Environmental Chemistry

Environmental Chemistry
A Global Perspective

Gary W. vanLoon

*Department of Chemistry and School of Environmental Studies,
Queen's University, Kingston, Ontario*

and

Stephen J. Duffy

Department of Chemistry, Brock University, St Catharines, Ontario

OXFORD
UNIVERSITY PRESS

OXFORD

UNIVERSITY PRESS

Great Clarendon Street, Oxford OX2 6DP

Oxford University Press is a department of the University of Oxford.
It furthers the University's objective of excellence in research, scholarship,
and education by publishing worldwide in

Oxford   New York

Athens   Auckland   Bangkok   Bogotá   Buenos Aires   Calcutta
Cape Town   Chennai   Dar es Salaam   Delhi   Florence   Hong Kong   Istanbul
Karachi   Kuala Lumpur   Madrid   Melbourne   Mexico City   Mumbai
Nairobi   Paris   São Paulo   Singapore   Taipei   Tokyo   Toronto   Warsaw

with associated companies in Berlin   Ibadan

Oxford is a registered trade mark of Oxford University Press
in the UK and in certain other countries

Published in the United States
by Oxford University Press Inc., New York

A catalogue record for this book is available from the British Library

Library of Congress Cataloging in Publication Data
(Data applied for)

ISBN 0 19 856440 6 (Pbk)

Typeset by Newgen Imaging Systems (P) Ltd., Chennai, India

Printed in Great Britain on acid-free paper by Bath Press Ltd.,
Bath, Avon

Preface

During the past decade, environmental chemistry has come into its own as a respected subdiscipline in the field of chemical science—a subject that occupies an important place in both teaching and research activities of many academic institutions.* In its early phases, environmental chemistry was essentially a catalogue or description of chemical properties of the natural world and of concentrations of contaminants. As the subject matured, however, it has come to encompass challenging studies of highly complex systems, and not just in a static way. Current research focuses on the processes that operate within and between various environmental compartments and the ways in which human activities interact with the natural processes. Studies in environmental chemistry use information from all the traditional subdisciplines, building on this and contributing new knowledge in highly specific ways. Beyond the specialized research, environmental chemistry also involves attempts to integrate the particular ideas into a comprehensive picture of how the natural environment functions and responds to stresses. In this book, we have tried to introduce the basic concepts of the subject and to capture its vitality and relevance to key issues of global concern.

In preparing *Environmental Chemistry—a Global Perspective*, we have kept several ideas in mind.

- The book deals with *chemical principles* operating in the natural and altered environment.
- It builds on the fundamentals of physical, organic and inorganic chemistry. As such, it is directed toward students at the second- or third-year level in an undergraduate chemistry program.
- It presents a descriptive approach to the most important topics within the overall subject of environmental chemistry; at the same time, we give an introduction to some basic quantitative calculations.
- The subject is considered in a global context. Examples are chosen from all the continents, emphasising the world-wide interconnectedness of all environmental issues.

The writing of the book builds on many years of research and teaching (for both of us) within the broad areas of environmental chemistry. We are indebted to students who have asked questions and to students who have found answers through their involvement in research projects. We also acknowledge support from colleagues in the Department of Chemistry and the School of Environmental Studies at Queen's University, as well as at Brock University. Finally, and most important, we wish to offer thanks to our family members for their support of our work over many years.

*Glaze, W. H., Environmental Chemistry Comes of Age, *Environ. Sci. Technol.*, **28(4)** (1994), 169A.

It is hoped that, in a small way, this book may contribute to understanding that is needed to maintain and restore this good Earth.

'. . . and God saw everything that he had made, and it was very good . . .'

Genesis 1.31a

December 1999

Gary W. vanLoon
Stephen J. Duffy

Contents

Part B The hydrosphere

Part C **The terrestrial environment**   367

Environmental Chemistry

A Global Perspective

We have to visualize the Earth as a small, rather crowded spaceship, destination unknown, in which humans have to find a slender thread of a way of life in the midst of a continually repeatable cycle of material transformations. In a spaceship, there can be no inputs or outputs. The water must circulate through the kidneys and the algae, the food likewise, the air likewise .... In a spaceship there can be no sewers and no imports.

Up to now the human population has been small enough so that we have not had to regard the Earth as a spaceship. We have been able to regard the atmosphere and the oceans and even the soil as inexhaustible reservoirs, from which we can draw at will and which we can pollute at will. There is writing on the wall, however .... Even now we may be doing irreversible damage to this precious little spaceship.

K. E. Boulding, 1966

Environmental chemistry

THE history of the Earth can be said to have begun more than 4.6 billion years ago.

For largely unexplained reasons, a cloud of molecular particles—mostly hydrogen—rotating through the galaxy began to contract, and spin with increasing velocity. As the gravitational energy increased, contraction continued to accelerate and massive amounts of heat were generated. Initially the heat was radiated out into space, but eventually it became trapped within the confines of the central body—the *protostar*—and its core became extremely dense and hot. The massive energy release caused hydrogen within the hottest regions to become ionized. The hydrogen nuclei became fuel for self-sustaining thermonuclear fusion reactions that maintained an interior temperature far in excess of 1 000 000 K.

The luminous sphere of gas that formed in this way could have been any typical star, but this particular one is known to us as the Sun. The rapidly rotating core of matter that had contracted to form the Sun left on its periphery other matter that took the shape of a disc, known as the *solar nebula*. As nebula particles remote from the Sun cooled, gases in that part of the solar system began to interact to form compounds. Some atoms and molecules condensed to form more particles, and collisions amongst them, over time, gradually drew them together into solid bodies known as planetesimals. Eventually, with further coalescence, the small planetesimals grew to such a size, now planets, that they could retain an atmosphere. Reactions occurred within and between the atmosphere and the solid/liquid phases of the young planets. The elements that were present, and the changing affinities between these elements as the system cooled, determined the molecular species that were created.

One of these planets was the Earth.

In this earliest period of the Earth's life, the solid materials present in its core consisted of iron and alloys of iron, while the mantle and crust of the Earth were in large part made up of oxides and silicates of metals. The major gases in the primeval atmosphere were dihydrogen, dinitrogen, carbon monoxide, and carbon dioxide. Over time, large amounts of the atmosphere were lost into space whereas continued volcanism brought other gases to the surface where reactions formed additional new gas species. Oxygen was abundant but there was no free oxygen gas. In its entirety, this element was present in combined form—associated with metals or in the atmosphere as carbon dioxide.

Very early in the Earth's history, water was formed most likely by reactions such as

$$3H_2 + CO_2 \rightarrow CH_4 + H_2O \tag{1.1}$$

$$H_2 + CO_2 \rightarrow CO + H_2O \tag{1.2}$$

To occur to any significant extent, the two reactions require the presence of catalysts and these were available in the form of metal oxides on the surface of the primordial Earth.

Water making up the early seas may have been acidic, due in part to dissolved carbon dioxide as well as hydrochloric acid and sulfur species which were trace components of the early atmosphere. The acids, thought to be concentrated enough to generate an aqueous pH of about 2, and the warm temperatures of the early oceans were sufficient to cause significant dissolution of components in the associated rocks. Dissolution is a neutralizing process and the pH of the seas rose to a value (pH $\sim$ 8) near that of the present-day oceans. At the same time, concentrations of metals in the water increased, sometimes exceeding the solubility products of secondary minerals. For one, the presence of dissolved aqueous carbonate species led to the formation of early sedimentary deposits of calcite ($CaCO_3$) and other carbonate minerals:

$$Ca^{2+}(aq) + CO_3^{2-}(aq) \rightarrow CaCO_3 \tag{1.3}$$

Also significantly affecting oceanic chemistry were continued underwater releases of gases and volcanic eruptions.

Volcanic activity, folding and uplift of rocks under pressure from the movement of tectonic plates, chemical and physical erosion and sedimentation all changed the nature of the crust of the Earth over long periods of the Earth's early history.

Because there was no free oxygen in the atmosphere, no ozone could be formed. The atmosphere was then transparent to a broad flux of solar radiation, including a large input of ultraviolet (UV) light. This highly energetic radiation and the presence of catalysts made it possible for simple organic compounds like methanol and formaldehyde to be synthesized:

$$CO + 2H_2 \rightarrow CH_3OH \tag{1.4}$$

$$CO + H_2 \rightarrow HCHO \tag{1.5}$$

Very early in the Earth's history these and other species—including HCN, NH_3, H_2S and many others—were formed. Some of the small molecules reacted further to produce more complex compounds, even including amino acids and simple peptides.

Very primitive forms of life are known to have developed as early as 4 billion years ago. The first cells used simple inorganic molecules as starting material for their synthesis and they, of course, lived in an environment devoid of free oxygen. With increasing complexity, around 3.5 billion years ago, some cells developed an ability to carry out photosynthesis—a reaction that released oxygen into the atmosphere:

$$CO_2 + H_2O \rightarrow CH_2O + O_2 \tag{1.6}$$

At first, the free oxygen was removed as quickly as it formed, by reaction with terrestrial materials. As the amount of aquatic plant life increased, however, free oxygen began to build up and by about 2 billion years ago the environment at the Earth's surface could be described as essentially oxidizing. Carbon dioxide gradually became a minor gas in the

Table 1.1 Some important physical properties of the present-day Earth

	Atmosphere	Oceans	Land
Mass/kg	5.98×10^{24}		
Radius/m	6.38×10^{6}		
Density/kg m^{-3}	5520		
Distance from Sun/km	1.5×10^{8}		
Surface temperature/K	290		
Mass/kg	5.14×10^{18}	1.37×10^{21}	
Surface area/m^2		3.61×10^{14}	1.48×10^{14}
Approximate density/kg m^{-3}	1.3 (at Earth's surface, 0 °C)	1030	2700 (surface rocks)
Major components	N_2, O_2, H_2O, Ar	H_2O, dissolved species Na^+, Cl^-, SO_4^{2-}, Mg^{2+}	Si, O, Al, Fe, Ca (as silicates, oxides, carbonates, etc.)

atmosphere. The presence of free oxygen led to the synthesis of ozone, which acted to partially shield the highly energetic components of solar radiation from reaching the Earth's surface. This opened the possibility for terrestrial life to emerge.

It was the development of life and an oxidizing atmosphere that dominated the change from the primitive to the present environment, and in the past billion years many features of the Earth's composition have remained relatively constant. Yet we should not leave the impression that geological and life processes have remained static during that period. On the contrary, the Earth is a dynamic system where processes such as volcanism, movement of tectonic plates, weathering, erosion, sedimentation, and the continuing evolution of life interact to provide the environment in which we now live. As changes occur in one compartment—through interactions and feedback—changes occur over the Earth as a whole.

Nevertheless, from about 1 billion years ago to the present, the average composition of the atmosphere, the oceans, and the land, in their major components, has remained relatively constant. Table 1.1 lists some important physical features of our present-day Earth.

1.1 The subject matter of this book

In the terminology of thermodynamics, we define the *universe* as consisting of a *system* and its *surroundings*. The system is that portion of the universe under direct investigation while the surroundings comprise everything beyond the system. We can apply this concept, for example, to consider an industrial chemical process such as the production of the wood preservative pentachlorophenol (PCP). The system—essentially what goes on in the factory reactor—is subject to investigation by chemists who develop and optimize appropriate synthetic reactions and by engineers who design and set up the manufacturing facility itself. Historically, much effort has been expended on examining the properties of systems but increasingly there is concern about surroundings. In our example, when scientists move outside the factory and focus attention on the impact of the industrial process on its surroundings—such as release of the product PCP or byproducts like dioxins—then the subject becomes one of environmental chemistry.

We may continue our example by moving away from manufacturing to the other end of the scientific spectrum where we examine mechanisms by which chemicals such as dioxins enter a living organism, the biochemical transformations they undergo, their molecular mode of action, and their elimination. The target organism has then become the system and is a subject for study by biologists, biochemists, and toxicologists. What goes on in the surroundings outside the organism—transport of dioxins, associative reactions with soil and water, degradation process, and so on—is the subject matter of environmental chemistry.

To express it simply, environmental chemistry is the chemistry of surroundings—the universe minus the system.

Given this very broad concept of environmental chemistry we must be clear about the physical extent of these surroundings. Beginning where humans live, we can move inward toward the centre of the Earth. We will find that below a relatively thin layer of the Earth's crust, few chemical processes affect the environment, at least over a timescale of years or even thousands of years. The thin layer may be as little as a metre when considering many soil processes, to tens of metres for lakes, or a few kilometres for oceans and when considering the disposal of nuclear wastes. Even at its greatest, the layer is only a very small fraction of the 6380 km radius of the Earth (Fig. 1.1).

Moving outward into the atmosphere above the Earth's surface, there occur complex processess supported by and supporting Earth-bound reactions. Many of these processes take place at low altitudes, but we know that chemical reactions at heights of 30 km or higher are critical to maintaining the Earth as it is.

Therefore, out of the total surroundings of the vast universe, a thin shell, perhaps 50 km in thickness, on and above the Earth's surface is the subject for most of what we say on the theme of environmental chemistry. Even these specified surroundings are very large and very complex—especially compared with the small controlled systems with which chemists are usually involved in a laboratory. We will therefore frequently zero in on a more limited,

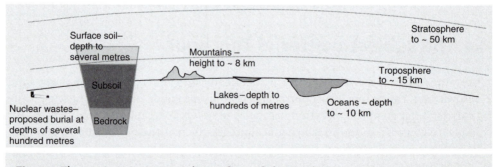

Fig. 1.1 The environment near the surface of the Earth.

defined portion such as the atmosphere of a building, a small lake, or a particular layer of soil, but when we do this we are really turning that part of the surroundings into a new system. We should never lose sight of the fact that each small part of the environment is connected with other parts and together they make an interrelated whole.

We must also establish some limitations as to what we can do in this particular book which surveys environmental chemistry. There are several important subjects which the book will *not* cover.

The book is not about *environmental analysis*, though almost every environmental topic discussed is based on information obtained by analysis. Often the quality of the discussion depends on the quality of the available data. For this reason, analytical chemistry is central to our understanding of the environment. And analysis is by no means straightforward. Determining ozone or nitrogen oxides in the stratosphere at an altitude of 25 km requires sophisticated instrumentation and interpretation. Finding the concentration of mercury which is available for uptake by fish in a body of water requires careful measurement and elegant reasoning related to the significance of the results. Environmental analytical chemistry is such a large subject in its own right that it will only be mentioned in passing in this book.

The book is not about *environmental toxicology*. One of the reasons—but by no means the only one—why twentieth-century inhabitants of the Earth are so concerned about the environment is that its contamination and degradation inevitably has effects on human beings. The direct toxicological effects, the mechanisms by which they occur, and the quantitative aspects defining conditions under which organisms, particularly the human organism, are influenced by chemicals, are the subjects of toxicology. In general, this book will be concerned with chemical behaviour external to specific organisms, although from time to time a brief discussion of factors affecting uptake in the biosphere is relevant to our purposes.

The book is not about *environmental control*—about detailed technologies for preventing or eliminating pollution, or about standards and laws that set guidelines and limits on levels of contaminants. These, too, are essential subjects for consideration if environmental science is to be more than a theoretical study. We will be concerned with the chemistry that underlies regulatory decisions and engineering design for control of pollution, but we will not examine the decisions or designs in detail.

And finally, the book is not about *environmental science* in its broadest sense. Books on environmental science attempt to cover the whole range of a topic, and they call on a large body of material from the sciences of climatology, geology, biology, and so on. To put any subject in context, we will make mention of these and other subjects, but we will not dwell on them, important though they are.

Now, we can restate the purpose of our present exploration of environmental chemistry. The book has been written to provide the chemical basis for understanding our surroundings, the global environment. Emphasis will be on the composition of the natural environment, the processes that take place within it, and the kinds of changes which come about as a result of human activities. Many examples will be used to illustrate the principles being described and these will be chosen from situations throughout the world, because environmental chemistry is indeed a global subject. Nevertheless, there are many important specific types of problems which, inevitably, will not be mentioned. It is not the goal of this book to provide a complete collection of all environmental issues but our objective is to give a comprehensive chemical background so that one can have a basis for understanding such issues.

1.2 Environmental composition

In learning about the chemistry of a particular component of the environment, a logical starting point is to describe its composition. To do this we make use of a range of units, some of which are defined below. Being familiar with these and being able to interconvert them is basic to any quantitative study.

1.2.1 Aqueous solutions

One of the most important and fundamental ways used by chemists to express concentrations of solutes in aqueous solutions employs units based on molarity. Sea water contains approximately 1.99% w/v chloride (% weight/volume = mass (g) of chloride per volume (100 mL) of solution), and so it is a 0.561 mol L^{-1} solution of the ion.

For elements or compounds present at lower levels, it is convenient to give concentrations in micromolar or nanomolar units. Typical would be zinc values ranging from 15 nmol L^{-1} to 6.7 μmol L^{-1} in the Toyohira River in Japan,[1] with the higher values found in samples within the Sapporo city boundaries.

It is also common practice to express low concentrations such as the above value of zinc in units of parts per million (ppm), parts per billion (ppb), or parts per trillion (ppt). While widely used, there are several problems associated with the use of such units and these should be kept in mind.

When dealing with aqueous solutions, the unit ppm is considered to be equivalent to grams of solute per million millilitres of solution, which is the same as μg mL^{-1} or mg L^{-1}. These units imply that the density of water is 1.00 kg L^{-1}. This is usually a good assumption for lakes, rivers, and other sources of fresh water, but it is not true for the oceans where the density is 1.025 kg L^{-1} at 15 °C. It is therefore preferable to specify the concentration units in unambiguous terms of mass per volume or mass per mass. A study of metals in the Huanghe (Yellow) River in China[2] found zinc concentrations ranging from 60 ng kg^{-1} far inland to 352 ng kg^{-1} near the estuary in Bohai. By using such units there is no uncertainty related to changes in density when moving from fresh to salt water.

A second complication with 'parts per . . .' types of units arises from the uncertainty associated with choosing a particular species for mass calculations. An example will illustrate this point. A 10 ppm ammonium ion solution corresponds to a molar concentration of 5.56×10^{-4} mol NH$_4^+$ L^{-1}, calculated in the following way:

$$10\,\text{ppm} = 10\,\text{mg NH}_4^+\,\text{L}^{-1} = 10 \times 10^{-3}\,\text{g NH}_4^+\,\text{L}^{-1} = 10 \times 10^{-3}/18.0\,\text{mol NH}_4^+\,\text{L}^{-1}$$
$$= 5.56 \times 10^{-4}\,\text{mol NH}_4^+\,\text{L}^{-1}$$

By an important reaction called nitrification—we shall study this reaction in detail later—the ammonium can be oxidized to form nitrate. The stoichiometry is such that one ammonium ion becomes one nitrate ion. In units of ppm, this becomes

$$5.56 \times 10^{-4} \, \text{mol} \, NH_4^+ \, L^{-1} = 5.56 \times 10^{-4} \, \text{mol} \, NO_3^- \, L^{-1}$$
$$= 5.56 \times 10^{-4} \times 62 \times 10^3 \, \text{mg} \, NO_3^- \, L^{-1} = 34 \, \text{ppm} \, NO_3^-$$

Obviously, the larger ppm value for nitrate compared to ammonium is not because any additional nitrogen-containing species has been created, but arises simply because of the differences in molar mass of the two nitrogen species. Uncovering such apparent anomalies may be avoided by expressing concentrations in both cases in terms of nitrogen alone. The two values are then both reported in terms of N as 7.7 ppm. For example,

$$34 \, \text{ppm} \, NO_3^- = 34 \, \text{mg} \, NO_3^- \, L^{-1} = 34 \times 14/62 \, \text{mg} \, N \, L^{-1} = 7.7 \, \text{mg} \, N \, L^{-1} = 7.7 \, \text{ppm} \, N$$

Of course, there will never be any confusion of this kind if molar units are used—the concentration of nitrogen would remain at $0.56 \, \text{mmol} \, L^{-1}$ whether as ammonium or as nitrate.

Yet another limitation with the use of units such as parts per million is that they give no indication of the concentration of reactive groups. For example, the concentration of total organic carbon (TOC) in a forest stream may be 9.0 ppm ($= 9.0 \, \text{mg} \, L^{-1}$). If the organic material being measured is largely dissolved humic material from the soil, then it will contain functional groups such as carboxylic acids which are capable of forming complexes with metals and therefore enhancing metal solubilization (a subject discussed in Chapter 12). One study concludes that the concentration of carboxylate functional groups per gram of humic material dissolved in the water is about $4 \, \text{mmol} \, g^{-1}$ of humic material. Using a carbon percentage in humic material of 50%, the 9.0 ppm TOC could be expressed in the following way:

$$9.0 \, \text{mg} \, C \, L^{-1} = 9.0 \times 10^{-3} \times 100/50 \, \text{g humic material} \, L^{-1} \, \text{water}$$
$$= 9.0 \times 10^{-6} \times 100/50 \, \text{g humic material} \, L^{-1} \times 4$$
$$\times 10^{-3} \, \text{mol carboxylate groups} \, g^{-1} \, \text{humic material}$$
$$= 8 \times 10^{-3} \, \text{mol carboxylate groups} \, L^{-1} \, \text{water}$$

This way of describing the concentration of available carboxylate groups could be used conveniently to describe aspects of the acid–base and complexing behaviour of the water.

1.2.2 Solids

Mass fractions of various kinds are invariably used for solids, and there should be no ambiguity in this—for example, the global average concentration of iron in the Earth's crust is 4.1% by weight. For vanadium the corresponding concentration is approximately $160 \, \mu g \, g^{-1}$ which is simply and correctly expressed as 160 ppm. In this usage, ppm stands for grams of vanadium per million grams of the solid and is equivalent to $\mu g\,g^{-1}$ or $mg\,kg^{-1}$. One minor modification of mass fraction calculations is the practice of Earth scientists to express the concentration of major elements in geological materials in terms of percentage of the appropriate oxide. A rock made up in part of 27.6% Si and 6.31% Al would then be described as containing 59.0% SiO_2 and 11.9% Al_2O_3. This does not imply that each element is actually present in the oxide form. However, since oxygen is frequently the only significant anionic species present in rocks, the sum of percentages of all major element oxides should add up to near 100%.

1.2.3 Gases

For gases which make up a large proportion of the atmosphere, fractional or percentage concentrations are used, but these are usually expressed on a molar, not mass, basis. The Avogadro relationship then defines that this would also be equivalent to a volume ratio or a ratio of partial pressures. A general term used to express atmospheric concentration is 'mixing ratio'. As an example, the mixing ratio for nitrogen in the dry troposphere is 0.7808 or 78.08%. This means that, for every 100 mol of all gases, 78.08 mol are nitrogen. It also means that at atmospheric pressure ($P° = 101\,325\,Pa$), the partial pressure of nitrogen can be calculated as shown:

$$0.7808 \times 101\,325\,Pa = 7.911 \times 10^4\,Pa$$

For gases present at low concentrations, the 'parts per ...' family of units is frequently used to express mixing ratios. As with the fractional and percentage concentrations, equivalent calculations are done on a molar, pressure, or volume basis. We then express concentrations as, for example, parts per million by volume for which the symbol is ppmv. The mixing ratio of methane in the dry troposphere is approximately 1.7 ppmv. This means that there are 1.7 μmol of methane for every 1.0 mol of the components of air. Using an average molar mass of air of $29\,g\,mol^{-1}$, a 1.7 ppmv concentration is the same as 0.94 ppm by mass. The latter concentration unit is rarely used in atmospheric studies.

Units of mass per volume, moles per volume, or molecules per volume are frequently employed to express concentration of gases and particulates in the atmosphere. At $0\,°C$ and 101.3 kPa, the molar volume of a gas is 22.4 L. For oxygen, whose mixing ratio in air is 20.95%, the concentration in $mg\,L^{-1}$ is calculated as follows:

1L of air contains $\dfrac{1}{22.4} = 0.0446$ mol gas of which 20.95 % is oxygen

The concentration of O_2 is then $0.0446 \times 20.95/100 = 9.35 \times 10^{-3}\,mol\,L^{-1}$.

In mass, this is equivalent to $32.0 \times 9.35 \times 10^{-3} = 0.299\,g\,L^{-1}$,

since 1 mol of oxygen has a mass of 32.0 g.

A related set of units giving molecules per volume is commonly used for some important atmospheric species that are present in very small concentrations. For solid or liquid particles in air, mass per volume units are invariably employed to measure concentration.

1.2.4 Species distribution

In many cases, it is insufficient to know only the total concentration of a particular element or compound in an environmental sample. Most substances can exist in more than one form, and a description of the species distribution for such substances is an important aspect of describing composition. In some instances, sophisticated analytical methods are used to distinguish between species in environmental samples. It is also possible to make use of distribution diagrams, of which there is a large variety of types, as an aid to evaluating species distribution. Figures 1.2 and 1.3 give examples of such diagrams for carbonate species and mercury in water. In the case of carbonate, the distribution is plotted as a function of pH of the water, while for mercury, its composition is plotted against the chloride ion concentration. The diagrams are therefore limited by the variables that are chosen for their calculation.

Distribution diagrams are based in part on analytical data, both for concentrations of particular species in environmental samples and also for thermodynamic equilibrium constants of a variety of types. The carbonate diagram, for example, depends on knowing the association constant for aqueous carbon dioxide and water, as well as the acid dissociation constants for carbonic acid. The validity of a distribution diagram therefore depends on the availability and quality of appropriate thermodynamic data, and also on the very important (and problematic) assumption that a particular natural system is at thermodynamic equilibrium.

On Fig. 1.2 the vertical line at pH 7 allows us to estimate the fraction of species at that pH. Aqueous carbon dioxide makes up about 16% of all carbonate species, hydrogen carbonate is 84%, while carbonate is a negligible fraction.

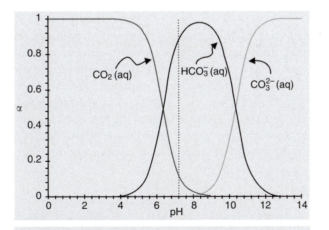

Fig. 1.2 Distribution of carbonate species as a function of pH. The α value is the fraction of a particular species: $\alpha = $ [individual species]/ [all species].

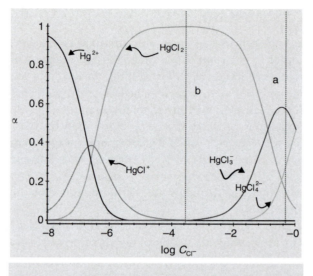

Fig. 1.3 Distribution of mercury chloro species in water as a function of chloride ion concentration, C_{Cl^-}(mol L^{-1}). The α value is the fraction of mercury in the form of a particular complex.

The vertical line (a) on Fig. 1.3 refers to the oceans which have a total chloride concentration of 0.56 mol L^{-1}. The HgCl$_4^{2-}$ is seen to make up 28% of mercury species, with HgCl$_3^-$ at 56% and HgCl$_2$ at 16%. Line (b) corresponds to well water with a chloride ion concentration of 9.5 ppm $= 2.7 \times 10^{-4}$ mol L^{-1}, where the only significant species of mercury is HgCl$_2$.

The species distribution of an element controls its behaviour in the environment and may be a major factor affecting biological availability. For example, in the aqueous environment mercury exists as inorganic species in the 0, +1, and +2 oxidation states depending on redox and other conditions. However, the major species found in fish is partially methylated mercury, CH$_3$Hg$^+$, which is produced in sediments by a variety of microbiological processes. This species is toxic both to the fish and to other animals (including humans) which may consume the fish. It is clearly important, then, to be aware not only of how much mercury is present in a sample, but also of the distribution of forms of the element in that sample.

1.3 Chemical processes

Knowledge of the composition of a particular compartment of the environment is a starting point for a description of environmental chemistry. However, if we stop there, we imply that the system is static, which of course is not true. There are many processes—physical, chemical, and biological—which operate within and connect the various components.

The processes may be completely 'natural' and, in fact, over the geological timescale, it is such processes that have contributed to making the Earth the way it is. Therefore, a second phase of developing an understanding of environmental chemistry is to learn about the chemical reactions that are a part of the environmental processes.

In order to summarize the broad features of reactions involving environmental species, we will find it useful to think of the environment in terms of four principal compartments— the atmosphere (the gaseous environment), the hydrosphere (the liquid, essentially aqueous environment), the terrestrial (solid) environment, and the biosphere (the living environment). At first glance, these categories appear to be quite clear cut; however, it will become evident that there are many overlapping areas. For example, we usually think of soil as part of the terrestrial environment, but chemical behaviour of soil solutions and soil gases plays a major role in determining the environmental characteristics of soil itself. Concerning the compartments, we can then describe chemical processes within each, and also reactions which bring about transitions from one to the other.

Figure 1.4[3] shows a simple diagrammatic representation of the water system and its relation to the various compartments. Strictly speaking, most of the processes shown in this example are physical in that they involve phase changes, not chemical reactions, but the form of the diagram is similar to those constructed to show environmental chemical relationships.

The figure allows us to obtain an overview of the relations between the various forms of the substance. It also shows the cyclic nature of many natural processes. In the steady state, the parts of the cycle are balanced so that concentrations remain constant. This enables us to calculate residence times. In the water cycle, the total mass of water at any time in the atmosphere is approximately 1.3×10^{16} kg. The inward flux is 4.23×10^{17} and 7.29×10^{16} kg y^{-1} by evaporation from oceans and land respectively. This is balanced by outward fluxes of 3.86×10^{17} and 1.10×10^{17} kg y^{-1} precipitation on to the ocean and land. Therefore the total inward and outward flux is 4.96×10^{17} kg y^{-1} and the residence time of water in the atmosphere is determined by

$$\tau = \text{residence time} = \frac{\text{steady state amount in the atmosphere}}{\text{flux (in or out)}}$$

$$= \frac{1.3 \times 10^{16} \text{ kg}}{4.96 \times 10^{17} \text{ kg y}^{-1}}$$

$$= 0.0262 \text{ y}$$

$$= 9.6 \text{ days}$$

This is the average time that a molecule of water spends in the atmosphere.

The box model cycle illustrated in Fig. 1.4 provides an overview of the major processes interconnecting compartments of the global water ecosystem, and may also allow for identification of important reactions within a single compartment. But there are many details involving specific reactions which are essential to completing the picture. As an example, a comprehensive description of water resource distribution in India is provided in Fig. 1.5. This shows details of the fate of water supplied by precipitation each year in that part of the subcontinent.

The information contained in such a diagram along with appropriate chemical data could be used, for example, in developing a complete picture of water used for irrigation.

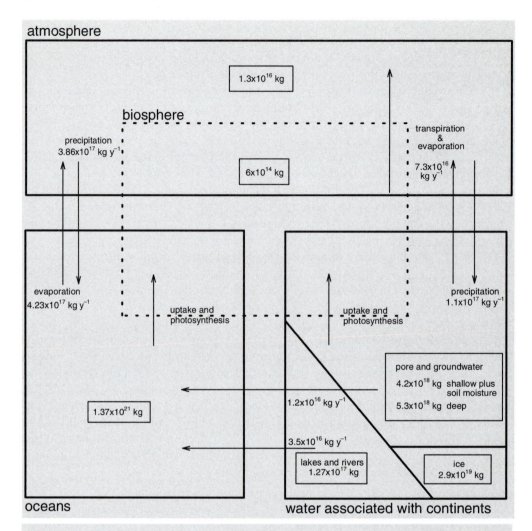

Fig. 1.4 The water cycle. Boxed values in kg are total amounts in the given compartment. Values in kg y^{-1} are fluxes or movement from one compartment to another. (Values taken from a number of sources and reported in Berner, E. K. and R. A. Berner, *The Global Water Cycle*, Prentice Hall, Inc., NJ; 1987.)

The combined quality and quantity of data allow for estimating fluxes of individual chemicals—a requirement for evaluating sustainability of the water resource in agriculture.

Generating a valid description of environmental processes involves moving back and forth between the general, broad picture with its global context, and the particular detailed delineation of specific chemical reactions. Studies in both these areas contribute to our knowledge of the subject. One of the challenges of environmental chemistry is to bring together the general and the particular.

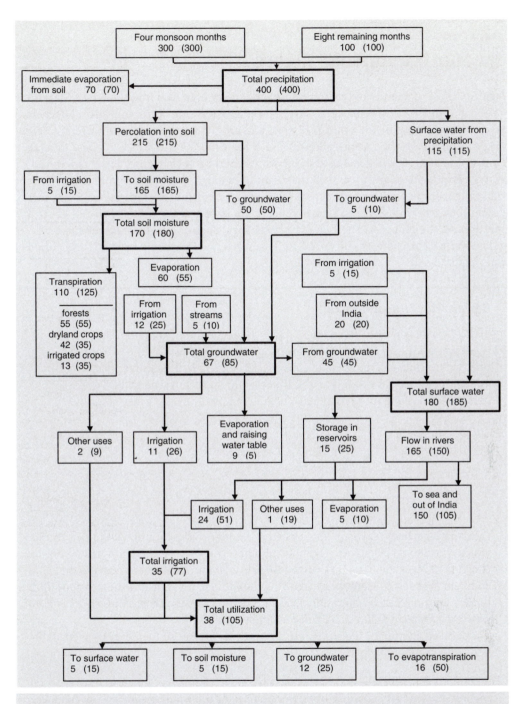

Fig. 1.5 Water resource flux in India. Values given are in million hectare metres – a hectare metre is the volume of water required to cover 1 ha (10 000 m²) to a depth of 1 m. This is a volume of 10^4 m³ or a mass of approximately 10^7 kg. The first numerical figure of each pair is a value for 1974; the second, in parentheses, is an estimate for 2025. (Centre for Science and Environment, *The State of India's Environment: A Citizen's Report*, India; 1978.)

1.4 Anthropogenic effects

A third aspect of a study of environmental chemistry is to examine the effects of human (anthropogenic) activities on the natural processes which are occurring. These effects may be catastrophic, usually in a localized area, and are then referred to as disasters. A recent tragic example was the uncontrolled release of toxic methyl isocyanate gas near Bhopal, in Central India in 1984 (Box 1.1). Other environmental perturbations may be more gradual and the effects may show up only in the medium or long term. This does not imply that the consequences are any less serious. For example, the possibility of global warming due to build-up of greenhouse gases in the troposphere is of great concern with respect to life throughout the planet in the twenty-first century. Should adverse consequences due to global warming become a reality, its reversal, even if possible, could take as long a period of time as its advent.

Box 1.1 The Bhopal disaster

In the early morning of 3 December 1984 just outside the city of Bhopal in central India there was a massive release of methyl isocyanate ($CH_3-N=C=O$) gas from a storage tank at the Union Carbide plant where the chemical had been manufactured since 1980. The dense gas floated across the surrounding landscape killing at least 3000 persons and leaving more than 300 000 affected by exposure—some severely.

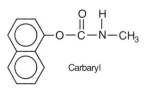

Carbaryl

Methyl isocyanate (MIC) is a starting material for the manufacture of the widely used carbamate pesticide carbaryl and was stored under refrigeration in an underground tank at the Bhopal plant site. A combination of technological problems—failure of the cooling unit, leakage of water into the tank, loss of nitrogen pressure above the MIC, and failure of several safety devices—was the immediate cause of the accident. But there were political, organizational, and human factors which provided a setting in which the multiple failures could occur simultaneously.

A detailed account of the story of this worst industrial crisis in history is given in the book *Bhopal, Anatomy of a Crisis* (P. Shrivastava, Ballinger Publishing Co., Cambridge MA; 1987).

In considering the potential effects of anthropogenic inputs into the environment, we make use of our knowledge of both the composition and the process data.

An example comes from consideration of the possible problems associated with the use of sewage sludge from municipal waste-water treatment facilities as an amendment/ fertilizer on soils used for agricultural purposes. This is a widely followed practice which has beneficial aspects as it supplies organic matter and small amounts of major and minor nutrients to the soil. One of the concerns regarding the practice is that potentially toxic concentrations of certain metals that are present in the sludge can be taken up by plants and incorporated in the food chain. Cadmium is one element of interest in this regard. Soils have a range of natural cadmium concentrations with a typical value being approximately $0.8 \, \mu g \, g^{-1}$. Of the cadmium, it is thought that most is present as inorganic mineral species associated with the clay mineral phase of the soil. Sewage sludge is a variable and heterogeneous material but a typical cadmium concentration is $80 \, \mu g \, g^{-1}$, some one hundred times greater than the soil itself. In considering whether addition of sludge to soil initiates a cadmium toxicity problem, the amount of sludge added becomes an issue as it will define the final concentration. Consideration of the form of cadmium in the sludge—it will probably be bound with the organic matrix—is also important. Finally, the geochemical and biochemical reactions which the added cadmium must undergo will be considered. These will include interconversions between various inorganic and organic species of the element in the soil, leaching of soluble forms, immobilization by ion exchange or other adsorption processes on the solid soil phases, and biological uptake by micro- or macro-organisms (Fig. 1.6).

The ultimate fate of the added cadmium depends on the extent of the individual reactions, how each process affects the others, and the ways in which a changing environment can alter the balance.

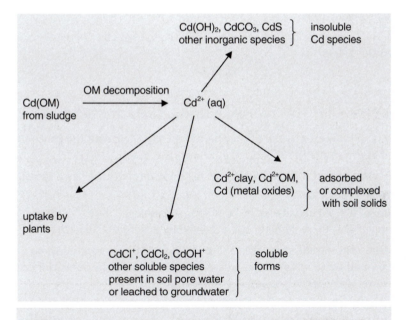

Fig. 1.6 Cadmium biogeochemical reactions in soil after addition as a trace component in sewage sludge. OM = organic matter.

The main points

The subject of this book is the chemistry of surroundings, the environment on and near the Earth's surface. These surroundings include air (the atmosphere), water (the hydrosphere), and land (the terrestrial environment) and we will deal with them in that order. In considering particular environmental subjects, we focus on the basic chemistry and to do this it is necessary to examine some or all of the three factors—composition, chemical processes, and perturbations caused by natural or anthropogenic activities—that we have discussed above. Although a systematic approach requires that individual topics be studied in isolation, the overlap, the connections, and the interdependence must always be kept in mind. Perhaps most important, as citizens of the Earth, while we are studying this vast subject, we should remember that we share together a single global environment.

Additional reading

1 Baird, C., *Environmental Chemistry*, W. H. Freeman & Co., Salt Lake City, Utah; 1995.

2 Bunce, N. J., *Environmental Chemistry*, Wuerz Publishing Ltd., Winnipeg; 1990.

3 Manahan, S. E., *Environmental Chemistry*, 5th edition, Lewis Publishers, Inc., Chelsea, Michigan; 1991.

4 Fifield, F. W. and Haines, P. J., *Environmental Analytical Chemistry*, Blackie Academic and Professional, London; 1995.

5 Zakrzewski, S. F., *Principles of Environmental Toxicology*, American Chemical Society, Washington DC; 1991.

Problems

1 The mixing ratio of oxygen in the atmosphere is 20.95%. Calculate the concentration in $mol\,L^{-1}$ and in $g\,m^{-3}$ at P° (101 325 Pa, 1.00 atm) and 25 °C.

2 What is the mass of air (at P° and 20 °C) contained in a room that has dimensions of $5.5\,m \times 7.0\,m \times 2.8\,m$?

3 The solubility of oxygen in water at 25 °C is approximately $8.5\,mg\,L^{-1}$. At P° and the same temperature, what is the volume of gas occupied by 8.5 mg of oxygen?

4 Average ozone concentrations in Jakarta, Indonesia have been reported to be $0.015\,mg\,m^{-3}$ and those in Tokyo, Japan are 20 ppbv. What is the approximate ratio of these two values, when expressed in the same units?

5 The concentration of titanium in the South Pacific Ocean, near the island nation of Fiji, has been found to be approximately $3.0 \times 10^{-9}\,mol\,L^{-1}$. Calculate the concentration in ppm, ppb, or ppt as appropriate.

6 The principal anion and cation in Lake Huron, one of the *Great Lakes* in North America, are respectively hydrogen carbonate and calcium. The concentration of the former is

approximately 1.05 mmol L^{-1}. Calculate the mass of solid calcium carbonate that would remain if 250 mL of Lake Huron water is evaporated to dryness.

7 Use Fig. 1.4 to estimate the residence time of water in the oceans. Indicate limitations in interpreting this result.

8 The concentration of copper in the surface organic layer of a forest soil is 37 ppm, and in the underlying mineral layer it is 17 ppm. The bulk densities of these two layers are 0.36 and 1.22 g mL^{-1} respectively. The very low density of the surface material is because it consists in large part of partially degraded organic material and there is very little of the heavier mineral matter. Which of these two layers has the larger amount of copper per unit volume?

9 The concentration of cadmium in the top 15 cm of soil in a field (often called the plow layer) can be estimated by taking a representative sample, dissolving the soil and analysing it by atomic absorption spectroscopy with electrothermal atomization. The concentration is found to be 0.78 ppm. Suppose that dewatered (solid) sewage sludge containing 22 ppm cadmium is added at the rate (mass per area) of 3 t ha^{-1}. Assuming sludge is well mixed within the plow layer, calculate the new average concentration of cadmium within this part of the soil. The bulk density of the soil is 1.1 g mL^{-1}.

Notes

1 Sakai, H., Y. Kojima, and K. Saito, Distribution of heavy metals in water and sieved sediments in the Toyohira river. *Wat. Res.* **20** (1986), 559.

2 Zhang, J., W.W. Huang, Dissolved trace metals in the Huanghe: the most turbid large river in the world. *Wat. Res.* **27** (1993), 1.

3 The form of this diagram will be used for other systems. For example see Fig. 14.10.

The Earth's atmosphere

When you understand all about the Sun
and all about the atmosphere
and all about the rotation of the Earth,
you may still miss the radiance of the sunset.

A. N. Whitehead, 1926

The Earth's atmosphere

2

O<small>F</small> the planets in the solar system Mercury, nearest the Sun, has almost no atmosphere. The next three planets, Venus, Earth, and Mars, lost whatever gases were present at their formation and the atmospheres they now possess are due to gases released from their interiors, and reactions these have undergone. The outer planets have extremely deep atmospheres made up mostly of hydrogen and helium, little changed from the original composition at the time of formation.

The Earth's atmosphere is a thin shell of gases surrounding the globe. Its unique chemistry, including compounds such as oxygen (molecular oxygen or dioxygen) and carbon dioxide which support the processes upon which all forms of life depend, distinguishes this atmosphere from that of other planets in the solar system. Table 2.1 shows the relative amounts of the four most abundant gases in the dry atmosphere.

The mixing ratios of the major gases remain relatively constant up to an altitude of about 80 km. The constancy is because the kinetic energy of the gas molecules is sufficient to overcome any gravitational forces that would lead to settling. Because the mixing ratios

Table 2.1 Major components of the atmosphere near the surface of the Earth

Component	Mixing ratio
Nitrogen	78.08%
Oxygen	20.95%
Argon	0.93%
Carbon dioxide	0.0365%

Mixing ratios are calculated on a dry atmosphere basis. The water content is a fifth major component, but its concentration is variable, ranging from 0.5 to 3.5%.

are constant, the average molar mass, $\overline{M}_a$, of the atmosphere can be calculated:

$$\overline{M}_a = M_{N_2} \times f_{N_2} + M_{O_2} \times f_{O_2} + M_{Ar} \times f_{Ar} + M_{CO_2} \times f_{CO_2} \qquad (2.1)$$

where M and f refer to the molar mass and fractional abundance of each component. Applied to the lower regions of the Earth's atmosphere,

$$\begin{aligned}
\overline{M}_a &= 28.01\,\text{g mol}^{-1} \times 0.7808 + 32.00\,\text{g mol}^{-1} \times 0.2095 + 39.95\,\text{g mol}^{-1} \times 0.0093 \\
&\quad + 44.01\,\text{g mol}^{-1} \times 0.00036 \\
&= 28.96\,\text{g mol}^{-1}
\end{aligned}$$

Above about 80 km altitude, the concentrations of the major species do begin to change significantly, due to photochemical processes that cause dinitrogen and especially dioxygen to dissociate. These processes will be discussed below.

In contrast to the major species, the composition of some trace gases is not constant in the lower troposphere, as a consequence of the various sources and removal processes that operate in different regions, on both a horizontal and a vertical scale.

Atmospheric pressure, P°, is measured by the force of gravity due to the atmosphere, divided by the total surface area of the Earth's surface, as calculated by eqn. 2.2:

$$P^\circ = \frac{M_{atm}g}{4\pi r^2} \qquad (2.2)$$

where P° is the pressure at the Earth's surface (sea level) $= 101\,325$ Pa; M_{atm} is the mass (in kg) of the atmosphere); g is the acceleration due to gravity $= 9.81$ m s^{-2} (this is approximately constant since most of the atmosphere is near the Earth's surface); and r is the radius of the Earth $= 6.37 \times 10^6$ m.

This equation can then be rearranged to calculate the total mass of the atmosphere. Solving for M_{atm} gives a mass of 5.27×10^{18} kg. We will make frequent use of this value.

2.1 Regions of the atmosphere

The atmosphere can be conveniently divided into four sections based on the direction of temperature change as one proceeds from lower to higher altitudes (Fig. 2.1). Beginning at the Earth's surface where the temporally and spatially averaged temperature is approximately 14 °C (287 K), the atmospheric temperature falls steadily to a value of about −60 °C at an altitude of approximately 15 km. This part of the atmosphere closest to the Earth's surface, in which humans live and most biological activity occurs, is called the troposphere.

The troposphere, a region of intense convective mixing, contains approximately 85% by mass of the entire atmosphere. The upper boundary of the troposphere, the tropopause, marks the altitude at which the direction of temperature change reverses. Above this, the region is called the stratosphere and increasing altitude brings increasing temperature up to approximately −2 °C at 50 km, the stratopause; such a temperature profile is called an inversion. Due to the rising temperature and decreasing density with increasing altitude, there is little convective mixing and the stratosphere is a relatively stable region. At the stratopause, there is a second temperature reversal and in the mesosphere, temperature

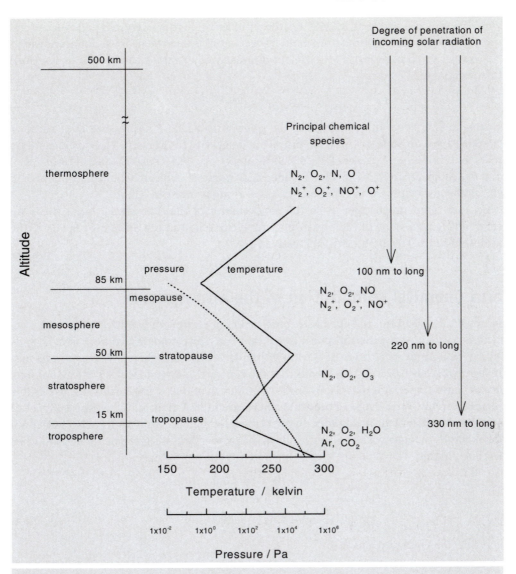

Fig. 2.1 The regions of the atmosphere showing temperature and pressure variations, principal chemical species, and penetration of incoming solar radiation. Solid line is temperature, broken line is pressure.

decreases with altitude to −90 °C at 85 km, the mesopause. Continuing above this height, the temperature once more begins to increase in a region called the thermosphere, reaching a value of about 1200 °C at 500 km. These, and other features of the atmosphere, are shown in Fig. 2.1.

In rarified atmospheres it is important to be aware that we are referring to *thermodynamic* temperatures, which are measures of the kinetic energy of molecules. At high altitudes, if we were to use a traditional mercury thermometer to measure temperature, the reading would be much lower than 1200 °C, as the number of (highly energetic) collisions of gas molecules with the thermometer would be very small under the near-vacuum conditions.

Pressure also changes with altitude but it undergoes a nearly steady decrease described by the equation

$$P_h = P° e^{-\overline{M}_a gh/RT} \tag{2.3}$$

where, in this equation, P_h = pressure at given altitude/Pa; $P°$ = pressure at sea level = 101 325 Pa; h = altitude/m; g = acceleration due to gravity = 9.81 m s^{-2}; $\overline{M}_a$ = average molar mass of atmospheric molecules = 0.0290 kg mol^{-1}; R = gas constant = 8.314 J mol^{-1} K^{-1}; and T = temperature/K. Note that an identical numerical result would be obtained if, simultaneously, altitude is expressed in km and molar mass in g mol^{-1}.

Because temperature appears in the denominator of the exponential term, there are small changes in slope of the pressure/altitude curve within the different regions of the atmosphere and this is shown graphically in Fig. 2.1.

2.1.1 Chemical composition of the atmosphere

In order to understand the chemical composition of regions in the atmosphere it is convenient to begin at the thermosphere and proceed in a direction toward the Earth. There is a near vacuum at an altitude of 100 km (in the lower thermosphere), where the pressure is approximately 0.025 Pa. This means that the concentration (molecules per unit volume) of all chemical species is only about one four millionth of that at the Earth's surface. As this region is on the outer fringe of the atmosphere, atoms and molecules are exposed to a full solar spectrum, including radiation in the ultraviolet region. This is high energy radiation; for example, radiation of wavelength 100 nm has associated energies calculated as follows. For one photon,

$$E = \frac{hc}{\lambda} = \frac{6.6 \times 10^{-34} \text{ J s} \times 3.0 \times 10^8 \text{ m s}^{-1}}{100 \times 10^{-9} \text{ m}}$$

$$= 2.0 \times 10^{-18} \text{ J per photon} \tag{2.4}$$

and thus for 1 mol of photons,

$$E = 2.0 \times 10^{-18} \times 6.0 \times 10^{23} = 1200 \text{ kJ mol}^{-1}$$

Such high energies are capable of bringing about dissociation of dinitrogen and dioxygen into their constituent atoms.

$$N_2 + h\nu \ (\lambda < 126 \text{ nm}) \rightarrow 2N \qquad \Delta H° = 945 \text{ kJ mol}^{-1} \tag{2.5}$$

$$O_2 + h\nu \ (\lambda < 240 \text{ nm}) \rightarrow 2O \qquad \Delta H° = 498 \text{ kJ mol}^{-1} \tag{2.6}$$

As shown, the energy required for dissociation in these two cases corresponds to electromagnetic radiation with $\lambda < 126$ nm and $\lambda < 240$ nm respectively.

Some dissociated atoms remain in the atomic form while others may recombine, and species such as NO are also produced in the thermosphere. The relative proportion of atoms to molecules increases with altitude and at 120 km the concentration of oxygen atoms approximately equals that of dioxygen molecules. Less nitrogen is present as atoms. Because a significant portion of dioxygen and dinitrogen gases are in the atomic form, the average molar mass, $\overline{M}_a$, becomes smaller than the value of 29.0 g mol^{-1} which obtains in the lower atmosphere.

Solar energy is also capable of ionizing both molecules and atoms. For this reason, the region above the mesopause is alternatively referred to as the ionosphere.

$$N_2 + h\nu \ (\lambda < 80 \ \text{nm}) \rightarrow N_2^+ + e^- \qquad \Delta H^\circ = 1500 \ \text{kJ mol}^{-1} \qquad (2.7)$$

$$O + h\nu \ (\lambda < 91 \ \text{nm}) \rightarrow O^+ + e^- \qquad \Delta H^\circ = 1310 \ \text{kJ mol}^{-1} \qquad (2.8)$$

(Note the use of standard enthalpy rather than free energy to describe the energetics of atmospheric gas-phase reactions. This is convenient because temperature and pressure have less influence on enthalpy compared with free energy changes. Also, it allows for description of the energetics of bond breaking and formation aside from statistical considerations of the bulk system.)

In regions of the atmosphere where energy is absorbed and causes ionization, the reverse endothermic electron-capture reactions also occur to some extent, releasing energy as kinetic energy. This is the reason for the high thermodynamic temperatures in the thermosphere. Proceeding closer to the Earth's surface but still in this region, there is a smaller flux of less highly energetic radiation to be absorbed and so the temperature and the population of atoms and ions decrease.

In the mesosphere, solar radiation begins to encounter new types of chemical species. Ozone (O_3), which has a particularly high concentration in the stratosphere, is present to some extent above the stratospause and is capable of absorbing solar radiation of lower energy than that required for the dissociation and ionization of more stable species. This absorption of longer wavelength radiation leads to an increase in temperature when moving to lower altitudes in the mesosphere.

The absorption of radiation by ozone causes it to dissociate, producing an oxygen molecule and atom, both in an excited state as indicated by the asterisks:

$$O_3 + h\nu \ (\lambda \simeq 230-320 \ \text{nm}) \rightarrow O_2^* + O^* \qquad (2.9)$$

The temperature decline, as one moves down through the stratosphere, has an explanation that is analogous to that given for the thermosphere. Radiation within the energy range of ozone absorption has been removed in the higher regions, producing elevated temperatures there; it therefore cannot penetrate further.

Little additional radiation is absorbed as solar photons pass through the troposphere. However, when the remaining spectrum of radiation strikes the Earth's surface it is partly absorbed and then remitted as lower energy infrared (IR) radiation. Some of the IR radiation is absorbed by certain gases in the troposphere—the two principal ones being water vapour and carbon dioxide. Absorption causes heating near the Earth's surface, an effect that declines with increasing altitude as there is less radiation remaining to be absorbed and the concentration of absorbing gases is also declining. The heating of the

Earth's lower atmosphere in this manner is the well known 'greenhouse effect'. It is a key factor in supporting life as we know it on Earth, but anthropogenic perturbations could seriously alter the present balance. Much more will be discussed about this topic in Chapter 8.

It is important to be aware that the Earth's atmosphere is not an equilibrium system. If that were so, then all of the oxygen would exist combined with other elements, and none would be free to support life. Ultimately, the Sun provides energy for many otherwise energetically unfavourable processes, including photosynthesis, which allow plants to make use of carbon dioxide and produce oxygen.

The constant supply of energy from the Sun is either absorbed, stored by chemical reactions, or reflected back into space by the Earth. The combination of these factors maintains energy relations of the Earth in delicate balance.

2.1.2 The troposphere

Most of the material concerning atmospheric chemistry discussed in this book relates to the troposphere. Here, the temperature profile—warmer, lighter air at the Earth's surface and cooler, denser air at higher altitudes—is such that convection currents and winds cause constant movement of the air. A molecule released at the Earth's surface would typically be swept up to the top of the troposphere in one or two days. Therefore, any gaseous matter which has a long residence time is well mixed, making the overall composition of the troposphere homogeneous. This includes all the major gases listed in Table 2.1. Local variations in concentration are characteristic of physically or chemically reactive species—water vapour and many trace species being good examples.

Other examples of how tropospheric air changes with location are given in Table 2.2. The table highlights particular characteristics of tropospheric air from a number of regions, including over the open ocean, over large continental areas, urban areas, the tropics, and the Arctic.

Table 2.2   Comparison of tropospheric atmospheres from various regions

Location	Atmospheric characteristic
Oceans	Sea salt aerosol (sodium, calcium, magnesium, chloride, sulfate)
Land (dry)	Airborne dust (soil related, plant pollens, etc.)
Urban	High levels of pollutants (smoke, dust, primary and secondary smog chemicals)
Arid tropics	Low humidity, intense solar radiation
Humid tropics	High humidity, natural volatile organics, intense solar radiation
Arctic	Sunlight period variable on a yearly cycle, Arctic haze (including sulfate aerosols, soot, and metals)

2.2 Reactions and calculations in atmospheric chemistry

Before beginning a detailed look at chemistry involved in the troposphere and stratosphere, we will examine some common features of atmospheric chemical reactions and introduce the types of calculations used in a quantitative description of the reactions. The discussion will make use of specific examples in order to illustrate important general principles, but the principles apply to all types of atmospheric reactions including 'natural' processes such as ozone creation and destruction in the stratosphere, and also anthropogenic ones where a particular pollutant generated by human activity interacts with other atmospheric components. Some reactions involve major gases and occur on a global scale, while others deal with trace species in a localized setting. In discussing the composition of the troposphere, we made note of the variability in mixing ratios of reactive minor components. While these gases are present in very small concentrations, many of them have short residence times, which is a reflection of their high reactivity. Frequently, then, a study of atmospheric chemistry centres on the reactions in which the trace species are involved.

2.2.1 Thermodynamic calculations

In some situations, thermodynamic calculations are appropriate for estimating concentrations of species produced in gas phase reactions. For example, nitric oxide is generated during combustion (in forest fires, industrial and domestic heating, internal combustion engines, lightning discharges, etc.) as a result of the combination of nitrogen and oxygen from the air:

$$N_2 \ (g) + O_2 \ (g) \rightarrow 2NO \ (g) \tag{2.10}$$

This reaction has a large positive free energy of formation at 25 °C (for NO (g), $\Delta G° = 2\Delta G_f° = 2 \times (+86.55) = +173.1 \text{ kJ mol}^{-1}$). Nitric oxide is therefore not formed to a significant extent at ambient temperatures. However, the reaction also has a positive entropy change and at the high temperatures associated with combustion, this causes the $\Delta G°$ value to become sufficiently small so that significant quantities of nitric oxide are produced.

For example, in a cylinder of an internal combustion engine at the time of ignition the temperature may reach approximately 2500 °C (=2773 K). In order to calculate the free energy change at the elevated temperature, we begin with data for the standard enthalpies of formation and absolute entropies, $\Delta H_f°$ and $S°$ respectively. The standard values are obtained at 25 °C (=298 K) and we assume that the $\Delta H_f°$ values are unaffected by temperature. The values of $\Delta H_f°$ for nitrogen and oxygen in the standard state are, by definition, zero. $\Delta G_{2773}°$ is then determined by

$$\Delta G_{2773}° = 2\Delta H_{f(NO)}° - T(2S_{NO}° - S_{N_2}° - S_{O_2}°)$$
$$= 2 \times 90.25 \text{ kJ mol}^{-1} - 2773 \text{ K} (2 \times 0.211 - 0.192 - 0.205) \text{ kJ mol}^{-1} \text{ K}^{-1}$$
$$= 111.2 \text{ kJ mol}^{-1}$$

From this final value, the equilibrium constant is readily calculated:

$$\ln K_p = \frac{-\Delta G_T^\circ}{RT} = \frac{-111\,200\,\text{J mol}^{-1}}{8.314\,\text{J mol}^{-1}\,\text{K}^{-1} \times 2773\,\text{K}} = -4.82$$

$$K_p = 0.0080$$

The partial pressure of nitric oxide in the exhaust gases produced under these conditions is evaluated assuming the equilibrium condition

$$K_p = \frac{(P_{NO}/P^\circ)^2}{(P_{N_2}/P^\circ)(P_{O_2}/P^\circ)}$$

In this equation, note that all pressures are relative to P°, 101 325 Pa. Consider a situation where most of the oxygen has been burned and the compressed cylinder gas includes $P_{N_2} = 650\,\text{kPa}$ and $P_{O_2} = 1.0\,\text{kPa}$ with temperature at 2500 °C. After reaction of N_2 and O_2, assume each has reacted so as to lose partial pressure of x kPa; $P_{NO} = 2x$ kPa. Therefore,

$$\frac{[(2_x)/101.3]^2}{[(650 - x)/101.3][(1.0 - x)/101.3]} = 0.0080$$

and

$$\frac{4x^2}{(650 - x)(1.0 - x)} = 0.0080$$

Assume $x \ll 650$ kPa and thus

$$\frac{4x^2}{650(1.0 - x)} = 0.0080$$

giving the quadratic

$$4x^2 + 5.2x - 5.2 = 0$$
$$x = 0.66$$

The assumption that $x \ll 650$ kPa is valid and

$$P_{NO} = 2x = 1.4\,\text{kPa}$$

The nitric oxide mixing ratio in the hot cylinder is then

$$\frac{1.4\,\text{kPa}}{650\,\text{kPa}} \times 10^6\,\text{ppmv} = 2200\,\text{ppmv}$$

Assuming no catalytic decomposition of nitric oxide in the exhaust system, it would be released to the atmosphere.

Concentrations of this order[1] have been observed in the exhaust of automobiles built without emission control. For example, measurements on the exhaust of a 1966 Valiant (Chrysler Corporation) V8 mid-size car running at 2000 rpm showed 1200 ppmv nitric oxide under no-load conditions and 2500 ppmv when the engine was operating with a 50 horsepower load.

We can attempt to go further with calculations concerning nitric oxide by determining what its concentration would be when the gases are cooled and diluted after exiting the exhaust system. The question we are asking is 'How much of the nitric oxide will revert back to dinitrogen and dioxygen?'

Suppose a concentration of 2000 ppmv is diluted in the open air by a factor of 20 000, giving an atmospheric concentration of 0.100 ppmv = 100 ppbv. Assume that the ambient temperature is 25 °C. Using a similar method of calculation for K_p we obtain a new equilibrium concentration of nitric oxide for the lower temperature:

$$\ln K_p = \frac{-173\,100}{8.314 \times 298}$$
$$K_p = 4.73 \times 10^{-31}$$
$$= \frac{(P_{NO}/P^\circ)^2}{(P_{N_2}/P^\circ)(P_{O_2}/P^\circ)}$$

The atmospheric pressures of nitrogen and oxygen are approximately 79 and 21 kPa respectively. Let x be the amount of nitrogen (and oxygen) produced as reaction 2.10 goes from right to left:

$$4.73 \times 10^{-31} = \frac{[(0.0101 - 2x)/101.3]^2}{[(79 + x)/101.3][(21 + x)/101.3]}$$

Making assumptions that $x \ll 79$ and 21, the equation is easily solved and $x \simeq 5 \times 10^{-3}$. Clearly, the assumptions were valid. P_{NO} is equal to $0.0101 - 2x$ in the numerator term. Therefore

$$P_{NO} = 2.8 \times 10^{-14} \text{ kPa}$$

We had assumed at the outset that the atmospheric concentration of nitric oxide when it was released was 100 ppbv. However, according to our thermodynamic calculation, the equilibrium partial pressure of nitric oxide in cool air near a combustion source should be extremely low, corresponding to a mixing ratio of

$$\frac{2.8 \times 10^{-14} \text{ kPa}}{101 \text{ kPa}} \times 10^9 \text{ ppbv} \simeq 3 \times 10^{-7} \text{ ppbv}$$

in the ambient atmosphere. This original concentration is almost one billion times higher than the calculated equilibrium value. In other words, we have shown that the nitric oxide released in the exhaust should quantitatively revert back to dinitrogen and dioxygen.

While our calculation has indicated that an almost negligible concentration of nitric oxide should remain in the atmosphere, in fact substantial amounts are frequently found (see, for example, Fig. 4.2(b)) near combustion sources. Therefore, unlike the calculation for nitric oxide produced within the engine, in this instance thermodynamics does not correctly predict the atmospheric situation.

2.2.2 Kinetic calculations

The reason for the major discrepancy is that the reverse of reaction 2.10 is *extremely slow* at ambient temperatures. During combustion, there is sufficient thermal energy that equilibrium is rapidly established but when cooling occurs, reactants and products are retained near levels corresponding to the high temperature equilibrium.

The second order rate constant for the reaction (reverse of reaction 2.10)

$$2NO \ (g) \rightarrow N_2 \ (g) + O_2 \ (g) \tag{2.11}$$

is given by

$$k_2 = 2.6 \times 10^6 e^{-3.21 \times 10^4/T} \, m^3 \, mol^{-1} \, s^{-1}$$

where T = temperature/K. At 25 °C, the value of k_2 is therefore $4.3 \times 10^{-41} \, m^3 \, mol^{-1} \, s^{-1}$.

The mixing ratio of 100 ppbv estimated above for nitric oxide corresponds to a concentration $[NO]_i$ in mol m^{-3} calculated as follows:

$$[NO]_i = \frac{n}{V} = \frac{P}{R \times T} = \frac{100 \times 10^{-9} \times 101\,325 \, Pa}{8.314 \, J \, mol^{-1} \, K^{-1} \times 298 \, K}$$
$$= 4.1 \times 10^{-6} \, mol \, m^{-3}$$

Therefore, the initial rate of decomposition of nitric oxide at ambient temperature is

$$k_2[NO]_i^2 = 7.2 \times 10^{-52} \, mol \, m^{-3} \, s^{-1}$$

The reaction is obviously very slow and the half-life, $t_{1/2} = 1/k_2[NO]_i$, is equal to

$$t_{1/2} = \frac{1}{4.3 \times 10^{-41} \, m^3 \, mol^{-1} \, s^{-1} \times 4.1 \times 10^{-6} \, mol \, m^{-3}}$$
$$= 5.7 \times 10^{45} \, s$$
$$= 1.8 \times 10^{38} \, y$$

which is more than 10^{28} times as long as the age of the Earth. It is obvious that the nitric oxide generated during combustion is thermodynamically unstable, but it is kinetically extremely inert at 25 °C with respect to reaction 2.11.

In this example then, thermodynamics predicted complete reversion of nitric oxide to dinitrogen and dioxygen. A kinetic evaluation for the same reaction predicted essentially the opposite result—the nitric oxide is completely stable with respect to these two products. Experimental observations indicate that actual atmospheric nitric oxide concentrations fall between these two extremes.

The observations imply that other reactions must be involved in controlling the atmospheric concentration of this gas and we must consider these simultaneously with reaction 2.11. The most important is the oxidation of nitric oxide to produce nitrogen dioxide. With dioxygen as oxidant (reaction 2.12), once again it can be shown that thermodynamics fails us, as a simple calculation shows that the ratio P_{NO_2}/P_{NO} should always be greater than 10^6:

$$2NO \ (g) + O_2 \ (g) \rightarrow 2NO_2 \ (g) \tag{2.12}$$

This is far different from ratios normally found in the natural atmosphere. Furthermore, kinetics still predicts that the oxidation should proceed slowly.

Consider the situation in a polluted urban atmosphere where the nitric oxide concentration in the morning might be approximately 150 ppbv (0.15 ppmv = 1.5×10^{-5} kPa, see Fig. 4.2(b)). Assuming reaction 2.12 to be an elementary process then the rate equals $k_3 (P_{NO})^2 (P_{O_2})$. The third order rate constant, k_3, has been evaluated to be $2.4 \times 10^{-3} \, kPa^{-2} \, s^{-1}$

at 25 °C. If we assume that oxygen has its normal partial pressure of approximately 21 kPa, the initial rate of oxidation would be

$$
\begin{aligned}
\text{rate} &= 2.4 \times 10^{-3}\,\text{kPa}^{-2}\,\text{s}^{-1}(1.5 \times 10^{-5}\,\text{kPa})^2(21\,\text{kPa}) \\
&= 1.1 \times 10^{-11}\,\text{kPa}\,\text{s}^{-1} \\
&= \frac{1.1 \times 10^{-11} \times 10^9}{101.3} \times 3600 \times 24 \\
&= 9.7\,\text{ppbv day}^{-1}
\end{aligned}
$$

This initial rate of oxidation is slow compared with the known concentration and cannot explain the rapid build-up of nitrogen dioxide observed during a smog event (see Fig. 4.2(a) and (b)). We shall see in Chapter 4 that dioxygen is not the most important oxidant in this reaction. Rather it is a combination of ozone (O_3), peroxy radicals (ROO·) and oxy radicals (RO·)—also present in a tropospheric smog—which have the major effect on oxidizing nitric oxide. A general form of the reaction with peroxy radical species is

$$
\text{ROO·} + \text{NO} \rightarrow \text{RO·} + \text{NO}_2 \tag{2.13}
$$

Unfortunately, detailed calculations for this general reaction are difficult because of the variety of peroxy and oxy radical species and their variable and frequently unknown concentrations.

2.2.3 Photochemical reactions

If we now consider the means by which nitrogen dioxide is decomposed and reverts to nitric oxide we encounter yet another feature characteristic of atmospheric reactions—the fact that many of these reactions have a photochemical component. By this we mean that the absorption of (usually solar) electromagnetic energy by species in the atmosphere is required to stimulate certain reactions. The reactant is excited to a higher energy state, thus enhancing bond-breaking or bond-making processes. The first step in a photochemical reaction is then

$$
\text{XY} + h\nu \rightarrow \text{XY}^* \qquad \text{absorption} \tag{2.14}
$$

The asterisk here and elsewhere is used to denote an excited state species. Absorption of energy is followed by further reactions which may take a variety of pathways such as

$$
\text{XY}^* \rightarrow \text{X} + \text{Y} \qquad \text{decomposition} \tag{2.15}
$$

or

$$
\text{XY}^* + \text{other reactant(s)} \rightarrow \text{products} \tag{2.16}
$$

The rate constant for a photochemical reaction, symbolized as f to distinguish it from the thermal rate constant k, is determined by several factors (eqn 2.17):

$$
f = \int_{\lambda_1}^{\lambda_2} J_\lambda \sigma_\lambda \phi_\lambda \, \mathrm{d}\lambda \tag{2.17}
$$

In this expression, λ_1 and λ_2 are the wavelength 'limits' of the radiation being considered. For example, at the Earth's surface the values for solar radiation would be approximately

300 and 800 nm respectively, while the range would extend to lower wavelengths in the stratosphere. The other terms in the expression all depend on the range of wavelengths under consideration. J is the radiative flux; σ is the absorption cross-section, which is a measure of the ability of the molecule of interest to absorb radiation of the type being considered; ϕ is the quantum yield, which is the ratio of the number of molecules undergoing the specific reaction of types, such as 2.15 or 2.16, to number of quanta of radiation absorbed.

When every absorption event leads to the particular reaction of interest, ϕ has a value of unity. Where the excited species is deactivated by 'quenching' through collisions with other gases (M) (reaction 2.18) or by transferring energy to other molecules leading to their excitation (reaction 2.19), which occurs to a large extent, ϕ is very small.

$$XY^* + M \rightarrow XY + M + \text{kinetic energy} \qquad \text{quenching} \qquad (2.18)$$

$$XY^* + AB \rightarrow XY + AB^* \qquad \text{intermolecular transfer} \qquad (2.19)$$

As another extreme, the quantum yield is greater than one when a single activated molecule sets off a chain of events leading to a sequence of reactions.

One important tropospheric photochemical reaction is the decomposition of nitrogen dioxide:

$$NO_2 + h\nu \rightarrow NO + O \qquad (2.20)$$

The atomic oxygen produced in this reaction is in the ground state ($O\,(^3P)$) and, as an aside, it can be mentioned that $O\,(^3P)$ reacts with molecular dioxygen to generate ozone—the only significant pathway to produce ozone in the lower troposphere. The photolytic quantum yield for reaction 2.20 depends on the wavelength of the electromagnetic radiation and is near 1 for $\lambda < 360$ nm (near the high energy end of the visible region of the spectrum) but falls off to 0 at about $\lambda > 440$ nm. It is commonly said that the minimum energy required to effect the process is associated with radiation of 400 nm. The rate of the reaction is given by

$$\text{rate} = f_1[NO_2]$$

As usual, the photochemical first order rate constant, f_1, depends on the intensity and wavelength of the incident radiation. Values of f_1 range from about $5.6 \times 10^{-3}\,\text{s}^{-1}$ in intense sunlight to 0 at night. In sunlight, therefore, nitrogen dioxide would have a half-life, $t_{1/2}$, of $\ln 2/5.6 \times 10^{-3}\,\text{s}^{-1} \simeq 120$ s, indicating that the photolysis is relatively rapid. Decomposition of nitrogen dioxide is balanced by synthesis, as shown in reaction 2.13, also a kinetically rapid process. The steady-state situation established between decomposition and synthesis is known as a photostationary state and is described by equating the rates of the production and consumption reactions:

$$\text{production (reaction 2.13)} = \text{consumption (reaction 2.20)}$$

$$k_2[ROO\cdot][NO] = f_1[NO_2]$$

Therefore

$$\frac{[NO_2]}{[NO]} = \frac{k_2[ROO\cdot]}{f_1}$$

(Note that ROO· is a composite of several species including ozone, and k_2 is a composite rate constant.) In any case, the final equation determines the ratio of nitrogen dioxide to nitric oxide, a ratio which is established quickly and may vary from a small value in bright sunlight when f_1 is large, to a higher night-time value.

The example we have used to illustrate the importance of photochemistry in atmospheric reactions is one which occurs in the troposphere. There are many other reactions and, as we move upward in altitude to the stratosphere and above, we find that photochemistry becomes even more important. Obviously, this is because of the availability of previously unabsorbed higher energy radiation in the rarified upper atmosphere. There, photolysis of high binding-energy molecules like dioxygen and dinitrogen becomes possible as well as ionization of molecules and atoms that are 'normally' stable.

As we have noted, where kinetic properties determine the behaviour of chemical species in the atmosphere, we frequently use the term half-life, $t_{1/2}$, to describe the progress of the reaction. A similar, alternative description is given by the term residence time, τ (also called atmospheric lifetime) which we defined in Chapter 1. Whereas $t_{1/2}$ represents the time taken for half of the reaction to occur, τ is the time during which the original concentration decreases to $1/e$ (37%) of its original value. For a first order or psuedo first order reaction, the residence time is equal to the reciprocal of the rate constant.

2.2.4 Reactions involving free radicals

We cannot leave an introduction to atmospheric chemical reactions without discussing the importance of free radicals[2] in many reactions. We will encounter a variety of radical species, but none is more important than the hydroxyl free radical (·OH), a species that plays a central role in many atmospheric chemistry reactions. Neutral hydroxyl (not to be confused with the negatively charged hydroxide ion, OH^-, which is so pivotal in aqueous solution chemistry) is formed in the troposphere by a variety of means but the most important is a four-step process (including two photochemical steps):

$$NO_2 + h\nu \rightarrow NO + O \tag{2.20}$$

$$O + O_2 + M \rightarrow O_3 + M \tag{2.21}$$

In this second step, the combining of O and O_2 takes place on a 'third body', represented as M in 2.21. Since dinitrogen and dioxygen are the most common species throughout the Earth's atmosphere, the third body is usually one of these molecules.

$$O_3 + h\nu \rightarrow O_2^* + O^* \tag{2.22}$$

$$O^* + H_2O \rightarrow 2 \cdot OH \tag{2.23}$$

As noted above, the atomic oxygen produced in reaction 2.20 is ground state or triplet $(O\,(^3P))$ oxygen while that in reaction 2.22 is in the excited state or singlet $(O\,(^1D))$ oxygen. For the former reaction, radiation in the visible region supplies sufficient energy to give a large value for the rate constant, f_1, while for the latter, higher energy $(\lambda < 325\,nm)$ ultraviolet radiation is required. Excited state oxygen atoms are essential in order to form hydroxyl by reaction with atmospheric (gaseous) water according to reaction 2.23. In competition with this reaction are quenching processes of the type

$$O^* + M \rightarrow O + M + \text{kinetic energy} \qquad (2.24)$$

Because of the availability of many M species (O_2, N_2, etc.) in air, reaction 2.24 proceeds readily and this results in a low quantum yield, ϕ, for reaction 2.23.

Because of its low concentration and highly reactive nature, it is extremely difficult to measure[3] concentrations of hydroxyl radicals but in many situations, levels have been estimated to be between 2.5×10^5 and 1×10^7 molecules per cm^3, with values being higher at low altitudes and latitudes, in daytime, and in areas of heavy pollution. In the tropics, an average concentration of 2×10^6 molecules per cm^3 may be a reasonable estimate. The factors determining actual concentrations of the hydroxyl radical relate to eqns 2.20 to 2.24.

We will later encounter a number of different kinds of atmospheric reactions in which hydroxyl radicals take part. However, in a quantitative sense the two most important are

$$\cdot OH \ + \ CO \ \rightarrow \ \underset{\text{hydrogen radical}}{\cdot H} \ + \ CO_2 \qquad (2.25)$$

$$\cdot OH \ + \ CH_4 \ \rightarrow \ \underset{\text{methyl radical}}{\cdot CH_3} \ + \ H_2O \qquad (2.26)$$

Reactions with carbon monoxide and methane account for approximately 70% and 30% respectively of all reactions involving hydroxyl radical in an unpolluted atmosphere. In the course of both reactions, other highly reactive radical species (the hydrogen and methyl radicals) are formed and these, in turn, react further:

$$\cdot H \ + \ O_2 \ + \ M \ \rightarrow \ \underset{\text{hydroperoxyl radical}}{HOO\cdot} \ + \ M \qquad (2.27)$$

$$\cdot CH_3 \ + \ O_2 \ + \ M \ \rightarrow \ \underset{\text{peroxymethyl radical}}{CH_3OO\cdot} \ + \ M \qquad (2.28)$$

The peroxy radical products are important oxidants, as we have already noted in the case of oxidation of nitric oxide (reaction 2.13).

In regions affected by either natural or anthropogenic gaseous emissions, other processes frequently contribute in a significant way to the loss of the hydroxyl radical. Most of the reactions may be classified in two categories. The first category is hydrogen abstraction as in the reactions with toluene (reaction 2.29) and formaldehyde (reaction 2.30)

$$\cdot OH \ + \ \text{HCHO} \ \longrightarrow \ H\dot{C}{=}O \ + \ H_2O \qquad (2.30)$$

Similar to the reaction with methane, in 2.29 and 2.30 new highly reactive radical species are produced and they too react further with oxygen as in reactions 2.31 and 2.32, forming peroxy radical species:

$$(2.31)$$

$$(2.32)$$

The second category of reactions is addition across a multiple bond, as illustrated by the reactions with ethene (reaction 2.33) and benzene (reaction 2.35). Again, new highly reactive species are produced which typically begin a further reaction sequence by adding oxygen (Reactions 2.34 and 2.36):

$$\cdot OH \quad + \quad H_2C{=}CH_2 \longrightarrow H_2\dot{C}{-}CH_2OH \qquad (2.33)$$

$$(2.34)$$

$$(2.35)$$

$$(2.36)$$

There is a great variety of organic species present in the atmosphere at low concentrations. These include ones of biogenic origin emitted by living or decomposing organisms, and the almost limitless range of chemicals released to the atmosphere from many industrial processes. Even present in buildings, organic species are given off by the many natural and synthetic materials used in construction. The possible chemical reactions with which these chemicals are involved are also limitless, but a surprising number are initiated by hydroxyl or other radical species. In many cases, the first step has been adequately described, often involving principles such as those just shown; rate constants are also available so that the rate of destruction can be calculated. Subsequent reaction steps are often less clear and the mechanisms leading to ultimate destruction are not known.

Box 2.1 The hydroxyl radical as an industrial chemical

Recently there has been a number of reports in which the hydroxyl radical is employed as an oxidant to destroy residual organic chemicals in air or water. This requires being able to generate the radical in an appropriate manner and at sufficient concentrations. An example of the application is illustrated by recent developments in technology for the decontamination of groundwater using photocatalysis (*Chemical and Engineering News*, **70** (1992) 25–8). A parabolic trough radiation collector has been designed to focus solar radiation on a titanium dioxide semiconductor catalyst which is immersed in a groundwater stream that had been pumped to the surface for decontamination. The ultraviolet component of the sunlight activates the catalyst, producing hydroxyl radicals which are able oxidatively to decompose 'stable' chlorinated compounds such as the widely used solvents trichloroethylene and perchloroethylene. The decomposition products are carbon dioxide, water, and hydrochloric acid. To remove soluble carbonate species, the water is previously acidified to pH 5, causing evolution of gaseous carbon dioxide, and after irradiation it is subsequently neutralized to pH 7 with sodium hydroxide. In this way up to 400 000 litres of contaminated water could be treated per day.

The main points

1 The Earth's atmosphere is a thin shell of gases with pressure of 10^5 Pa at the surface, decreasing to less than 10^{-2} Pa at an altitude of 100 km. In the lower altitudes, nitrogen and oxygen make up about 80% and 20% respectively of the gaseous species. While other gases are present in much smaller, even trace amounts, they play an important role in processes that influence life on the Earth.

2 The atmosphere may be divided into four regions—the troposphere, stratosphere, mesosphere, and thermosphere—on the basis of the direction of temperature change with increasing altitude.

3 For any given atmospheric chemical process, either thermodynamic or kinetic control is possible, with the latter being frequently the case. Therefore thermodynamic calculations alone often lead to erroneous predictions about concentrations, at least in the short term.

4 Many important atmospheric processes are, partly or completely, photochemical processes. The energy of solar radiation increases with increasing altitude so that photochemistry becomes even more important in the upper atmosphere.

5 Free radicals play a dominant role in many atmospheric reactions and the hydroxyl radical is especially prominent in this regard.

Additional reading

1 Chamberlain, J. W., and D. M. Hunten, *Theory of Planetary Atmospheres, An Introduction to their Physics and Chemistry*, 2nd edition, Academic Press, London; 1987.

2 Finlayson-Pitts, B. J., and J. N. Pitts Jr, *Atmospheric Chemistry*, John Wiley, Chichester; 1986.

3 Wayne, R. P., *Chemistry of Atmospheres, An Introduction to the Chemistry of the Atmospheres of Earth, the Planets and their Satellites*, 2nd edition, Clarendon press, Oxford; 1991.

4 Atkins, P. W., *Physical Chemistry*, 6th edition, Oxford University Press, Oxford; 1997.

Problems

1 Calculate the atmospheric pressure at the stratopause. What are the concentrations ($mol\,m^{-3}$) of dioxygen and dinitrogen at this altitude? How do these concentrations compare with the corresponding values at sea level?

2 What is the total mass of the stratosphere (the region between 15 and 50 km above the Earth's surface)? What mass fraction of the atmosphere does this make up?

3 If the mixing ratio of ozone in a polluted urban atmosphere is 50 ppbv, calculate its concentration (a) in $mg\,m^{-3}$ and (b) in molecules per cm^3.

4 The gases from a wood-burning stove are found to contain 1.8% carbon monoxide at a temperature of 65 °C. Express the concentration in units of $\mu g\,m^{-3}$.

5 Using Fig. 8.1 (p. 155) compare the mixing ratio (as a percentage) of water in the atmosphere for a situation in a tropical rain forest in Kinshasa, Zaire ($T = 36$ °C, relative humidity = 92%) with that in Denver, USA ($T = -8$ °C, relative humidity = 24%).

6 Calculate the maximum wavelength of radiation which could have sufficient energy to effect the dissociation of nitric oxide (NO). In what regions of the atmosphere would such radiation be available?

7 The average distance a gas molecule travels before colliding with another (mean free path, s_{mfp}) is given by the relation

$$s_{mfp} = \frac{kT}{\sqrt{2}P\sigma_c}$$

where T and P are the temperature (K) and pressure (Pa), k is Boltzmann's constant $= 1.38 \times 10^{-23}\,J\,K^{-1}$, and σ_c is the collision cross-section of the molecule (m^2). Calculate the mean free path of a dinitrogen molecule at the Earth's surface ($P°$ and $T = 25$ °C) and at the stratopause. The value of σ_c for dinitrogen is 0.43 nm^2. What does this indicate regarding gas phase reaction rates in these two locations?

8 Calculate the maximum wavelength of radiation required to bring about dissociation of (a) a dinitrogen molecule and (b) a dioxygen molecule? Account qualitatively for the difference.

9 Carbon dioxide in the troposphere is a major greenhouse gas. It absorbs infrared radiation which causes changes in the frequency of carbon–oxygen stretching vibrations. What are the ranges of wavelength (μm), frequency (cm^{-1}), and energy (J) associated with this absorption?

10 The stability of compounds in the stratosphere depends on the magnitude of the bond energies of the reactive part of the molecules. Using thermochemical tables, calculate bond energies of HF (g), HCl (g), and HBr (g), in order to determine the relative ability of these molecules to act as reservoirs for the respective halogen atoms.

11 Use the tabulated bond energy data in a first year chemistry text to estimate the enthalpy change of the gas-phase reaction between hydroxyl radical and methane.

12 Which of the following atmospheric species are free radicals?

$$OH, O_3, Cl, ClO, CO, NO, N_2O, NO_3^-, N_2O_5$$

13 In an indoor atmosphere, for NO_2 the value of the first order rate constant has been estimated to be $1.28\,h^{-1}$. Calculate its residence time.

14 If the rate laws are expressed using $mol\,L^{-1}$ for concentrations and Pa for pressure, what are the units of the second and third order rate constants, k_2 and k_3? Calculate the conversion factor for converting k_2 values obtained in the units above to ones using molecules per cm^3 for concentration and atm for pressure.

15 For the reaction

$$NO + O_3 \rightarrow NO_2 + O_2$$

the second order rate constant has a value of $1.8 \times 10^{-14}\,molecules^{-1}\,cm^3\,s^{-1}$ at 25 °C. The concentration of NO in a relatively clean atmosphere is 0.10 ppbv and that of O_3 is 15 ppbv. Convert these two concentrations into units of molecules cm^{-3}. Calculate the rate of the NO oxidation using units of molecules $cm^{-3}\,s^{-1}$. Show how the rate law may be expressed in pseudo first order terms and calculate the corresponding pseudo first order rate constant.

16 At a particular temperature, the Arrhenius parameters for the reaction

$$\cdot OH + H_2 \rightarrow H_2O + \cdot H$$

are $A = 8 \times 10^{10}\,s^{-1}$ and $E_a = 42\,kJ\,mol^{-1}$. Given that the concentration of hydroxyl in the atmosphere is 7×10^5 molecules cm^{-3} and that of H_2 is 530 ppbv, calculate the rate of reaction (units of molecule $cm^{-3}\,s^{-1}$) for this process.

17 The equilibrium constant for the reaction

$$N_2O_4 \rightleftharpoons 2NO_2$$

has a value of $K_c = 4.65 \times 10^{-3}\,L\,mol^{-1}$ at 25 °C and the corresponding standard enthalpy change is $\Delta H^\circ = -57\,kJ$. Calculate the equilibrium concentration of N_2O_4 in an atmosphere where the concentration of NO_2 is 200 μg m^{-3}. If the temperature were to increase would you expect the relative concentration of N_2O_4 to increase or decrease?

18 It is postulated[4] that the hydroxyl radical concentration in the tropical atmosphere is directly proportional to the rate constant for photolysis of O_3 to O^*, and to the concentration of O_3 and H_2O. It is inversely proportional to $(C_{CO} + 0.03 \times C_{CH_4})$; all concentrations are expressed as mixing ratios. Use this hypothesis to predict the relative hydroxyl concentrations in two tropical atmospheres with the following conditions:

Air temp/°C	25–30	20–25
Cloudiness/%	50–75	25–50
O_3/ppbv	10	19
H_2O/kPa	>25	20
CO/ppbv	38	50
CH_4/ppmv	1560	1580
$J(O_3$ photolysis)/s^{-1}	1.3	0.68

Notes

1 Gould, R. F., ed., Advances in Chemistry Symposium Series, No 143. *Catalysts for the Control of Automotive Pollutants*, American Chemical Society, Washington DC; 1975.

2 Free radicals are atoms or molecules that have one or more unpaired electrons. They are usually highly reactive species. In most cases, we will denote radicals using a dot (•) beside or above the radical species. While some compounds containing nitrogen or chlorine atoms have an unpaired electron, we make an exception and will not usually show the unpaired electron in such compounds, unless some other radical species is involved in the reaction.

3 Elsele, F. L., and J. K. Bradshaw, The elusive hydroxyl radical: measuring OH in the atmosphere. *Analyt. Chem.* **65** (1993), 927A–39A.

4 Data taken from Rodhe, H., and R. Herrera, *Acidification in Tropical Countries* (Scope 36), John Wiley and Sons, Chichester; 1988.

3

Stratospheric chemistry—ozone

Iɴ studying the chemistry of our atmospheric surroundings, for the most part we will limit ourselves to the troposphere—that part of the atmosphere where humans live and where most other life processes also occur. We make an exception in this chapter, however, as we examine important aspects of stratospheric chemistry. This is because gas molecules in the stratosphere act as absorbing centres, moderating the transmission of solar radiation to the Earth. The qualitative and quantitative effect of this is an important determining factor with respect to life processes. Furthermore, unlike the Sun itself which continues to emit its characteristic radiation unaffected by human activities, the extent to which stratospheric gases interact with radiation can be significantly altered by anthropogenic processes. It is for these reasons that Earth-bound environmental chemists should have some knowledge of the chemistry of the stratosphere.

The stratosphere is the region of space running between approximately 15 km and 50 km above the Earth's surface (Fig. 2.1). It is distinguished from the troposphere below and the mesosphere above by having a temperature inversion—that is, the temperature increases with altitude. Similar to the troposphere, the principal constituent gases are dinitrogen and dioxygen, but the higher energy radiation striking these gases, oxygen in particular, leads to chemical reactions different from those found nearer the surface of the Earth.

The most important of these reactions relate to the synthesis and decomposition of ozone. Ozone is a naturally occurring gas found throughout the atmosphere, with a maximum mixing ratio at altitudes ranging from 15 to 30 km above the Earth. This region is frequently called the 'ozone layer'. Figure 3.1 plots the ozone distribution in two ways and it is important to note the distinction between graphs expressed as mixing ratio and those as the logarithm of number density. In the latter case, you can see that the concentration of ozone at the Earth's surface actually increases to within an order of magnitude of that in the ozone layer. Near the Earth, elevated levels can be toxic to both plants and animals. We will discuss aspects of the tropospheric chemistry of ozone in Chapter 4.

In the stratospheric ozone layer, the mixing ratio of the gas is in the part per million range, so, even in that region, it makes up only a small fraction of all molecules. However, this relatively small concentration of molecules is of great importance in screening out harmful ultraviolet (UV) radiation from the Sun, which would otherwise reach the surface

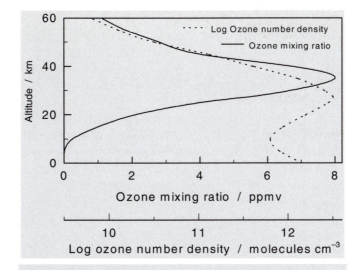

Fig. 3.1 Concentration profile of ozone in the lower atmosphere, shown both as the mixing ratio (solid line) and as the log of number density (broken line). (Data from Wayne, R. P., *Chemistry of Atmospheres*, Clarenden Press; Oxford; 1991. Reprinted with permission.)

of the Earth. In the stratosphere, ozone is an effective filter capable of absorbing ultraviolet radiation with wavelengths between 200 and 315 nm.

The Sun emits radiation over a broad region of the electromagnetic spectrum corresponding to wavelengths of more than 1000 μm to those less than 250 nm. The maximum intensity occurs at 550 nm in the yellow region of the visible spectrum. Longer wavelength photons provide heat and light, defining the global climate and all manner of biological activity. The shorter wavelength (ultraviolet) radiation comprises only a small portion of the total flux, yet is of interest for a number of reasons—one being that it can be harmful, even lethal, to living organisms. In a biological classification,[1] ultraviolet radiation is subdivided into three categories:

- UV-A, 315 to 400 nm (the near ultraviolet ranging into the visible, 7% of the total solar flux) which is not particularly harmful to living species;
- UV-B, 280 to 315 nm (1.5% of the total flux) which can be harmful to both plant and animal species, especially after prolonged exposure; and
- UV-C < 280 nm (0.5% of the total flux) which rapidly damages biota of all types.

The ultraviolet portion of the Sun's emission spectrum is shown in Fig. 3.2. The small flux of UV-C radiation penetrates into the upper atmosphere but is efficiently and completely absorbed by ozone and other atmospheric species before it reaches the Earth's surface. On the other hand, the 1.5% of radiation that is in the UV-B region is only partially absorbed. Ozone is the species responsible for intercepting ultraviolet photons but, as shown in

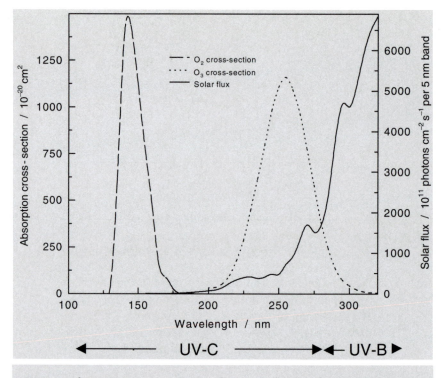

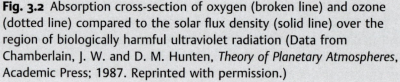

Fig. 3.2 Absorption cross-section of oxygen (broken line) and ozone (dotted line) compared to the solar flux density (solid line) over the region of biologically harmful ultraviolet radiation (Data from Chamberlain, J. W. and D. M. Hunten, *Theory of Planetary Atmospheres*, Academic Press; 1987. Reprinted with permission.)

Fig. 3.2, it has a relatively small absorption cross-section in the UV-B wavelength range. Where concentrations of ozone in the stratosphere are substantially reduced, dangerous levels of UV-B radiation can penetrate into the troposphere.

One type of unit associated with ozone analysis is called the Dobson unit (DU). Measurement involves determining the transmission of ultraviolet radiation at two wavelengths—where ozone does and does not absorb—and relating the values to the amount of ozone in the light path. Because measurements are made from ground level looking up through the atmosphere, they give an integrated value of total ozone in the column of air. One hundred DU are equivalent to a 1 mm thick layer of pure ozone at 0 °C and $P°$. A world average value is approximately 300 DU, ranging from 250 in the tropics to 450 in more northern and southern regions of the planet. The term 'ozone hole' is used to describe the thinning of a region of the ozone layer such as has been observed in recent years, in the spring, over the Antarctic (October) and more recently the Arctic (March). 'Ozone hole' is a partial misnomer as ozone is still present in the polar regions of the stratosphere, though at much reduced concentrations. For example, measurements through the Antarctic 'ozone hole' have been found to be lower than 150 DU.

Two issues are important with respect to ozone concentrations in the stratosphere. The first relates to the global average concentrations which, incidentally, vary from month to month and region to region, and the factors which affect these values. The second issue concerns concentrations in specific regions of the globe, especially the Antarctic and Arctic areas. It is in these latter two locations in particular that human activities have had the greatest effect on the natural variations in stratospheric ozone concentrations. Both of these issues will be discussed in further detail later in this chapter, but we will first investigate some of the basic features of 'natural' ozone chemistry.

3.1 Oxygen-only chemistry—formation and turnover of ozone

Both synthesis and decomposition of ozone may be described in terms of chemistry involving only oxygen-containing species. This was first done by Chapman[2] and involves four fundamental reactions. The enthalpies are a combination of energy involved in making or breaking bonds between atoms and, in two of the reactions, energy derived from the Sun. In keeping with thermodynamic conventions, the solar energy is always expressed as having a negative value.

$$\Delta H^\circ / \text{kJ}$$

Synthesis

$$O_2 + h\nu\,(\lambda < 240\,\text{nm}) \rightarrow O + O \qquad \text{Slow} \qquad -E(h\nu) + 498.4 \qquad (3.1)$$

$$O + O_2 + M \rightarrow O_3 + M \qquad \text{Fast} \qquad -106.5 \qquad (3.2)$$

Decomposition

$$O_3 + h\nu\,(\lambda \simeq 230-320\,\text{nm}) \rightarrow O_2^* + O^* \qquad \text{Fast} \qquad -E(h\nu) + 386.5 \qquad (3.3)$$

$$O + O_3 \rightarrow O_2 + O_2 \qquad \text{Slow} \qquad -391.9 \qquad (3.4)$$

In reaction 3.2, M is a neutral third body, usually N_2 or O_2, the predominant species in the stratosphere. $E(h\nu)$ is the energy of the photon in excess of that required to bring about the reaction.

Looking at the oxygen-only reactions, we begin with the synthesis process. The photochemical reaction 3.1 is slow and results in the production of two 'odd oxygen' species—that is, species containing an odd number (in this case one) of oxygen atoms. The amount of energy and therefore the wavelength of radiation required for this process to occur is calculated as follows. We assume that the enthalpy change is independent of temperature.

$$\Delta H^\circ(\text{reaction 3.1}) = 2\Delta H_f^\circ(O\,(g)) - \Delta H_f^\circ(O_2\,(g))$$
$$= 2 \times 249.2 - 0$$
$$= 498.4\,\text{kJ}\,\text{mol}^{-1}$$

The formation of 2 mol of O (g) from one mol of O_2 (g) therefore requires 498.4 kJ of energy. The wavelength of light associated with this amount of energy is calculated from

$$\lambda = \frac{hcN_A}{E}$$

$$= \frac{6.626 \times 10^{-34} \, \text{J s} \times 2.998 \times 10^8 \, \text{m s}^{-1} \times 10^9 \, \text{nm m}^{-1} \times 6.022 \times 10^{23} \, \text{mol}^{-1}}{498\,400 \, \text{J mol}^{-1}}$$

$$= 240.0 \, \text{nm}$$

It is evident therefore that radiation having wavelengths greater than 240 nm would not possess sufficient energy to cause this reaction to occur, and would penetrate further into the atmosphere.

Reactions 3.2 (synthesis) and 3.3 (dissociation) rapidly interconvert single oxygen atoms with ozone, another odd oxygen species. The $\Delta H°$ (reaction 3.2) for formation of ozone from ground state oxygen atoms and dioxygen molecules is calculated as shown below.

$$\Delta H°(\text{reaction 3.2}) = \Delta H_f°(O_3 \, (\text{g})) - \Delta H_f°(O \, (\text{g})) - \Delta H_f°(O_2 \, (\text{g}))$$

$$= 142.7 - 249.2 - 0$$

$$= -106.5 \, \text{kJ mol}^{-1}$$

Reaction 3.3 appears to be the reverse of reaction 3.2 but a careful examination illustrates an important concept that must be taken into account. When ozone is photodissociated, according to spin conservation theory the products dioxygen and atomic oxygen must both be in ground (triplet) or excited (singlet) states. The ground state formation of the two species requires energy of only 106.5 kJ mol^{-1} (calculation is analogous to that above for reaction 3.2), corresponding to a wavelength of 1123 nm, in the infrared region. However, as shown in eqn 3.5, the rate constant of a photochemical reaction is dependent on the absorption cross-section (σ_λ) of the molecule, and the ability to absorb depends on the wavelength of the radiation:

$$f = \int_{\lambda_1}^{\lambda_2} J_\lambda \sigma_\lambda \phi_\lambda \, \mathrm{d}\lambda \tag{3.5}$$

If the curve for ozone in Fig. 3.2 were extrapolated to 1123 nm, it would be seen that the absorption cross-section is very small at that wavelength. Therefore, because ozone molecules do not significantly absorb 1123 nm photons, the value of the rate constant, f, for dissociation to ground state products approaches zero.

On the other hand, calculation of the energy to produce the two excited species, $O_2(^1\Delta_g)$ and $O(^1D)$ takes into account the excitation energies (E_e) of O_2 (g) and O (g) which are approximately 90 and 190 kJ mol^{-1} respectively.

$$\Delta H° \, (\text{reaction 3.3}) = \Delta H_f°(O \, (\text{g})) + E_e(O) + \Delta H_f°(O_2 \, (\text{g})) + E_e(O_2) - \Delta H_f°(O_3 \, (\text{g}))$$

$$= 249.2 + 190 + 0 + 90 - 142.7$$

$$= 387 \, \text{kJ mol}^{-1}$$

The calculated energy corresponds to a wavelength of 309 nm and radiation of this wavelength or less is capable of effecting the dissociation. The region between 309 and 200 nm is the part of the spectrum where ozone absorbs photons strongly. Absorption with consequent decomposition is the process that protects the Earth from harmful ultraviolet radiation.

Reaction 3.4 describes a mechanism for the destruction of odd oxygen species. This is an exothermic process $(\simeq -400\,\text{kJ}\,\text{mol}^{-1})$ but the relatively large activation energy $(18\,\text{kJ}\,\text{mol}^{-1})$ determines that it is a slow process in the stratosphere.

3.1.1 The ozone layer

The existence of a layer of ozone in the lower stratosphere is explained qualitatively in terms of the reactions above. In the upper stratosphere the intensity of high energy solar radiation is strong, so that dioxygen has a short lifetime in terms of dissociation into oxygen atoms (at 50 km the lifetime of dioxygen is approximately one hour). Therefore the ratio, O to O_2, is relatively large, but because the absolute concentration of oxygen is low, ozone production is limited by reaction 3.2. At low altitudes, there is a plentiful supply of molecular oxygen, but little radiation sufficiently energetic to cause it to dissociate and so the limitation is reaction 3.1. At 20 km the lifetime of dioxygen is about 5 years. In the intermediate range there are both dioxygen and intense high-energy radiation sufficient to maximize the rate of oxygen atom (and therefore ozone) production. The maximum ozone concentration is therefore predicted and found to be at an altitude of approximately 23 km (Fig. 3.1). The actual height of the maximum varies with time of year and latitude on the Earth.

Although oxygen-only chemistry correctly describes the shape and altitude of the ozone profile, it predicts an absolute concentration of ozone which is high by a factor of about two. This indicates that other processes contribute to ozone destruction. It has been suggested that reaction 3.4 accounts for nearly 20% of the odd oxygen removal from the stratosphere. The remainder is due to catalytic reactions involving other trace species.

3.2 Catalytic decomposition processes of ozone

In the previous section, we have seen that there are 'natural' chemical and photochemical processes that lead to the continuous formation and destruction of ozone in the strato-sphere. The balance between these processes would result in a steady-state situation with a well defined and stable ozone layer. However, oxygen-only chemistry cannot adequately explain the present situation. Other chemical species also play a role in determining actual concentrations and, in particular, can lead to enhanced ozone destruction. Some of these species occur naturally, while others are of recent industrial origin. The amount of some of the natural species moving into the stratosphere has also been augmented in recent years by a variety of human activities. We will look at several catalytic routes which have been shown to take part in additional removal of ozone in the stratosphere. Many of the processes share a general mechanism, using X as the reacting species, as follows:

$$X + O_3 \rightarrow XO + O_2 \tag{3.6}$$

$$XO + O \rightarrow X + O_2 \tag{3.7}$$

net reaction $\quad\overline{O + O_3 \rightarrow 2O_2} \tag{3.8}$

where reaction 3.8 is the sum of reactions 3.6 and 3.7.

We are interested in the nature, origin and relative importance of species X. The most common have been identified to be free radicals and are symbolized in three categories:

$$HO_x \ (\cdot H, \cdot OH, HOO\cdot),$$

$$NO_x \ (\cdot NO, \cdot NO_2),$$

$$ClO_x \ (\cdot Cl, ClO\cdot)$$

Depending on altitude and the mixing ratio, each species has varying ability to destroy ozone. For example, near the stratopause the HO_x radicals account for as much as 70% of the total mechanism of ozone destruction including the oxygen-only processes. Lower in the stratosphere around 30 km, the NO_x catalytic decomposition cycle dominates the removal mechanism of ozone. The NO_x cycle in the region just above the tropopause is usually considered to account for 70% of the destruction of ozone, although some recent work indicates that this figure may be high.[3] There are also questions about the relative involvement of the other catalytic cycles in the region near 30 km; they are generally thought to account almost equally for the remaining destruction.

3.2.1 The hydroxyl radical (HO_x) cycle

The radicals are generated by either of two photochemical processes:

$$O(^1D) + H_2O \rightarrow 2 \cdot OH \tag{3.9}$$

$$H_2O + h\nu \rightarrow \cdot H + \cdot OH \tag{3.10}$$

They can catalytically decompose ozone according to the usual cycle:

$$\cdot OH + O_3 \rightarrow HOO\cdot + O_2 \tag{3.11}$$

$$HOO\cdot + O \rightarrow \cdot OH + O_2 \tag{3.12}$$

$$\text{net reaction} \quad \overline{O + O_3 \rightarrow 2O_2} \tag{3.13}$$

Production of hydroxyl radicals is largely due to the 'natural' processes shown and increases in relative importance with increasing altitude.

The hydrogen radical (formed in reaction 3.10) also can participate in ozone removal cycles by reacting with ozone to form a hydroxyl radical and molecular oxygen. In this way the end result is the same as in reaction 3.9, where one molecule of water contributes two hydroxyl radicals, which then participate in the cycle of ozone destruction. Similar to the other radicals, the hydrogen radical can also be shown in an ozone destruction catalytic cycle.

$$\cdot H + O_3 \rightarrow \cdot OH + O_2 \tag{3.14}$$

$$\cdot OH + O \rightarrow \cdot H + O_2 \tag{3.15}$$

$$\text{net reaction} \quad \overline{O + O_3 \rightarrow 2O_2} \tag{3.16}$$

3.2.2 The nitric oxide (NO_x) cycle

The nitric oxide cycle takes the same form as other catalytic cycles, and contributes in a significant way to the 'natural' destruction of stratospheric ozone in the lower region of the stratosphere. The catalytic process is

$$NO + O_3 \rightarrow NO_2 + O_2 \tag{3.17}$$

$$NO_2 + O \rightarrow NO + O_2 \tag{3.18}$$

net reaction $\quad O + O_3 \rightarrow 2O_2 \tag{3.19}$

In the upper atmosphere, nitric oxide is produced by reactions involving atomic and ionic species of nitrogen and oxygen—species produced by the highly energetic radiation that penetrates that part of the atmosphere. At altitudes much greater than 30 km photochemical decomposition of dinitrogen is possible, producing both an electronically excited $N(^2D)$ and a ground state (but with high translational energy) $N(^4S)$ atom. The ground state nitrogen subsequently reacts with dioxygen, resulting in another natural source of nitric oxide:

$$N_2 + h\nu \, (\lambda < 126 \, nm) \rightarrow N(^4S) + N(^2D) \tag{3.20}$$

$$N(^4S) + O_2 \rightarrow NO + O \tag{3.21}$$

At altitudes of less than 30 km, additional nitric oxide is generated *via* a reaction between nitrous oxide and atomic oxygen (reaction 3.22). Nitrous oxide originates principally as a result of denitrification in soils high in nitrate ion (see p. 340). The molecule does not have unpaired electrons and is a stable species with a tropospheric residence time estimated to be 160 y. The principal mechanism for its removal is by migration into the stratosphere where it undergoes the reaction with excited-state oxygen:

$$N_2O + O(^1D) \rightarrow 2NO \tag{3.22}$$

In large part, the production of stratospheric nitric oxide *via* the various processes can be attributed to natural causes, although the mixing ratio of nitrous oxide in the troposphere is steadily increasing. Much of the increase is associated with greater use of nitrogenous fertilizers. The nitric oxide that has been generated by combustion near ground level, or even at altitudes where subsonic aircraft fly, is thought not to be a major contributor to the stratospheric burden, as it has a very short ($\sim$4 day) tropospheric residence time.

The advent of the use of supersonic aircraft, which fly in the stratosphere at altitudes of about 17–20 km, in the 1960s raised fears that they would contribute to considerable ozone destruction. Such aircraft typically consume about 20 000 kg h^{-1} of fuel and in the process generate 160 kg of NO$_x$. Twenty-five thousand kilograms of water are simultaneously released and it is possible that this could contribute to increased production of hydroxyl radicals (reactions 3.9 and 3.10) which would then also enhance ozone decomposition. The fact that these species are injected directly into the stable stratosphere is of particular concern. However, the fleet of these aircraft is very small and up to the present time it is thought that their contribution to enhanced ozone destruction is negligible.

The involvement of nitrogen oxides in ozone chemistry goes beyond that described by the catalytic cycle. Nitric oxide also reacts with hydroxyl radicals to produce nitrous acid:

$$\cdot NO + \cdot OH + M \rightarrow HNO_2 + M \tag{3.23}$$

Since both reactants (but not the product) are potential catalysts for ozone destruction, the consequence of their reaction together is reduced removal of ozone. This is one example of the fact that catalytic effects of individual species are not necessarily additive.

3.2.3 The chlorine radical (ClO$_x$) cycle

Chlorine atoms catalyse the destruction of ozone and at 30 km the natural contribution to ozone destruction by chlorine accounts for about 10% of the total. The only significant natural precursor of atomic chlorine is methyl chloride (CH_3Cl), which is produced biogenically from the oceans, from the burning of vegetation and from volcanic emissions. Once in the atmosphere this compound photolyses to release reactive chlorine:

$$CH_3Cl + h\nu \rightarrow \cdot CH_3 + \cdot Cl \tag{3.24}$$

Methyl chloride is partially removed by reaction with hydroxyl:

$$CH_3Cl + \cdot OH \rightarrow \cdot CH_2Cl + H_2O \tag{3.25}$$

Other natural chlorine sources are various forms of chloride. Such sources as hydrochloric acid released by volcanoes or chloride ion arising from sea-salt spray are short-lived in the troposphere. Most is washed out through rainfall before it reaches the stratosphere.

Similar to the previous cases, an ozone destruction mechanism involving chlorine can be described as follows:

$$\cdot Cl + O_3 \rightarrow ClO\cdot + O_2 \tag{3.26}$$
$$ClO\cdot + O \rightarrow \cdot Cl + O_2 \tag{3.27}$$
$$\text{net reaction} \quad \overline{O + O_3 \rightarrow 2O_2} \tag{3.28}$$

There are other catalytic cycles involving chlorine:

$$HOO\cdot + ClO\cdot \rightarrow HOCl + O_2 \tag{3.29}$$
$$HOCl + h\nu \rightarrow \cdot OH + \cdot Cl \tag{3.30}$$
$$\cdot Cl + O_3 \rightarrow ClO\cdot + O_2 \tag{3.31}$$
$$\cdot OH + O_3 \rightarrow HOO\cdot + O_2 \tag{3.32}$$
$$\text{net reaction} \quad \overline{2O_3 \rightarrow 3O_2} \tag{3.33}$$

and

$$BrO\cdot + ClO\cdot \rightarrow \cdot Br + \cdot Cl + O_2 \tag{3.34}$$
$$\cdot Br + O_3 \rightarrow BrO\cdot + O_2 \tag{3.35}$$
$$\cdot Cl + O_3 \rightarrow ClO\cdot + O_2 \tag{3.36}$$
$$\text{net reaction} \quad \overline{2O_3 \rightarrow 3O_2} \tag{3.37}$$

In reactions 3.29–3.31 bromine can replace chlorine in the cycle and together these reactions account for nearly 60% of the routes by which ozone is destroyed by halogens.

Although the natural sources produce small amounts of atomic chlorine in the stratosphere, of greater significance to the destruction of the ozone is chlorine derived from the anthropogenically produced chlorofluorocarbons (CFCs).

3.2.4 Anthropogenic sources of chlorine

Chlorinated fluorocarbons (CFCs) are a class of compounds initially developed in the 1930s; they have properties that make them particularly useful in a number of applications.

The desirable properties include low viscosity, low surface tension, low boiling point, and chemical and biological inertness which accounts for their being non-toxic and non-flammable. Applications that take advantage of these properties include use as refrigerants, solvents for cleaning electronic and other components, and blowing agents for polymer foams.

To understand the environmental importance of CFCs, we begin by looking at a few examples and how they are named. The nomenclature for CFCs is as follows:

CFC-*xyz* ⟵ number of C atoms −1 (usually omitted if $x = 0$)
number of F atoms
number of H atoms +1

The balance of atoms required for a saturated carbon is then made up of chlorine.

A simple way of determining the chemical formula from the numerical symbol for a specific CFC is to add 90 to the number. The modified three-digit number then gives the number of carbons, hydrogen, and fluorine in sequence. As an example, the chemical formula for CFC-115 would be determined in the following manner:

$$115 + 90 = 205$$

The number 205 then indicates that there are 2 carbons, 0 hydrogens, and 5 fluorines. The remaining one atom is chlorine. The chemical formula for CFC-115 is therefore CF_3CF_2Cl. Postscript letters are used to indicate different structural isomers. CFC-114a is $CFCl_2CF_3$ to distinguish it from CF_2ClCF_2Cl.

Table 3.1 lists some properties and environmental characteristics of common CFCs. One of the important properties is the ozone depletion potential (ODP), which is defined as the ratio of the impact on ozone from a specific chemical to the impact from an equivalent mass of CFC-11, the standard by which all others are calculated. The ODP values generated are then useful as a guide to predict the overall impact on the ozone layer from certain chemical species. The ODP takes into account the reactivity of the species, its atmospheric lifetime, and its molar mass. The amount of chlorine in the chemical species is also important. CFC-114 (CF_2ClCF_2Cl) contains two chlorine atoms and has an ODP of 1.0 while the similar compound CFC-115 (CF_2ClCF_3) with a single chlorine has a substantially lower ODP of 0.6. Similarly, carbon tetrachloride (CCl_4), containing four atoms of chlorine per molecule, has a high (1.1–1.2) ODP. Most CFCs have ODP values between 0.1 and 1.0, while hydrochlorofluorocarbons (HCFCs) have ODP values which are about ten times lower (0.01–0.1). Hydrofluorocarbons, which contain no chlorine, have a zero ODP value.

Despite the fact that CFC molecules are much heavier than those of air, they tend to be well mixed and homogeneously distributed throughout the troposphere in the gas phase due to the strong convective mixing. As noted above, an important property of CFCs is that they are almost completely inert both biologically and chemically in the Earth's environment, including in the troposphere. Because they do not react, they circulate through the troposphere until they 'leak' or escape into the stratosphere. While unreactive in the troposphere, at higher altitudes they are able to undergo ultraviolet photolytic decomposition, as a consequence of being exposed to the intense flux of energetic

Table 3.1 Properties of common CFCs

	Formula	Atmospheric lifetime/y	ODP[a]	Release rate/ 10^6 kg y^{-1}	Concentration/pptv 1977	Concentration/pptv 1993	Contribution to O$_3$ loss[b]/%
CFC-11	CFCl$_3$	60	1.0	281	140	272	31
CFC-12	CF$_2$Cl$_2$	195	1.0	370	255	519	36
CFC-113	CF$_2$ClCFCl$_2$	101	0.8	138	–	–	14
CFC-114	CF$_2$ClCF$_2$Cl	236	1.0	–	–	–	–
CFC-115	CF$_2$ClCF$_3$	522	0.6	–	–	–	–

[a]Ozone depletion potential (ODP).
[b]The percentage contribution to ozone depletion is based on the major halogen-containing species only. ODP values were obtained from the US EPA's Stratospheric Protection Division; CFC concentrations and from Environment Canada (SOE Bulletin No 94–6, Fall 1994); and the remaining values are from Wayne, R. P., *Chemistry of Atmospheres*, Clarendon Press, Oxford; 1991.

ultraviolet radiation:

$$CFCl_3 + h\nu(\lambda < 290\,nm) \rightarrow \cdot CFCl_2 + \cdot Cl \qquad (3.38)$$

The released chorine radical is now able to take part in the catalytic cycles shown in reactions 3.26–3.28. A second and possibly third chlorine radical could also be produced by further decomposition of the remnant $\cdot CFCl_2$ radical, generating additional potential for ozone depletion.

A large portion of the CFCs produced over the past 60 years have been released and have migrated into the stratosphere. Therefore, although legislation to phase out the use of these compounds, accepted by many nations in the Montreal Protocol of 1989 and further amendments in London and Copenhagen, may be significant in reducing the rate of stratospheric accumulation of CFCs, the 'memory effect' of past releases will persist for many decades.

There has not been unanimous acceptance of the provisions of the Montreal Protocol. Chlorinated fluorocarbons are very useful materials and are less expensive than substitutes which have been developed. Low-income nations in particular have argued that they need to take advantage of this well established low-cost technology. Current research is aimed at developing replacement compounds that retain the desirable properties of the original CFCs, but will not be involved in stratospheric ozone decomposition. In a way, there is a contradiction here, as the essential property of chemical inertness is also the property which leads to tropospheric stability and slow leakage into the stratosphere. Furthermore, the fact that both new and old compounds are greenhouse gases must also be taken into consideration.

Current research is centred on modifying the relative amounts of fluorine, chlorine, and hydrogen in new compounds. A common approach has been to incorporate hydrogen in the structure—then called hydrochlorofluorocarbons (HCFCs)—or replace chlorine altogether and produce what are known as hydrofluorocarbons (HFCs). Increasing the hydrogen content reduces inertness, in this way making the tropospheric lifetime shorter.

The presence of hydrogen means that the compounds are more reactive and flammable and the higher reactivity is a disadvantage in some applications. Increasing the proportion of fluorine at the expense of chlorine tends to produce a highly stable compound. The C–F bond has a large bond energy (441 kJ mol^{-1} compared with 328 kJ mol^{-1} for the C–Cl bond) and stratospheric photolysis does not occur to any significant extent. However, such compounds are excellent and persistent greenhouse gases. For compounds containing no chlorine, obviously no Cl atoms can be released, so that there is no potential to deplete ozone. Increasing chlorine content is undesirable both because the ability to destroy ozone is intrinsically higher and because compounds containing chlorine and hydrogen tend to be more toxic.

Table 3.2 lists some CFC alternatives and indicates the outlook as far as regulation in the USA is concerned.

From a combined industrial–environmental perspective, one of the more promising new compounds is HFC-134a (CF_3CH_2F). This HFC has intermediate stability; it is oxidized by the hydroxyl radical in the troposphere, and has a residence time of 18 y. It is not very flammable and, having no chlorine, will have no ozone depletion potential. Unfortunately it is difficult and expensive to manufacture. Another CFC replacement is HCFC-123 (CF_3CHCl_2),

Table 3.2 CFC alternatives, applications, and regulations

Substance	Formula	ODP[a]	Major uses	Regulatory outlook
HCFC-22	$CHClF_2$	0.055	Foams, air-conditioning, refrigeration, aerosols	Clean Air Act bans aerosol use in new equipment after 2005
HCFC-142b	CH_3CClF_2	0.065	Foams, refrigerants	EPA likely to ban use in new equipment after 2005
HCFC-141b	CH_3CCl_2F	0.11	Foams, solvents	EPA likely to approve for foams use only and ban use in new equipment after 2005
HCFC-123	$CHCl_2CF_3$	0.02	Foams, air-conditioning, fire fighting	EPA likely to approve only air-conditioning use; Clean Air Act bans use in new equipment after 2015
HFC-134a	CH_2FCF_3	0.0	Refrigeration, air-conditioning	No restrictions anticipated
HCFC-124	$CHClFCF_3$	0.022	Refrigeration, sterilant	Clean Air Act bans use in new equipment after 2005
HFC-125	CHF_2CF_3	0.0	Refrigeration	No restrictions anticipated
HFC-22	CH_2F_2	0.0	Refrigeration, air-conditioning	No restrictions anticipated

[a] Estimates from United Nations Environment Program's 'Scientific Assessment of Ozone Depletion: 1991'. Estimates depend on chlorine content and atmospheric lifetime. Potentials are set relative to CFC-11, which is assigned a value of 1.0.
Reproduced with permission from Zurer, P. S., *Industry*, consumers prepare for compliance with pending CRC ban, *Chem. Eng. News*, **70** (1992), 7–13.

which contains chlorine but has only about one-tenth the ozone depleting potential of CFC-11, largely because of its relatively short tropospheric lifetime.

Halons, the bromine analogues of CFCs, are widely used in fire extinguishers because of their high density (they settle over and smother a fire at ground level) and because the bromine atom terminates the radical chain reactions which propagate the combustion process. Methyl bromide also finds application as a soil fumigant. Chemicals containing bromine are much more reactive than the chlorine analogues in terms of destroying ozone; as a result some of the ODP values for bromine compounds are very high. For example, the halons bromochlorodifluoromethane, CF_2ClBr (1211), bromotrifluoromethane, CF_3Br (1301), and dibromotetrafluoroethane $CBrF_2CBrF_2$ (2402) have ODP values of 3.0, 10.0, and 6.0 respectively. The nomenclature for the halons is straightforward. The first digit indicates the number of carbon atoms; the second, the number of fluorine atoms; the third, the number of chlorine atoms; and the fourth, the number of bromine atoms. Additional atoms needed for a saturated carbon are assigned to hydrogen. At the 1995 world conference on Ozone Depleting Substances in Vienna, Austria, agreement was reached to completely phase out the use of methyl bromide by the year 2010.

3.2.5 Kinetic calculations

We have presented the main cycles describing stratospheric ozone decomposition, and made general comments on their relative importance. The percentage involvement for each catalyst given earlier is based on estimates from kinetic data.

As an example, the overall rate of destruction of ozone by the NO_x cycle (reactions 3.17–3.19) at an altitude of 20 km ($T \simeq 220$ K) is given here. The rate expressions for reactions 3.17 and 3.18 are

$$\text{rate} = k_{17}[NO][O_3]$$
$$\text{rate} = k_{18}[NO_2][O]$$

where k_{17} and k_{18} represent the respective second order rate constants.

The overall rate of reaction is determined not by the sum of the rates of reaction but by the rate determining step. In order to determine which of the two is rate limiting, we must take into account the concentrations of the species involved and calculate the rate constants under conditions obtaining in a particular region of the atmosphere.

The rate constants can be calculated using the Arrhenius expression

$$k = Ae^{-E_a/RT}$$

and values at 220 K are given in the Table below:

	E_a/kJ mol^{-1}	A/cm^3 molecule^{-1} s^{-1}	k/cm^3 molecule^{-1} s^{-1}
Reaction 3.17	11.4	1.8×10^{-12}	3.5×10^{-15}
Reaction 3.18	0	9.3×10^{-12}	9.3×10^{-12}

At an altitude of 20 km the number concentrations of the four species involved in the two reactions can be approximated as follows:[4]

$$[O] = 2.0 \times 10^7 \text{ molecules cm}^{-3} \qquad [O_3] = 3.0 \times 10^{12} \text{ molecules cm}^{-3}$$
$$[NO] = 2.0 \times 10^9 \text{ molecules cm}^{-3} \qquad [NO_2] = 8.0 \times 10^9 \text{ molecules cm}^{-3}$$

The rates of the individual reactions are then determined to be, for reaction 3.17,

$$\begin{aligned}
\text{rate}_{3.17} &= k_{17}[NO][O_3] \\
&= (3.5 \times 10^{-15}) \times (2.0 \times 10^9) \times (3.0 \times 10^{12}) \\
&= 2.1 \times 10^7 \text{ molecules cm}^{-3}\text{s}^{-1}
\end{aligned}$$

and, for reaction 3.18,

$$\begin{aligned}
\text{rate}_{3.18} &= k_{18}[NO_2][O] \\
&= (9.3 \times 10^{-12}) \times (8.0 \times 10^9) \times (2.0 \times 10^7) \\
&= 1.5 \times 10^6 \text{ molecules cm}^{-3} \text{ s}^{-1}
\end{aligned}$$

The overall rate of reaction for the NO_x catalytic cycle is limited by this latter value at 1.5×10^6 molecules cm^{-3} s^{-1}. It is interesting that reaction 3.18, in spite of having a lower activation energy than reaction 3.17, is rate limiting.

The same calculation repeated at an altitude of 40 km (T $\simeq$ 250 K) indicates that reaction 3.17 becomes rate limiting with $k = 1.7 \times 10^6$ molecules cm^{-3} s^{-1}. Rate parameters for other catalytic cycles are shown in Table 3.3.

3.3 Null and holding cycles

Calculations of rates of individual reactions are readily carried out but their accuracy depends on the quality of the analytical data and rate constants. Furthermore, as was indicated earlier by reaction 3.23, there are other processes that occur in the stratosphere and are in competition with the catalytic cycles. Their relative importance further complicates our ability to make predictions about the extent of ozone destruction that may occur under various conditions. Null and holding cycles are two other types of reaction sequences that prevent species from taking part in catalytic processes.

Null (do nothing) cycles interconvert the species X and XO while effecting no net odd oxygen removal. Null cycles involving nitrogen oxides are shown below.

$$NO + O_3 \rightarrow NO_2 + O_2 \qquad (3.39)$$

$$\underline{NO_2 + h\nu \rightarrow NO + O} \qquad (3.40)$$

$$\text{net reaction} \quad O_3 + h\nu \rightarrow O_2 + O \qquad (3.41)$$

This sequence competes with the catalytic cycle and is important only during daytime as it requires radiation in the near-ultraviolet region. While the net effect is ozone photolysis, ozone is rapidly and stoichiometrically resynthesized by reaction 3.2.

Table 3.3 Kinetic data for various reactants involved in the catalytic decomposition of ozone calculated at 235 K

$X + O_3$				
X	Concentration/ molecules cm^{-3}	A	$E_a/kJ \, mol^{-1}$	k^{235}
O	1.0×10^9			
H	2.0×10^5	1.4×10^{-10}	3.9	1.9×10^{-11}
OH	1.0×10^6	1.6×10^{-12}	7.8	3.0×10^{-14}
NO	5.0×10^8	1.8×10^{-12}	11.4	5.3×10^{-15}
Cl	Very small	2.8×10^{-11}	21	9.6×10^{-12}

$XO + O$				
XO	Concentration/ molecules cm^{-3}	A	$E_a/kJ \, mol^{-1}$	k^{235}
O_2	5.0×10^{16}	8×10^{-12}	17.1	1.3×10^{-15}
HO	1.0×10^6	2.3×10^{-11}	0	2.3×10^{-11}
HO_2	2.5×10^7	2.2×10^{-11}	−0.1	2.3×10^{-11}
NO_2	5.0×10^9	9.3×10^{-12}	0	9.3×10^{-12}
ClO	2.0×10^7	4.7×10^{-11}	0.4	3.8×10^{-11}

Units of A and k^{235} are $cm^3 \, molecule^{-1} \, s^{-1}$. The concentration of species shown are for an altitude of 30 km, with the exception of ClO which is for 35 km. The ozone concentration at 30 km is 2.0×10^{12} molecules cm^{-3}. No value is given for chlorine. The mixing ratio of chlorine-containing compounds throughout the stratosphere is about 3 ppbv at all altitudes. This corresponds to a concentration of 5.1×10^{13} molecules cm^{-3} at 30 km. Only a small fraction of this is free chlorine. Concentrations are derived from plots in Chamberlain, J. W. and D. M. Hunten, *Theory of Planetary Atmospheres*, Academic Press, London; 1987. The Arrhenius parameters are from Wayne, R. P., *Chemistry of Atmospheres*, Clarendon Press, Oxford; 1991.

Another reaction involving NO_2 results in production of NO_3 and the establishment of a second type of null cycle:

$$NO_2 + O_3 \rightarrow NO_3 + O_2 \qquad (3.42)$$

$$\underline{NO_3 + h\nu \rightarrow NO_2 + O} \qquad (3.43)$$

$$\text{net reaction} \quad O_3 + h\nu \rightarrow O_2 + O \qquad (3.44)$$

Some of the NO_3 reacts in a three-body process to produce N_2O_5:

$$NO_3 + NO_2 + M \rightleftharpoons N_2O_5 + M \qquad (3.45)$$

The N_2O_5 is a relatively stable species and is itself not a catalyst for ozone destruction. It therefore behaves as an unreactive reservoir of NO_x, at any one time making up 5 to 10% of the total NO_x budget. Formation of N_2O_5, however, does not constitute a permanent loss of odd-nitrogen species as it ultimately decomposes back to NO_2 and NO_3. Reaction 3.45 therefore acts as a *holding cycle*, temporarily limiting the availability of NO_x for catalysing ozone decomposition in the stratosphere.

Nitric acid (as shown before) and hydrochloric acid are formed in the stratosphere and also serve as important reservoirs for ozone depleting nitric oxide and chlorine species.

$$\cdot NO_2 + \cdot OH + M \rightarrow HNO_3 + M \qquad (3.46)$$

$$\cdot Cl + CH_4 \rightarrow HCl + \cdot CH_3 \qquad (3.47)$$

Almost 50% of NO_x is stored in the nitric acid reservoir, while 70% of the stratospheric chlorine is present as hydrochloric acid. The nitric acid photolyses in daylight, producing nitrogen dioxide in the reverse of reaction 3.46, and the hydrochloric acid releases chlorine and water after reaction with hydroxyl radical.

In addition, several lesser known species have recently been identified as reservoirs of chlorine and NO_x in the stratosphere. Reactions producing these species include

$$ClO\cdot + HOO\cdot \rightarrow \underset{\text{hypochlorous acid}}{HOCl} + O_2 \qquad (3.48)$$

$$HOO\cdot + \cdot NO_2 + M \rightarrow \underset{\text{pernitric acid}}{HO_2NO2} + M \qquad (3.49)$$

$$ClO\cdot + \cdot NO_2 + M \rightarrow \underset{\text{'chlorine nitrate'}}{ClONO_2} + M \qquad (3.50)$$

These compounds serve to store reactive species until they are released as active catalysts or leak back into the troposphere. A consequence of their release is the development in recent years of the Antarctic and Arctic 'ozone holes'.

3.4 Antarctic and Arctic 'ozone hole' formation

Because ozone has long been recognized as an important chemical species with respect to the absorption of biologically damaging solar radiation, concentrations in the stratosphere have been monitored world wide since the mid 1950s. One of the monitoring sites associated with the British Antarctic Survey is at the Halley Bay Station, Antarctica, and in the early 1980s, a noticeable decline in ozone levels in the early spring at that location was observed. By 1984 it became evident that declining ozone was an annual event recurring each year in late October. During that month the average ozone column thickness in the period between 1956 and 1966 had been 314 DU, while from 1983 to 1993 the same average dropped to 182 DU. The observations led to a major research effort for the purpose of understanding this unexpected phenomenon.

In the Antarctic (and Arctic), conditions exist where spring-time ozone depletion events can occur, resulting in a major loss of ozone over a relatively short time. The conditions include a complex combination of climatic factors and the accumulation of reservoir species which, at the onset of the first daylight of the polar spring, react vigourously to bring about a major loss of ozone.

During the long dark winter in the Antarctic, due to the intense cold and the Earth's rotation, a stream of air is drawn toward the South Pole, creating a giant vortex (Fig. 3.3). The area within the vortex acts like a self-contained chemical reactor in which important and unique chemical processes occur. For one thing, inside the reactor, stratospheric clouds form as a result of exceptionally low temperatures that exist under conditions of no sunlight. The clouds are categorized in two types. Type 1 (the most common) polar stratospheric clouds (PSCs) form at about 193 K and consist of approximately 1 μm diameter particles of nitric acid and water in a ratio of 1 to 3. Type 2 PSCs occur when the temperature has dropped to 187 K. Type 2 PSCs occur when the temperature has dropped to 187 K. They are made up of relatively pure particles of water-ice up to 10 μm in size. Also present within the vortex are the accumulated gases—reservoir chlorine- and nitrogen-containing species, principally hydrochloric acid (reaction 3.47), hypochlorous acid (reaction 3.48), and chlorine nitrate (reaction 3.50). During the winter, on the surface of the PSCs, these species take

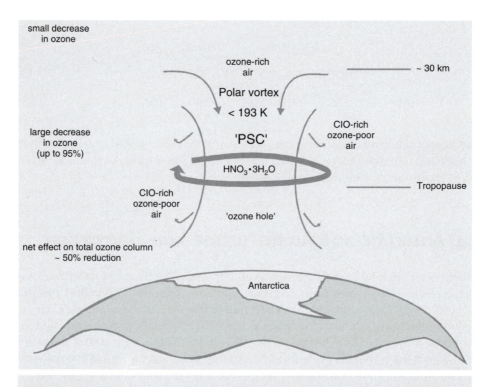

Fig. 3.3. The Antarctic 'ozone hole' illustrating the polar vortex and the location and relative amounts of ozone loss during the polar sunrise. (Redrawn with permission from R. P. Wayne, *Chemistry of Atmospheres*, Clarendon Press, Oxford; 1991.)

part in heterogeneous reactions which release chlorine molecules (reaction 3.51) and hydrogen hypochlorite (reaction 3.52):

$$HCl + ClONO_2 \overset{\text{on PSCs}}{\rightarrow} Cl_2 + HNO_3 \qquad (3.51)$$

$$H_2O + ClONO_2 \overset{\text{on PSCs}}{\rightarrow} HOCl + HNO_3 \qquad (3.52)$$

This relatively stable situation persists until after the onset of sunlight in late October. At that time, the solar radiation provides energy for photolysis of chlorine and hydrogen hypochlorite to produce the chlorine radical species:

$$Cl_2 + h\nu \rightarrow 2 \cdot Cl \qquad (3.53)$$

$$HOCl + h\nu \rightarrow \cdot Cl + \cdot OH \qquad (3.54)$$

The chlorine radical is then available to deplete ozone by the standard catalytic cycle or according to reactions 3.55–3.59. Destruction of ozone occurs rapidly so that in a matter of days, ozone levels fall dramatically to half or less of their winter value.

Similar to the cycles discussed earlier involving chlorine, the presence of atomic oxygen is unnecessary for depletion to occur:

$$2 \cdot Cl + 2O_3 \rightarrow 2ClO \cdot + 2O_2 \qquad (3.55)$$

$$ClO \cdot + ClO \cdot \rightarrow ClOOCl \qquad (3.56)$$

$$ClOOCl + h\nu \rightarrow ClOO \cdot + \cdot Cl \qquad (3.57)$$

$$\underline{ClOO \cdot \rightarrow \cdot Cl + O_2} \qquad (3.58)$$

$$\text{net reaction} \qquad 2O_3 + h\nu \rightarrow 3O_2 \qquad (3.59)$$

This situation persists until the air temperature rises, causing the vortex to break up and the polar stratospheric clouds to dissipate. When that occurs in mid to late spring, the chlorine radicals again become tied up by the formation of hydrochloric acid and chlorine nitrate, and the ozone level begins to recover to 'pre-hole' levels. There is concern that the air mass, while low in ozone concentration, will extend out over the most southern land masses, exposing people in the southern parts of South America, Australia, and New Zealand to unusually high levels of UV-B radiation.

Events such as those described have been observed through measurements made in the southern polar region since 1984. More recently, analagous observations of 'ozone layer thinning' have been made in the Arctic. Figure 3.4 shows ozone number density measurements obtained from balloon-borne sensors above Spitsbergen, Norway (79° North) in March of (a) 1992 and (b) 1995. The 1995 data at 18 km altitude gave concentrations that are only about half their 'normal' values. At other Arctic locations, there is similar evidence of major spring-time ozone loss. Nevertheless, the decreases are not as great as those near the South Pole. There, the polar vortex is stronger and more persistent, and the temperatures within it are as much as 10 K lower than in the Arctic.

Recently in the Arctic atmosphere, as a result of the eruption of Mt Pinatubo in the Phillipines during 1991, sulfate was found to be present at higher concentrations than

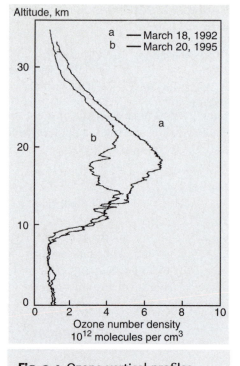

Fig. 3.4 Ozone vertical profiles determined by balloon-borne sensors above Spitsbergen, Norway (79° N). (Reprinted with permission from von der Gathen, P., Complexities of ozone loss continue to challenge scientists, *Chem. Eng. News*, 73 (1995), 24.)

usual. Stratospheric sulfate is usually in the form of an aerosol (see Box 6.1, p. 123) and acts as a catalyst for the removal of N_2O_5 gas by forming nitric acid:

$$N_2O_5 + H_2O \overset{\text{sulfate}}{\underset{\text{aerosol}}{\rightarrow}} 2HNO_3 \tag{3.60}$$

The ultimate consequence of reaction 3.60 is that it becomes a means by which NO_x species are removed from the polar stratosphere; thus eliminating one of the species that is associated with holding cycles that tie up chlorine monoxide radicals (reaction 3.50). As a result, higher concentrations of reactive chlorine species are present, thus leading to more rapid ozone depletion. As well, there is concern that this reaction could occur to a large extent throughout the stratosphere and not just at the poles, resulting in a lowering of the worldwide ozone concentration.

The main points

1 Ozone plays a major role in regulating the flux of ultraviolet radiation that reaches the Earth's surface. This allotrope of oxygen occurs naturally, with its synthesis and destruction being controlled by oxygen concentrations and photolysis reactions. These factors are such that a maximum in concentration of ozone occurs in a layer at approximately 23 km altitude—the so-called ozone layer.

2 In addition to the oxygen-only chemistry, other radical species are involved in ozone destruction. The presence in the atmosphere of two of these—the chlorine radical and nitric oxide—is to a large extent due to human activities, and their elevated levels in the stratosphere have caused a general global reduction in ozone concentrations, estimated at about 5 to 7%. Even greater reductions are observed in the northern mid-latitudes over heavily populated areas of Asia, Europe, and North America.

3 A more acute problem has been observed since the mid 1980s in the polar regions, where climatic and chemical factors combine to bring about a very large seasonal loss in ozone concentrations—up to 50% or more of the original concentration—during the polar spring time.

4 Both these phenomena are ones where human activities such as the release of stable, volatile chlorinated hydrocarbons and the excessive use of nitrogen-containing fertilizers affect the global environment. Significantly, whether these activities are local or widespread, our Earth's surroundings are altered over broad regions of space and time.

Additional reading

1 Brasseur, G. and C. Granier, Mt Pinatubo aerosols, *chlorofluorocarbons, and ozone depletion*, *Science.*, **257** (1992), 1239.

2 Elkins, J., T. Thompson, T. Swanson, J. Butler, B. Hall, S. Cummings, D. Fisher, and A. Raffo, Decrease in growth rates of atmospheric chlorofluorocarbons 11 and 12, *Nature*, **364** (1993), 780.

3 Gleason, J., P. Bhatia, J. Herman, R. McPeters, P. Newman, R. Stolarski *et al.*, Record low global ozone in 1992. *Science*, **260** (1993), 523.

4 Solomon, S., Progress towards a quantitative understanding of Antarctic ozone depletion. *Nature*, **347** (1990), 347.

5 Zuger, P.S., Ozone depletion's recurring surprises challenge atmospheric scientists. *Chem. Eng. News*, **71** (1993), 9–18.

Problems

1 Draw Lewis structures for ozone and for dioxygen. Using the data given below, qualitatively compare the bond enthalpies, bond orders, and bond lengths of these two compounds.

$$O_2 (g) \rightarrow 2O (g) \qquad \Delta H^\circ = +498 \, kJ$$

$$O (g) + O_2 (g) \rightarrow O_3 (g) \qquad \Delta H^\circ = -105 \, kJ$$

2 It has been suggested that the loss of ozone in the stratosphere could lead to a negative feedback which might allow more ozone to be produced. Explain why such a feedback is possible. (This 'self-healing' does in fact, occur, but only to a very small extent.)

3 Using data from Table 3.3, identify the rate-determining step in the catalytic cycle involving hydrogen and hydroxyl radicals, and determine the overall rate of ozone destruction as a consequence of this cycle. (Note that the calculation applies to reactions occurring at 30 km only.)

4 A catalyst may be defined as a substance that enhances the rate of a chemical reaction without being consumed in the process. By this definition, a catalyst would have an infinite lifetime. The ozone decomposition catalysts, however, have finite lifetimes. What are possible sinks for removal of the stratospheric catalysts, NO and Cl?

5 Using bond energies, explain the reaction sequence of ozone-destroying capability in the stratosphere of hydrocarbons containing the halogens: $Br > Cl > F$.

6 HCFC-123 has been proposed as a substitute for CFC-11. How would you expect the following environmental properties to compare:

(a) tropospheric lifetime
(b) combustibility
(c) ozone depletion potential (ODP)
(d) greenhouse gas properties (this can be answered after reading Chapter 8)?

7 CFC-114 has a tropospheric lifetime of 236 y. Would you expect CFC-115 to have a longer or shorter lifetime? Why?

8 A proposal to repair the ozone layer has been made. The suggestion is to inject 'negative charges' into the lower stratosphere, and these would react with CFCs to produce harmless products. From your knowledge of basic chemistry, indicate whether this process would be theoretically possible, and discuss the practical requirements of it. (*Chem. Eng. News*, May 23, 1994, p. 36; and *Phys. Rev. Lett.*, **72** (1994), 3124.

Notes

1 Miller, D. H., *Energy at the Surface of the Earth*, Academic Press Inc., New York; 1981.

2 Chapman, S. A., A theory of upper-atmosphere ozone, *Mem. Roy. Meteorol. Soc.* **3** (1930), 103.

3 Wennberg, P.O., *et. al.*, Removal of stratospheric O_3 by radicals: in situ measurements of OH, HO_2, NO, NO_2 ClO, and BrO, *Science*, **226** (1994), 398.

4 Values from plots in Chamberlain, J. W. and D. M. Hunten, *Theory of Planetary Atmospheres*, Academic Press, London; 1987.

Tropospheric chemistry—smog

I N the next several chapters, we will be looking at aspects of tropospheric chemistry. Because the troposphere is where we live, chemical reactions involving gases in this region of the atmosphere directly and immediately affect human life and the total environment around us.

Urban smog is a highly visible problem encountered in major metropolises on every continent of the Earth. We will begin by examining the nature of smog, particularly photochemical smog; the chemical processes involved in creating this type of smog event are well known. Moreover, the same reactions that produce smog operate to a more limited degree widely throughout the globe, even in regions we would identify as being free of pollution. Consequently, in a general sense, the topic of this chapter could be described broadly as gas-phase tropospheric chemistry of organic compounds.

4.1 Smog

Smog is a general term referring to forms of air pollution in which atmospheric visibility is partially obscured by a haze consisting of solid particulates and/or liquid aerosols. The name, in its original definition, is derived from a combination of smoke plus fog. Within the range of smog characteristics are two well defined types—the so-called *classical* or *London smog* and the *photochemical* or *Los Angeles smog*.

Classical smog, so named because it is associated with the use of the traditional fuel, coal, is characterized by a high concentration of unburned carbon soot as well as elevated levels of atmospheric sulfur dioxide. Due to the presence of sulfur dioxide, a mild reducing agent and the precursor of a weak acid, the overall chemical properties are reducing and acidic. Where the atmosphere is humid, the carbon particles may serve as nuclei for condensation of water droplets, forming an irritating fog.

The classical smog situation was encountered in many nineteenth century heavily populated industrial centres such as London, England. In these cities high-sulfur coal was used both for domestic heating and as an energy source for industry. There were no pollution controls and, in many cases, the emissions were released near ground level.

Depending on climatic conditions, smog episodes were frequent and continued to occur well into the present century. As late as 1952, in London a severe smog lasting several weeks led to the death of more than 4000 persons, mostly by the acute aggravation of pre-existing respiratory problems. Over the years, technical improvements and strict legislation have combined to reduce the problem in London and elsewhere.

Unfortunately, there are still present-day examples where classical smog situations persist. Much of Eastern Europe is the site of coal-burning industries which operate without emissions control. Added to the dust and soot from these sources are domestic heating emissions and those from automobiles and other vehicles, many of which use two-stroke engines that emit several times higher concentrations of volatile hydrocarbons as well as substantially more carbon monoxide than conventional four-stroke engines. One area where the problem is especially acute is in Poland's Upper Silesia region adjacent to the city of Krakow. This industrial region is home to some four million people and it has been determined that airborne dust, carbon monoxide, sulfur dioxide, nitrogen oxides, and lead all exceed the safe limit, sometimes by factors of ten or more.

In contrast to classical smog, photochemical smog is based on emissions from petroleum combustion, principally from motor vehicles, followed by a sequence of chemical and photochemical reactions occurring under specific conditions. The smog contains elevated levels of oxidants and carbon-containing reaction products. Photochemical smog is a twentieth-century phenomenon as it requires the presence of unburned gaseous hydrocarbons and nitrogen oxides, both of which are emitted by internal combustion engines. Specific climatic conditions are also required (Fig. 4.1). Some of the smog-producing

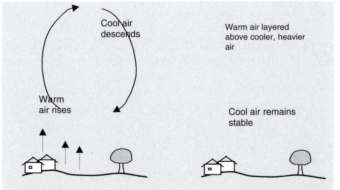

Fig. 4.1 A well mixed troposphere occurs when warm air at the surface of the Earth rises and is replaced by cooler air descending from above. In a less common situation, the troposphere is stable during an inversion event, when the surface air is cooler (and more dense) than that above.

reactions are thermal reactions and are therefore favoured by warm temperatures; others are photochemical, so sunlight is a requirement. A stable atmosphere ensures that the released gases remain in the same location where they are able to react. Stability is achieved when there is a temperature inversion in the troposphere. During an inversion, cooler air remains close to the surface of the Earth. Having greater density than the air above, it does not rise and there is little convective mixing. A variety of topographical and climatic features can produce an inversion situation.

The requisite conditions are found frequently in the automobile-rich city of Los Angeles but the phenomenon is now very widespread, appearing during hot, stable weather in many North American and Western European conurbations. In other major urban centres— large cities such as Mexico City, Cairo, Lagos, Jakarta, and Beijing—air pollution results from a complex suite of emissions. For example, Delhi has a high density of motorized vehicles operating under minimal pollution control standards. At the same time a significant portion of the domestic energy supply comes from coal, charcoal, and a range of biomass sources. Usually these fuels are burned in low efficiency units at ground level. In addition, during the nine dry months from October through June, the atmosphere is laden with dust from the clay-rich alluvium which makes up the soil in the city and its surroundings. The combination of gaseous and particulate pollutants together forms a uniquely characteristic haze which is especially evident during hot, dry summer evenings. In this dusk period from 6 pm to 8 pm, the heavy smokey haze hangs over the entire city.

4.2 Photochemical smog

The chemistry involved in the formation of photochemical smog has been studied in detail and the processes can be summarized graphically as in Fig. 4.2. Part(a) of the figure depicts an *idealized* photochemical smog event while part(b) plots *actual data* for one such occurrence in Toronto, Canada on 21 May 1992. The most obvious features are as follows.

Beginning at about 6 am on what is to be a sunny, hot day as the morning traffic takes to the streets, a simultaneous increase in the atmospheric concentrations of volatile hydrocarbons and nitric oxide is observed. The nitric oxide concentration rapidly reaches a maximum and then decreases while, at the same time, nitrogen dioxide levels begin to rise. Later in the morning, both hydrocarbon and nitrogen dioxide concentrations fall off and elevated levels of oxidizing agents and aldehydes are detected. Some aspects of the pattern are repeated on a smaller scale during the period of heavy evening traffic, but there is a general lowering of concentrations of all the aforementioned species to background levels which remain constant during the night.

The smog, consisting of a mixture of partially oxidized hydrocarbons, ozone, and other oxidants, is observable from midday to late afternoon. In addition to producing a haze, it also causes eye and other membrane irritation, can adversely affect plant growth, and is implicated in other serious ecotoxicological problems.

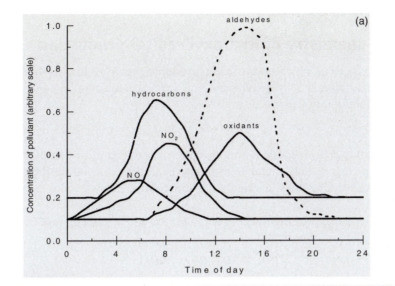

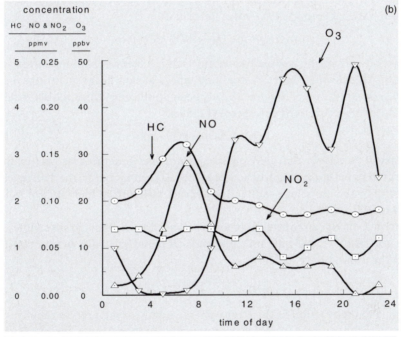

Fig. 4.2 (a) Sequence of chemical species appearing during a photochemical smog event. Idealized plot based on results from laboratory studies in a smog chamber. (b) Sequence of chemical species during a photochemical smog event. Actual data obtained from measurements on the corner of Bay and Grosvenor Streets in Toronto, Canada on 26 May 1992. This was a cloudless, still day with maximum temperature of 26 °C. (Data from the Ministry of Environment and Energy, Government of Ontario.)

4.2.1 The chemistry of hydroxyl radical production

The chemical reactions which lead to smog formation centre around the hydroxyl radical. As we showed in Chapter 2, hydroxyl formation is due to a reaction sequence which begins with the production of nitric oxide:

$$N_2 + O_2 \rightleftharpoons 2NO \tag{4.1}$$

Because the left-to-right reaction is endothermic and endergonic, nitric oxide is produced to a significant extent only under high-energy conditions, including those provided by internal combustion engines. A calculation illustrating this was done in Section 2.2 of Chapter 2.

When it is exhausted to the open atmosphere the nitric oxide is oxidized to nitrogen dioxide by oxygen or other oxidants:

$$2NO + O_2 \rightarrow 2NO_2 \tag{4.2}$$

The three-body reaction with oxygen proceeds only very slowly and cannot account for all the NO_2 generated. A second mechanism for oxidation is

$$NO + O_3 \rightarrow NO_2 + O_2 \tag{4.3}$$

The reaction with ozone is rapid but ozone itself is a byproduct of the formation of NO_2 (see reaction 4.6 below). It therefore does not appear in significant quantities until after a substantial concentration of NO_2 has already been produced. A third route for oxidation of nitric oxide involves reactions with peroxyl radicals:

$$ROO\cdot + NO \rightarrow RO\cdot + NO_2 \tag{4.4}$$

Later, it will be shown that peroxyl species are also generated as part of the overall sequence of steps in the oxidation of hydrocarbons. Like nitric oxide, the hydrocarbons are a product of vehicular emissions—either due to evaporation from fuel tanks or as unburned species in the exhaust.

The nitrogen dioxide produced in reaction 4.4 absorbs visible and ultraviolet radiation from the sunlight ($\lambda < 400$ nm) and this leads to photolysis with production of atomic oxygen in the ground state ($O(^3P)$):

$$NO_2 \overset{h\nu, \lambda < 400\,\text{nm}}{\rightarrow} NO + O \tag{4.5}$$

Analogous to the reaction in the stratosphere, atomic oxygen reacts rapidly with molecular oxygen in the presence of a third body–usually another O_2 or N_2 molecule–to produce ozone:

$$O + O_2 + M \rightarrow O_3 + M \tag{4.6}$$

Ozone then photolyses due to ultraviolet radiation from sunlight:

$$O_3 \overset{h\nu, \lambda < 315\,\text{nm}}{\rightarrow} O_2^* + O^* \tag{4.7}$$

The products are an oxygen molecule ($O_2(^1\Delta g)$) and an oxygen atom ($O(^1D)$), both in an excited state. A large fraction of the excited-state oxygen atoms is deactivated by collisions with ground-state dioxygen or dinitrogen but that which retains its additional energy can react with water vapour to produce hydroxyl radicals:

$$O^* + H_2O \rightarrow 2 \cdot OH \tag{4.8}$$

Accordingly, a single nitrogen dioxide molecule is able to create two hydroxyl radicals. The sum of reactions 4.5 to 4.8 is

$$NO_2 + H_2O \rightarrow NO + 2 \cdot OH \tag{4.9}$$

Quantitatively, this is the most important means by which the hydroxyl radical is produced.

A second mechanism of generating hydroxyl arises from reactions involving nitrogen dioxide as follows:

$$NO + NO_2 + H_2O \rightarrow 2HONO \tag{4.10}$$

$$2HONO \xrightarrow{h\nu, \lambda < 400\,nm} 2NO + 2 \cdot OH \tag{4.11}$$

Reactions 4.10 and 4.11 are especially important in heavily polluted atmospheres and it is evident that two hydroxyl radicals are again produced by one nitrogen dioxide molecule in the process.

Therefore, once again, the sum of reactions 4.10 and 4.11 is equal to the following overall reaction:

$$NO_2 + H_2O \rightarrow NO + 2 \cdot OH \tag{4.12}$$

It is again worth emphasizing that the accelerated synthesis of hydroxyl radicals in the urban atmosphere is substanitially a consequence of production of nitrogen oxides in an internal combustion engine. The actual concentration of hydroxyl is very small and difficult to measure, but has been estimated to be in the order of 10^7 molecules cm^{-3} in a polluted urban environment in contrast to a concentration of about 2.5×10^5 molecules cm^{-3} in a relatively clean rural area in a temperate zone. Other factors enhancing hydroxyl radical concentration are the temperature and the intensity of sunlight, so values tend to be higher in tropical compared to temperate regions.

4.2.2 Oxidation of hydrocarbons

Besides being a source of nitrogen oxides, vehicle engines simultaneously emit unburned volatile hydrocarbons and these are oxidized through reactions initiated by the highly reactive hydroxyl radical. Considering a generic aliphatic hydrocarbon, one version of the hydroxyl radical initiated oxidation sequence is summarized in the following way:

$$\cdot OH + RCH_3 \rightarrow \underset{\text{alkyl}}{R\dot{C}H_2} + H_2O \tag{4.13}$$

$$\underset{}{R\dot{C}H_2} + O_2 + M \rightarrow \underset{\text{peroxyalkyl}}{RCH_2OO\cdot} + M \tag{4.14}$$

$$RCH_2OO\cdot + NO \rightarrow \underset{\text{alkoxyl}}{RCH_2O\cdot} + NO_2 \tag{4.15}$$

$$RCH_2O\cdot + O_2 \rightarrow \underset{\text{aldehyde}}{RCHO} + \underset{\text{hydroperoxyl}}{HOO\cdot} \tag{4.16}$$

$$HOO\cdot + NO \rightarrow NO_2 + \cdot OH \tag{4.17}$$

Note that each step in the sequence produces a new radical. The sum of the reactions is

$$RCH_3 + 2O_2 + 2NO \rightarrow RCHO + 2NO_2 + H_2O \tag{4.18}$$

If we incorporate the fact that nitrogen dioxide, besides being a precursor of ozone, is the principal source of hydroxyl radical in the atmosphere through the overall reaction 4.9, and include this with reaction 4.18, the oxidation of the hydrocarbon is now written as

$$RCH_3 + 2O_2 + H_2O \rightarrow RCHO + 4 \cdot OH \tag{4.19}$$

In this overall reaction, the hydroxyl radical as well as other radicals play a catalytic role (there is actually the potential for net production) so that a very small amount of the radical species produces a large amount of product. As was noted in Chapter 2, the central role of hydroxyl in tropospheric chemistry cannot be overemphasized.

Were it not for various means of removing hydroxyl radicals from the atmosphere, its concentration would continue to increase and the rate of oxidation of hydrocarbons would accelerate. Several termination reactions serve to remove hydroxyl, its precursor nitrogen dioxide, and the hydroperoxyl radical:

$$\cdot OH + \cdot NO_2 + M \rightarrow HNO_3 + M \tag{4.20}$$
$$2HOO \cdot \rightarrow H_2O_2 + O_2 \tag{4.21}$$
$$\cdot OH + HOO \cdot \rightarrow H_2O + O_2 \tag{4.22}$$

The products of each of these reactions are relatively stable. The nitric acid and hydrogen peroxide are water soluble and are removed from the atmosphere by precipitation.

Keep in mind also that hydroxyl production involves two photochemical steps. Consequently, night-time also brings the smog-forming reactions to a close.

4.2.3 Secondary reactions

There are secondary reactions which occur simultaneously with the oxidation of hydrocarbons. The first two have been noted previously:

$$NO_2 \xrightarrow{h\nu, \lambda < 400\,nm} NO + O \tag{4.23}$$
$$O + O_2 + M \rightarrow O_3 + M \tag{4.24}$$

Other very important reactions involve the aldehyde product of oxidation of hydrocarbons. Using acetaldehyde as an example, the reactions[1] are:

$$CH_3CHO + \cdot OH \rightarrow CH_3\dot{C}O + H_2O \tag{4.25}$$
$$CH_3\dot{C}O + O_2 + M \rightarrow CH_3C(O)OO \cdot \tag{4.26}$$
$$\text{acetylperoxy}$$
$$CH_3C(O)OO \cdot + \cdot NO_2 \rightleftharpoons CH_3C(O)OONO_2 \tag{4.27}$$
$$\text{peroxyacetic nitric anhydride}$$
$$\text{(PAN)}$$

Peroxyacetic nitric anhydride and related compounds, PANs,[2] are the major eye irritants in a photochemical smog. The importance of PANs as reaction products is that they act as reservoirs for nitrogen oxide species. Peroxyacetic nitric anhydride is a relatively stable molecule, especially at low temperatures, and therefore may be transferred over long distances by air currents. In warmer locations often remote from the source, PAN breaks down *via* the reverse process of reaction 4.27 so that nitrogen dioxide is released with the potential of producing additional ozone and hydroxyl radicals. This allows for carry-over of smog conditions through both space and time.

4.2.4 The nature of photochemical smog

From this complex series of thermal and photochemical reactions it is seen that a number of chemicals would be expected to be present at elevated atmospheric concentrations during a photochemical smog event. These include the nitrogen oxide and hydrocarbon precursors as well as products such as aldehydes and oxidants, ozone, and PANs. Some of the chemicals are gases but others, particularly aldehydes, are present as small liquid droplets in the form of an aerosol. This is the cause of the hazy appearance present during an intense smog. The yellowish colour is due to nitrogen dioxide. The temporal order of reactions described throughout this section is also consistent with the sequence as depicted in Fig. 4.2.

Typical values of some smog chemicals found under polluted and non-polluted conditions confirm expectations regarding higher concentrations of particular chemicals (Table 4.1). A summary of the reactions which we have been discussing is given in Fig. 4.3.

Table 4.1 Typical atmospheric concentrations of selected species characteristic of a photochemical smog

Species	Concentration/μg m^{-3}	
	Polluted area	Unpolluted area
Carbon monoxide	10 000–30 000	< 200
Nitrogen dioxide	100–400	< 20
Hydrocarbons (excluding methane)	600–3000	< 300
Ozone	50–150	< 5
PANs	50–250	< 5

Most values are estimates based on data in *Air Quality in Ontario* 1991, Environment Ontario, Queen's Printer for Ontario; 1992.

4.2.5 Volatile organic compounds and their oxidation

In discussing a reaction sequence for the oxidation of organic compounds in the gas phase and the generation of photochemical smog, our example of a hydrocarbon substrate was a general aliphatic saturated hydrocarbon. Of course, there are many other types of volatile organic compounds (VOC) which are released into the atmosphere from both anthropogenic and natural sources. For vehicular emissions, the list of compounds is long and variable depending on the fuel, type of engine, and operating conditions. Hydrocarbons

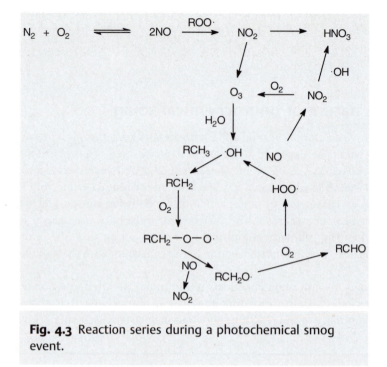

Fig. 4.3 Reaction series during a photochemical smog event.

such as ethene, ethyne, higher aliphatic hydrocarbons, benzene, toluene, and xylenes are important emissions in almost all cases. Each of these compounds can be released unreacted or can undergo oxidation reactions.

In a major city, the atmospheric levels of fuel-derived gases can be very high. For example, measurements of personal exposure to volatile organic compounds, while commuting by motorcycle in Taipei city (Taiwan), showed large concentrations of several hydrocarbons, especially aromatic compounds. The ten hydrocarbons found in highest concentration are listed in Table 4.2.

In their unreacted state, the compounds may have undesirable ecotoxicological properties. For example benzene, besides causing annoying physiological reactions such as dizziness and membrane irritation, is known to be a human carcinogen. Furthermore, each of the compounds can undergo oxidation producing a range of products directly and by related secondary reactions.

Considering the various hydrocarbon reactants, there are two principal mechanisms by which hydroxyl radicals initiate oxidation (see pp. 36–7). The first mechanism is the type that we have already seen in the case of saturated aliphatic hydrocarbons. It begins when a hydroxyl abstracts a hydrogen to form water and an organic radical. The abstraction reaction can occur whenever there is a hydrogen atom available to be removed but the rate depends on the strength of the carbon–hydrogen bond. The general order of strength of carbon–hydrogen bonds is primary > secondary > tertiary carbons and so the rate of hydrogen abstraction is in the reverse order.

Table 4.2 Mean concentration of volatile organic compounds in the atmosphere of Taipei city

Compound	Atmospheric concentration/μg m^{-3}
Toluene	980
m,p-xylene	910
o-xylene	510
Benzene	370
Ethylbenzene	310
1,3,5-trimethylbenzene	230
1-ethyl,4-methylbenzene	200
Hexane	150
Heptane	130
1-ethyl,2-methylbenzene	120

From Chan, C.-C., S.-H. Lin, and G.-R. Her, Student's exposure to volatile organic compounds while commuting by motorcycle and bus in Taipei city. *Air and Waste*, **43** (1993), 1231–8.
Measurements were made in the breathing zone of cyclists and pedestrians in three parts of the city frequented by commuters.

4.2.6 Alkenes

The second type of hydroxyl radical initiation reaction is expressed with olefins where the electrophilic hydroxyl adds to a double or other multiple bond, a region of high electron density (reaction 4.28):

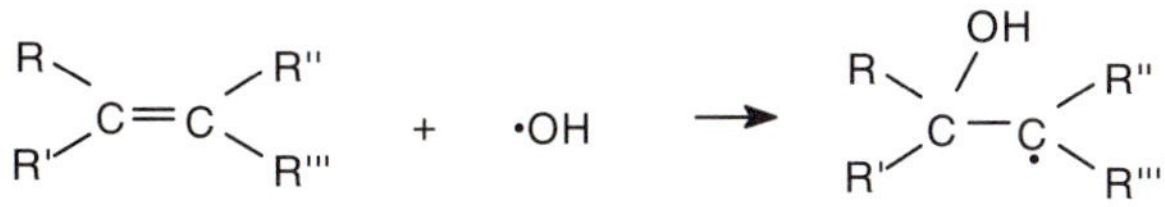

$$(4.28)$$

Following this step, one mechanism continues with the addition of dioxygen to the hydroxylated species. The β-hydroxy peroxyl compound then transfers an oxygen atom to nitric oxide and the resultant β-hydroxyalkoxyl radical usually undergoes decomposition. Finally a dioxygen molecule abstracts a hydrogen from the alkoxyl radical (one of the decompostion products), forming a hydroperoxyl radical and a ketone. The RC(OH)R$'$ species reacts further with oxygen (reaction 4.32) and, in the overall reaction, two ketones are produced. For the common situation of a straight chain alkene (R$'$ and R$''$ = H), instead of ketones the products are aldehydes. The entire process is shown below in its most general form.

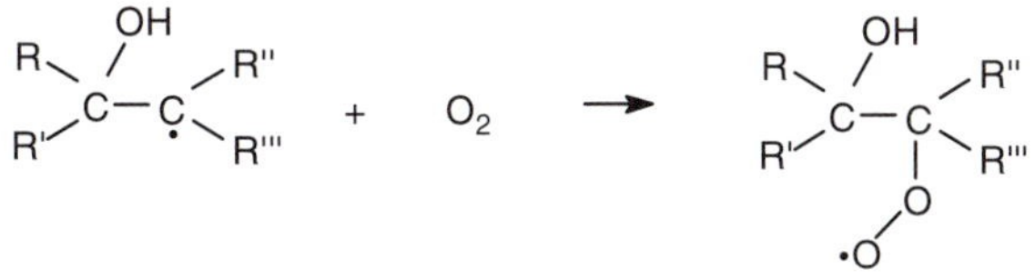

$$(4.29)$$

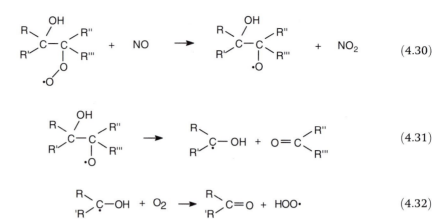

4.2.7 Alkynes

For alkynes, the initial step is also addition of hydroxyl, in this case to the electron-rich triple bond. For acetylene, after further reaction, the principal products are glyoxal $(CHO)_2$ and formic acid (HCOOH).

4.2.8 Aromatics

Aromatic compounds are also oxidized in reactions initiated by the hydroxyl radical. Where the aromatic is alkyl substituted, a hydrogen atom is first abstracted and the sequence continues in the same way as for other alkyl compounds. The final products are aldehydes.

An alternative route which can also apply where there is no alkyl substituent involves addition of hydroxyl radical to the ring. The reaction is shown below.

$$(4.33)$$

By reaction with molecular oxygen, the hydroxybenzene radical eventually forms phenol. When the initial substrate is a substituted aromatic compound, corresponding phenols and other types of products are also formed.

$$(4.34)$$

In addition to this sequence, it is observed that ring cleavage of the peroxidized aromatic also occurs to a significant extent, producing various poorly characterized products.

An environmentally significant class of compounds is the fused-ring polynuclear aromatic hydrocarbon (PAH) class. Except for the simplest PAH, naphthalene and its derivatives, these species are stable to oxidation and therefore have long atmospheric lifetimes.

The PAHs form an important constituent of some atmospheric aerosols and more will be said about them in Chapter 6.

4.2.9 Aldehydes and ketones

We have seen that aldehydes (and ketones) are important products of several of the oxidation schemes. These are somewhat stable species, but they do undergo reactions of their own. As shown in reactions 4.25 to 4.27, one possibility begins with yet another reaction initiated by hydroxyl radical, followed by the addition of dioxygen and then reaction with nitrogen dioxide to form a member of the PAN family of compounds. A second possibility also starts with reactions 4.25 and 4.26. The acetylperoxy radical then gives up an oxygen to nitric oxide followed by cleavage of the C–C bond to form a methyl radical and carbon dioxide (reactions 4.35 and 4.36):

$$CH_3C(O)OO\cdot + NO \rightarrow CH_3C(O)O\cdot + NO_2 \tag{4.35}$$

$$CH_3C(O)O\cdot \rightarrow \cdot CH_3 + CO_2 \tag{4.36}$$

A third important route for decomposition is by photolysis. Aldehydes are capable of absorbing UV radiation longer than 290 nm and this leads to their photochemical degradation. For acetaldehyde, observed photolytic reactions are

$$CH_3CHO \overset{h\nu,\lambda\simeq290\,nm}{\rightarrow} \cdot CH_3 + H\dot{C}O \tag{4.37}$$

$$CH_3CHO \overset{h\nu,\lambda\simeq290\,nm}{\rightarrow} CH_4 + CO \tag{4.38}$$

The lifetime of aldehydes in the atmosphere is in the range of 24 hours.

4.2.10 Methane

We have left a discussion about the tropospheric oxidation of methane to the end. Quantitatively methane, with a mixing ratio of 1.7 ppmv, is present in the global atmosphere at the largest concentration of any hydrocarbon. At most locations, higher molar mass hydrocarbons and partially oxidized species are present in much smaller concentrations. However, there are situations in urban areas when non-methane hydrocarbon (NMHC) mixing ratios may reach values of 5–10 ppmv C as a result of vehicular and other emissions.[3] As has been noted, in cities these compounds derive mostly from unburned petroleum-based fuel.

In the countryside, the hydrocarbon background is provided by emissions of terpenes and other volatile organic compounds wherever there are trees, especially in heavily forested regions. In total, this background level of hydrocarbons other than methane is typically 10–20 ppbv C, two orders of magnitude lower than that of methane.

There are several natural and anthropogenic sources of methane. It is produced biogenically under anaerobic conditions in submerged soils and landfills and it is released during extraction, production, and transport of natural gas. Methane is a greenhouse gas and its role in absorbing infrared radiation is described in Chapter 8.

All of the hydrocarbons are susceptible to oxidation *via* sequences initiated by the hydroxyl radical, but the rate of oxidation depends on the particular substrate. The oxidation reaction rate of methane is considerably slower than that of most other hydrocarbons. Therefore, the presence of methane provides a background, but local emissions of other hydrocarbons can have a larger effect on the daily cycle of hydrocarbon oxidation.

The background reactions involving methane are as follows:

$$CH_4 + \cdot OH \rightarrow \cdot CH_3 + H_2O \tag{4.39}$$

$$\cdot CH_3 + O_2 \rightarrow CH_3OO\cdot \tag{4.40}$$

$$CH_3OO\cdot + NO \rightarrow CH_3O\cdot + NO_2 \tag{4.41}$$

$$CH_3O\cdot + O_2 \rightarrow CH_2O + HOO\cdot \tag{4.42}$$

Up to this point, the oxidation of methane has produced formaldehyde and the sum of the reactions is

$$CH_4 + \cdot OH + 2O_2 + NO \rightarrow CH_2O + H_2O + HOO\cdot + NO_2 \tag{4.43}$$

The formaldehyde is involved in additional reactions, with photolysis being most important.

$$CH_2O \overset{h\nu,(\lambda<330\,nm)}{\rightarrow} H\dot{C}O + \cdot H \tag{4.44}$$

$$H\dot{C}O + O_2 \rightarrow HOO\cdot + CO \tag{4.45}$$

$$CO + \cdot OH \rightarrow CO_2 + \cdot H \tag{4.46}$$

The hydrogen and the hydroperoxyl radical then undergo subsequent reactions:

$$2(\cdot H + O_2 \rightarrow HOO\cdot) \tag{4.47}$$

$$2(HOO\cdot + NO \rightarrow \cdot OH + NO_2) \tag{4.48}$$

In sum, the net reaction is

$$CH_4 + 5O_2 + 2H_2O \rightarrow 2HOO\cdot + 6\cdot OH + CO_2 \tag{4.49}$$

Therefore, the oxidation of methane ultimately produces carbon dioxide as the final stable carbon compound. The reaction is initiated by hydroxyl radical but in the process additional radicals are generated. Clearly, these reactions are self-sustaining but the variety of termination processes serves to keep them in balance.

Consider an urban situation with the following atmospheric concentrations. The mixing ratio of methane has the usual background value of 1.7 ppmv. This is equivalent to a concentration of 4.6×10^{13} molecules cm^{-3}. The concentration of one of the higher hydrocarbons, hexane, is 100 µg m^{-3} which equals 7.0×10^{11} molecules cm^{-3}. The concentration of hydroxyl radical is 2.0×10^6 molecules cm^{-3}. The rate constants for the second order reactions between hydroxyl and the hydrocarbons are 8.36×10^{-15} and 5.61×10^{-12} cm^3 molecule^{-1} s^{-1} for methane and hexane respectively (see ref. 1, Additional reading).

Using the atmospheric data and rate constants given, it can be calculated that the instantaneous rate of disappearance of hydroxyl is 7.7×10^5 molecules cm^{-3} s^{-1} due to methane and 7.8×10^6 molecules cm^{-3} s^{-1} due to hexane. Note that these are instantaneous values and assume a steady state concentration of hydroxyl radical (implying its very rapid synthesis). Under these circumstances, hexane is a more important reactant for hydroxyl than methane, even though its concentration is two orders of magnitude lower. This is a consequence of the rate of reaction being much greater for hexane. In a non-polluted atmosphere, where the hexane concentration would be several orders of magnitude lower than the urban value, it would make a negligible contribution to hydroxyl radical destruction compared with methane.

4.2.11 General principles of volatile organic compound oxidation

In this survey of hydroxyl-initiated reactions of volatile organic compounds, we have observed a number of important general steps.

1. The initiation begins with hydrogen abstraction or hydroxyl addition.

2. The radical produced by step 1 adds an oxygen molecule, forming a peroxyl species, or in the case of aromatics, the dioxygen abstracts a hydrogen.

3. The peroxyl species transfers an oxygen atom to a molecule of nitric oxide.

4. The product molecule now loses a hydrogen atom to another oxygen molecule, or it splits into two smaller species. In either case, aldehydes (or, less commonly, ketones) are formed. The hydroperoxyl radical is another product.

5. The aldehydes react with nitrogen dioxide to form PANs, undergo further hydroxyl-initiated oxidation, or photochemically decompose.

6. The decomposition products are again subject to oxidation and the ultimate stable products are carbon dioxide and water.

7. The reactions indicated in this section are a summary of some of the more important reactions occuring in the troposphere. The situation is highly complex and many alternative processes occur. For those who model atmospheric behaviour in a polluted environment, hundreds of simultaneous reactions must be taken into account, rate constants measured, and concentrations estimated. Assumptions are involved at every step and the natural climatic variations make quantitative calculations even more difficult.

4.3 Exhaust gases from the internal combustion engine

The point has been made that a major anthropogenic source of organic chemicals and other atmospheric pollutants is from vehicles of various kinds. In many countries, especially Australia, Japan, and those in Europe, North America, and parts of Asia, the gasoline-powered four-stroke internal combustion engine is the predominant type of vehicle engine. Diesel engines are used more commonly for larger vehicles such as buses, trucks, locomotives, ships, and in some power generators, for example those used in mines. Two-stroke gasoline-powered engines are less common, but are important sources of power for chainsaws, lawn mowers, mopeds, motorcycles, and in outboard motors for boats.

In low income countries, there is a smaller proportion of gasoline-powered automobiles, while highly developed public transportation systems involving diesel buses and trains are very important. Scooters, motorcycles, and auto-rickshaws are also common and, invariably, are powered by small two-stroke engines. In Shanghai, China, for example, there are estimated to be around 650 000 scooters operating without significant pollution controls.

The nature and quantity of emissions from each of these engine/fuel combinations are different. However, in a quantitative sense, it is very difficult to compare emissions from the various engine types. The size of vehicles on which each type is used is usually different. Some engines have catalysts which effectively remove a part of the harmful gases before they are sent to exhaust. Furthermore, in practice, many vehicles operate under far less than ideal conditions. Laboratory data obtained using controlled loads and a well tuned engine are therefore not reliable indicators of the real situation on the road. We will look briefly at the nature of emissions emanating from engines of the three types.

4.3.1 Gasoline-powered four-stroke engines

The Otto cycle (Fig. 4.4) describes the sequence followed in the cylinders of a gasoline-powered four-stroke engine. It is similar to, but not identical with, the well known Carnot cycle.

When the piston is at the bottom of its intake stroke, it is at position 'a' on the pressure–volume diagram. Compression takes place as the piston rises rapidly and there is little time for heat exchange, so this may be considered an adiabatic process, one in which no heat enters or leaves the system. This is represented by the line 'ab'. At 'b', the spark ignites the fuel/oxidant mixture and there is almost instantaneous combustion of the contents of the cylinder. In this short period of time, the piston has moved to a negligible extent so that there is a constant volume addition of heat corresponding to line 'bc'. The result of the explosive combustion is that the piston begins to move downwards in the power stroke. Again there is negligible heat transfer and the process is represented by the adiabatic curve 'cd'.

At this point, the exhaust valve opens and the hot gases are expelled followed by intake of fresh, cool gases. This complicated sequence of processes may be represented in simplified form by the line 'da'. The cycle then repeats itself. The efficiency of the Otto engine is given by the relation

$$\frac{W_{\text{cycle}}}{Q_{\text{bc}}} = 1 - \frac{T_{\text{d}}}{T_{\text{c}}} = 1 - \left(\frac{V_{\text{c}}}{V_{\text{d}}}\right)^{R/C_v} \tag{4.50}$$

In this relation, W_{cycle} is the work produced by the engine and Q_{bc} is the heat generated in the cylinder during combustion. Their ratio as given in the left-hand side of the equation gives the thermodynamic efficiency of the engine. R is the gas constant and C_v is the heat capacity of the gas at constant volume. T_{c} and T_{d} are the temperatures at the beginning and end of the power stroke and V_{c} and V_{d} are the corresponding volumes. The ratio of cylinder

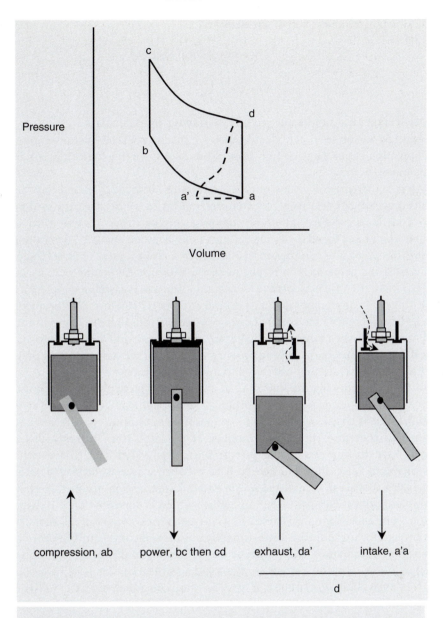

Fig. 4.4 The Otto cycle. A description of the sequence of steps is given in the text.

volume at the bottom to the top of the stroke is

$$\frac{V_d}{V_c} = \text{compression ratio}, r_c$$

$$\therefore \frac{W_{cycle}}{Q_{bc}} = 1 - \left(\frac{1}{r_c}\right)^{R/C_v} \tag{4.51}$$

There is environmental significance to this thermodynamic calculation.

From eqn 4.51 we can see that high efficiencies in the internal combustion engine require high compression ratios. High compression ratios lead to engine knocking, which can be overcome in two ways.

One way is to increase the octane number of gasoline. Straight chain hydrocarbons tend to produce knocking at relatively low compression ratios, while aromatic and branched chain hydrocarbons combust quietly at somewhat higher ratios. An empirical scale has therefore been devised based on binary mixtures of n-heptane and 2,2,4-trimethylpentane ('isooctane') in which the compression ratio where a certain level of knocking occurs is plotted against the percentage of isooctane in the mixture. An octane number of 87 corresponds to an 87 : 13 ratio of isooctane to n-heptane. An actual gasoline sample is a mixture of a wide range of hydrocarbons and if its octane number is defined as 87, this means that it behaves as does a binary mixture of the above composition. Higher octane numbers correspond to mixtures with more branched chain, oxygenated, and aromatic molecules and these tend to be relatively expensive to produce. Addition of substances like methyl tertiary butyl ether to gasoline are therefore used to increase the octane number.

A second way of reducing the knocking of an engine (effectively increasing the octane number of gasoline) is through the addition of tetraethyl lead ($(C_2H_5)_4Pb$). The addition of this substance at a rate of about $1 \, g \, L^{-1}$ of gasoline substantially increases anti-knock properties and therefore the octane number of the fuel. Until recently this additive was widely used to improve engine performance. The added lead reacts with halogenated compounds (also added to the fuel) to produce a variety of lead halides which are sufficiently volatile to be emitted in the gaseous combustion products, but condense in the ambient atmosphere and form an aerosol which is deposited on the surrounding vegetation, soil, and water. Legislation in many countries now requires that tetraethyl lead not be used for this purpose. Unfortunately most low-income countries have not been able to enact such proposals, and vehicular lead emissions continue. In Nigeria, a country with major high-quality oil resources, the 1990 specification for lead in gasoline was $0.74 \, g \, L^{-1}$. In 1985 about 20 000 000 L of gasoline were consumed each day in Nigeria, corresponding to approximately 5400 t of this element being deposited throughout the country, mostly in areas close to traffic corridors. Deposition rates are especially high in a city like Lagos which has over one million vehicles and where congestion results in peak hour average operating velocities of less than 10 km per hour. Excessive concentrations of carbon monoxide and major smog episodes have also been observed.

It is clear that the compression ratio–octane rating issue has a well established technological solution, yet it persists as an economic problem. In addition, there is the emission of carbon monoxide and smog-producing chemicals for which partial technical solutions have been developed. More efficient technology, capable of effecting better removal of carbon monoxide, nitric oxide, and unburned hydrocarbons, is being sought.

The complete combustion of octane may be described in simple overall form as

$$C_8H_{18} + 12.5O_2 \rightarrow 8CO_2 + 9H_2O \tag{4.52}$$

For this or any other saturated hydrocarbon, the stoichiometric oxygen : fuel ratio (on a mass basis) is about 3.5 : 1, corresponding to an air to fuel ratio of 17 : 1. With insufficient oxygen, incomplete combustion occurs and the reaction products include carbon monoxide and a range of partially reacted or unreacted hydrocarbons. The former compound is a well known toxic gas and some jurisdictions have established stringent maximum allowable levels, typically 10 to 30 ppmv in the atmosphere. The latter compounds, as we have seen, are precursors in the photochemical smog production sequence of reactions. (It should be noted that losses of hydrocarbons also occur to a significant extent by evaporation from the fuel tank or from the carburettor. The losses are especially severe when the ambient temperature is very high.) During combustion, to minimize emissions of fuel hydrocarbons, the obvious solution is to use a lean fuel–oxidant mixture, that is, one in which the ratio of air to gasoline is larger than the stoichiometric requirement. This is consistent with elevating the combustion temperature as a means of increasing the compression ratio and maximizing fuel efficiency. Unfortunately there is a negative environmental side-effect which results from this practice.

The increased supply of air not only elevates the combustion temperature but also augments the concentration of nitrogen and supplies excess oxygen in the combustion mixture. All of these factors, while reducing emissions of hydrocarbons, increase the formation and release of nitric oxide.

We have seen that nitric oxide product is an essential component in photochemical smog formation and is also a precursor for atmospheric generation of nitric acid, one of the constituents of acid rain.

Initial attempts at emission control involved using lower temperature combustion and lower compression ratios along with recycling partially burned exhaust gases as a diluent in the fuel–air mixture. While this practice lowered emission levels of nitrogen oxides, it also reduced fuel efficiency. Therefore a number of more complex strategies of exhaust gas clean-up have been developed. Current technology centres around the use of a *three-way catalyst* system capable of reducing levels of non-methane hydrocarbons, carbon monoxide, and nitric oxide, which involves two catalysts in series following a near-stoichiometric combustion process. The catalysts are mounted on a high-surface-area ceramic material through which the exhaust gases pass. The first catalyst is typically rhodium, which facilitates the reduction of the nitrogen oxides to form dinitrogen; in this case, the reducing agent is one of the incompletely oxidized components of the exhaust gas stream. Unreacted hydrocarbons, hydrogen gas, and carbon monoxide can all react in this way:

$$2NO + 2CO \rightarrow N_2 + 2CO_2 \tag{4.53}$$

As the process proceeds, some ammonia is cogenerated due to reaction of nitric oxide with hydrogen, which is present as a result of decomposition of some components of the hydrocarbon fuel:

$$2NO + 5H_2 \rightarrow 2NH_3 + 2H_2O \tag{4.54}$$

To oxidize the residual hydrocarbons, carbon monoxide, and ammonia, air is supplied to the system and it is passed over a second catalyst, either palladium or platinum or nonstoichiometric oxides such as Fe_2O_3 or $CoO \cdot Cr_2O_3$, to favour oxidation of the reduced components.

$$RH + O_2 \rightarrow CO_2 + H_2O \tag{4.55}$$

$$CO + \tfrac{1}{2}O_2 \rightarrow CO_2 \tag{4.56}$$

$$2NH_3 + \tfrac{3}{2}O_2 \rightarrow N_2 + 3H_2O \tag{4.57}$$

The use of efficient three-way catalyst systems serves to reduce emissions of the three problem gases by 80% or more. There are, however, outstanding technical problems that need to be solved in order to achieve even further reductions. Principal among these is the need to reduce emissions during the 'warm-up' period of vehicle operation, when the temperature in the catalytic chamber is not high enough to bring about efficient reaction of the exhaust gases.

Box 4.1 Reformulated gasolines

One of the complex environmental issues centred on gasoline use in urban areas is related to new types of fuel formulations. Certain additives, particularly methyl *t*-butyl ether (MTBE), methanol, or ethanol, are oxygen suppliers and reduce emissions of carbon monoxide during fuel combustion in an automobile engine. For this reason, legislation in the United States (The Clean Air Act, 1990) has been enacted to require year-round use of reformulated oxygen-enriched gasoline in nine major American conurbations which are locations of severe air pollution problems. These areas are Greater Connecticut, New York City, Philadelphia, Baltimore, Chicago, Milwaukee, Houston, Los Angeles, and San Diego. Other regions are subject to less stringent restrictions. One formulation recommended in compliance with the Act includes 10% ethanol in the fuel.

While the ethanol contributes to improvements in terms of carbon monoxide emissions, it also increases the volatility of the fuel by about 7 kPa measured as 'Reid vapour pressure' (RVP). The RVP is the vapour pressure measured at a temperature of 34.4 °C, which is 100 °F. The increased volatility operates against other sections of the Clean Air Act that require reductions in volatility from levels of 62 kPa for conventional gasoline to 56 kPa in northern cities and 50 kPa in the south. These regulations were introduced because volatility of fuel is a major contributor to release of volatile organic carbons (VOC) which are, along with carbon monoxide significant factors in urban air pollution. Appropriate legislation then requires weighting of the carbon monoxide and VOC emission factors. As in many other situations, a simple solution in response to a particular problem opens up an array of interrelated issues. Aside from the complexities of local urban pollution, one must also consider the effects of devoting vast areas of prime agricultural land to continuous production of corn through high-intensity agriculture. Corn is the major feedstock in ethanol-production plants and its production at high yields depends on heavy application of synthetic fertilizer whose manufacture in turn depends on fossil fuels as well as other raw materials. The environmental consequences of producing and using such chemicals is yet another factor which must be included in any calculation of problems and benefits of gasoline reformulation strategies. A comparison of emissions from regular and reformulated gasolines is given in *Environ. Sci. Technol.* **26** (1992), 206.

4.3.2 Gasoline-powered two-stroke engines

The two-stroke engine is used to power small vehicles and other labour-saving devices such as chainsaws and power mowers. In simplified form, the cycle is shown in Fig. 4.5.

In part a of the figure, the engine is near the bottom of its stroke and the piston is about to move upward, compressing the fuel/air mixture. As it move up to position b, fresh fuel/air mixture is drawn in below the piston while the mixture above is compressed. At sufficiently high compression, a spark ignites the fuel and, in the power stroke, the piston is forced down (part c). As it nears the bottom, exhaust gases begin to escape through the exhaust vent. This continues to occur while fresh fuel/air mixture is forced into the cylinder simultaneously. The cycle then repeats itself.

The engine described here is a simple two-stroke design, and many design modifications have been made to improve efficiency and other characteristics. Nevertheless, it should be clear that one feature of the design—namely the simultaneous introduction of fuel and release of exhaust gases—can lead to problems of loss of unburned hydrocarbons. The problem is enhanced because two-stroke engine fuel normally consists of a mixture of gasoline and a petroleum lubricant; the latter material has higher average molar mass and is less efficiently oxidized during the combustion process than the lighter gasoline.

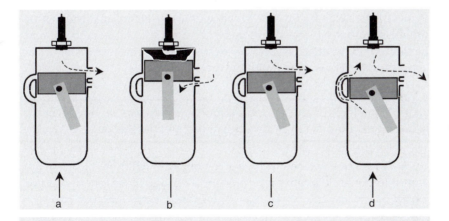

Fig. 4.5 The two-stroke engine cycle. A description of the steps in the cycle is given in the text.

Table 4.3 Exhaust gas output of CO, NO_x and hydrocarbons, from 7.3 kW two- and four-stroke engines operating under the same conditions[4]

	Output/10^{-8}g J^{-1}		
	CO	NO_x	Hydrocarbons
Two-stroke engine	165	0.3	89
Four-stroke engine	127	0.7	7

Another aspect of the engine design is that combustion takes place at a lower temperature than in the conventional four-stroke engine. Taken together, these features lead to an exhaust gas composition that is relatively low in nitric oxide and high in oxygen and hydrocarbons, even when the engine is operated under stoichiometric conditions.

The nature of the hydrocarbons emitted from a 7.3 kW marine outboard engine burning gasoline is similar to that released from road vehicles burning the same fuel.[4] The components are dominated by aromatic hydrocarbons, especially alkyl aromatics. From an outboard motor, the exhaust is released directly into water; although the solubility of these hydrophobic molecules is small, their ecotoxicological effects on aqueous organisms could be significant.

Two- and four-stroke outboard engines with the same power output are available and therefore it is possible to make comparisons of emissions under similar operating conditions. Carrying out this kind of test using 7.3 kW engines burning gasoline,[4] exhaust was found to have composition as shown in Table 4.3.

The two previously noted effects are confirmed in these data—that is, that two-stroke engines emit relatively large amounts of hydrocarbons, and relatively small amounts of nitrogen oxides. Emissions catalysts are used to reduce the hydrocarbon emissions, and clearly the catalyst must be of the oxidative variety. Because most two-stroke engines are physically small, gases are emitted close to the combustion chamber and tend to be very hot under running conditions. It may therefore be necessary to cool them to a temperature at which the catalyst can be effective and one means of doing this is to introduce additional cooling air just prior to passage through the catalyst system.

Table 4.4 The effects of engine optimization and catalysts on release of CO, NO_x, and hydrocarbons from a 125 cc two-stroke motorcycle engine

	Output/g km^{-1}		
	CO	NO_x	Hydrocarbons
Production engine	21.7	0.01	16.9
Optimized engine	1.7	0.03	10.4
Engine with catalyst	0.8	0.02	1.9
Swiss standards	8	0.1	3

Catalysts are able successfully to reduce emissions to a much lower level. Table 4.4 gives data for a 125 cc high-performance motorcycle engine operating in three formats—as a standard production engine, with optimization of the carburettor and output gas scavenging conditions, and then with the addition of a catalyst to the exhaust system. In the final configuration, the emissions met the strict standards of the Swiss government[5] (Table 4.4).

4.3.3 Diesel-powered engines

Diesels are usually efficient four-stroke engines which operate with higher compression ratios than the standard Otto engine. The large pressure in the cylinder results in a high temperature so that the fuel mixture spontaneously burns in the absence of a spark ignition source. The cycle takes the following form shown in Fig. 4.6.

In the cylinder filled with air, adiabatic compression takes place (ab) with a compression ratio of about 15 : 1. At the end of step ab, fuel is introduced. Compression to a very small volume results in a temperature increase to about 500 °C in the cylinder and the fuel ignites spontaneously and instantaneously, burning continuously as the piston is forced downwards in the power stroke (bc and cd). The time taken for fuel injection and combustion is sufficiently long that cylinder pressure remains nearly constant in the initial stage

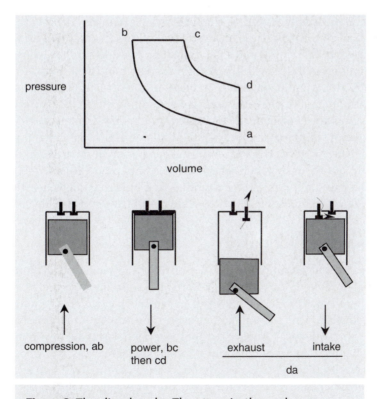

Fig. 4.6 The diesel cycle. The steps in the cycle are described in the text.

(bc) of expansion. Finally, exhaust of the combustion products occurs, followed by intake of fresh air (da).

The diesel engine uses a lower grade higher-boiling fuel than the gasoline employed in the previously described engines. The high compression ratio leads to a more efficient power source (eqn 4.51) but also requires that the engine be heavily constructed to withstand the pressures generated in the cylinders. Because fuel is injected at the completion of the compression stroke, to some extent mixing of fuel and air is incomplete. Consequently, one of the principal problems associated with diesel engines is release of particulates, which are made up of unburned carbon particles (soot) and a soluble organic fraction (SOF). The SOF has at least two components–unburned or partially burned fuel which forms a liquid aerosol, and other components which are sorbed on to the soot particles. Gaseous hydrocarbon emissions are less than from gasoline-powered engines because of the higher boiling nature of diesel fuel. Inorganic sulfates are also present in the aerosol as the sulfur content of diesel fuel is somewhat higher (typically 0.1 to 0.3%) than that found in most gasolines.

The other important constituent of exhaust gas is nitric oxide, which is present as a consequence of the high combustion temperature and the lean conditions that exist for at least a part of the combustion process.

Soot traps are one means of controlling emission of the aerosol particulates. However, as the soot accumulates, resistance to flow increases and it is necessary to regenerate the filter by burning off the accumulated material. The technology for accomplishing regeneration is complex, usually requiring an energy source, a second trap to be used during regeneration, and a control package. An alternative approach is to use an oxidation catalyst which can accommodate liquid and solid emissions as well as unburned gaseous hydrocarbons. One type of catalyst consists of platinum or palladium deposited on a high-surface-area silica substrate. Alumina is less desirable as a support because of its strong specific interaction with sulfate, which is also a component of the exhaust gas. The usual reduction catalysts are suitable for removing nitric oxide from the gas stream.

While removal of pollutant gases after their production in the engine remains a requirement, an even better approach is to prevent the generation of the gases in the first place. This is difficult, but some success has been achieved by engine design wherein recirculation of exhaust gases, altering injection timing, and ensuring good fuel–air mixing, are used to bring about complete combustion while minimizing nitric oxide production.

4.3.4 Ozone production

In the mix of gases emitted as products of combustion in any engine are a number of species with undesirable, sometimes toxicological properties. Secondary reactions bring about chemical changes which produce other gases and aerosol particulates. Among these, ozone is of considerable concern as it can have adverse effects on both plants and animals. Furthermore, in photochemical smog situations, elevated levels of ozone are frequently observed. In Chapter 7, we will note some urban areas where excessive atmospheric ozone levels have been recorded.

We have seen that ozone is one of the byproducts of the sequence of reactions by which a hydrocarbon is oxidized in reactions initiated by the hydroxyl radical. In order to reduce the extent of ozone production in an urban area then, it is necessary to reduce the atmospheric concentration of the reactants, nitric oxide (the precursor of hydroxyl), and the

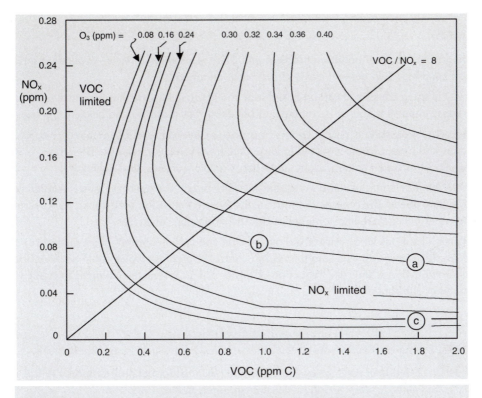

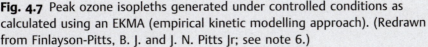

Fig. 4.7 Peak ozone isopleths generated under controlled conditions as calculated using an EKMA (empirical kinetic modelling approach). (Redrawn from Finlayson-Pitts, B. J. and J. N. Pitts Jr; see note 6.)

hydrocarbons themselves. Ideally, one would wish to reduce the concentration of both types of gases. We have seen, however, that this is a technological challenge, and it is possible to reduce ozone synthesis by considering which of the two reactants is the *limiting reagent*. Fig. 4.7 plots ozone isopleths (contours of constant concentration) as a function of both nitric oxide and hydrocarbon concentration.[6]

Consider a situation where the number of two-stroke engines is large compared to other vehicles and emissions are therefore highly enriched in hydrocarbon content. In this location, represented by point 'a', a region of large hydrocarbon : nitrogen oxides ratio on the plot, reducing the hydrocarbon emissions by 50% would change the conditions to point 'b', a point at which little improvement in ozone concentration would be observed. On the other hand, a reduction in the already low nitric oxide levels might alter the situation to point 'c' on the plot, with significant reduction in ozone production. This is due to the fact that the amount of ozone produced is limited by the availability of hydroxyl radical which, in turn, depends on a plentiful supply of nitrogen oxides.

It is important to understand these factors but, at the same time, they apply only to the production of a single pollutant gas–ozone. It may also be important to reduce emissions of hydrocarbons because, in themselves, compounds like benzene have harmful properties.

The main points

1 Smog is an urban air pollution problem and there is a variety of types depending on the local situation. Two general classes have been identified.

2 Classical smog consists of carbon-based soot and other solid particulates and sulfur dioxide. Its most important source is combustion of coal. It has reducing and acidic properties.

3 Photochemical smog is produced by atmospheric chemical reactions between nitrogen oxides and hydrocarbons emitted in large part from vehicular exhaust. The smog itself consists of nitrogen oxides, ozone and other organic oxidants, and aldehydes.

4 Photochemical smog oxidation reactions are initiated by the hydroxyl radical (produced in large part due to the presence of nitric oxide from combustion emissions). Hydrocarbons and other volatile organic compounds are the oxidizable substrates.

5 Engines of various kinds release the nitric oxide and carbon compounds essential to generating photochemical smog. Emission control strategies involve catalytic oxidation of unburned hydrocarbons and reduction of nitric oxide. Current technology can reduce the emissions substantially.

Additional reading

1 Guderian, R. (ed.), *Air Pollution by Photochemical Oxidants: Formation, Transport, Control, and Effects on Plants*, Ecological Series, Vol. 52, Springer, Berlin; 1985.

2 Fenn, J. B., *Engines, Energy, and Entropy*, W. H. Freeman & Co, New York; 1982.

3 Atkinson, R., Gas-phase tropospheric chemistry of organic compounds: a review. *Atmospheric Environment*, **24A** (1990), 1–41.

4 Farrauto, R. J., R. M. Heck, and B. K. Speronello, Environmental catalysts. *Chem. Eng. News*, **70** (1992), 34–44.

Problems

1 A general formula for gasoline is C_7H_{13}. Calculate the air to gasoline mass ratio required for stoichiometric combustion.

2 Write out a general sequence for the photochemical oxidation of butane.

3 Suppose propene ($CH_2 = CH–CH_3$) is the hydrocarbon that reacts with the hydroxyl radical ·OH. Write the set of chemical reactions which ultimately produce an aldehyde. What is this final aldehyde?

4 Peroxyacetic nitric anhydride (PAN) can be thought of as related to hydrogen peroxide. Draw its structure and indicate why you suppose that it is a powerful oxidizing agent.

5 The atmospheric ratio of PAN : PAN + inorganic nitrate varies from less than 0.1 to 0.9. High values of the ratio are associated with 'photochemically aged' air masses, situations where precipitation has recently occurred, and situations where unusually high night-time nitrogen oxide concentrations are found (Roberts, J. M., The atmospheric chemistry of organic nitrates. *Atmospheric Environment*, **24A**(2) (1990), 243–287. What do these observations tell us about factors affecting the relative rates of formation and removal of nitrogen oxide compounds?

6 Nitrogen oxide is formed at night by dissociation of dinitrogen pentoxide. For the first order reaction

$$N_2O_5 \rightarrow NO_2 + NO_3$$

the rate constant is 3.14×10^{-2} s^{-1} at 25 °C, and 1.42×10^{-3} s^{-1} at 55 °C. Calculate the half life of this molecule at the two temperatures. For a concentration of N_2O_5 of 3.6 ppbv, calculate the length of time it could take for the concentration to be reduced to 1.0 ppbv at a constant temperature of 25 °C. Calculate the Arrhenius parameters, A and E_a, for the reaction.

7 Use the data in Table 4.2 to calculate the concentration of atmospheric carbon compounds in ppbv C.

Notes

1 The brackets in the formula of some of the species indicate that both the bracketed atom and the next one are bonded to the adjacent carbon. In these particular formulae, the single oxygen in brackets is joined to carbon by a double bond. This allows us to write the formulae in a linear fashion.

2 Peroxyacetic nitric anhydride is the name we use here. Peroxyacetyl nitrate is the normal name used, but does not conform to IUPAC conventions. Using the IUPAC rules it should be called ethane peroxoic nitric anhydride, and a compromise name is peroxyacetic nitric anhydride (PAN). Other carbonyl precursors result in the production of related compounds–for example, from propionaldehyde, peroxypropionic nitric anhydride (PPN) is produced. The entire class of compounds is usually referred to as PANs.

3 When considering non-methane hydrocarbons as a group, it is useful to report the mixing ratio in ppbv C. This is done by multiplying the normal mixing ratio (ppbv) of an individual compound by the number of carbon atoms. Therefore, one ppbv ethane is equivalent in carbon content to two ppbv methane. For conversion of NMHC concentrations in units such as μg m^{-3} to mixing ratios in ppbv C, the number of μmoles of carbon is estimated by dividing the mass (μg) by 14, equivalent to the mass of a methylene group.

4 Juttner, F., D. Backhaus, U. Matthias, U. Essers, R. Greiner, and B. Mahr, Emissions of two- and four-stroke outboard engines–I. *Quantification of gases and VOC. Water Research*, **29** (1995), 1976–82.

5 Laimbock, F. as in Blair, G. P., *The basic Design of Two-Stroke Engines*, Society of Automotive Engineers, Inc., warrendale, Pa; 1990; p. 328.

6 Finlayson-Pitts, B. J. and J. N. Pitts Jr, Atmospheric chemistry of tropospheric ozone formation: *scientific and regulatory implications, Air and Waste*, **43** (1993), 1091–1100.

5

Tropospheric chemistry— precipitation

Rain and snow are the forms of precipitation that are most familiar to all of us. In an environmental context, however, the term precipitation has a broader definition and can be subdivided into two categories. Precipitation by *wet deposition* refers to deposition on the Earth's surface of water-based particles in liquid or solid form—rain, snow, and various types of ice such as sleet and hail. In its wider meaning, precipitation also includes surface deposition of *dry* species, including gases such as sulfur dioxide and solid particles such as various components of dust. Without specifically using the term *dry deposition*, in the next chapter we will discuss the chemical and physical behaviour of dry species in the atmosphere. In this chapter, the emphasis is on wet deposition.

When water condenses to form clouds, various chemical species are incorporated into the cloud droplets and sometimes further transformed there. During a precipitation event, the accumulated species are carried to the Earth's surface. As a sink for atmospheric chemicals, accumulation in clouds is referred to as *rainout*, while *washout* is the term used when chemicals are taken up by water droplets as they pass through air below the clouds.

5.1 The composition of rain

If rain water is in equilibrium with gaseous species in the atmosphere, it contains soluble forms of the gases with concentrations determined by Henry's law. Amongst the gases is carbon dioxide whose atmospheric mixing ratio in the northern hemisphere averaged approximately 365 ppmv in 1995. Dissolved carbon dioxide is a weak acid and an aqueous solution in equilibrium with the unpolluted atmosphere has a pH of 5.7. The significance of this value will become clear when we carry out calculations of gas solubilities in water and the effect on pH in Chapter 11. Rain water containing dissolved carbon dioxide is therefore slightly acidic, with a hydronium ion concentration approximately 20 times greater than that of pure water. Since most of the carbon dioxide is of 'natural' origin, it is frequently suggested that the pH of unpolluted rain is 5.7. In fact, other natural acid-producing species, including organic acids and sulfur compounds which are produced by microbiological processes from living and dead biomass, can be a source of enhanced acidity in rain in areas remote from human influence. Still other 'natural' chemicals such as dust containing calcium carbonate, when incorporated in rain, can result in a mildly alkaline solution. Depending on the setting therefore, a variety of anionic and cationic species accumulate in rain and the distinction between natural and anthropogenic origin is not always clear. As a consequence of the combined factors, the pH of unpolluted rain might range from below 4.5 to 8 or higher. In any case, rain and snow are not pure water and their chemistry is determined by a range of complex, interrelated factors. The concentration of major species in samples of rain at a variety of locations is given in Table 5.1.

The same dissolved species as those listed in the table are also important components of precipitation occurring at other locations around the globe. While individual species cannot be assigned with certainty to a specific origin, sodium and chloride are, to a large extent, derived from the oceans (an example is the St Georges, Bermuda rainfall). Accompanying sodium and chloride are somewhat smaller concentrations of other marine ions including potassium, calcium, magnesium, and sulfate. In less common situations,

Table 5.1 The composition of rain from several locations

	Urban Guiyang, Guizhou, PRC[a]	Birkenes, Southern Norway[b]	Katherine, Northern Territiories, Australia[c]	Pune, Maharashtra State, India[d]	St Georges Bermuda[c]
	Concentration/μmol L^{-1}				
H$^+$	112 (pH = 3.95)	57 (pH = 4.2)	16.6 (pH = 4.8)	0.04 (pH = 7.4)	16.2 (pH = 4.8)
Cl$^-$		58	11.8	155	175
NO$_3^-$	10.3	38	4.3	18	5.5
SO$_4^{2-}$	222	68	6.3	11	36.3
Ca^{2+}	128	9	2.5	55	9.7
Mg^{2+}		13	2.0	35	34.5
Na$^+$		56	7.0	150	147
K$^+$		4	0.9	36	4.3
NH$_4^+$	57	38	2.4	28	3.8

The sites in China and Norway are considered to contain anthropogenic-source chemical species. The others are influenced to a smaller or negligible extent by human activity.
[a]Dianwu, Z. and X. Jiling, Acidification in southwestern China, in ref. 4. No data are provided for Cl, Mg, Na, and K.
[b]Overrein, L. N., H. M. Seip, and A. Tollan, Acid precipitation—effects on forest and fish. Final report of the SNSF project, Norwegian Ministry of the Environment, 1972–1980, Oslo, 1980.
[c]Legge, A. H. and S. V. Krupa, *Acid Deposition, Sulfur and Nitrogen Oxides*, Lewis Publishers, Chelsea MI; 1990.
[d]Khemani, L. T., G. A. Momin, M. S. Naik, P. W. Prakasa Rao, P. D. Safai, and A. S. R. Murty, Influence of alkaline particulates on pH of cloud and rain water in India. Atmospheric Environment, **24** (1987), 1137–45.

elevated concentrations of sodium and/or calcium and chloride in the atmospheric aerosol originate from salts used to melt ice and snow on highways during the winter in countries like Russia and Canada. From dust, calcium and magnesium are incorporated into water droplets, and consequently high levels of these elements are usually associated with a terrestrial origin. An example is the Pune, India sample. Ammonium ion originates from biological processes involving nitrogen from plant and animal residues and from inorganic fertilizers. As well as the principal species, smaller concentrations of other elements are incorporated into rain, especially in heavily industrialized regions.

Since the sodium, potassium, calcium and magnesium cations and the chloride anion are all common, of natural origin, and generally benign species at the μmol L^{-1} concentrations found in rain, their presence has relatively minor environmental significance. It is the sulfur and nitrogen species and associated acidity that are of particular concern and together create the phenomenon known as acid precipitation. To a large extent these species derive from anthropogenic sources. Table 5.2 lists sources and estimates of mixing ratios and residence times for sulfur and nitrogen species present in the atmosphere.

As with other compounds, distinctions between natural and anthropogenic sources are not clear, but there is no doubt that a major contribution of both sulfur and nitrogen atmospheric species arises from combustion and a variety of industrial processes. It is these emissions which ultimately contribute to enhanced production of the strong acids found in wet and dry precipitation.

Table 5.2 Nitrogen and sulfur species present in the atmosphere

	Approximate atmospheric mixing ratio/ppbv[a]	Approximate residence time/days	Source
Nitrogen species			
Nitrogen oxides, NO_x	1– >10 (urban) 0.1–1 (remote)	0.2 (urban, summer) to 10 (remote, winter)	Fossil fuel, biomass, combustion; lightning; microbiological release
Ammonia, NH_3	0.1–1	2–70	Animal excreta, fertilizers, microbiological release
Sulfur species			
Sulfur dioxide, SO_2	0.01–0.3	3–5	Fossil fuel, biomass combustion; sulfide ore smelting
Hydrogen sulfide, H_2S	0.05–0.3	1–2	Submerged soils, wetlands
Carbon disulfide, CS_2	0.02–0.5	50	Submerged soils, wetlands
Dimethyl sulfide, $(CH_3)_2S$	0.01–0.07	1	Oceans
Carbonyl sulfide, COS	0.3–0.5	200–400	Oceans, soils
Methyl mercaptan, CH_3SH			Oceans, soils
Dimethyl disulfide, CH_3SSCH_3			Oceans, soils

[a] Mixing ratios are for unpolluted areas unless otherwise noted. Data are from various sources.

5.2 Atmospheric production of nitric acid

The principal reaction sequence contributing to production of nitric acid is relatively straightforward and has been discussed in Chapters 3 and 4. It begins with nitric oxide emissions primarily occurring during combustion processes. For high temperature combustion, most of the nitrogen originates from the atmosphere, but some can also be derived from organic nitrogen compounds in fuels such as wood. Smaller amounts of nitric oxide are released as a byproduct of microbial nitrification in soil, a process that is enhanced in the high temperature tropical environment. Lightning, which is also most frequent in the tropics, adds a further small input to the nitric oxide budget.

5.2.1 Daytime chemistry

Nitric oxide is oxidized by O_2, O_3, or $ROO\cdot$ (where R is an alkyl group). For example

$$NO + O_3 \rightarrow NO_2 + O_2 \tag{5.1}$$

The nitrogen dioxide produced in this way subsequently contributes to ozone and hydroxyl radical production and therefore is responsible (in part) for the initiation of a photochemical smog sequence. In the process, nitric oxide is regenerated and is therefore available to once again contribute to additional ozone and smog production. The principal mechanism for removal of nitrogen oxides from the atmosphere is through conversion to nitric acid via oxidation of nitrogen dioxide by the hydroxyl radical

$$\cdot NO_2 + \cdot OH + M \rightarrow HNO_3 + M \tag{5.2}$$

where M is a third body. The pseudo second order rate constant for the reaction has a value of $1.2 \times 10^{-11}\ (T/298)^{-1.6}$ cm^3 molecule^{-1} s^{-1}. Because reaction 5.2 involves starting materials formed in part through photochemical processes, it is largely a daytime reaction.

5.2.2 Night-time chemistry

After sunset, an alternative sequence becomes more important for producing nitric acid. It involves the nitrate radical which is formed during both day and night but accumulates only at night-time because it is destroyed by photolysis. The reaction for nitrate radical formation is

$$\cdot NO_2 + O_3 \rightarrow \cdot NO_3 + O_2 \tag{5.3}$$

The radical can take part in a number of reactions. To some extent, it is destroyed by reactions with the NO_x compounds:

$$NO_3 + NO_2 \rightarrow NO + NO_2 + O_2 \tag{5.4}$$
$$NO_3 + NO \rightarrow 2NO_2 \tag{5.5}$$

In the same manner as hydroxyl, nitrate radical is able to add to the double bond of olefins:

$$\cdot NO_3 + C_n H_{2n} \rightarrow \cdot C_n H_{2n} NO_3 \tag{5.6}$$

Because the product is also a radical, it is susceptible to further reactions. Typically, addition of the nitrate radical is followed by rapid addition of dioxygen.

Also like hydroxyl, nitrate radical initiates reaction sequences by first abstracting a hydrogen. In these cases, nitric acid acid is formed:

$$\text{with aldehydes} \qquad \cdot NO_3 + RCHO \rightarrow R\dot{C}O + HNO_3 \qquad (5.7)$$

$$\text{with alkanes} \qquad \cdot NO_3 + RH \quad \rightarrow \cdot R + HNO_3 \qquad (5.8)$$

In both reactions 5.7 and 5.8, R is an alkyl group. The alkyl radicals that are produced take part in further reactions such as the addition of dioxygen.

A kinetically rapid pair of reactions involving the nitrate radical and nitrogen dioxide results in the formation of dinitrogen pentoxide and then nitric acid:

$$NO_3 + NO_2 \rightleftharpoons N_2O_5 \qquad (5.9)$$

$$N_2O_5 + H_2O \rightarrow 2HNO_3 \qquad (5.10)$$

In many environmental situations, most of the NO_x present in the atmosphere at night-time proceeds through reactions 5.3, 5.9, and 5.10 and this is an important means of producing nitric acid. A smaller fraction of NO_x is finally converted to PAN by the reaction

$$RC(O)OO\cdot + \cdot NO_2 \rightarrow RCOO_2NO_2 \qquad (5.11)$$
$$\text{PAN}$$

5.2.3 Removal of nitric acid

Nitric acid is removed from the atmosphere by either wet or dry deposition, and is one of the main contributors to precipitation acidity. To some degree it also reacts with ammonia, whose principal source is volatilization from urea in animal urine and other organic reduced nitrogen sources:

$$NH_3 + HNO_3 \rightarrow NH_4NO_3 \qquad (5.12)$$

The ammonium nitrate can act as a condensation nucleus for the formation of a water droplet or it is deposited as part of the solid aerosol.

5.3 Atmospheric production of sulfuric acid

5.3.1 Oxidation of reduced sulfur species

The production of sulfuric acid is a more complex process than that of nitric acid. For one thing, the reaction sequence can begin with a wide range of reduced as well as partially oxidized sulfur compounds. Hydrogen sulfide, carbonyl sulfide, carbon disulfide, methyl mercaptan, dimethyl sulfide, and dimethyl disulfide all contain sulfur in its lowest (-2) oxidation state. These compounds are released from the oceans and from soils under reducing conditions as a consequence of various microbiological processes. High temperatures favour microbial activity and so the release of reduced sulfur compounds is especially significant in the tropics. In the atmosphere, hydrogen sulfide, carbon disulfide, and carbonyl sulfide are oxidized *via* hydroxyl giving the thionly radical (·SH) as an initial product:

$$H_2S + \cdot OH \rightarrow H_2O + \cdot SH \tag{5.13}$$

$$CS_2 + \cdot OH \rightarrow COS + \cdot SH \tag{5.14}$$

$$COS + \cdot OH \rightarrow CO_2 + \cdot SH \tag{5.15}$$

Of the three compounds, carbonyl sulfide, released directly from the oceans or produced by oxidation of carbon disulfide, is kinetically relatively stable to the oxidation reaction shown. It therefore has a long tropospheric lifetime estimated to be between 0.5 and 1 year. As a consequence, diffusion of this particular sulfur-containing species into the stratosphere is an important removal process. In the stratosphere, it can undergo photochemical oxidation to produce sulfur dioxide and ultimately sulfate anion which is an important component of the stratospheric aerosol. Volcanic activity can be a means by which sulfur dioxide is injected directly into the stratosphere, increasing the density of the aerosol there and leading to significant depression of the global temperature by physically blocking solar radiation. The sulfur compounds in the stratosphere also take part in other chemical processes having diverse environmental consequences (Box 5.1).

Box 5.1 The Mount Pinatubo volcano

Mount Pinatubo is a volcanic mountain located about 100 km north-west of Manila in the Philippines. After being quiescent for 635 years, it erupted dramatically in June of 1991, with peak releases on 14 and 15 June. Approximately 7 km³ of magma was expelled as lava, and as ash into the atmosphere. There was heavy ashfall up to 40 km from the volcano, and the typhoon Yunya, which occurred shortly after the volcanic eruption, carried solids, depositing them as far away as Thailand and Singapore. The ash—a calc-alkaline pumice—contained phenocrysts of anhydrite ($CaSO_4$), indicating that there were high concentrations of sulfur in the magma.

Besides ash, there was release of large quantities of gases with composition in the carbon–oxygen–hydrogen–sulfur family. Principal gases included water vapour and carbon dioxide along with about 20 Mt of sulfur dioxide (this amount, ~ 10 Mt expressed as sulfur, is about one tenth of the annual global anthropogenic release of sulfur dioxide; see Table 5.5). The gases were injected into the stratosphere at an altitude of 20 to 30 km, and the sulfur dioxide over time converted to a sulfuric acid aqueous aerosol. The aerosol cloud drifted to the north-west and could be observed as far away as the Greenland–Iceland area in early January of 1992. Various processes caused the cloud to dissipate over a 1–3 year period.

There have been a number of environmental effects attributed to the aerosol. The dispersed cloud particles blocked solar radiation and a measureable, but not uniform, average global cooling was observed in the 2 years following the eruption. There is also evidence that ozone loss was accelerated, especially within the polar vortexes. This is attributed to enhanced conversion of dinitrogen pentoxide to nitric acid (reaction

5.10), a reaction that readily occurs on the surface of the sulfuric-acid-ice crystals. By removing NO$_x$ from the stratosphere there was less tendency for nitrogen dioxide to react with chlorine monoxide (reaction (3.50)). This, in turn, augmented the chlorine catalytic cycle for ozone destruction and caused a reduction of ozone concentrations.

The amount of sulfur dioxide released from Pinatubo was too large to have occurred only by exsolution (release out of solution) from the magma at the time of the eruption; it is believed that pre-eruption releases of large amounts of vapour also occurred. Such emissions occur regularly at other terrestrial and marine sites, and are not always associated with catastrophic volcanic eruptions. As we noted at the beginning of the book, this type of release of gases has occurred throughout the entirety of Earth history. This led to the formation of our planet's unique atmosphere. Much, if not all, of the water on Earth was derived in this way.

Further oxidation of thionyl leads to production of sulfur dioxide:

$$\cdot SH + O_2 \rightarrow SO + \cdot OH \tag{5.16}$$

$$\cdot SH + O_3 \rightarrow SHO\cdot + O_2 \tag{5.17}$$

$$SHO\cdot + O_2 \rightarrow SO + HOO\cdot \tag{5.18}$$

$$SO + \left. \begin{array}{c} O_2 \\ O_3 \\ NO_2 \end{array} \right| \rightarrow SO_2 + \text{other products} \tag{5.19}$$

Phytoplankton living in surface waters of the oceans produces a large amount of dimethyl sulfide—one of the most important reduced sulfur compounds released to the atmosphere. The hydroxyl radical reacts with dimethyl sulfide by hydrogen abstraction or by addition:

$$(CH_3)_2S + \cdot OH \left. \begin{array}{c} \rightarrow \\ \\ \rightarrow \end{array} \right| \begin{array}{c} CH_3 S\dot{C}H_2 + H_2O \\ \\ CH_3 \dot{S}(OH)CH_3 \end{array} \rightarrow \text{further oxidation products} \tag{5.20}$$

The further oxidation products include dimethylsulfoxide and methane sulfonic acid, both of which have been found in the marine atmospheric aerosol. Sulfur dioxide is also formed; the mechanisms for these reactions have not been clearly mapped out.

By means of these processes, any reduced sulfur compounds are able to be oxidized, with one of the principal products being sulfur dioxide. This is ultimately converted to sulfuric acid. In areas of the world remote from human activity the emitted dimethyl sulfide, in particular, is a major cause of rainfall or aerosol pH being less than the value of 5.7 as defined by the solubility of atmospheric carbon dioxide.

Sulfur dioxide is also released in large quantities directly into the atmosphere from sulfide ore smelting and fossil fuel combustion. It is this excess anthropogenic source that gives rise to even higher acidity in some continental regions of the globe which are influenced by intense industrial activity.

5.3.2 Oxidation of sulfur dioxide by homogenous reactions

Sulfuric acid production from sulfur dioxide takes place by at least two distinct sets of processes. The first sequence occurs homogeneously in the gas phase and most frequently begins with reaction (5.21) as the rate-determining step:

$$SO_2 + \cdot OH + M \rightarrow HO\dot{S}O_2 + M \tag{5.21}$$

The reaction as shown is third order, but in the lower troposphere, where the concentration of M—mostly dinitrogen and dioxygen—is large, it becomes pseudo second order. The second order rate constant therefore would decrease with decreasing pressure moving to higher altitudes in the troposphere and stratosphere. However, a second factor must be considered in determining the rate constant. As with many other radical–radical and ion–molecule reactions involving simple species, the activation energy for this association

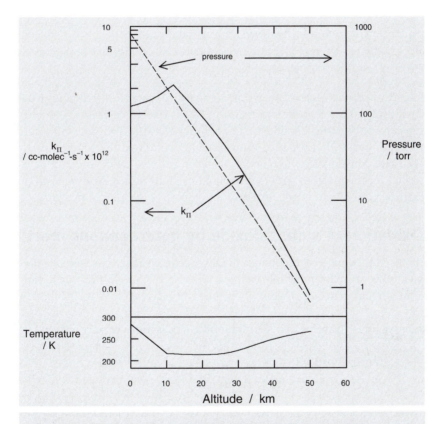

Fig. 5.1 Variation in the second order rate constant for the production of HO$\dot{S}$O$_2$ as a function of altitude in the troposphere and stratosphere. (Redrawn from J. G. Calvert, and W. R. Stockwell, Mechanisms and rates of the gas-phase oxidations of sulfur dioxide and nitrogen oxides in the atmosphere, in ref. 2.)

reaction is negative. Therefore, the decrease in temperature with altitude in the troposphere by itself would result in a corresponding increase in the rate constant for the reaction. Combining both temperature and pressure factors, best estimates of the second order rate constant are shown in Fig. 5.1.

The $HO\dot{S}O_2$ radical can undergo a number of relatively rapid reactions some of which result in sulfuric acid production. The simplest and most important acid-producing process is

$$HO\dot{S}O_2 + O_2 + M \rightarrow HOO\cdot + SO_3 + M \tag{5.22}$$

This is followed by dissolution in water to form sulfuric acid:

$$SO_3 + H_2O \rightarrow H_2SO_4 \tag{5.23}$$

The hydroperoxyl radical produced in the sequence also reacts with nitric oxide:

$$NO + HOO\cdot \rightarrow NO_2 + \cdot OH \tag{5.24}$$

and the nitrogen dioxide and hydroxyl radical take part in the nitric-acid-generating sequence

$$\cdot NO_2 + \cdot OH + M \rightarrow HNO_3 + M \tag{5.2}$$

described in the previous section. A portion of the hydroxyl radicals reacts with additional sulfur dioxide and so the set of reaction 5.21 to 5.24 is a self-accelerating sequence.

Starting again with sulfur dioxide, there are other quantitatively less important homogeneous reaction series that produce sulfuric acid, including direct reaction with atomic oxygen. The rate constant for the reaction between sulfur dioxide and atomic oxygen is similar to that for the reaction with hydroxyl radical, but the atomic oxygen tropospheric mixing ratio is approximately two orders of magnitude below that of hydroxyl.

5.3.3 Oxidation of sulfur dioxide by heterogenous reactions

In Chapter 3, we encountered a sequence of heterogeneous reactions which leads to large-scale losses of stratospheric ozone during the polar spring. Sulfuric acid can also be produced in a heterogeneous process when the required reactants are available in cloud droplets.

Beginning again with sulfur dioxide, the following reactions occur:

$$SO_2(g) \rightleftharpoons SO_2(aq) \qquad\qquad K_H = 1.81 \times 10^{-5}\,\text{mol}\,L^{-1}\,Pa^{-1} \tag{5.25}$$

$$SO_2(aq) + 2H_2O \rightleftharpoons HSO_3^-(aq) + H_3O^+(aq) \quad K_{a1} = 1.72 \times 10^{-2}\,\text{mol}\,L^{-1} \tag{5.26}$$

$$HSO_3^-(aq) + H_2O \rightleftharpoons SO_3^{2-}(aq) + H_3O^+(aq) \quad K_{a2} = 6.43 \times 10^{-8}\,\text{mol}\,L^{-1} \tag{5.27}$$

Like that of carbon dioxde, the aqueous solubility of sulfur dioxide is pH dependent, but it is much larger. If the atmospheric mixing ratio of sulfur dioxide is 10 ppbv at P° the solubility is $2.2 \times 10^{-6}\,\text{mol}\,L^{-1}$ when the aerosol has pH of 4, and is $2.2 \times 10^{-3}\,\text{mol}\,L^{-1}$ when the pH is 7. The solubility is one of the factors affecting the rate of the heterogeneous reaction. Oxidation of sulfur species takes place within the water droplets. The most important oxidant is hydrogen peroxide, a chemical which has an atmospheric mixing ratio of around 1 or 2 ppbv and is readily soluble in water ($K_H = 7.0 \times 10^{-1}$ mol L^{-1} Pa^{-1}).

$$HSO_3^-(aq) + H_2O_2(aq) \underset{k_2}{\overset{k_1}{\rightleftharpoons}} HOOSO_2^-(aq) + H_2O \tag{5.28}$$

Peroxymonosulfite, $HOOSO_2^-$, has a structure

and rapidly rearranges to form hydrogen sulfate, HSO_4^-, whose structure is

In protonated form, hydrogen sulfate is sulfuric acid. The combined rearrangement and protonation reaction is therefore

$$HOOSO_2^-(aq) + H_3O^+(aq) \overset{k_3}{\rightarrow} H_2SO_4(aq) + H_2O \tag{5.29}$$

In the steady state condition, the rate of production of sulfuric acid by hydrogen peroxide oxidation of sulfur dioxide is calculated in the following way. Based on reaction 5.29, the rate of production of sulfuric acid is given by

$$\frac{d[H_2SO_4]}{dt} = k_3[HOOSO_2^-][H_3O^+]. \tag{5.30}$$

Assuming a steady state concentration of HSO_4^-,

$$\frac{d[HOOSO_2^-]}{dt} = 0 = k_1[HSO_3^-][H_2O_2] - k_2'[HOOSO_2^-] - k_3[HOOSO_2^-][H_3O^+] \tag{5.31}$$

In eqn 5.31, $k_2' = k_2[H_2O]$ is a pseudo first order rate constant since $[H_2O] \gg [HOOSO_2]$.

$$k_1[HSO_3^-][H_2O_2] = [HOOSO_2^-](k_2' + k_3[H_3O^+]) \tag{5.32}$$

$$[HOOSO_2^-] = \frac{k_1[HSO_3^-][H_2O_2]}{k_2' + k_3[H_3O^+]}. \tag{5.33}$$

Substituting eqn 5.33 in 5.30,

$$\text{rate} = \frac{d[H_2SO_4]}{dt} = \frac{k_1 k_3 [HSO_3^-][H_2O_2][H_3O^+]}{k_2' + k_3[H_3O^+]}. \tag{5.34}$$

The values of the rate constants are[1] $k_1 = 5.2 \times 10^6$ L mol^{-1} s^{-1} and $k_2'/k_3 = 10^{-1}$. When the pH is greater than 2, $k_3[H_3O^+] \ll k_2'$ and the rate is given by

$$\text{rate} = \frac{k_1 k_3}{k_2'}[HSO_3^-][H_2O_2][H_3O^+]. \tag{5.35}$$

Using the equations for K_H and K_{a1},

$$[H_3O^+][HSO_3^-] = K_H K_{a1} P_{SO_2} \qquad (5.36)$$

$$\text{rate} = \frac{k_1 k_3 K_H K_{a1}}{k_2'}[H_2O_2]P_{SO_2} \qquad (5.37)$$

$$= k'[H_2O_2]P_{SO_2}. \qquad (5.38)$$

This rate law applies between approximately pH 2 and 5 where hydrogen sulfite (HSO_3^-) is the principal aqueous sulfur (IV) species. Within this range, the oxidation of sulfur dioxide by hydrogen peroxide is the dominant mechanism and the reaction rate is approximately independent of pH. Below pH $\sim$ 2 the rate decreases, as reflected by an increase in the denominator of eqn 5.34. At higher pH values, sulfite (SO_3^{2-}) becomes the dominant sulfur species and because it does not react with hydrogen peroxide, the rate of oxidation *via* this oxidant again decreases.

A second heterogeneous pathway involves ozone as the oxidant; in this case both hydrogen sulfite and sulfite ions are oxidizable.

$$HSO_3^-(aq) + O_3 \rightarrow SO_4^{-2}(aq) + H_3O^+(aq) + O_2 \qquad (5.39)$$

$$SO_3^{2-}(aq) + O_3 \rightarrow SO_4^{2-}(aq) + O_2 \qquad (5.40)$$

The rates of these reactions are such that, taken together with the previous processes, below pH 5.5, hydrogen peroxide is the primary oxidant while above that value, ozone becomes more important. Depending on availability of oxidants, the heterogeneous reactions can make a greater contribution to sulfur dioxide oxidation than do the gas-phase processes.

5.3.4 Catalytic enhancement of oxidation of sulfur dioxide

By itself, molecular oxygen oxidizes sulfite only very slowly in aqueous solution. However, the presence of small amounts of some metal ions catalyses the reaction. Metals that have been shown to increase the rate of the reaction include iron (II) and (III), manganese (II), copper (II), and cobalt (III). In acidified water, small concentrations of these metals are soluble and the soluble species are the catalytic agents. However, even in higher pH situations where metals such as iron (III) are very insoluble, catalysis still occurs. Other solids including carbon particles have also been found to increase the rate of oxidation by molecular oxygen. Even when catalysed, however, oxidation of sulfur dioxide by dioxygen in aqueous solution probably makes a relatively small contribution compared to other oxidation routes. Metal ions may also enhance the rate of heterogeneous reaction of sulfur dioxide by hydrogen peroxide and ozone.

5.4 Acidifying agents in rain

The major ions present in precipitation are well known and are listed in Table 5.1. As noted sodium, potassium, calcium, and magnesium ions are cations of strong bases and chloride, nitrate, and sulfate are anions of strong acids. As such, all these species are themselves neutral and therefore the only major ions which perturb the acid–base balance of the water are ammonium and hydronium ion itself. The activity of hydronium ion, of course, directly determines the solution pH. Since ammonium is a very weak acid ($pK_a = 9.25$), in the presence of even a small excess of hydronium ion, it is unable to donate protons and has a negligible effect on the precipitation pH. This is not to say that ammonium lacks acid-producing capability. On the contrary, when deposited in soil or water, under aerobic conditions, microbial oxidation of ammonium generates two hydronium ions for each ammonium molecule:

$$NH_4^+(aq) + 2O_2 + H_2O \xrightarrow{microorganisms} NO_3^-(aq) + 2H_3O^+(aq) \tag{5.41}$$

In this indirect way, ammonium in precipitation is a potent contributor to acidification. More will be said about this very important reaction in Chapter 18 when we deal with the chemistry of soil processes.

We have indicated that the hydronium ion in rain is associated with either nitric or sulfuric acid produced by mechanisms described above. If these two components were the only sources, there should be a good correlation between hydronium ion concentration and that of sulfate and/or nitrate. Of the numerous studies designed to examine the issue, many do show excellent correlations (correlation coefficient > 0.8) for one or other of the relations. But there are a number of exceptions, indicating that additional factors including meteorological anomolies, other emissions, and proximity to source, overrule any simplified relation. A good example is in the prairie regions of western North America where hydronium ion is best correlated—in an inverse fashion—to calcium ion, indicating terrestrial control of precipitation acid/base balance through alkaline soil minerals.

The relative importance of nitric and sulfuric acids depends on distance from source because the rate of conversion of NO_x to nitric acid, and its deposition velocity is greater than corresponding rates for sulfuric acid. This can be shown by the following calculation which considers the rate-determining step for the two principal homogeneous reaction processes.

Consider a situation where the atmospheric concentrations of nitrogen dioxide and sulfur dioxide are respectively 50 and $25\,\mu g\,m^{-3}$. These are typical concentrations observed in heavily industrialized areas such as in western Europe and eastern North America. A reasonable average 24 h value for the concentration of hydroxyl radical during the summer months is 1.7×10^6 molecules cm^{-3}. The relevent pseudo second order rate constants at the Earth's surface for reactions (5.2) and (5.21) are 1.2×10^{-11} cm^3 molecule^{-1} s^{-1} and 1.2×10^{-12} cm^3 molecule^{-1} s^{-1}, both estimates made at P° and 25 °C.

The atmospheric concentration of nitrogen dioxide is

$$50\,\mu g\,m^{-3} = \frac{50 \times 10^{-6}\,g\,m^{-3}}{46\,g\,mol^{-1}} \times 6.02 \times 10^{23}\,\text{molecules mol}^{-1} \times 10^{-6}\,m^3\,cm^{-3}$$

$$[NO_2] = 6.5 \times 10^{11}\,\text{molecules cm}^{-3}\,NO_2$$

For sulfur dioxide the concentration is

$$25\,\mu g\,m^{-3} = \frac{25 \times 10^{-6}\,g\,m^{-3}}{64\,g\,mol^{-1}} \times 6.02 \times 10^{23}\,\text{molecules mol}^{-1} \times 10^{-6}\,m^3\,cm^{-3}$$

$$[SO_2] = 2.4 \times 10^{11}\,\text{molecules cm}^{-3}\,SO_2$$

For nitrogen dioxide, the rate of oxidation

$$
\begin{aligned}
= \frac{-d[NO_2]}{dt} &= k_{NO_2}[NO_2][OH] \\
&= 1.2 \times 10^{-11} \, cm^3 \, molecules^{-1} \, s^{-1} \times 6.5 \times 10^{11} \, molecules \, cm^{-3} \\
&\quad \times 1.7 \times 10^6 \, molecules \, cm^{-3} \\
&= 1.3 \times 10^7 \, molecules \, cm^{-3} \, s^{-1} \\
&= 4.8 \times 10^{10} \, molecules \, cm^{-3} \, h^{-1}
\end{aligned}
$$

This last figure represents a loss rate of approximately 7% of the original concentration of nitrogen dioxide in one hour.

For sulfur dioxide, a similar calculation is as follows:

$$
\begin{aligned}
\text{Rate of oxidation} = \frac{-d[SO_2]}{dt} &= k_{SO_2}[SO_2][OH] \\
&= 1.2 \times 10^{-12} \, cm^3 \, molecule^{-1} \, s^{-1} \times 2.4 \times 10^{11} \, molecules \, cm^{-3} \\
&\quad \times 1.7 \times 10^6 \, molecules \, cm^{-3} \\
&= 4.9 \times 10^5 \, molecules \, cm^{-3} s^{-1} \\
&= 1.8 \times 10^9 \, molecules \, cm^{-3} h^{-1}
\end{aligned}
$$

The initial loss rate of sulfur dioxide is therefore approximately 0.7% of the original concentration in one hour.

Note that there are several assumptions and approximations in this calculation, and we have only considered the homogeneous oxidation processes.

Taking into account these and other oxidation pathways, it is found that when both sulfur and nitrogen oxides are emitted from an industrial region, and the emissions along with transformation products move downwind together, the molar ratio of sulfate to nitrate increases away from the source. For example, the ratio is approximately 1 : 1 to 1.5 : 1 in the Netherlands, a level indicative of the industrial heartland of western Europe. Moving north-east with prevailing winds, the ratio becomes 2 : 1 in southern Scandinavia and is as high as 5 : 1 in northern Scandinavia.

5.5 Rain, snow, and fog chemistry

5.5.1 Rain

The chemical composition of rain is highly variable depending on the geographic location and the influence of natural and anthropogenic chemical processes on the atmosphere in that region. We have emphasized the role of nitrogen and sulfur compounds in determining acidity of rain and other precipitation forms. Table 5.1 reported the major element composition of rain at several sites. Additional elements (including metals) are present in rain in trace amounts, again depending on location. To some extent the trace components are derived from soil and other dust particles which act as nuclei around which water condenses to form cloud droplets. The solubility of metals from such sources depends on the nature of the metal and the original form in which it was present. Metals associated with a silicate mineral matrix are almost completely insoluble. Where insoluble iron (III) and aluminium hydrous oxides (also commonly of terrestrial origin) are found in water droplets, they act as scavengers by adsorbing various species on their surface. Elevated amounts of these solids therefore suppresses the solubility of other metals. The rain water pH is another factor which controls metal solubility, with most metals becoming more soluble under more acid conditions.

5.5.2 Fog

Water droplets in fog and mist also contain chemical species accumulated from the atmosphere. Composition is similar to that of rain, but concentrations tend to be higher in fog because of its location near the Earth's surface where levels of contaminating gases and other species are usually greater. One of the foggiest areas in the world is the Bay of Fundy on the Atlantic coast of eastern Canada between Nova Scotia and New Brunswick. During the summer season from April to October, warm air is drawn into the Bay from the south. Passing over the cold ocean water, it is chilled, causing condensation and frequent heavy fog conditions. During this season, areas adjacent to the Bay of Fundy are subjected to fog for 12 to 30% of the time, sometimes for 3 to 5 days continuously at a stretch.

There is concern that birch trees growing in the forests adjacent to the Bay would be adversely affected by prolonged exposure to the acid-containing fog. In one study, analysis of fog composition along a 37.5 km transect inland from the coast has been carried out. The volume-weighted mean concentrations (over the 1987 growing season) of major constituents at all five sites are given in Table 5.3.

These average concentrations are greater than have been measured in many rainfall samples (Table 5.1). By following the trend to values beginning at the coast and moving inland, it was found that concentrations of most soluble species increased steadily and this was attributed to evaporation of the aqueous solvent. Increases were largest for hydrogen ion and for sulfate, probably an indication of an additional cause—rapid heterogeneous oxidation of dissolved sulfur dioxide in the aqueous aerosol during the time taken for the fog to drift away from the ocean. The oxidant may have been ozone or alkyl peroxides, both produced by reactions involving the hydrocarbon emissions from the forest.

Table 5.3 Concentrations of major constituents of fog near the Bay of Fundy, Canada

Species	H_3O^+	Na^+	K^+	Ca^{2+}	Mg^{2+}	NH_4^+	Cl^-	NO_3^-	SO_4^{2-}
Conc./μmol L^{-1}	330 (pH 3.5)	78	31	13	11	50	61	160	245

Volume-weighted mean values from five sites, taken between April and October, 1987.
From Cox, R. M., J. Spavold-Tims, and R. N. Hughes, Acid fog and ozone: their possible role in birch deterioration around the Bay of Fundy, Canada. *Water, Air and Soil Pollution*, **48** (1989), 263–76.

5.5.3 Snow

The chemistry of snow must be considered in two aspects. First is the nature of the *snowfall* as precipitation—that is, its compositon at the time it is deposited on the surface of the Earth. Second, because snow frequently remains on the ground for extended periods, it

Table 5.4 Composition of two fresh snow samples (bracketed values are standard deviations)

	H_3O^+	Na^+	K^+, Ca^{2+}	NH_4^+	Mg^{2+}	Cl^-	NO_3^-	SO_4^{2-}
	Concentration/μmol L^{-1}							
Location								
Antarctica surface snow	1.52 (0.60) pH 5.82	0.64 (0.26)	Not detected	0.11 (0.04)	0.073 (0.030)	0.84 (0.31)	0.82 (0.35)	0.26 (0.09)
Ciste Mhearad Scotland	279 (31) pH 3.55					13 (5)	23 (9)	86 (25)

The Antarctic results are from 14 samples of surface snow taken along a transect moving inland from 100 to 430 km in Terre Adelie, from Legrand, M. and R. J. Delmas, Spatial and temporal variations of snow chemistry in Terre Adelie (East Antarctica), *Annals of Glaciology*, **7** (1985), 20–25.

The limited data from Scotland are based on 15 samples taken along a 700 m transect, from Brimblecombe P., M. Tranter, P. W. Abrahams, I. Blackwood, T. D. Davies, and C. E. Vincent, Relocation and preferential elution of acidic solute through the snowpack of a small remote, high-altitude scottish catchment, *ibid*, pp. 141–47.

is subject to further inputs from the atmosphere by wet and dry deposition processes. Therefore we must also consider the chemistry of the *snowpack* as an accumulated deposit.

Table 5.4 shows concentrations of ionic species in freshly fallen snow in the Antarctic and in Scotland. The extremely low values found in the Antarctic samples are indicative of a location remote from anthropogenic sources of these ions. In both cases, for samples relatively close to one another, considerable variability in concentrations was observed in spite of there being no obvious local influences which should affect the results. The inhomogeneity indicates different rates of atmospheric scavenging of ions over space and time and/or the effects of lateral movement of snow due to wind action. At other locations around the Earth, there is even more variability in the chemical composition of snow samples.

Snow remaining on the ground over the winter season is subject to chemical alteration due to inputs from dry and wet deposition influenced by urban or industrial sources as well as natural organic debris—the latter especially important in forested areas. The additional deposits contribute to spatial variability in physical and chemical composition and perhaps, more importantly, can influence the melting processes and the nature of the meltwater.

Aside from the additional deposits, during the winter season, snow undergoes metamorphosis with individual particles coalescing and recrystalizing into larger grains.[2] As part of this process, the solute ions are partially excluded from the ice crystal lattice and tend to migrate to the crystal surfaces. In winter it is also normal that there be periods of partial melting of the snowpack due to higher temperatures and/or to exposure to intense sunlight. During these periods, the early melt fractions encounter the surface impurities and dissolve them in the meltwater, leaving reduced concentrations in the remaining snow. Such events may occur several times before complete melting in the *spring run-off*. The total amount of solute available for dissolution declines as winter proceeds but during each event, an initial flux of high-concentration solution is produced. For the major precipitation anions, preferential elution occurs in the order sulfate > nitrate > chloride. This means that early meltwater is enriched in sulfate. The snowpack is generally depleted of ions but is *relatively* enriched in chloride—which will then be carried away in the final meltwater.

The combination of additions of organic and inorganic species to the snowpack by wet and dry deposition, and removal of chemical species by mid-winter melting means that snowpack chemistry inevitably changes over the season. For any particular species, depending on location and winter climate, the snow concentration may show an increase or a decrease.

Later (Chapters 11 and 18), we will discuss the ability of water and soil to neutralize inputs of acid. Where acid is provided continously and slowly over extended periods, it may be neutralized. However, a sudden release of acids into the aqueous or terrestrial environment can lead to a phenomenon know as *acid shock*. This term is used to refer to the large flux of water that passes through the soil when the final spring thaw occurs. Although meltwater released at that time may not have as high a concentration of some species as was present in earlier releases, the great volume of water ensures that the soil, rivers, and lakes which receive the water are also receiving large amounts of cations (including hydronium ion), anions, and organic species; the sudden influx can have a major effect on water, soil, and biota.

5.6 The global picture—sources and sinks

Quantitative estimates of the sources and sinks of nitrogen and sulfur compounds found in the atmosphere are very difficult to make, and amounts measured and calculated by different researchers show large variability. The amounts from various sources reported in Table 5.5 are recent estimates for the Earth as a whole and give some sense of the major processes affecting atmospheric concentrations. The ultimate sinks for these compounds are through deposition onto water and soil.

Sea salt particles contain sulfur that is mostly in the form of alkali and alkaline earth metal sulfates. These particles act as condensation nuclei for water and play an important role in cloud and fog formation. For the most part, however, they are chemically unreactive species. Dust composition is highly variable and may affect precipitation pH as noted in the case of rain over western India and the North American prairies.

Of the compounds that are precursors of sulfuric and nitric acids, about half is emitted as a result of human activities, mostly related to energy production, while the other half arises from naturally occuring geochemical and biogenic processes. There are two major anthropogenic sulfur dioxide sources—emissions from the smelting of sulfide-based ores and the combustion of fossil fuels. The former include production of copper, nickel, lead, and zinc, which are all frequently found as metal sulfide minerals. One well known case is the high grade nickel–copper ores at Sudbury, Ontario, Canada. In the last century and the

Table 5.5 Major sources and sinks of nitrogen and sulfur compounds in the atmosphere

$N/10^{12}$ g y^{-1}		$S/10^{12}$ g y^{-1}	
Nitrogen compounds		Sulfur compounds	
NH_3			
Biogenic volatilization	122	Solid species, mostly SO_4^{2-}	
NO_x		Sea salt	44
From stratosphere	1	Dust	20
Atmospheric oxidation of NH_3	1	Reduced sulfur	
Lightning	5	Biogenic (oceans and land)	98
Biogenic	8	Partially oxidized sulfur	
Biomass combustion	12	Volcanoes (average)	5
Fossil fuel combustion	20	Fossil fuel combustion/smelting	104

Nitrogen data based on Jaffe, D.A., The nitrogen cycle. In *Global Biogeochemical Cycles*, ed. S. S. Butcher, R. J. Charlson, G. H. Orians, and G. V. Wolfe, Academic Press, London; 1991.
Sulfur data based on Scriven, R., What are the sources of acid rain? In *Report of the Acid Rain Inquiry*, Scottish Wildlife Trust, Edinburgh; 1985.

early part of this one, refining of the ore was done in open roasting beds releasing huge quantities of sulfur dioxide at ground level. The ambient sulfur dioxide and the acid generated from it destroyed much of the vegetation in the Sudbury area and the barren, shallow soils were eroded from the underlying bedrock. In recent years, stringent controls have reduced the emissions substantially. There are no low level emissions but the smaller amounts of sulfur dioxide are now released 400 m above ground level through a 'super-stack'. Of course, while the emissions that are now released have minimal effect in the local area, they still contribute to the regional and global budget.

All fossil fuels contain some sulfur. Coal is considered to be the major source, with sulfur contents ranging from fractions of a percent to 10% in some cases. Smaller amounts are present in liquid fuels. Gasoline may have 10 to 500 ppm sulfur depending on its origin and the refining process. There can be large amounts of sulfur compounds in some natural gas (methane) supplies, but these are substantially removed in the refining process.

Combustion-produced nitrogen originates from atmospheric dinitrogen whenever the burning temperature is very high, as in internal combustion engines or in large industrial scale units which burn fossil fuels. When biomass is burned as in forest fires or for domestic heating and cooking purposes, the temperature is usually too low to oxidize substantial amounts of atmospheric nitrogen. In these cases, the nitrogen oxides are derived almost exclusively from the fuel itself. The amount released is then equal to

$$N = \text{mass of biomass} \times \text{fraction of N in biomass} \times \text{conversion efficiency} \qquad (5.42)$$

Table 5.6 gives values estimated for annual nitrogen oxide emissions associated with agricultural practices in various regions in the tropics. The total nitrogen oxides produced

Table 5.6 Annual emissions of nitrogen oxides from biomass burning associated with agricultural production in various tropical areas, based on fuel N and conversion efficiencies

	Total area/10^{12} m^2	Biomass burned/10^{15} g dry matter y^{-1}	Fuel N content/%	Conversion efficiency/%	Nitrogen oxide emission/10^{12} g N y^{-1}	Nitrogen oxide flux from total area/g N m^{-2} y^{-1}
Tropical forest	15.9	0.8–2.0	1	13	1.0–2.6	~0.1
Tropical woodland, shrubland and grassland	22.1	2.0–3.7	0.6	10	1.2–2.2	~0.1
Agricultural land	17.6	1.7–2.1	0.6	10	1.0–1.3	~0.1

From Galbally, I. E. and R. W. Gillett, Processes regulating nitrogen compounds in the tropical atmosphere. In *Acidification in Tropical Countries*, ed. H. Rodhe, and R. Herrera, John Wiley and Sons, Chichester; 1988.

are calculated to be between 3.2 and 6.1 Tg y^{-1}, a significant part of the 12 Tg y^{-1} attributed to biomass combustion (Table 5.5).

Most atmospheric ammonia is derived from biogenic sources (see Chapter 6); nevertheless some of these sources have origins closely related to humans. The cattle industry—both in terms of milk and meat production—has grown with the human population and ammonia produced from manure makes a large contribution to the atmospheric budget. A small portion of the volatilized ammonia is oxidized in the atmosphere and the rest (about 8 Tmol y^{-1}) is neutralized by atmospheric nitric and sulfuric acids of which 4 and 5.5 Tmol y^{-1} respectively are potentially produced from the atmospheric nitrogen and sulfur compounds. These acids are sufficient to completely neutralize 4 (*via* nitric acid) + 2 × 5.5 (*via* sulfuric acid) = 15 Tmol y^{-1} of ammonia, and therefore, taking the Earth as a whole, there is excess acid that contributes to precipitation acidity.

If all anthropogenic releases of acidic and basic substances were eliminated, natural production of both might be in closer balance and the global average pH could be near the 5.7 value.

The release of nitrogen and sulfur compounds varies from country to country around the Earth. About 60 or 70% of the anthropogenic emissions of both elements come from sources in Europe and North America. As a consequence, rainfall chemistry also varies and some of the most severe acid precipitation problems occur in these continents. The average pH in parts of the highly industrialized world is as low as 4.0, and individual events containing even more acidic precipitation have been observed. Figure 5.2 shows mean rainfall pH values in regions around the globe, some individual values for less well characterized locations, and places where problems associated with acidic depositions have been reported. Such maps provide a simple means of displaying perturbations in precipitation chemistry but it is important to be aware that pH is not the only precipitation property of concern. Other cations and especially anions can also adversely affect the properties of the receiving water and soil. Some of these effects will be discussed in later chapters.

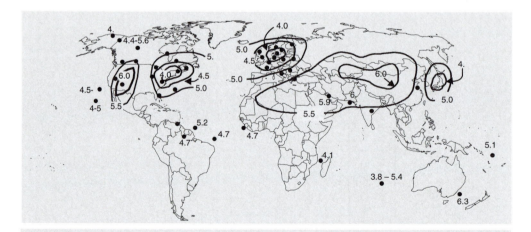

Fig. 5.2 Mean precipitation pH throughout the Earth (Redrawn with permission from Park, C. C., *Acid Rain*, Routledge, London; 1987).

5.7 Control of anthropogenic nitrogen and sulfur emissions

We have seen that the anthropogenic sources of nitrogen oxides and gaseous sulfur compounds centre around energy-related activities. Four approaches for reducing these emissions are possible—decreasing energy use by various efficiency measures, producing energy *via* non-combustion processes, preventing emissions of the problem gases, or removing the gases after they have been generated. All of the approaches are technically possible and there are philosophical, political, and economic arguments to be made for and against each one. In Chapter 4, we considered catalytic methods of reducing nitric oxide emissions from vehicles. Here, using coal combustion as an example, we will look very briefly at technology related to ways of minimizing emissions of both nitrogen and sulfur compounds. Simultaneously, other environmental effects of the modified technology must be considered.

5.7.1 Fluidized-bed combustion

New types of combustion chambers have been designed to enhance the efficiency of coal combustion and of heat transfer and therefore to minimize fuel use. Fluidized-bed combustion (FBC) is one such technique (Fig. 5.3).

In the FBC combustion chamber, preheated air is forced upward through a bed of powdered coal. The passage of air, as well as convection generated by the hot gases from the burning finely divided particles of coal, creates a fluid-like suspension. In this configuration, uniform and complete combustion occurs with reduced emissions of carbon monoxide. The temperature of combustion is somewhat lower than that in a static bed, thus

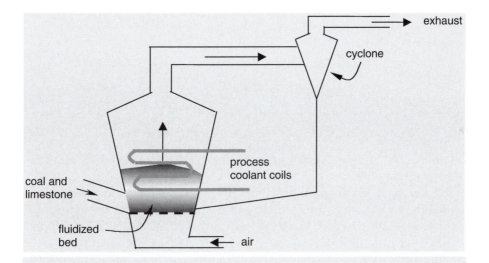

Fig. 5.3 A fluidized-bed combustion unit with a cyclone device for removal of particulate material in the flue gases.

reducing the amount of nitric oxide produced. A modification of the method allows for the simultaneous injection of powdered limestone into the bed and generates a reaction between sulfur dioxide and lime (calcium oxide):

$$CaCO_3 \rightarrow CaO + CO_2 \tag{5.43}$$

$$CaO + SO_2 + \tfrac{1}{2}O_2 \rightarrow CaSO_4 \tag{5.44}$$

At the high bed temperature, unburned coal particles are mechanically separated from the flue gas by a centrifugal cyclone device and fed back into the combustion bed. Heavy ash, including calcium sulfate, settles through a grid under the fluidized bed and finer particles, carried upward in the flue gases, are trapped by an electrostatic precipitator or a fabric filter. Although conventional precipitators are capable of removing well over 99% (by mass) of the airborne particulates emitted during coal combustion, the efficiency may be much less (about 30%) when calculated on the basis of number of particles. In particular, these systems are much less effective in controlling particles with diameters less than approximately 5 μm and it is these colloids that are potentially most hazardous to human health. More will be said about atmospheric particulates and their control in the next chapter.

As a means of removing sulfur dioxide, this is a highly efficient process, effecting around 90% recovery. Nitrogen oxides emission is reduced by about 50% due to the controlled combustion conditions which also enhance conversion of carbon monoxide to carbon dioxide.

5.7.2 Retrofitted flue gas desulfurization

For an existing conventional coal-fired plant, retrofitted flue gas desulfurization may be applied. Amongst the many processes used are lime and limestone slurry scrubbers in which the combustion gas passes through an aqueous slurry where reactions to form calcium sulfite take place:

$$\text{with hydrated lime slurry } Ca(OH)_2 + SO_2 \rightarrow CaSO_3 + H_2O \tag{5.45}$$

$$\text{with limestone slurry } CaCO_3 + SO_2 \rightarrow CaSO_3 + CO_2 \tag{5.46}$$

The insoluble calcium sulfite may be oxidized downstream to produce $CaSO_4 \cdot 2H_2O$:

$$CaSO_3 + \tfrac{1}{2}O_2 + 2H_2O \rightarrow CaSO_4 \cdot 2H_2O \tag{5.47}$$

which settles in disposal ponds or may be recovered for use in applications such as manufacturing plaster or plasterboard.

This technology efficiently desulfurizes the gas stream (~90%) but vast quantities of water and lime (or limestone) are required and the resulting wastes are correspondingly large. For a 1000 MW coal-fired electricity-generating plant, about 10 000 t of coal are burned daily, producing (for 10% ash coal) 1000 t of ash. The daily volume of water required for the scrubber slurry is around 7 000 000 L and the limestone requirement is about 2000 t. Over 3000 t gypsum is produced in the process for each day of operation.

An obvious alternative to sulfur removal by these processes is the use of low-sulfur coal. The content of this element in coal ranges from a few tenths of a percent to 5% or more. The decision about reduction of sulfur emissions then becomes one of economics—whether it is less expensive to transport low-sulfur coal to the plant site or to equip the facility with the requisite.control system.

5.7.3 The SONOX process

A recently developed process for control of acid gas emission from power plants is called SONOX. This process, developed in Canada by Ontario Hydro at its small $640\,MJ\,h^{-1}$ Combustion Research Facility, is soon to be tested on full-sized boilers.

The control system involves in-furnace injection of an aqueous slurry of a calcium-based sorbent—usually powdered limestone—and a nitrogen-containing additive—usually urea—at temperatures ranging between 900 and 1350 °C. The following reactions occur in the high temperature atmosphere of the furnace reactor:

$$CaCO_3 \overset{\text{heat}}{\rightarrow} CaO + CO_2 \qquad (5.43)$$

$$CaO + SO_2 + \tfrac{1}{2}O_2 \rightarrow CaSO_4 \qquad (5.44)$$

$$NH_2CONH_2 + 2NO + \tfrac{1}{2}O_2 \rightarrow 2N_2 + CO_2 + 2H_2O \qquad (5.48)$$

Spraying the additives into the furnace *via* a high pressure nebulizer ensures rapid evaporation of the solvent and efficient 'cracking' of the calcium carbonate and urea to produce calcium oxide and the reactive amidogen radical ($NH_2O\dot{N}H$) respectively. The process is shown schematically in Fig. 5.4.

The capture efficiency for sulfur dioxide and nitrogen oxides depends on the nature of the additive, the rate of addition, the spray characteristics, the reactor temperature, and

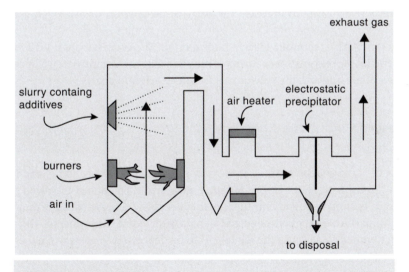

Fig. 5.4 The SONOX process for removal of nitrogen and sulfur oxides from stack gases.

the sulfur content of the coal. Optimum conditions include a furnace temperature of ~1150 °C, concurrent injection, mean droplet size of ~6.6 μm. Sulfur dioxide sorbent is ~90% porous limestone and 10% dolomite or hydrated lime, and the solids are present at a concentration of 40% in an aqueous slurry concentration, calcium : sulfur ratio 2.5 to 3 : 1. The nitrogen oxides sorbent is urea or ammonium carbonate in aqueous solution at a stoichiometric molar ratio, additive : nitrogen oxide = 1.7 to 2.0 : 1. Under these conditions up to 85% sulfur dioxide and 85 to 95% nitrogen oxides removal can be effected. When urea is used as an additive, the concentration of nitrous oxide in the flue gases is augmented from a value of 10–25 ppm (without additive) to 50–150 ppm. More studies may result in methods for reducing these concentrations.

The solid waste obtained after slurry injection is about double in mass that which results from combustion of coal without emission control. (The actual amount depends on the ash content of the coal as well as the amount of sulfur dioxide sorbent used. This latter amount depends, in turn, on the sulfur content of the coal.) The additional material in the solid waste consists of unreacted calcium oxide and calcium sulfate and consideration must be given to disposal of these mixed residues.

5.7.4 Conversion of coal to gaseous and liquid forms

Finally, the environmental consequences of using coal to manufacture synthetic gaseous and liquid fuels, often called *synfuels*, must be considered. The conversion of coal to gaseous or liquid forms is carried out in order to create energy commodities that may be transported *via* pipelines, are readily stored in containers, are clean, and are suitable for use in small-scale facilities, particularly in vehicular engines. Conversion, especially of the poorer grades of coal, to liquid or gaseous forms also allows for the upgrading of energy content of the fuel. The fuel components of coal are principally carbon and hydrogen, and the basic principle for conversion is to increase the relative proportion of hydrogen compared to carbon.

We will look at these technologies in a later section. The processes, besides achieving the goal of producing more convenient forms of fuel, can also incorporate steps for the removal of sulfur and other undesirable substances from the original coal.

The main points

1 The forms of liquid and solid water in the atmosphere include rain, snow, and fog. These and other atmospheric forms of precipitation always contain various dissolved species derived from both natural and anthropogenic processes. Of these substances, acids produced from the oxides of nitrogen and sulfur are of widespread concern. Their presence in excessive amounts can cause the rainfall pH to be depressed to values as low as 4 or even less.

2 Natural processes producing acid-generating nitrogen and sulfur compounds include lightning, volcanic activity, and a variety of biogenic reactions. The released compounds are transformed in the atmosphere to nitric and sulfuric acids. Some degree of acid/base

balance is achieved due to the simultaneous release of basic substances, particularly ammonia, to the atmosphere. In the atmosphere, dust particles derived from soils containing minerals like calcite (calcium carbonate) may also function to neutralize acids in precipitation.

3 Areas where acid precipitation is endemic are usually in regions of intense industrialization – Europe and North America being prime examples. Anthropogenic sources of both nitrogen and sulfur oxide emissions are associated with combustion processes. Sulfur dioxide is also released during the smelting of sulfide minerals.

4 Technologies for control of nitrogen and sulfur oxide industrial emmisions are available. Economics of the control processes and problems associated with disposal of the waste materials are two factors limiting application of these technologies.

Additional reading

1 Legge, A. H. and S. V. Krupa, eds, *Acidic Deposition: Sulphur and Nitrogen Oxides*, Lewis Publishers Inc., Chelsea, Michigan; 1990.

2 Calvert, J. G., ed., *SO$_2$, NO and NO$_2$ Oxidation Mechanisms: Atmospheric Considerations*, Butterworth, Boston; 1984.

3 Lindberg, S. E., A. L. Page, and S. A. Norton, eds, *Acidic Precipitation. Volume 3: Sources, Deposition, and Canopy Interactions*, Springer, New York; 1990.

4 Rodhe, H. and R. Herrera, eds, *Acidification in Tropical Countries*. John Wiley and Sons, Chichester; 1988.

Problems

1 The tropospheric processes discussed in this chapter cannot be considered as independent reactions. Discuss the relationship between the production of nitrogen oxide species and the formation of 'acid rain' and the role played by carbon monoxide, methane, and the hydroxyl radical.

2 In a particular 3000 km^2 region of southern Sweden, the annual rainfall averages 850 mm, its mean pH is 4.27, and 66% of the hydrogen ion is associated with sulfuric acid; the remaining 34% is derived from nitric acid. Calculate whether soils of this region are subject to excessive sulfate loading if the only source of sulfate is rainfall and if the recommended maximum is set at 20 kg SO_4^{2-} ha^{-1}.

3 The mean monthly pH values of rainfall in Guiyang city in Guizhou province in southern China in 1984 were as follows:

Jan   3.9, 4.0, 3.8, 4.1, 4.0, 4.5, 4.5, 4.1, 3.7, 3.8, 3.7, 3.4   Dec.

For the same year, the measurements at Luizhang, an adjacent rural area were

Jan   4.3, 4.4, 4.4, 4.2, 4.5, 4.9, 4.9, 4.6, 4.8, 4.3, 5.4, 5.4   Dec.

(a) Calculate the mean monthly pH of the rain at the two locations.
(b) What is the ratio of the mean hydrogen ion activity at the two sites?
(c) What is likely to be the most important anion in the rainfall at the urban site?

(From Dianwu, Z. and X. Jiling, Acidfication in southwestern China, in ref. 4.)

4 The ionic composition (in units of $ng\ m^{-3}$) of an atmospheric aerosol in a tropical rain forest is SO_4^{2-}, 207; NO_3^-, 18; NH_4^+, 385; K^+, 180; Na^+, 247. The pH of the aerosol is 5.22. Use these data to calculate the total positive and negative charge 'concentration' ($mol\ m^{-3}$) in the aerosol and suggest reasons which might account for any discrepancy in anionic and cationic charge.

5 The tropospheric mixing ratios of carbon monoxide are higher in the northern hemisphere than in the southern hemisphere. However, it has been observed that there has been a general global decline in carbon monoxide concentration everywhere in the 1990s. Two reasons have been suggested for this—one is the eruption of Mount Pinatubo and the other is the occurrence of several relatively dry years in the tropics. Comment on these two possibilities in terms of tropospheric and stratospheric processes.

6 The estimated atmospheric carbon dioxide concentration in the northern hemisphere in 1950 was 310 ppmv. It may be predicted with some certainty that the concentration in 2000 will be 370 ppmv. Calculate the pH of *pure* rain which would be in equilibrium with the carbon dioxide in each of the two years cited, and comment on the contribution which carbon dioxide makes towards precipitation acidity.

Notes

1 Martin, L. R., *Kinetic Studies of Sulfite Oxidation in Aqueous Solution, in SO₂, NO and NO₂ Oxidation Mechanisms: Atmospheric Considerations*, ed. J.G. Calvert, Butterworth, Boston; 1984.

2 Jeffries, D. S., Snowpack storage of pollutants, release during melting, and impact on receiving waters. In *Acidic Precipitation*, Vol. 4, ed. S. A. Norton, S. E. Lindberg, and A. L. Page, Springer, New York; 1989.

6

Atmospheric aerosols

A N aerosol is a suspension of particles in a gas and an atmospheric aerosol consists of particles that remain aloft in the air.

When we refer to particles in an aerosol, in the definition we are including both solids and liquids. Particles are distinguished from smaller gas molecules or molecular clusters by their ability to cause incoherent scattering of visible light and therefore to interfere with light transmission. As a consequence, the presence of a high concentration of aerosol is indicated by a hazy appearance in the atmosphere. To scatter visible light, particles must have dimensions comparable with or larger than the wavelengths of that light—say, within an order of magnitude (for example, at least one tenth of 400 nm, that is 0.04 μm).

There are a number of factors that determine how long particles can stay suspended in the air. Large particles readily settle out. Except for those of very low density, most particles with dimensions greater than 10 μm require strong air currents to keep them aloft. On the other hand, very small particles have limited lifetimes as independent entities, because they come together and coagulate to form larger ones. Particles within the size range between 0.01 to 1 μm are most likely to remain suspended for long periods, sometimes up to a month or even longer.

Aerosol particles are known by a variety of names, depending on the source and nature of the particles. Many of the names are common; we are all familiar with terms like dust, smoke, fly ash, and pollen (solids in gas) and cloud, mist, fog, and smog (liquids in gas). Figure 6.1 documents some properties of various types of atmospheric aerosols.

Many aerosol particles—dust, pollen, smoke—are released or injected into the air as preformed entities. There is another class, however, that results from chemical processes involving gaseous species reacting together in the atmosphere to form liquid or solid particles. Clouds and mist are simple examples of these *condensation aerosols*. Another condensation process is the generation of photochemical smog as described in Chapter 4.

While aerosols are present everywhere, excessive concentrations and/or the presence of particular chemical species may lead to human health problems, and so they are regularly monitored at many locations. A variety of parameters are determined to

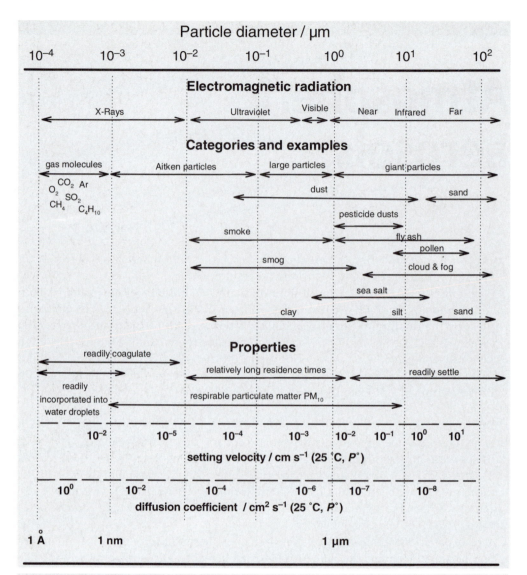

Fig. 6.1 Classification and properties of atmospheric particulates. For purposes of comparison, dimensions of gaseous molecules and wavelengths of electromagnetic radiation are also given.

assess the concentration of particulates. Typical of these parameters are the following operationally defined aerosol measurements that are regularly made in Ontario, Canada.

- Coefficient of haze (COH) is determined by drawing 300 linear metres of air through a porous filter tape, after which the optical absorbance of the tape is compared with standards. The coefficient of haze is equal to the numerical value of $100 \times$ absorbance, and a result of 6 or greater indicates air that may cause adverse symptoms to persons suffering from respiratory problems such as asthma. The method of sampling favours retention of solid particles in the size range 5–10 μm.

- Total suspended particulate (TSP), expressed in units of $\mu g\, m^{-3}$, is determined by gravimetrically measuring the quantity of particulates that have been obtained after filtering air at a rate of $1.4\, m^3\, h^{-1}$. An average concentration of individual measurements made over a 1 year period of greater than $60\,\mu g\, m^{-3}$ is considered excessive. Values of about $10–30\,\mu g\, m^{-3}$ are observed in many locations and individual readings greater than $500\,\mu g\, m^{-3}$ have been observed in the core of cities like Toronto. Inhalable particulates (IP), also expressed in units of $\mu g\, m^{-3}$, are those particulates that are smaller in dimension than 10 μm. This component of the aerosol is often referred to as the PM_{10} (PM = particulate matter) fraction and is considered important because smaller particles are the agents of many serious respiratory problems. An observed approximate relation is

$$IP = 0.45\ TSP \tag{6.1}$$

- Using a passive technique, total dustfall (TDF) is measured in $g\, m^{-2}\, month^{-1}$ by weighing dust that settles into an open-topped container over a 30 day period. A value greater than $7.0\, g\, m^{-2}\, month^{-1}$ (mass per unit area per time) of settled dust is considered excessive, and this value is frequently found to be exceeded in industrial cities such as Hamilton, Ontario.

As would be expected, it is very difficult to determine quantitatively the input rate of particulates to the atmosphere even from point sources. When all natural and anthropogenic sources around the world are considered together, estimates become exceedingly problematic. The natural sources are particularly difficult to quantify because many of them are diffuse releases occurring in small quantities over vast areas. Table 6.1 lists the major sources and gives approximate ranges for the yearly input. For obvious reasons, the table does not include essentially aqueous aerosols such as fog and clouds.

The total annual global production of aerosol particles then appears to be between 2500 and $4000\, Tg\, y^{-1}$. Although this range is only a highly approximate estimate, it is

Table 6.1 Estimated ranges of yearly input fluxes of particles that make up atmospheric aerosols

Aerosol	Natural (N) or anthropogenic (A)	Annual flux/$Tg\, y^{-1}$
Sea spray	N	1000–1500
Dust	N,A	100–750
Forest fires	N,A	35–100
Volcanic emissions	N	50 (highly variable)
Meteors	N	1
Anthropogenic combustion	A	50
Condensation	N,A	1500

Values obtained from various sources. $1\, Tg = 10^{12}\, g$.

consistent with many reported values. We will look at aspects of the formation and properties of some of the common aerosol materials.

6.1 Sea spray

At any given time, whitecaps (waves with broken and foaming crests) cover approximately 2% of the ocean surface. The vigorous wind-driven action of the waves generates numerous extremely small bubbles that form and collapse at a rate of greater than 10^6 events $m^{-2}s^{-1}$. These include not only the bubbles that foam and make the whitecaps white, but also many that are small (anywhere from 5 to 500 μm in diameter) and individually invisible. Figure 6.2 illustrates the initial processes of sea-spray aerosol formation. Within milliseconds after a bubble has formed, the pressure of surrounding water causes it to collapse on itself, contorting and then rupturing the surface film, producing typically 1–10 very small water droplets. A few larger drops are also formed when the collapsing bubble ejects a stream of water that breaks apart as it falls under the force of gravity. The film disintegration droplets are typically between 5 and 25 μm in diameter and contain a mass of sea salt between 2 and 300 pg, while the central jet droplets are about 25 to 500 μm in diameter with 300 pg to 2 μg of salt in each.

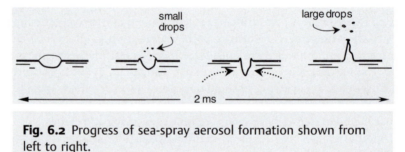

small drops   large drops

2 ms

Fig. 6.2 Progress of sea-spray aerosol formation shown from left to right.

After forming, many of the larger droplets fall back into the ocean, but smaller ones are wafted into the atmosphere where the water rapidly evaporates creating a solid aerosol having a sea salt chemical matrix. The particle composition reflects that of the soluble matter in sea water, but because the bubbles are produced at the surface of the sea, the composition shows any anomalies associated with surface water. In fact, the surface microlayer of a natural water body is enriched in surface active components. Many of these are amphiphilic organic macromolecules that can sequester both neutral organic solutes and metal ions. Therefore, the salt matrix frequently is enriched in species other than the sodium, chloride, and other major solutes in bulk sea water. The enrichment is reflected in the composition of the sea-salt aerosol and its extent is described by a chemical concentration factor (CCF) that is defined in relation to one of the major elements, usually sodium, of the ocean:

$$CCF = \frac{(C_X/C_{Na})_{aerosol}}{(C_X/C_{Na})_{sea}}. \tag{6.2}$$

In eqn 6.2, C_X and C_{Na} in the numerator are the concentrations of the element of interest and of sodium in the aerosol, while the denominator terms are the respective concentrations in sea water itself.

Ocean-derived CCF values that are greater than 100 have been observed for some elements, especially ones like mercury, lead, and cadmium that have an atmospheric source. These elements also tend to form complexes with carboxylic acid- and nitrogen-containing ligands such as some organic macromolecules found in the oceans. Large CCF values have also been measured for certain organic species.

6.2 Dust

Aerosol dust is produced by simple physical processes whereby solid materials having very small diameter, such as clay minerals and finely divided organic matter, are lifted into the atmosphere by wind currents. When air currents drive these particles along the surface, they collide with other solids on the ground causing fragmentation into still smaller particles that may then become airborne, adding to the aerosol. It is found that the chemical composition of dust reflects the general composition of the solid surface from which it was derived. For example, desert dust storms consist of mostly siliceous material, and sometimes this is carried for hundreds or even thousands of kilometres to remote locations. In West Africa, northerly winds, called the *Harmattan*, coming off the Sahara between December and February, carry dust at concentrations of up to $1000\,\mu g\,m^{-3}$ which settles over vast areas from the Cameroons to Liberia. The dust has a composition similar to that of the sands in the great desert to the north-west. Similarly dust from the Gobi Desert in the north-western China and Mongolia is carried far to the south and east, setting as far away as Japan and the Pacific Ocean.

Chemical concentration factors analogous to those for the sea-spray aerosol have also been determined for dust. For the land-based material, a reference element that is a major component of the crust is used. Silicon is an obvious choice, but because of analytical difficulties it is rarely used. Aluminium, the most abundant metal in the geosphere, is easier to determine and makes a good reference element. Factors affecting species enrichment in dust are more varied than in the marine situation. They include anomalies in the surface soil composition, many a result of human activities such as agriculture, and a wide range of urban influences.

Urban dust has been the subject of a good number of studies, but usually the investigations are done on settled dust that favours larger particulates but also including particles that are sufficiently large and/or heavy so that they were never a part of the aerosol. Besides soil components, city dust contains vegetative plant fragments, cement, tire and brake lining particles, solid aerosols from vehicle exhaust, and many other synthetic and natural materials in smaller amounts. The urban aerosol contains the same substances in the form of fine particles, which can legitimately be termed dust, but it contains other non-dust components too. Smoke, pollen, and condensation products are intermixed with dust, and identification of individual sources is very difficult.

6.3 Combustion products

Combustion products are formed as a consequence of a wide variety of natural and human-influenced activities ranging from forest fires to fossil-fuel-fired power production plants. In most cases, the principal products are carbon dioxide and water and they are accompanied by other gases in smaller amounts. Depending on the fuel and the manner of burning, particulate matter is simultaneously emitted. Where combustion of carbon-based fuels is incomplete, elemental carbon is given off as a black smoke. This is the case when the ratio of fuel to oxidant is larger than the stoichiometric value and/or the combustion temperature is relatively low. We have seen that there are instances when the cylinders of diesel engines operate under these conditions making them a major source of carbon emissions. On the other hand, a white plume from a factory stack is usually due to condensed water vapour (also a combustion product) and is then evidence of efficient and complete reaction.

Even when combustion of the carbonaceous material is complete, other particulates derived from minor constituents originally present in the fuel are released. Coal invariably contains non-combustible ash that is mostly siliceous in nature. Some of the minor elements remain in the combustion chamber along with the siliceous residue as *bottom ash*, while the finer particles are released into the atmosphere as *fly ash* unless otherwise removed by post-combustion methods. Table 6.2 gives ranges of fly ash composition resulting from coal combustion. Unburned carbon is not included in the listing.

Water extracts of fly ash are generally alkaline, giving pH values as high as 11. This suggests that ash collected in precipitators has potential value as a low-grade lime amendment for acidic soils. However, trace elements are frequently present in significantly large concentrations so that disposal of ash collected in precipitators becomes a problem. Metals like lead and mercury appear to be deposited on the surface of the very fine material. This supports a theory that low-boiling metals present in fuels volatilize during combustion and then condense on to ash particle surfaces in cooler regions of the furnace or stack. Being surface deposits they are readily leachable and also available for biological uptake. We will say more in Chapter 19 about ash produced during combustion of urban soild waste materials.

One combustion product class that has been widely studied and is of particular concern because of its carcinogenicity is the class of compounds called polyaromatic hydrocarbons (PAHs). These are produced when wood, coal, or other carbon-based fuels are burned in the presence of limited amounts of oxygen. Along with elemental carbon the PAHs are products of incomplete combustion under relatively low temperature conditions. Large amounts are also released during the electrolytic refining of aluminium using carbon electrodes. Typical PAH compounds include angular, pericondensed, and linear compounds as shown from left to right below.

chrysene   pyrene   anthracene

Table 6.2 Range of composition of fly ash from coal combustion

Component	Range/%	Mean/%
Si	9.0 to 28	21
Al	4.6 to 15	11
Fe	2.5 to 18	7.6
Ca	0.7 to 22	6.2
K	0.3 to 2.5	1.4
S	0.1 to 6.4	1.3
Mg	0.2 to 4.2	1.1
Na	0.1 to 6.3	0.9
Ti	0.1 to 1.0	0.7
P	0.1 to 1.0	0.3
	Range/μg g^{-1}	Mean/μg g^{-1}
Zn	27 to 2900	450
V	$<$ 95 to 650	270
Cr	37 to 650	250
Pb	21 to 2100	170
As	8 to 1400	160
U	11 to 30	19
Cd	6 to 17	12

Most of these data are based on measurements of US fly ash as reported by Ainsworth, C. C. and D. Rai, *Chemical Characterization of Fossil Fuel Combustion Wastes, EPRI EA*-5321, Electric Power Research Institute, Pao Alto, California; 1987. Where sulfur emissions from coal combustion are controlled by a lime slurry scrubbing process, the ash has elevated calcium sulfite or sulfate levels.

Altogether more than 150 PAH compounds containing two to seven rings, and some with alkyl or other substitution, have been identified as products present in smoke. Although these compounds are termed aromatic, application of the Huckel rule indicates that a species like pyrene does not meet the '$4n + 2\pi$-bonding electrons' criterion for aromaticity. The criterion is met, however, if one considers only the periphery of the compound. Nevertheless, we will treat all PAHs as aromatic and draw structures accordingly. Molar masses run from 128 to over 300. The smaller species are somewhat volatile while those with larger molar mass are present as solids, usually as surface deposits on soot and other combustion product particles. Chrysene (molar mass = 228) has been observed to be distributed approximately equally between the vapour and solid phases.

Air measurements made in the industrial city, Hamilton, Ontario, have indicated about 10 ng m^{-3} PAH in summer and 30 ng m^{-3} in winter.[1] The PAH compounds were found to be associated with soot particles, 80% of which were less than 3.3 μm in diameter. While we would expect to identify PAH compounds in an industrial urban centre, we might not expect that they could also be found in regions remote from major sources of combustion. However, concentrations of $1\,\text{ng m}^{-3}$ or more have been detected in North American Arctic regions. This is evidence of their non-reactivity so that they persist while being carried by wind currents from distant sources in Eurasia and North America. The time taken to travel these distances may be more than one year. The resultant behaviour has been described in terms of a *grasshopper effect*. During the summer season, southerly winds and high temperatures favour gas-phase transport in a northward direction, but in winter the solid phases are favoured so that there is less transport by the northerly air currents and the compounds remain essentially stationary. In a subsequent warm season they 'hop' further toward the pole.

6.4 Condensation aerosols

6.4.1 Ammonium sulfate aerosols

There are many and varied chemical processes involving gaseous reactants that can produce liquid or soild condensation aerosol particles in the atmosphere. Two major condensation-formed components of both the continental and oceanic aerosol are ammonium hydrogen sulfate (NH_4HSO_4) and ammonium sulfate (($NH_4)_2SO_4$). Both of these solid species are produced as a result of two parallel series of reactions. The first series begins with reduced (oxidation state of -2) organic forms of sulfur such as dimethyl sulfide, $(CH_3)_2S$.

As has been described in detail in the previous chapter, oxygen and the hydroxyl radical serve as oxidizing agents to additional amounts convert reduced sulfur-containing gases into sulfur dioxide. This sulfur dioxide, augmented by released directly from other, mostly industrial, sources, is further oxidized to form sulfuric acid.

At the same time, ammonia from several sources is released to the atmosphere.

During microbial degradation of decaying biomass and organic matter in soil and water, nitrogen compounds like proteins are ammonified to release ammonia/ammonium ion into the surroundings. An important source of nitrogen-rich biomass that undergoes these reactions is animal excreta deposited on soil or maintained in manure piles. One of the nitrogen compounds in animal urine is urea and it hydroyses to produce ammonia and carbon dioxide:

$$CO(NH_2)_2 + H_2O \rightarrow CO_2 + 2NH_3 \tag{6.3}$$

The actual percentage of nitrogen released depends on temperature, moisture, soil texture and pH, and the nature of plants growing where the excreta are deposited. On a global scale, it may be that up to 20% of the nitrogen in animal waste is volatilized as ammonia.

Similarly where synthetic fertilizers containing reduced nitrogen (ammonium ion or urea) are used, it is possible that ammonia species are released. Gaseous ammonia evolves in an alkaline environment while ammonium ion is favoured under acidic or neutral conditions. The ammonium ion is soluble in the aqueous soil solution, and, assuming that the soil does not have an excessively high pH, it becomes *fixed* on to cation-exchange sites (Chapter 18) and/or oxidized to nitrate before being taken up by the growing crop.

There is little loss of volatile ammonia if the organic matter degradation occurs under well aerated conditions or if the fertilizer is incorporated into a surface soil and efficiently used by plants. On the other hand, when degradation or application of these same fertilizers takes place in submerged soils during rice culture, the reducing conditions maintain the nitrogen in the ammonia/ammonium form and, depending on soil solution pH, substantial loss of ammonia due to volatilization may occur.

Additional ammonia is released from a range of industrial processes. The ammonia and sulfuric acid react together to form ammonium hydrogen surface or ammonium sulfate (reactions 6.4 and 6.5) in the form of particles that are approximately 0.1 to 1 μm in diameter. Because sulfuric acid is usually in excess, the former compound is dominant in most situations.

$$NH_3 + H_2SO_4 \rightarrow NH_4HSO_4 \quad\quad (6.4)$$

$$2NH_3 + H_2SO_4 \rightarrow (NH_4)_2SO_4 \quad\quad (6.5)$$

The hydrophilic sulfate salt particulates act as nuclei around which water condenses to form clouds. Where there is a high concentration of particles, the clouds that form consist of a large number of very small droplets. Such clouds are whiter and more reflective than those containing fewer large drops. This is one way in which cloud reflectivity, an important regulator of global climate, is influenced by atmospheric aerosols.

Box 6.1 Arctic haze

The term 'Arctic haze' was first used 40 years ago by Mitchell ("Visual range in the polar regions with particular reference to the Alaskan Arctic" *J. Atmos. Terr. Phys.* Special supplement, pp. 195–211, 1956) to describe reduced visibility observed during weather reconnaissance flights in the Arctic. Today, the term conveys the notion of haziness, due to atmospheric pollution in a part of the Earth that is perceived to be a pristine environment. Since the early 1970s scientific investigation has helped us gain a better understanding of the cause of this phenomenon.

Arctic haze is the result of wind-blown dust and industrial emissions (predominantly sulfur dioxide, but also includes hydrocarbons, soot, metals, and other gases and particulates) that are carried to the Arctic from mostly Eurasian locations. There is a distinct seasonal variation in the haze. From December to April, as Arctic air covers areas far to the south, including some heavily industrialized areas in Eurasia and North America, the Arctic haze reaches a peak as a result of increased particulate and gaseous pollutant transport to the north. During the northern winter season, natural atmospheric cleansing processes are less efficient in the cold dry Arctic, which then leads to even higher aerosol levels. Another important factor in the production of the haze is the steep inversion layer that exists in the Arctic in late winter and early spring.

A temperature differential of up to 30–40 °C may exist between the cooler surface temperature and the warmer (albeit still quite cold) air temperature several hundred metres above.

Arctic haze consists mainly of particles of variable chemical and physical properties. The particle concentration ranges from 10–4000 particles cm^{-3} with a geometric average of 200–350 particles cm^{-3}. The most numerous particles are those in the size range of 0.005–0.2 μm. Particles in the range 0.1–1 μm consist mainly of sulfate-derived aerosols and are primarily responsible for the visible haze. There are also coarse particles in the range of 1–10 μm, and giant particles (> 10 μm) that result mainly from soil dust and sea spray. While the larger particles contribute a significant fraction to the aerosol mass, there are relatively few of these particles and they have little influence on the actual appearance of haze.

The haze spreads over a vast area, stretching across an expanse of 800–1300 km, and occupies altitudes below 9 km, with a maximum concentration at an altitude of 4–5 km. The January–April average Arctic haze has a reported composition consisting of 2 μg m^{-3} of SO_4^{2-}, 1.0 μg m^{-3} of organic compounds (unspecified), 0.3–0.5 μg m^{-3} of black carbon, a few tenths of a μg m^{-3} of other substances and a few μg m^{-3} of water.

The period between May and November is associated with conditions that are 20–40 times less hazy; at this time, the residual aerosols consist mostly of wind-blown dust and sea spray.

Barrie, L. A., Arctic air pollution: an overview of current knowledge, *Atmospheric Environment*, **20** (1986), 643–63.

6.4.2 Organic condensation nuclei

Smog-forming reactions are an example of complex condensation processes that lead to the formation of a liquid aerosol. Through similar reactions, some types of organic condensation nuclei also form *via* substantially natural processes. A good example is the haze that develops over heavily forested areas during warm summer days. Terpenes and related compounds are low molar mass chemicals synthesized within the leaves and stems of a variety of plants. These compounds are relatively volatile and so are emitted into the atmosphere, producing the attractive odours that characterize forests. One of the terpenes, α-pinene, is produced by coniferous species including pine, spruce, and fir trees, and has been measured in forest atmospheres at concentration levels between 0.1 and 50 ppbv (0.5 and 300 μg m^{-3}).

α-pinene

For deciduous species such as willow, oak, poplar, and aspen, a more characteristic emission product is isoprene and concentrations between 1 and 10 ppbv (3 and 30 μg m^{-3})

of this compound have been observed. Emission rates of these and other compounds depend on the tree species and are maximal during the daytime and in warm temperatures.

$$CH_2 = C - CH = CH_2$$
$$|$$
$$CH_3$$

isoprene

Both isoprene and the terpenes are highly reactive with respect to photochemical oxidation and undergo reactions which are very much analogous to the urban photochemical smog formation processes. Two oxidative pathways[2] have been proposed. The first involves the presence of NO_x species as a source of the hydroxyl radical. Reaction sequence 6.6 illustrates the hydroxyl initiated oxidation of isoprene:

$$k_{6.6,overall} \simeq 9 \times 10^{-11} \text{ cm}^3 \text{ molecule}^{-1} \text{ s}^{-1} \qquad (6.6)$$

There are alternative oxidative processes using the hydroxyl radical or ozone and these can follow several routes to produce a range of products. Some are shown in reaction 6.7.

While the oxidation of isoprene initiated by hydroxyl has an observed rate constant, $k_{6.6,overall}$, of approximately $10^{-10} \text{ cm}^3 \text{ molecule}^{-1} \text{ s}^{-1}$, the ozone-based sequence has a much smaller observed rate constant, $k_{6.7,overall}$, of $10^{-16} \text{ cm}^3 \text{ molecule}^{-1} \text{ s}^{-1}$. Yet both pathways may contribute comparably to the oxidation of isoprene. This is because the hydroxyl radical is present in smaller mixing ratios than ozone—10^{-5} ppbv compared to 30 ppbv would be typical levels of the two species in forested areas remote from urban influences.

The oxygenated products including aldehydes, ketones, and carboxylic acids make up some of the constituents of a photochemical haze that is characteristic of certain forests on otherwise clear sunny days. The Great Smokey Mountains of North Carolina in the USA are often cited as an example of a location where this natural phenomenon is evident. Nevertheless, it is possible that human activities can affect the otherwise natural process since reactants like ozone and NO_x, required to create the haze, are augmented by combustion and other processes. The particles in the haze aerosol typically are less than 0.3 μm in diameter, putting them in the 'large' and 'Aitken particle' size range. Total annual global emissions of gaseous organic species from forests have been estimated to be around 20 Tg y^{-1}.

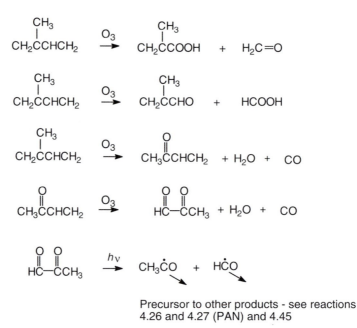

Precursor to other products - see reactions
4.26 and 4.27 (PAN) and 4.45

$$k_{6.7,\text{overall}} \simeq 10^{-16} \, \text{cm}^3 \, \text{molecule}^{-1} \, \text{s}^{-1} \tag{6.7}$$

6.5 Aerosol concentrations and lifetimes

Table 6.3 lists typical total aerosol concentration ranges. Values are invariably reported in units of number or mass of particles per volume of air. Concentrations involving moles (mixing ratios) cannot be used since the aerosol always consists of a mixture of ill-defined species.

While the aerosol particles consist of condensed-phase material, their mass concentration in the atmosphere may not be as large as that of the minor gaseous constituents. The atmospheric mean mixing ratio of methane gas is about 1.7 ppmv, which is equivalent to

Table 6.3 Total concentration ranges of aerosols in various geographic settings

Geographic setting	Typical concentration/μg m^{-3}
Open ocean	10 to 150
Sea shore	up to 500
Vegetated rural areas	10 to 50
Arid areas	up to 500
Urban atmosphere	up to 200

1200 μg of the compound in a cubic metre of air. This is much larger than the 10 to 100 μg m^{-3} that is typical of the total concentration for many aerosols in the atmosphere.

Two types of physical processes are most important in determining the lifetime of aerosol particles. The first of these is settling, which is the primary means of removal of larger particles from the atmosphere. Settling occurs due to the force of gravity and in a simple way is described by the Stokes relation. As given below, Stokes' law determines the terminal velocity or settling velocity (Fig. 6.1) of spherical particles falling under the force of gravity in a fluid:

$$v_t = \frac{(\rho_p - \rho_a)Cgd_p^2}{18\eta} \tag{6.8}$$

where v_t = terminal velocity of particles/m s^{-1}; ρ_p = density of particle/g m^{-3}; ρ_a = density of air = 1.2×10^3 g m^{-3} at $P°$ and 25 °C; C = the Stokes–Cunningham slip correction factor (see Table 6.4); g = 9.8 m s^{-2} = acceleration due to gravity; d_p = particle diameter/m; and η = viscosity of air = 1.9×10^{-2} g m^{-1} s^{-1} at $P°$ and 25 °C.

The dimensionless Stokes–Cunningham slip correction factor C accounts for the discontinuous nature of fluid interactions when the particle size is small compared with the molecular mean free path in air.

Consider the case of a dust particle of geological origin whose particle diameter is 10 μm and density is 2.5 g mL^{-1}(= 2.5×10^6 g m^{-3}):

$$
\begin{aligned}
v_t &= \frac{(2.5 \times 10^6 - 1.2 \times 10^3)1.016 \times 9.8 \times (10 \times 10^{-6})^2}{18 \times 1.9 \times 10^{-2}} \\
&= 7.3 \times 10^{-3} \, \text{m s}^{-1} \\
&= 0.73 \, \text{cm s}^{-1}
\end{aligned}
$$

For most particles, the density of air could be neglected in these calculations.

Opposing the downward movement is the natural convective upward movement of air. As a broad generalization, particles greater than 10 μm in size are considered to be settleable, while those smaller than 10 μm remain suspended until removed by other processes such as washout in rain. Estimates of terminal or settling velocities are included in Table 6.4 and Fig. 6.1. A number of approximations, including assumptions about particle density and shape, have been made in producing these estimates.

The other physical process that determines particle lifetime is referred to as coagulation. This process involves the coming together by Brownian diffusion of small particles to form larger ones.

The rate of coagulation in a monodisperse system of particles (i.e. an aerosol made up of particles having uniform size) with a particular composition and density is then given by

$$-\frac{dN}{dt} = 4\pi DCd_pN^2 \tag{6.9}$$

where N = particle concentration/m^{-3}; D = diffusion coefficient of the particles in air/ m^2 s^{-1}; C = Stokes–Cunningham slip correction factor (Table 6.4); and d_p = particle diameter/m. Assuming that D, C, and d_p are constant, the rate of the coagulation process is second order:

$$-\frac{dN}{dt} = k_2N^2 \tag{6.10}$$

Table 6.4 Aerosol transport properties assuming spherical particles, density 2.0 g cm^{-3}, in air at P° and 25 °C

d_p/μm	C	v_t/cm s^{-1}	D/m^2 s^{-1}	$t_{1/2}$
0.001	216		5.14×10^{-6}	1 min
0.005	43.6		2.07×10^{-7}	0.5 h
0.01	22.2		5.24×10^{-8}	2 h
0.05	4.95		2.35×10^{-9}	38 h
0.1	2.85	1.7×10^{-4}	6.75×10^{-10}	110 h
0.5	1.326	2.0×10^{-3}	6.32×10^{-11}	520 h
1.0	1.164	6.8×10^{-3}	2.77×10^{-11}	690 h
5.0	1.032	1.5×10^{-1}		
10.0	1.016	6.0×10^{-1}		
50.0	1.003	15		
100.0	1.0016	58		

For half-life calculations, particle number density in the atmosphere is taken as 10^9 m^{-3}.

The half-life (in seconds) is then given by

$$t_{1/2} = \frac{1}{k_2 N} = \frac{1}{4\pi D C d_p N} \tag{6.11}$$

For example, an aerosol consisting of particles, all 0.01 μm ($= 0.01 \times 10^{-6}$ m) in diameter, with a number concentration, N, of 10^9 per m^3, the half-life is estimated to be

$$t_{1/2} = \frac{1}{4 \times 3.14 \times 5.24 \times 10^{-8} \times 22.2 \times 0.01 \times 10^{-6} \times 10^9}$$

$$= 6800 \, \text{s}$$

$$\simeq 2 \, \text{h}$$

Other values of $t_{1/2}$ in Table 6.4 are calculated in a similar manner.

Where particle sizes approach molecular dimensions, the above equations based on collisions limited by Brownian diffusion do not apply and the rate of coagulation is better described using relations derived from the kinetic theory of gases.

Because diffusion of particles is inversely proportional to the square of diameter, coagulation processes are more significant for very small particles and become negligible when the diameter is greater than approximately 0.01 μm. Table 6.4 lists diffusion coefficients as well as slip correction factors, terminal settling velocities, and half-lives for particles in the atmosphere at P° and 25 °C.

As a consequence of settling and coagulation processes, large and small particles respectively have relatively short residence times in the atmosphere. As shown in Fig. 6.3, the longest-lived particles are those whose size is in the intermediate range.

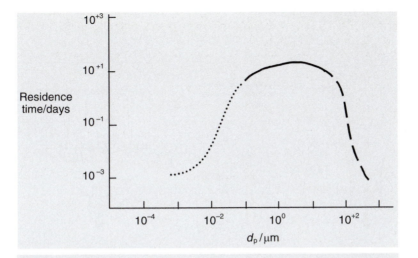

Fig. 6.3 Residence times of particles in the atmosphere. The dotted line indicates that coagulation is the principal removal process; the broken line indicates setting; and the solid line represents particles that are relatively long lived.

6.6 Air pollution control for particulate emissions

It is possible to minimize emissions of aerosol particles from point sources such as at thermal electrical generating stations or industrial smelting units. Obviously other point source emissions such as volcanic eruptions cannot be attenuated. Nor is it usually possible to control particulate emissions from non-point sources such as sea spray, wind-blown dust (e.g. the Harmattan, see Section 6.2), or forest or grassland fires. In the latter case, steps can be taken to prevent some uncontrolled fires, and this would lessen the amount of aerosol released to the atmosphere.

Elimination of much of the solid particulate material from the industrial gaseous emission stream can also be achieved by prevention, through methods such as improving the efficiency of the combustion process or using a better grade of fuel. However, in many cases some production and release of particulates is inevitable and containment is therefore required before the waste gases are released into the atmosphere.

Containment is accomplished using devices that remove the aerosol particles from the fast-moving stack-gas stream. Common collection methods include settling chambers, cyclones, fabric filters, scrubbers, and electrostatic precipitators, which are depicted in simple diagrammatic form in Fig. 6.4. Figure 6.5 compares the various methods in terms of the particle sizes that are most effectively removed using the different technologies.

Settling chambers are the simplest form of collection device and are one of the most widely applied methods of particulate control. The construction includes chambers with a variety of baffles and open spaces designed to allow the particles sufficient time to settle

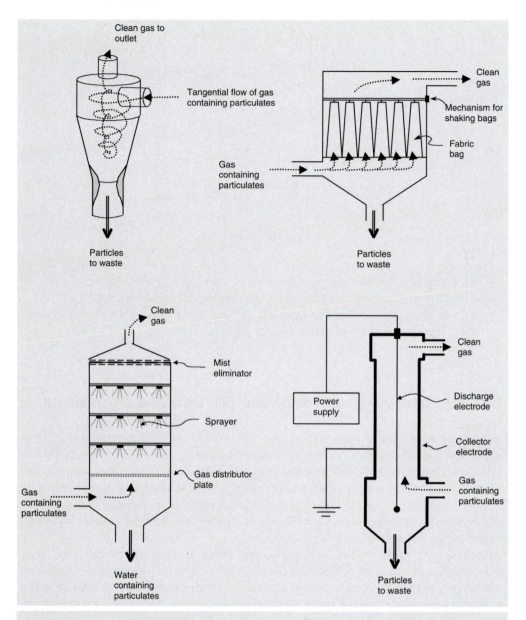

Fig. 6.4 Methods of air pollution (aerosol) control from point sources. Clockwise from upper left: cyclone, fabric filter, electrostatic precipitator, scrubber.

under the force of gravity. Because settling rates are limited by gravity, the method is most effective only for large particles ($> 10\,\mu m$).

Cyclones are a cone-shaped device that directs the waste gas stream to swirl rapidly in spiral fashion, causing larger particles to move towards the wall of the cone by centrifugal force. Once in contact with the wall, the particles slide down the inside of the cone to a collection container below. Stokes' law again determines the extent of removal of particles,

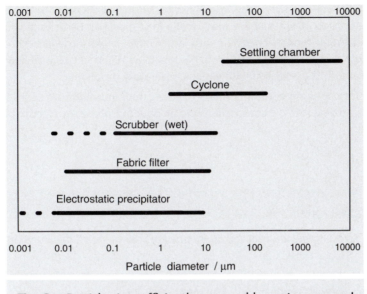

Fig. 6.5 Particle sizes efficiently removed by various aerosol control devices used in industry.

but the 'settling' rate can greatly be enhanced by the increased force due to the cyclone action. The effective removal range therefore extends to much smaller particles.

A bag or fabric filter operates in a manner that is similar to a vacuum cleaner—that is, the air stream is made to pass through a porous fabric material—and is effective for particulates in the 0.01 to 10 μm range. Bags or fabric filters are sensitive to temperature and humidity and because the fine pore filters become clogged during use, they must be cleaned periodically.

Scrubbers allow for contact between the gas stream and a fine mist or spray of water. By using relatively larger water droplets to capture small particles, their size is effectively increased so that they are able to settle more rapidly. Various scrubber designs have been developed.

An electrostatic precipitator causes the particles in a gas stream to become negatively charged by capturing electrons that were produced through an electrical discharge between two electrodes. Once the particles are charged they migrate to the positive electrode and are collected and removed from the emission stream.

The main points

1 Atmospheric aerosols consist of particles of solid or liquid in the air. The particles derive from both natural (sea spray, dust, etc.) and anthropogenic (industrial combustion, condensed organic vapours, etc.) sources.

2 Aerosol particles cover a size range from 1 nm to 100 μm in diameter, but it is particles in the range 0.01 to 10 μm that are most stable in suspension. Particles smaller than 0.01 μm in diameter tend to coagulate into larger units, while those larger than 10 μm readily settle out. The aerosol fraction that is less than 10 μm in size is of special concern because it is associated with various respiratory problems.

3 Technology for capturing aerosol particles released from industrial processes is based on filtration, settling rates, incorporation into an aqueous phase, or electrostatic precipitation.

Additional reading

1 Hidy, G. M., *Aerosols: An Industrial and Environmental Science*. Academic Press, Inc., Orlando; 1984.

2 Linak, W. P. and J. O. L. Wendt, Toxic metal emissions from incineration: mechanisms and control. Prog. Energy Combust. Sci., **19** (1993), 145–85.

3 Pye, K., *Aeolian Dust and Dust Deposits*, Academic Press, London; 1987.

Problems

1 Which of the two—cloud droplets or rain drops—would you expect to be more effective in scavenging gases from the atmosphere? Explain your reasoning.

2 A particular fog consists of 10 000 droplets of water per cm^3. The average diameter of the drops is 1.5 μm. Compare the mass of water in the liquid phase to that in the gaseous form if the temperature is 35 °C and the relative humidity is 100%.

3 Compare the amount of nitric oxide in air at a concentration of 300 ppbv with the amount of nitrate ion in a fog consisting of 10 000 droplets cm^{-3} having an average diameter of 2 μm and a concentration in the droplets of 3×10^{-5} mol L^{-1}.

4 A fly ash ($\rho = 1.8$ g mL^{-1}) aerosol consists of particles averaging 13 μm in diameter and with a concentration of 800 μg m^{-3}. Use the average diameter to calculate the settling velocity (cm s^{-1}) and settling rate (μg m^{-2} s^{-1}) of the particles in still air.

5 Suggest a possible sequence of reactions (beginning with oxidation *via* the hydroxyl radical) by which dimethyl sulfide ((CH_3)$_2$S) can be oxidized to produce a sulfuric acid aerosol.

6 Polyaromatic hydrocarbons are commonly associated with soot particles, as adsorbates on the surface. Calculate the actual and relative surface areas (actual values in m^2 g^{-1}) of soot particles ($\rho = 0.6$ g L^{-1}) having diameters 20 μm and 2 μm respectively, assuming that the particles are spheres. Speculate on the health implications with regard to human intake of PAH compounds.

7 In fly ash, the lead and chromium concentrations are measured for four particle diameter ranges. Results are as follows:

Particle diameter/μm	Lead conc./μg g^{-1}	Chromium conc./μg g^{-1}
>10	870	330
6–10	990	760
2–6	1100	1800
<2	1300	2700

Suggest reasons for these trends.

Notes

1 Isidorov, V. A., *Organic Chemistry of the Earth's Atmosphere*, Springer, Berlin; 1990.

2 Isidorov, V. A., Organic Chemistry of the Earth's Atmosphere, Springer, Berlin; 1990; and Hanst, P. L., J. W. Spence, and E. O. Eddy, *Atmos. Environ.*, **14** (1980), 1077.

7

Chemistry of urban and indoor atmospheres

IN previous chapters, we have examined some of the important chemical processes that occur in the troposphere. This has allowed us to consider specific types of atmospheric pollution including pollution due to sulfur dioxide, particulates, carbon monoxide, and a range of oxidants such as ozone and organic peroxides.

Here we will make use of this and additional information to describe the chemical composition of air in places where people live. We begin with an examination of air quality in major urban areas around the world. We will see that there are serious problems in some cities and to a very large extent such problems are a direct and indirect consequence of energy use. In particular, the combustion of fossil fuels in motor vehicles, in space heating and cooling, in power generation and industrial processes, and the incineration of waste materials are all major contributors to atmospheric pollution. Use of petroleum products, especially in motor vehicles, results in ground-level emissions of carbon monoxide, volatile hydrocarbons, nitrogen oxides, and sometimes lead. Wherever these compounds are released, aldehydes and other secondary pollutants are also formed. The combustion of biomass and coal produces substantial concentrations of solid particulate matter along with nitrogen oxides, PAH compounds, and in the case of much of the world's coal, sulfur dioxide as well. Open burning of refuse is an important cause of air pollution in many countries and this is a source of a wide variety of volatile organic carbon compounds (VOC) and solid particulate matter (SPM). Particulates are also provided from non-fuel sources—notably dust thrown up by movement of people and vehicles and wind-blown dust, especially in cities adjacent to deserts or dry, barren areas.

A second part of the chapter will examine related issues concerning the chemistry of atmospheres inside buildings. Depending on the building design, the same combustion products that are found in the outdoor urban atmosphere are again a source of concern indoors. Construction, furnishing, and cleaning materials are additional potential sources of other atmospheric pollutants.

7.1 Pollutants in the urban atmosphere

In many countries or regions, data have been proposed that set standards for air quality. One such set, the guidelines recommended by the World Health Organization (WHO) of the United Nations is summarized in Table 7.1 (see ref. 3).

Considering the WHO guidelines, studies of air quality in 20 of the world's 'megacities' have been carried out through the WHO and the United Nations Environment Program (UNEP). The material in the present section is based on the WHO–UNEP report, referenced above. In this context, a 'megacity' is defined as an urban agglomeration with a projected population of ten million or more by the year 2000. These cities are found on every continent except Australia and Antarctica and represent a variety of climatic, cultural, and technological situations. Table 7.2 summarizes, in a semi-objective manner, air quality properties in the megacities.

Table 7.1 Summary of World Health Organization (WHO) air quality guidelines

Pollutant	Maximum time-weighted average concentration	Averaging time[a]
Sulfur dioxide/$\mu g\,m^{-3}$	500	10 min
	350	1 h
	100–150	24 h
	40–60	1 y
Carbon monoxide/$mg\,m^{-3}$	30	1 h
	10	8 h
Nitrogen dioxide/$\mu g\,m^{-3}$	400	1 h
	150	24 h
Ozone/$\mu g\,m^{-3}$	150–200	1 h
	100–120	8 h
Suspended particulate matter (SPM)/$\mu g\,m^{-3}$		
Black smoke	100–150	24 h
	40–60	1 y
Total suspended particulates	150–230	24 h
	60–90	1 y
Respirable particulates (PM_{10})	70	24 h
Lead/$\mu g\,m^{-3}$	0.5–1	1 y

[a] The averaging time refers to the period during which the weighted-average value should not exceed the specified guideline concentration. Reproduced with permission.

Table 7.2 Summary of air quality in twenty megacities from all continents around the globe

	SO_2	SPM	CO	NO_2	O_3	Pb
Bangkok	−	+ +	−	−	−	+
Beijing	+ +	+ +	na	−	+	−
Bombay	−	+ +	−	−	na	−
Buenos Aires	na	+	na	na	na	−
Cairo	na	+ +	+	na	na	+ +
Calcutta	−	+ +	na	−	na	−
Delhi	−	+ +	−	−	na	−
Jakarta	−	+ +	+	−	+	+
Karachi	−	+ +	na	na	na	+ +
London	−	−	+	−	−	−
Los Angeles	−	+	+	+	+ +	−
Manila	−	+ +	na	na	na	+
Mexico City	+ +	+ +	+ +	+	+ +	+
Moscow	na	+	+	+	na	−
New York	−	−	+	−	+	−
Rio de Janeiro	+	+	−	na	na	−
São Paulo	−	+	+	+	+ +	−
Seoul	+ +	+ +	−	−	−	−
Shanghai	+	+ +	na	na	na	na
Tokyo	−	−	−	−	+ +	na

  −   Low pollution; WHO guidelines normally met. Short-term guidelines may be exceeded occasionally.
  +   Moderate to heavy pollution; WHO guidelines exceeded by up to a factor of two. Short-term guidelines may be exceeded regularly in some locations.
+ +   Serious pollution; WHO guidelines regularly exceeded by a factor of more than two.
 na   No data or insufficient data.
Data reproduced with permission.

7.1.1 Suspended particulate matter (SPM)

The problem of excessive concentrations of atmospheric particulates is a severe one in at least 12 of the megacities. The average levels in these cities range from 200 to 600 μg m^{-3} and peak concentrations may exceed 1000 μg m^{-3}. The human health consequences associated with high values depend on the nature of the particulates; those derived from coal and those in the PM_{10} category have been shown to be particularly hazardous.

In some cities including Beijing, Shanghai, and Seoul, the elevated SPM levels are largely due to domestic heating while Beijing, Cairo, Delhi, Karachi, and Mexico City have high natural loadings of wind-blown particulates. Of the 20 cities, only London, New York, and Tokyo consistently met WHO guidelines with respect to particulates.

7.1.2 Carbon monoxide

In urban areas, concentrations of carbon monoxide vary substantially across short distances, and depend on proximity to regions of high traffic density. Not surprisingly cities with large automobile populations, such as Mexico City, London, New York, and Los Angeles, have areas and times where excessive carbon monoxide levels are observed.

7.1.3 Sulfur dioxide

Several cities including Beijing, Mexico City, and Seoul have ambient levels of sulfur dioxide whose annual average concentration exceeds WHO guidelines by a factor of up to three. Peak daily concentration is sometimes even greater than the higher value specified as the 10 minute limit.

   In contrast, 12 cities appear to maintain sulfur dioxide concentrations that are well within the defined range, although brief high-level excursions are observed in London and São Paulo as well as in Calcutta during the dry season. Because coal use is now restricted in many cities, there has been a significant decline in sulfur dioxide concentrations over the past decade and further improvements are predicted.

7.1.4 Nitrogen dioxide

There are only limited data on atmospheric concentrations of this component of photochemical smog. Of the cities where regular measurements are made, Mexico City, Los Angeles, São Paulo, and Moscow all have shown short-term concentrations that exceeded guidelines. Where kerosene or natural gas is used for heating and cooking in households, higher levels of nitrogen oxides may be expected indoors (and have been measured in some cases), unless there is proper ventilation.

7.1.5 Ozone

Ozone concentrations were frequently high in Los Angeles, Mexico City, São Paulo, and Tokyo. Mexico City was the worst case with levels as high as $900\,\mu\mathrm{g}\,\mathrm{m}^{-3}$, four times the WHO guideline value. Less frequently, several other cities exceeded guidelines, while there were insufficient data in ten of the cities studied.

7.1.6 Lead

Airborne lead depends on the number density of motor vehicles, the concentration of lead additives in the fuel, and the availability of unleaded fuels. Concentrations in leaded gasoline vary between 0.1 and $2\,\mathrm{g}\,\mathrm{L}^{-1}$ although increasingly, using tetraethyl lead to augment the octane number is being abandoned. Cairo and Karachi atmospheric lead concentrations were consistently excessive and Bangkok, Jakarta, Manila, and Mexico City had somewhat high levels. These latter cities have plans to reduce or eliminate lead additives in gasoline. Where tetraethyl lead has already been eliminated, atmospheric lead concentrations are measurably smaller. This includes Brazilian cities where low-lead 'gasohol' is frequently used as a fuel. Residual high concentrations of lead persist, however, in roadside soils around the world.

7.2 Mexico City

Urban air pollution is a worldwide phenomenon, and each situation has individual characteristics related to the local conditions. Of all the great cities, a particularly complex situation is found in Mexico City and we will briefly examine its air chemistry characteristics.

It is evident from Table 7.2 that Mexico City represents a unique example of severe urban air pollution, suffering from excessive levels of most of the defined pollutants. The Metropolitan Area of Mexico City (MAMC) occupies an area of about 2500 km^2. Situated at an altitude of 2240 m, the region is surrounded by mountains. Within this contained territory, winds tend to be light and frequent inversions occur, leading to persistent accumulations of air pollutants. In 1990, the total population of the MAMC was near 20 million, giving a density of about 8000 persons per square kilometre. Many and varied industries are located within the area and there is a vast and complex transport system; the combined factors lead to a total consumption of energy estimated to be greater than 500 PJ (1 PJ (petajoule) $= 10^{15}$ J) each year.

Concentrations of all major air pollutants are regularly monitored and Figs 7.1 to 7.6 plot data obtained over several years up to 1991 along with the corresponding WHO guideline values. As can be seen, levels in excess of the guidelines occur frequently—in some cases continuously.

Suspended particulate matter (Fig. 7.1) has been found to be consistently above the guideline values, frequently by a factor of five or more. Part of the excessive amount is derived from 'natural' particulates (soil dust, biological particulates, etc.) but much results from the whole range of domestic, industrial, and transportation combustion processes. The PM$_{10}$ fraction makes up about half of the total SPM concentrations.

In Mexico city, mean carbon monoxide levels (Fig. 7.2) have generally exceeded the WHO guidelines. Very high values are especially common during times of heavy traffic in early morning and, to a reduced extent, in the evening. Being located at a high altitude, the partial pressure of oxygen is low and this enhances the possibility of incomplete combustion in furnaces and engines. The low pressure also contributes to adverse health effects associated with carbon monoxide.

The major sources of sulfur dioxide (Fig. 7.3) are power plants that burn high-sulfur (3.5% S) fuel oil, and buses and trucks that burn diesel oil (1.2% S). Future decreases in atmospheric concentrations are expected to occur as switch-over to natural gas fuel in power plants takes place.

More than three quarters of the nitrogen dioxide in the MAMC derives from transport vehicles of all types. Figure 7.4 shows that maximum hourly levels usually exceeded the WHO guidelines. Mean values were approximately one third of the maximum concentrations, and these average levels are generally considered to be 'safe'. However, ozone (Fig. 7.5), resulting in part from NO$_x$ emissions, is also frequently high. In the Pedregal

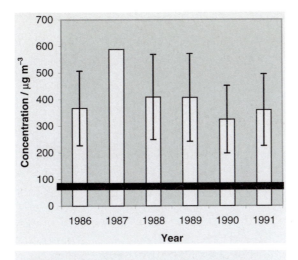

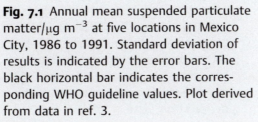

Fig. 7.1 Annual mean suspended particulate matter/μg m^{-3} at five locations in Mexico City, 1986 to 1991. Standard deviation of results is indicated by the error bars. The black horizontal bar indicates the corresponding WHO guideline values. Plot derived from data in ref. 3.

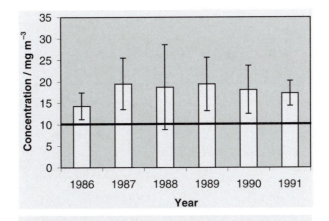

Fig. 7.2 Maximum 8 h mean carbon monoxide concentrations/mg m^{-3} at various locations in Mexico City, 1986 to 1991. Standard deviation of results is indicated by the error bars. The black horizontal line indicates the corresponding WHO guideline values. Plot derived from data in ref. 3.

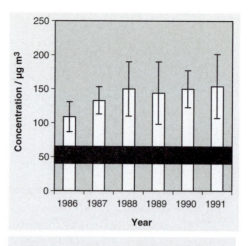

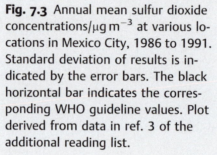

Fig. 7.3 Annual mean sulfur dioxide concentrations/μg m^{-3} at various locations in Mexico City, 1986 to 1991. Standard deviation of results is indicated by the error bars. The black horizontal bar indicates the corresponding WHO guideline values. Plot derived from data in ref. 3 of the additional reading list.

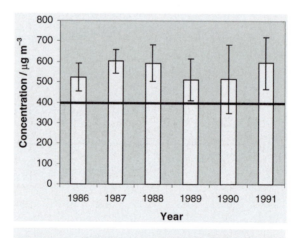

Fig. 7.4 Maximum hourly nitrogen dioxide concentrations/μg m^{-3} at various locations in Mexico City, 1986 to 1991. Standard deviation of results is indicated by the error bars. The black horizontal line indicates the corresponding WHO guideline value. Plot derived from data in ref. 3.

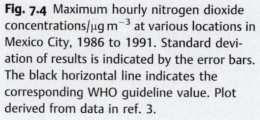

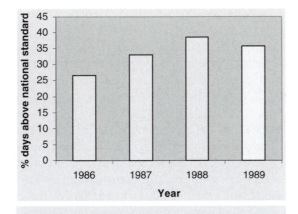

Fig. 7.5 Percentage of days when ozone levels exceed national standards in Mexico City, 1986 to 1989. Plot derived from data in ref. 3.

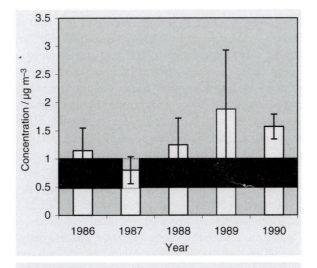

Fig. 7.6 Annual mean atmospheric lead concentrations/$\mu g\,m^{-3}$ at various locations in Mexico City, 1986 to 1991. Standard deviation of results is indicated by the error bars. The black horizontal bar indicates the corresponding WHO guideline values. Plot derived from data in ref. 3.

region occupying the south-western sector of the city, excessive ozone concentrations were observed on more than 60% of the days.

Atmospheric lead (Fig. 7.6) concentrations were observed to be excessive in some locations but appear to be decreasing after reaching peak values around 1980. Since that time, the maximum concentration of tetraethyl lead allowed in gasoline has been lowered substantially to $0.54\,g\,L^{-1}$, leading to a decline in atmospheric values in spite of increased use of fuel.

On all counts, the MAMC area has a very serious air quality problem. It is clear that this problem is a consequence of a high population in a contained area and an energy-intensive lifestyle. These factors are unlikely to change substantially and without urgent action, the situation will deteriorate even more rapidly. Plans for improvement in air quality will require use of clean fuel, improved combustion efficiency, and introduction of strict emission controls as well as fuel conservation measures. Again, we emphasize that similar kinds of air pollution problems have been identified and characterized in many other cities around the world.

7.3 Indoor air quality

Many of us spend the greater portion of our lives indoors, and the atmospheres we encounter there exhibit an extreme variety of conditions. They range from the highly purified, sterile atmosphere in a hospital operating room to the dusty environment of a poorly ventilated feed mill. Space does not permit us to deal with such specific and individual cases, but we will consider the nature of indoor atmospheres under 'normal' conditions such as in homes. Here too there are a number of possibilities. In some parts of the world where the climate is warm year round, a sizeable fraction of the population lives in houses constructed from clay-rich soils or other fresh or baked-earth materials. Because of the comfortable temperatures, these homes are open much of the time and air exchange is very rapid. However, sometimes heating (in cooler seasons) or cooking is done over an open fire in a room without a chimney and a variety of fuels are used so that a complex suite of gaseous and particulate emissions is produced. The range of air properties in these houses will be much different from those built where the climate is temperate or cold. In the latter regions, construction materials are typically a combination of brick, stone, wood, various plastics, and metals. Often the homes are well insulated and sealed so that there is little exchange between indoor air and that outside. Many activities take place inside the home, including cleaning, cooking, heating by open or enclosed fires, and smoking, to name a few. The result is a different set of variable air conditions.

Four major factors determine the quality of indoor air.

- The first is the nature of the ambient air, outdoors around the building. Whether it is rapid or slow, exchange inevitably does occur so that the indoor atmosphere is always influenced by air outside. In the case of some pollutants, the only sources are from outdoors. We have seen that ozone production is a photochemical process requiring ultraviolet radiation. Therefore it cannot normally be produced within a building (except by certain electrical devices such as copying machines and electrostatic precipitators used for dust control), yet indoor ambient levels of 5 to $15\,\mu\mathrm{g\,m^{-3}}$ are typically measured. In most cases all of the ozone present in a building originated in the atmosphere outside. As would be expected therefore, the ratio of C_{ozone} (indoors)/ C_{ozone} (outdoors) is usually substantially less than unity.

- The second factor is the infiltration rate or air exchange rate for a building. This relates to the site and design of the building. In the open design home of the tropics, there are many exchanges per hour. On the other hand, an insulated temperate-climate house in winter typically has an air exchange rate of $1\,\mathrm{h^{-1}}$ while tightly sealed 'energy efficient' buildings exchange air only every 2 or even 10 h, giving air exchange rates of 0.5 to $0.1\,\mathrm{h^{-1}}$ respectively. In these latter cases, the influence of the outside atmosphere on air quality inside is much less. A corollary is that any atmospheric chemical that originates indoors accumulates to a greater degree when air exchange is limited.

- A third influence on indoor air is the material present in the building as construction material or in other forms. We shall see that many modern polymers used for construction are sources of formaldehyde and their presence in the building provides a passive but continuous supply of this chemical. Likewise building materials derived from the Earth—clays, laterite, concrete—contain varying trace amounts of radioactive elements that generate the gas radon, ultimately a source of α radiation. Every material, just by being present, is a potential source for emission of a variety of chemicals.

- Finally, and perhaps most important, the activities that go on inside a building determine the nature of the indoor atmosphere. Combustion for heating or cooking, whether enclosed as in a high-efficiency natural gas furnace, or in an open fire, releases gaseous and/or particulate emissions. Smoking is another combustion process that has a major effect on the atmosphere of an enclosed area. Any cleaning activity, whether by mechanical means that raise dust, or with cleaning aids using volatile solvents, alters the indoor atmosphere. In fact, every activity inside a building inevitably has a small or large effect on the atmospheric composition within the enclosed space.

A general equation describing the steady-state behaviour of a stable compound with respect to indoor and adjacent outdoor concentrations is

$$R_{\mathrm{i}} = k_{\mathrm{e}}C_{\mathrm{i}} - k_{\mathrm{e}}C_{\mathrm{o}} \tag{7.1}$$

where R_{i} (units of concentration per time) is the net rate of production of the compound inside, C_{i} and C_{o} are the indoor and outdoor concentrations, and k_{e} is the first order rate constant for atmosphere exchange (defined above as the air exchange rate, with units of $\mathrm{time^{-1}}$). In the steady state the interior concentration is given by

$$C_{\mathrm{i}} = C_{\mathrm{o}} + R_{\mathrm{i}}/k_{\mathrm{e}} \tag{7.2}$$

For cases where the outdoor contribution is negligible,

$$C_{\mathrm{i}} = R_{\mathrm{i}}/k_{\mathrm{e}} \tag{7.3}$$

Where none of the chemical is produced indoors,

$$C_{\mathrm{i}} = C_{\mathrm{o}} \tag{7.4}$$

unless the chemical is lost by reaction as it infiltrates the building.

Each of these relations assumes a stable chemical, meaning that it is not lost either by reaction or deposition inside the building.

In the next sections, we will consider three kinds of air quality issues that are encountered in many indoor situations. The nature and concentration of the chemicals relate to all four of the factors affecting air quality, including the types of structural materials present in the building and the activities that are undertaken indoors.

7.3.1 Radioactivity

Inside a building, most radioactivity above the outdoor background level is associated with radon. Radon (Rn, element 86) is a dense, radioactive, noble gas that is itself the product of radioactive decay processes beginning with uranium-238 and thorium-232. The two parent isotopes are present at low concentration in many geological materials and have half-lives of 4.5 and 14 billion years respectively. Because of their long half-lives, trace quantities persist to the present day in soil that surrounds structures, in building materials derived from rock or other geological sources, and also in water that is in contact with the soil and rocks. Both uranium and thorium decay through complex radiochemical reaction series, emitting alpha, beta, and gamma radiation, as shown in Fig. 7.7. Intermediate daughter products in the sequences include radon-222 and radon-220.

High levels of radiation in indoor air are not usually associated directly with the emissions from uranium and thorium. When α particles are given off by elements in solid materials or water, the heavy matrix absorbs most before they reach the surrounding air. Only a small number of particles are able to escape out of the material surface and this forms part of the background radiation within a building. The heavy α particles are also readily absorbed by air and very few move more than 30 or 40 cm away from the solid or liquid. Where they do encounter biological tissue they are even more strongly absorbed and do not penetrate much beyond the surface.

The situation is different with the radon gas produced in the decay sequence of the two heavy elements. Being a gas, radon escapes from construction materials, the surrounding soil, and water; it penetrates through cracks in a building, and is released into the indoor atmosphere where it can be inhaled. Decay of radon-222 (half-life = 3.8 days) and radon-220 (half-life = 55.6 s) *via* α emission then takes place directly inside the lungs. The radioactive decay products include isotopes of metals such as polonium-218, -216, -214, and -212 that are deposited on the internal tissues. Further decay of the radioactive metals release more α particles that are then available to interact immediately with cell molecules. Of the two isotopes of radon, the radon-220 has a very short half-life giving it little time to escape into and accumulate in the ambient atmosphere. It is therefore a less abundant and less important form of the element than is radon-222.

It is instructive to consider a semi-quantitative relation between production and loss of radon in an enclosed air space—a relation that is analogous to eqn 7.1. We can assume a steady-state indoor rate of radioactive decay, A_i, over time—measured as activity in a 1 L volume of air, Bq L^{-1}, using the SI unit for radioactivity.[1] Note that this is a way of measuring 'concentration' of a radioactive element. Outside the building, the decay rate is also constant at A_o (Bq L^{-1}). In the present situation, two first order rate constants are required to describe the interchanges in radioactivity concentration; k_e (h^{-1}) is the air

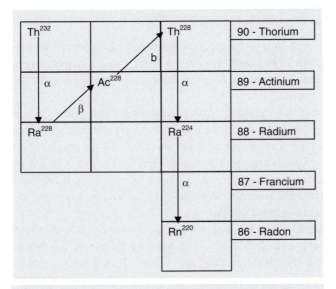

Fig. 7.7 Uranium-238 (above) and thorium-232 (below) radioactive decay schemes to produce radon-222 and radon-220 respectively. The product isotopes are gases and are themselves radioactive, decaying further to stable products.

exchange rate as defined earlier and k_d (h^{-1}) is the radioactive decay constant, in this case for radon. Equation 7.1 is therefore modified for this new situation:

$$R_i + k_e A_o = k_e A_i + k_d A_i \qquad (7.5)$$

The left-hand terms include internal and external sources of radon, and the right-hand terms are for loss by exchange and radioactive decay respectively. Rearranging eqn 7.5, the interior radioactivity concentration is given by

$$A_i = (R_i + k_e A_o)/(k_d + k_e) \qquad (7.6)$$

Radon-222 has a half-life of 3.83 days, equivalent to a decay constant, $k_d = \ln 2/t_{\frac{1}{2}}$, of 0.00754 h^{-1}. Consider a situation where the rate of production of radioactivity inside the building, R_i, is 1.0×10^{-2} Bq L^{-1} h^{-1}, and the external radioactivity concentration, A_o, is 4.0×10^{-3} Bq L^{-1}. The steady-state interior radioactivity concentration, A_i, would be 4.5×10^{-3} if the air exchange rate were 20 h^{-1} (open building, excellent air exchange), but it would be 0.097 Bq L^{-1} where the air exchange rate was 0.10 h^{-1} (tightly sealed, 'energy efficient' building). Clearly, the rate of production and the exterior activity of radon would also be important determinants affecting activity inside. Levels of radioactivity less than 0.01 Bq L^{-1} are considered low, around 0.1 Bq L^{-1} are normal, and greater than 4 Bq L^{-1} are high. In the above calculation, the value used for outdoor radioactivity associated with radon is the global average of about 4×10^{-3} Bq L^{-1}.

7.3.2 Volatile organic compounds

Volatile organic compounds (VOCs) present in indoor atmospheres arise from a number of sources. Many organic compounds are emitted from construction materials, furnishings, and consumer products in the building. Combustion processes generate an additional suite of VOCs.

Construction materials including wood composites made from chips or sawdust, insulating foams, floor tiles, carpet, and the adhesives used for installation all contribute to the VOC burden in an indoor atmosphere. Emissions from fresh material are particularly high in a new building; they decline rapidly over the first few months and then more slowly over extended periods. There are many aromatic and aliphatic compounds contributing to VOC concentrations, with chloroform, acetone, chlorinated compounds, and formaldehyde being predominant in many locations.

Consumer products used in homes contribute other VOCs to the atmosphere. For example, latex paint contains toluene, ethylbenzene, 2-propanol, and butanone. Cleaning agents, household solvents, detergents, and waxes all contain various small molar mass volatile organic species.

Small concentrations (typical values are between 10 and 200 $\mu g\ m^{-3}$ in new construction) of formaldehyde are observed in the air of many buildings and its behaviour has been extensively studied. Formaldehyde or its precursors are present in a wide variety of modern materials found in homes and other buildings. They are found in resins that are used to make plywood and various types of particle board. Anti-shrink, anti-wrinkle, colour-retention, and fire-retardant chemicals modify many fabrics; all of these may also be sources of the chemical. Some papers such as pre-pasted wallpaper contain formaldehyde resins. Urea–formaldehyde (UF) polymers, in several forms, are perhaps the largest source of formaldehyde. Where the polymers have been made into high-density moulded or extruded plastics, the surface area is relatively small and chemical reactions to release formaldehyde in gaseous form are slow. However, the UF polymers are also widely used in the form of foams that produce a sponge-like porous solid with an enormous surface area. Such UF foams are convenient and efficient insulating materials for structures built in areas of cold climate. However, UF foams have been shown to be sources of significant quantities of formaldehyde, with emission occurring by both rapid- and slow-release processes depending on whether the formaldehyde is in free or combined forms.

There is initial rapid evolution of free formaldehyde, and N-methylol end groups present in the resin can react rapidly, releasing additional product (reaction 7.7):

$$R-\underset{H}{N}-\overset{\overset{\displaystyle O}{\|}}{C}-\underset{H}{N}-CH_2-OH \longrightarrow R-\underset{H}{N}-\overset{\overset{\displaystyle O}{\|}}{C}-NH_2 + \overset{H}{\underset{H}{}}C=O \qquad (7.7)$$

When these sources are expended, continuous, nearly steady-state slow release occurs, mostly due to hydrolysis reactions of methylene bridge groups in the polymer backbone (reaction 7.8).[2] Release rates are enhanced at high temperatures and, because the reactions are due to hydrolysis, humid conditions also increase the rate during the steady-state period.

$$R-\underset{H}{N}-\overset{\overset{\displaystyle O}{\|}}{C}-\underset{H}{N}-CH_2-\underset{H}{N}-\overset{\overset{\displaystyle O}{\|}}{C}-\underset{H}{N}-R + H_2O \longrightarrow 2\,R-\underset{H}{N}-\overset{\overset{\displaystyle O}{\|}}{C}-NH_2 + \overset{H}{\underset{H}{}}C=O \qquad (7.8)$$

A qualitative picture of the pattern of predicted formaldehyde release to the indoor air of a newly constructed building is shown in Fig. 7.8.

Loss of atmospheric formaldehyde occurs by reactions that have been discussed in Chapter 4. The first step (reaction 7.9) of its decomposition is a photochemical process that requires energetic radiation having wavelengths at the lower end of the UV-A and in the UV-B regions of the spectrum. Fluorescent lamps emit significant amounts of UV-A radiation, while radiation from incandescent lights is mostly in the visible and near infrared regions.

$$HCHO \xrightarrow{h\nu\,(\lambda < 330\,nm)} H\dot{C}O + \cdot H \qquad (7.9)$$

The two radicals react further with oxygen, producing the hydroperoxy radical:

$$\cdot H + O_2 \rightarrow HOO\cdot \qquad (7.10)$$

$$H\dot{C}O + O_2 \rightarrow HOO\cdot + CO \qquad (7.11)$$

The hydroperoxy radical is a highly reactive species and is consumed by a number of possible reactions, one of which is the oxidation of nitric oxide:

$$HOO\cdot + NO \rightarrow \cdot OH + NO_2 \qquad (7.12)$$

Outdoor clean-air concentrations of formaldehyde are typically less than 10 ppbv, but in buildings that contain substantial formaldehyde-producing materials, concentrations

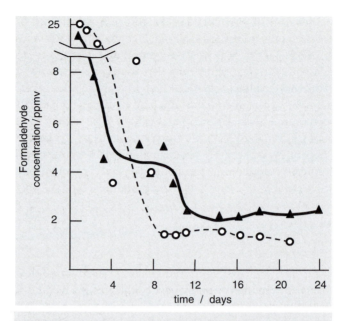

Fig. 7.8 Time sequence of formaldehyde release in a building containing formaldehyde-based polymers. Room maintained at 33 °C; measurements taken over a 30 minute sampling period. Filled triangles: high humidity conditions; open circles: low humidity conditions. (Figure redrawn from data provided in Gammage, R. B. and K. C. Gupta, Formaldehyde, in ref. 2, p. 113.)

between 100 and 500 ppbv have frequently been observed. Values several times higher than these are not unusual. A mixing ratio of 100 ppbv has been proposed as a level of concern.

In recent decades, in Europe and North America, there has been an increasing number of complaints about the indoor atmosphere from residents of some private homes and workers in offices and other commercial buildings. It is claimed that the climate in certain buildings produces unhealthy symptoms, commonly including mucosal and skin irritations, tiredness, dizziness, headaches, and general unease. Where this is a common occurrence, the building is often characterized as *sick* and the collection of symptoms are together referred to as *sick building syndrome*.

While the syndrome is not associated with a well defined source, there are common features related to materials used in construction and furnishings, inadequate ventilation, and features of the building's operation. Most probably, formaldehyde and other organic chemicals, as have been discussed, contribute to the unhealthy symptoms. The effects of these chemicals are enhanced in situations where they are able to accumulate by adsorption on surfaces within the building. This occurs when the building contains a large 'concentration' of high-surface-area materials such as books and papers on open shelves, carpets, draperies, and other textiles. Chemicals are sorbed by these high-surface-area materials and then can be re-emitted, especially when the temperature is elevated—often when people are present in the building. Many volatile organic compounds have been measured in the air of such buildings. Formaldehyde and phenol are two commonly encountered chemicals, but other aliphatic and aromatic hydrocarbons, aldehydes, ketones, acids, alcohols, and esters are frequently found in measurable concentrations in the atmosphere. Careful choice of building and furnishing materials and efficient ventilation are therefore design considerations required in the construction of a healthy building.

7.3.3 Emissions from indoor combustion

Combustion inside a building contributes to the concentration of VOCs and it also is a source of stable inorganic gases. Burning fuel indoors is a common practice—usually for purposes of heating or cooking. The nature of the combustion products depends on the fuel used as well as on the fuel : oxidant ratio and other combustion conditions. There are many types of fuel, but oxygen in air is invariably the oxidant. The extent to which combustion products are released into the building depends on the design of the heat production system. Such designs range from high-efficiency furnaces that draw in air and release the emission products outside the building, to heating or cooking devices that generate and release both gases and particulates inside a poorly ventilated room. Tobacco smoking is another combustion issue.

Combustion of carbon-based fuels always produces carbon dioxide leading to increased atmospheric mixing ratios of this gas. Even where ventilation is poor, however, carbon dioxide levels rarely exceed 1000 ppmv (recall that outdoor ambient levels are 365 ppmv), a mixing ratio that is usually considered to be acceptable. Carbon monoxide emissions are also associated with all burning activities but enhancement of indoor levels may be minimized by ensuring that combustion conditions favour complete oxidation to carbon dioxide. A reasonable air exchange rate is also important so that levels of carbon monoxide will not be higher than 10 ppmv. Nitric oxide (NO) is derived from the combined nitrogen in the fuel. Wood or other forms of biomass contain substantial amounts of this element, but there is very little in petroleum products or natural gas. An additional amount of nitric oxide is produced from the dinitrogen in air during high temperature combustion. Because the rate of generation depends on flame temperature, much more is produced by burning of natural gas ($T_{flame} > 2000\,°C$) than of biomass such as wood ($T_{flame} < 1000\,°C$). The postcombustion oxidation reaction $NO \rightarrow NO_2$ requires photochemically generated species such as peroxy radicals in order to progress rapidly. One of the routes for producing the hydroperoxy species was shown in reactions 7.9 to 7.11 above. Mixing ratios greater than 100 ppbv of either NO_x species are considered to be excessive. Except in the unusual case of indoor burning of high-sulfur coal, sulfur dioxide emissions are not a serious problem.

Volatile organic carbon compounds are other products of combustion, their nature and amount again depending on the fuel type and burning conditions. Coal, wood, and other forms of biomass tend to release more gaseous hydrocarbons and other partially oxidized VOCs than do petroleum and natural gas products. When the former fuels are burned in unvented appliances, elevated levels of these gases are found in the indoor air. Proper venting of exhaust gases minimizes this problem. Formaldehyde is a minor product of combustion of all types of fuel.

Tobacco smoking is a source of many VOCs including organic bases like nicotine, aldehydes, ketones, organic acids, and hydrocarbons. Cigarette smoke is yet another source of formaldehyde with an estimated amount of 2.4 mg being emitted by each cigarette.[3] The directly inhaled air drawn through a cigarette may contain formaldehyde concentrations more than 400 times greater than the level of concern. More will be said about the products of tobacco combustion in the next section.

7.3.4 Particulates

Much of the solid aerosol in a building is due to dust lifted into the air by human activities and air circulation. Release of particulate matter also accompanies the combustion of coal and biomass materials. Particle size of the aerosol derived from combustion of these solid fuels is in the PM_{10} range, with many particles having diameters less than 2 μm, making them able to penetrate deep into the respiratory passages. A concentration of 100 μg m^{-3} total solid particulates is a typical guideline value for an acceptable upper limit inside a building. (Refer to Table 7.1 for recommended outdoor values.)

In homes in rural Mexico, it has been found that very high levels of particulates were present during the (typically 2 to 3 h) period required for the preparation of tortillas. In one study,[4] four preparation methods were considered and the average particulate concentrations within the PM_{10} range were

- 1140 μg m^{-3} for use of biomass fuel in an unvented environment;
- 330 μg m^{-3} for unvented liquefied petroleum gas (LPG);
- 540 μg m^{-3} for a mixture of unvented biomass and unvented LPG;
- 430 μg m^{-3} for vented biomass.

All of these values, but especially those associated with biomass fuels, are higher than the recommended acceptable levels.

Smoking can be a significant contributor to respirable particulate matter inside a building. Figure 7.9 illustrates the estimated range of exposure to particulates for persons inside a building where smokers are present. The respirable suspended particulates (RSP) concentration is plotted against the volume density of persons in the enclosed area. In this theoretical calculation, it is assumed that at any time one third of the persons in the room are smokers and at any time during the day, for every three smokers, one cigarette burns constantly. Typical occupancy densities are 4 per 100 m^3 for office space, 25 per 100 m^3 for restaurants, and up to 50 per 100 m^3 for crowded places like auditoriums, trains, and aeroplanes. Values of the air exchange rate, k_e, often range from 1 h^{-1} for an unventilated building through 10, for a well ventilated commercial establishment, to 26 for a commercial aircraft with excellent air exchange. It is clear that high occupancy by smokers and poor air exchange, or a combination of the two factors, can lead to unacceptable levels of RSP within the confined areas.

In Chapter 6 we saw that polyaromatic hydrocarbons (PAHs) are associated with particulate and gaseous emissions from coal and biomass. While outdoor concentrations of

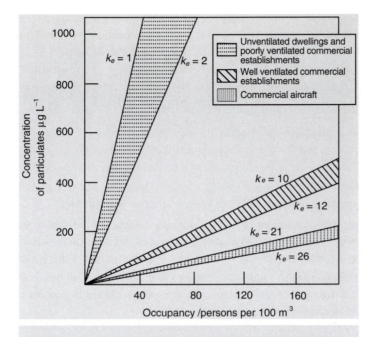

Fig. 7.9 Theoretical steady-state density of respirable suspended particulates from environmental tobacco smoke in an enclosed indoor space. Air exchange rate is $k_e\,h^{-1}$. (Redrawn from Repace, J. L. and A. H. Lowrey, Indoor air pollution, tobacco smoke, and public health. *Science*, 208 (1980), 464–71.)

PAHs are often below $1\,ng\,m^{-3}$, levels may be several times higher in communities where extensive biomass combustion takes place. Concentrations are further enhanced in geographic situations such as valley communities or during inversion events. There are cases of wood-burning communities in the United States where, under unfavourable meteorological conditions, outdoor levels of benzo(a)pyrene have been measured around $10\,ng\,m^{-3}$ and occasional levels above $100\,ng\,m^{-3}$ have been reported. (For analytical reasons, levels of benzo(a)pyrene are often reported as a surrogate for total PAH. This may be misleading, since other particular species may be present in much higher concentrations and, in any case, proportions of all species vary in each situation.)

Indoor concentrations of PAH compounds are greatly affected by the amount and conditions of combustion in the building. Where outdoor levels of benzo(a)pyrene are in the usual low range, residences with enclosed wood stoves can have concentrations of PAHs in a range around $5\,ng\,m^{-3}$ and those with open fireplaces (but a well-vented chimney) often have levels above $10\,ng\,m^{-3}$.

Tobacco smoking is a source of PAH compounds as well as of particulates in general. It has been shown that one unfiltered cigarette provides about 25 ng of benzo(a)pyrene to the smoker.[5] Using this figure and a breathing rate of $23\,m^3\,d^{-1}$, each cigarette provides as much of the compound as does continuous breathing over one day of an atmosphere containing $1 \, ng \, m^{-3}$. A 20-cigarette-a-day smoker is then exposed to the equivalent of an atmosphere with a constant level of benzo(a)pyrene of about $20 \, ng \, m^{-3}$.

The main points

1 Urban air pollution problems are frequently associated with the products of combustion in industry, in vehicles, and for domestic purposes. Suspended particulate matter, carbon monoxide, sulfur dioxide, nitrogen oxides, and ozone are common pollutants.

2 Many large cities around the globe exhibit excessive levels of one or more of these pollutants. Concentrations vary with the season and the time of day.

3 The indoor atmosphere is influenced by outdoor air quality, by the rate of air exchange with the outdoor atmosphere, by materials used in construction and for furnishing, and by activities—especially those involving combustion—that go on inside the building.

4 Radioactivity associated with radon gas, gaseous formaldehyde, and other organic compounds, and excessive levels of atmospheric particulates are important sources of air quality problems inside buildings.

Additional reading

1 Otson, R. and P. Fellin, Volatile organics in the indoor environment: sources and occurrence, in J. O. Nriagu, ed., *Gaseous Pollutants: Characterization and Cycling*. John Wiley and Sons, New York; 1992.

2 Walsh, P. J., C. S. Dudney, and E. D. Copenhaver, eds, *Indoor Air Quality*, CRC Press, Boca Raton; 1984.

3 World Health Organization and the United Nations Environment Programme, *Urban Air Pollution in Megacities of the World*, Blackwell, Oxford; 1992.

Problems

1 Atmospheric pollutants are sometimes classified into two categories:

(a) primary pollutants—those that are emitted directly from the source;
(b) secondary pollutants—those that are produced by reactions in the open atmosphere.

Which category(s) would each of the following fall into? Carbon monoxide, carbon dioxide, sulfur dioxide, nitric oxide, nitrogen dioxide, ozone, PAH compounds, formaldehyde.

2 The indoor/outdoor concentration ratios for carbon monoxide are usually near 1, for carbon dioxide are usually greater than 1, and for sulfur dioxide are usually less than 1. Give reasons for these normal situations and cite instances that might be exceptions to the values.

3 In the uranium-238 decay series, radium-226 is an immediate radioactive parent of radon-222. Consider its position in the periodic table and indicate its usual oxidation state, its probable mobility in water, and factors affecting its solubility in soil/water systems.

4 As an experiment, gas cooking units were activated until high levels of carbon monoxide and NO_x compounds were measured in a house and then the gas burners were turned off. The carbon monoxide levels returned to background values in 1.6 h, but the NO_x levels only required 0.7 h before they had reached the 'normal' concentration. Suggest an explanation.

5 In Chinese village homes, heating is often by an open wood or coal fire. Where outdoor PAH concentration is $0.60\,\text{ng}\,\text{m}^{-3}$, the rate of emission of PAH from indoor combustion is $3.5\,\text{ng}\,\text{m}^{-3}\,\text{h}^{-1}$ and the air exchange rate is $20\,\text{min}^{-1}$, estimate the indoor air concentration of PAH compounds. Assume that the only loss mechanism is by air exchange.

Notes

1 The becquerel (Bq) is defined as one disintegration per second. Another basic (non-SI) unit for radiation is the curie (Ci), which is equal to $3.7 \times 10^{10}\,\text{Bq}$.

2 Allan, G. G., J. Dutkiewicz, and E. J. Gilmartin, Long-term stability of urea–formaldehyde foam insulation. *Environ. Sci. Technol.*, **14** (1980), 1235–41.

3 Olander, L., J. Johansson, and R. Johansson, Tobacco smoke removal with room air cleaners. *Scand. J. Work Environ. Health*, **14** (1998), 390–7.

4 Brauer, M., K. Bartlett, J. Regalado-Pineda, and R. Perez-Padilla, Assessment of particulate concentrations from domestic biomass combustion in rural Mexico. *Environ. Sci. Technol.*, **30** (1996), 104–9.

5 National Research Council, *Particulate Polycyclic Organic Matter*, Committee on Biological Effects of Atmospheric Pollutants, National Academy of Sciences, Washington, DC; 1972 as reported in ref. 1.

8

The chemistry of global climate

IN the simplest manner, climate may be defined as average weather. At a particular geographical location, the local climate describes the average of day to day weather variations–temperature, wind, amounts and patterns of precipitation–over a period of many years. Global climate is a spatial average of all local climates around the world.

Climate is not a static phenomenon but rather shows changes–some unidirectional, some cyclical–over periods of geological time. The global climate and the changes it undergoes are determined by many factors, one very significant factor being the chemical nature of the Earth's atmosphere. Gases and aerosol species determine the degree to which the Sun's radiation is absorbed or reflected, and they do the same for radiation emitted back into space from the Earth itself. In a complex manner, then, temperature, precipitation, wind patterns, and other climatic features are affected by the chemical composition of the atmosphere. At the same time, the reverse effect is true as well. Changes in climate bring about changes in atmospheric composition and so it is frequently difficult to identify cause and effect.

We will consider how the chemistry of the Earth's atmosphere plays a role in determining its average temperature. We will also examine some of the human activities that are affecting the chemical composition in ways that can cause changes in global climate patterns.

8.1 Composition of the Earth's atmosphere

The average mixing ratio of the nine principal gases in the dry troposphere is given in Table 8.1. Because these gases have long residence times, the values are relatively constant at all locations on the Earth.

The most highly variable major gaseous component in the troposphere is water. Its residence time (calculated from the values in Fig. 1.4) is about 11 days–a period that is much smaller than the time required for complete mixing of the troposphere. It is for this reason that the mixing ratio of water vapour varies from day to day and place to place.

Table 8.1 Average composition of the dry troposphere	
Component	**Mixing ratio**
Nitrogen	78.08%
Oxygen	20.95%
Argon	0.93%
Carbon dioxide	365 ppmv
Neon	18 ppmv
Helium	5.2 ppmv
Methane	1.77 ppmv
Hydrogen	0.53 ppmv
Nitrous oxide	0.31 ppmv

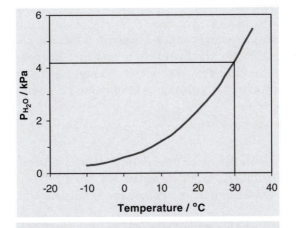

Fig. 8.1 The vapour pressure of water (P_{H_2O}) as a function of temperature. The value of (P_{H_2O}) at 30 °C is determined from the plot, or calculated to be 4.24 kPa.

The maximum mixing ratio of water that can be present in the atmosphere is related to temperature through the vapour pressure curve given in Fig. 8.1. Relative humidity is an expression of the percentage of this maximum (equibrium) value that obtains in a given situation. In the dry tropics, a temperature of 30 °C and relative humidity of 40% could be typical. This corresponds to a non-equilibrium P_{H_2O} of

$$0.40 \times 4.24 \, \text{kPa} = 1.7 \, \text{kPa}$$

At $P°$, the mixing ratio of H_2O is then equal to the ratio of its partial pressure to total pressure:

$$\text{mixing ratio}(\%) = \frac{P_{H_2O}}{P°} \times 100 = \frac{1.7\,\text{kPa}}{101\,\text{kPa}} \times 100 = 1.7\%$$

An equivalent calculation for a situation in a cooler region where the temperature is $-10°C$ and the humidity 100% gives a value of $P_{H_2O} = 0.260\,\text{kPa}$ and a mixing ratio of $0.0026 = 0.26\%$. These two examples illustrate that

- water is a quantitatively important component (ranked third or fourth according to mixing ratio) of the atmosphere;
- the atmospheric concentration of water in time and space is highly variable;
- the mixing ratio of water depends on both temperature and the extent of non-equilibrium as expressed by the relative humidity.

8.2 Energy balance

The ultimate source of most energy available on Earth is the Sun. The energy[1] that the Sun (or any other object, including the Earth) emits may be described in terms of *black-body radiation*. According to the Planck relation[2] for black-body radiation the emissive energy per unit wavelength of any material with a finite temperature is given by

$$M_\lambda = \frac{2\pi hc^2}{\lambda^5} \left(\frac{1}{e^{(hc/kT\lambda)} - 1} \right) \tag{8.1}$$

where M_λ = the emissive energy $W\,m^{-3}$ (the units signify watts per square metre of surface per metre of wavelength); h = Planck's constant = $6.63 \times 10^{-34}\,J\,s$; c = velocity of light = $3.00 \times 10^8\,m\,s^{-1}$; λ = wavelength/m; k = Boltzmann's constant = $1.38 \times 10^{-23}\,J\,K^{-1}$; and T = temperature/K.

A plot of M_λ vs λ is shown for different temperatures in Fig. 8.2. The area under the curves between any two wavelengths equals the total power per area ($W\,m^{-2}$) emitted between these two wavelengths. The spectrum at 5000 K corresponds closely to the situation for the Sun (whose surface temperature is about 5800 K). As is evident, solar radiation is given off as a broad continuum throughout parts of the near ultraviolet, visible, and infrared regions. A good approximation of the wavelength (in m) of maximum emission (for objects with temperatures greater than 100 K) is given by

$$\lambda_{max} = \frac{2.88 \times 10^{-3}}{T} \tag{8.2}$$

At 5800 K, the λ_{max} is then predicted to be about 500 nm, in the green region of the visible spectrum.

For reasons we have discussed in earlier chapters, much of the solar radiation is absorbed in the upper atmosphere. The total energy reaching the part of space occupied by the Earth (that is, the energy that would be received by a satellite orbiting above the Earth's

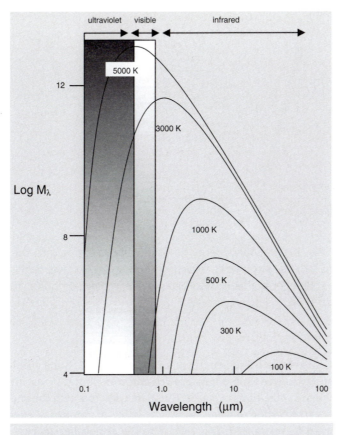

ultraviolet visible infrared

5000 K

3000 K

1000 K

500 K

300 K

100 K

Log M_λ

12

8

4

0.1 1.0 10 100

Wavelength (μm)

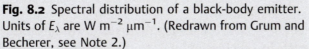

Fig. 8.2 Spectral distribution of a black-body emitter. Units of E_λ are W m^{-2} μm^{-1}. (Redrawn from Grum and Becherer, see Note 2.)

atmosphere) is called the solar flux, F_s, and averages 1368 W m^{-2}. However, only a portion of this energy can actually be absorbed at the Earth's surface. Figure 8.3 shows diagrammatically what happens to this solar energy as it flows into the global system. The numerical values are normalized to a value of 100 units corresponding to the total flux of incoming solar radiation.

Of the total solar flux, a portion is reflected back into space and does not contribute to the energy budget of the Earth. The reflected portion includes 6 units from the Earth's surface, 17 units due to clouds, and 8 units by aerosols including dust, salt from sea spray, smoke from fires, and volcanic ash. Thus the total reflectivity of the Earth is 31 units. This is commonly referred to in percentage terms as the Earth's albedo being 31% (or $A = 0.31$). Of the non-reflected 69 units of solar energy, 4 units are absorbed and thus retained within the Earth's atmosphere by water droplets in the clouds, and 19 units are absorbed by other aerosol particles and gaseous species such as ozone. Thus 23% of the solar energy reaching the Earth is absorbed in the atmosphere and only 46% is actually absorbed by land or water.

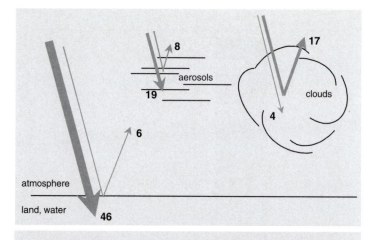

Fig. 8.3 Relative energy flows of solar radiation into the Earth's environment (based on 100 units). Arrows show solar energy that is absorbed or reflected by components of the Earth's environment.

Once absorbed, the relatively high energy, short wavelength (UV and visible) solar radiation is used as an energy source for biomass growth and contributes to warming of the Earth's surface, and to phase transitions of water. These are ways in which the energy is stored, but the situation is a steady-state one and, sooner or later, an amount of energy equivalent to that received must be re-emitted. This release to the atmosphere occurs as lower energy, longer wavelength (IR) radiation.

If the re-emitted energy were completely lost in space, we could predict the temperature of the Earth's surface in the following way. At any given time, solar radiation strikes half of the globe, but not directly—not at right angles to the entire surface (Fig. 8.4). The average value of the solar flux, F_s, over space and time is, as previously stated, 1368 W m^{-2}.

The integrated value of F_s at all angles over the spherical Earth is equivalent to the full flux continuously and directly striking an area normal to the solar beam $= \pi r^2$ where r is the Earth's radius. The total solar energy reaching the Earth is then given by $F_s \pi r^2$. The total amount of solar energy absorbed by the Earth is equal to this integrated flux minus the portion that is reflected back into space:

$$E_s = F_s(1 - A)\pi r^2 \qquad (8.3)$$

where E_s = total solar energy absorbed by the Earth/W; and r = radius of the Earth/m.

This absorbed energy can be compared with the average energy emitted from the Earth. To estimate emitted energy, we make use of Wien's law (eqn 8.4), which gives the total energy radiated from any object, a black body, at all wavelengths from one square metre of the body's surface. Equation 8.4 is obtained by integration of eqn 8.1:

$$F = \sigma T^4 \qquad (8.4)$$

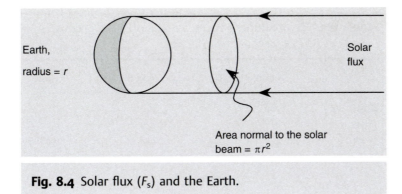

Fig. 8.4 Solar flux (F_s) and the Earth.

In referring to radiative emission from the Earth we will write eqn 8.4 as

$$F_e = \sigma T_e^4 \tag{8.5}$$

where F_e = radiant flux from the Earth/W m^{-2}; σ = the Stefan–Boltzmann constant:

$$\sigma = \frac{2\pi^5 k^4}{15 h^3 c^2}$$
$$= 5.67 \times 10^{-8}/\text{W m}^{-2}\text{K}^{-4}$$

and T_e = effective temperature of the Earth/K. The total radiative energy, E_e, emitted from the entire area of the Earth's surface (area = $4\pi r^2$) is then

$$E_e = 4\pi r^2 \sigma T_e^4 \tag{8.6}$$

where E_e = total energy emitted by the Earth/W.

Over any substantial time span, in the steady state, total energy absorbed from the sun (E_s) = total energy emitted by the Earth (E_e). Therefore

$$F_s(1 - A)\pi r^2 = 4\pi r^2 \sigma T_e^4 \tag{8.7}$$

Knowing the values of F_s, A, and σ, the temperature of the Earth is then predicted by

$$T_e = \left(\frac{(1 - A)F_s}{4\sigma} \right)^{\frac{1}{4}} \tag{8.8}$$

$$T_e = \left(\frac{(1 - 0.31) \times 1368 \, \text{W m}^{-2}}{4 \times 5.67 \times 10^{-8} \, \text{W m}^{-2}\text{K}^{-4}} \right)^{\frac{1}{4}}$$

$$= 254 \, \text{K}$$

$$= -19 \, ^\circ\text{C}$$

The calculation therefore leads us to predict an average global surface temperature of $-19\,^\circ$C. In fact the measured average surface temperature is known to be $+17\,^\circ$C, 36 $^\circ$C higher than the calculated value (the atmosphere just above the Earth's surface has an average temperature about 3 $^\circ$C less than this value). It is instructive to note the calculated and actual temperatures for other planets too (Table 8.2).

Table 8.2 Temperatures at the surface of three planets; $\Delta = T_{actual} - T_{calculated}$

	Calculated T/K	Actual T/K	Δ/K
Earth	254	290	+36
Mars	217	223	+6
Venus	227	732	+505

Why are temperatures on the planets greater than those predicted in this simple way? In each of the cases, the positive value of Δ ($= T_{actual} - T_{calculated}$) can be attributed to a 'greenhouse effect'. Recall that our simple calculation began with an assumption that all of the IR radiation left the Earth and went into space. In fact, a portion of the radiation re-emitted from the planet's surface is not immediately radiated into space. Rather it is absorbed by gases in the planetary atmosphere and in this way it contributes to enhanced warming. Life as we know it on Earth is compatible with the climate that results from these conditions.

On both Venus and Mars the principal gas responsible for greenhouse warming is carbon dioxide but the amount of this gas in the atmospheres of these two planets is greatly different. On Mars, the total pressure of carbon dioxide is only about 0.6 kPa; therefore the greenhouse effect is small, about 6 °C. In contrast, the Venusian atmosphere contains about 95% carbon dioxide and its partial pressure is greater than 9000 kPa, generating an enormous greenhouse warming effect of about 505 °C.

Coming back to the Earth, as we move upwards in the troposphere, we reach an altitude above which emitted IR radiation is absorbed only to a small extent and the radiation does go directly into space. This altitude is about 5.5 km, where the temperature is actually near the predicted value of 254 K. Observed from space, the Earth behaves as a black body having this temperature.

From the Earth's surface then, infrared radiation in the 'thermal IR' region from about 3 to 40 µm is given off, but not all of it leaves the region of the Earth. Figure 8.5 shows three emission spectra—those for black-body radiators at 240 K (a) and 280 K (b) and the actual emission spectrum from the Earth (c).

In the figure, we see that the actual IR emission spectrum from the Earth appears to track close to the 280 K spectrum with the predicted (eqn 8.2) maximum emission at 13 µm, near that actually observed. A major difference in the Earth's emission spectrum compared with the theoretical black-body spectrum is that the Earth's radiation is reduced in certain regions, especially those around 15 ± 2 µm, 9.5 ± 0.5 µm, and < 8 µm. This indicates that species capable of absorbing and attenuating radiation in these wavelength ranges must be present in the Earth's atmosphere. Shortly we will examine what these species are.

The greenhouse effect is a measure of the extent to which this absorption retains heat at the Earth's surface. The magnitude of this greenhouse effect can be estimated quantitatively, and a value in good agreement with the observed value is obtained, but the calculations are difficult and involve several simplifying assumptions.

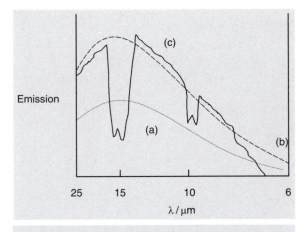

Fig. 8.5 Infrared emission spectra for black-body radiators at (a) 240 K, (b) 280 K, and (c) the Earth. (Redrawn from Wayne, R. P., *Chemistry of Atmospheres*, Clarendon Press, Oxford; 1991.

8.3 The greenhouse gases and aerosols

8.3.1 Water

For the Earth, water vapour is actually the most important of all greenhouse gases, and it absorbs IR in the ranges 2.5 to 3.5 μm, 5–7 μm, as well as over a broad range above 13 μm (Fig. 8.6). While the mixing ratio of water vapour is highly variable in time and space, the global average relative humidity is constant at about 1% and there are no anthropogenic activities that directly cause it to increase. Nevertheless gaseous water is involved in feedback processes: positive in that increased global warming means increased evaporation from ocean and land surfaces leading to higher ratios of atmospheric water—therefore enhanced warming; negative in that the troposphere becomes more cloudy leading to increased reflection and absorption in the atmosphere so that the solar flux at the solid/liquid surface of the Earth is reduced. At present, greenhouse warming associated with water vapour is estimated to be about 110 W m^{-2}.

8.3.2 Carbon dioxide

It is well established that carbon dioxide is a major contributor to greenhouse warming. It absorbs in the 14–19 μm range, and completely blocks the radiative flux between 15 and 16 μm; it also absorbs at wavelengths between 4 and 4.5 μm (Fig. 8.7).

There are many natural sources of carbon dioxide including animal and plant respiration and decay, combustion through forest and grassland fires, and volcanic activity. Regions in the oceans too are important sources; in the mid latitudes of the Pacific, upwelling of

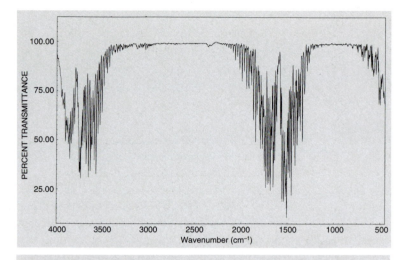

Fig. 8.6 Infrared absorption spectrum of water. The relation between wave number and wavelength is: wave number $(cm^{-1}) = 10\,000/$wavelength (μm) (© BIO-RAD Laboratories, Sadtler Division, 2000).

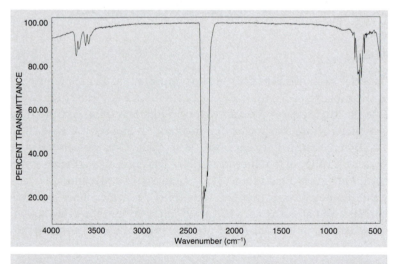

Fig. 8.7 Infrared absorption spectrum of carbon dioxide. The relation between wave number and wavelength is: wave number $(cm^{-1}) = 10\,000/$wavelength (μm) (© BIO-RAD Laboratories, Sadtler Division, 2000).

carbon-dioxide-rich waters causes release of large quantities of the gas. There are also natural sinks, including photosynthesis on land and by biological and abiotic processes in water. An 'ultimate' sink is the precipitation of calcite (limestone, $CaCO_3$) from oceans and lakes to form part of the sedimentary material.

Human activities have a significant effect on the global carbon cycle. The anthropogenic sources of atmospheric carbon dioxide include carbon release *via* combustion of fossil fuels and forest destruction and burning. Besides releasing carbon dioxide into the atmosphere, burning trees eliminates their contribution to carbon dioxide removal by the photosynthesis reactions:[3]

$$CO_2 + H_2O \rightarrow \{CH_2O\} + O_2 \tag{8.9}$$

It is estimated that about 7 Gt (as C) of additional carbon dioxide are released to the atmosphere each year. Three quarters are from fossil fuel combustion and the rest from changing land use in the tropics. About 2 Gt (as C) of this are assimilated by dissolution in oceans and the same amount is assimilated by increased plant growth rates. The remaining 3 Gt stay in the atmosphere. While the complex relations between sources and sinks are only partially understood, the net consequence of all the processes involving carbon dioxide is a steady annual increase of about 1.5 ppmv (about 0.5% of the 1995 concentration of 360 ppmv) in the atmosphere. The present greenhouse warming effect due to carbon dioxide is estimated to be approximately 50 W m^{-2}.

Water and carbon dioxide are the two most important greenhouse gases. Together, they absorb much of the radiation in the thermal infrared region below 7.5 μm and above 13 μm. Between these two wavelengths, very little IR radiation is absorbed by these gases, and this part of the spectrum can be considered to be a 'window, transparent to the radiation of interest. Other gases that absorb in that region, however, partially close the window and can have a major effect on heat retention in the region of the Earth.

8.3.3 Methane

The present (1995) tropospheric concentration of methane is 1.77 ppmv and it is increasing at a rate of about 0.5% per year. The tropospheric lifetime is about 7 years. Methane absorbs radiation in the wavelength ranges from 3 to 4 μm and 7 to 8.5 μm (in the window region) (Fig. 8.8).

Methane is produced where organic matter is subject to an oxygen-depleted highly reducing aqueous or terrestrial environment (Chapter 15). For example it is released from wetlands, including both natural swamps and cultivated paddy (rice) fields. The amount released is positively correlated with temperature, and is related to vegetation and soil type. Methane is also produced during extraction and inefficient combustion of fossil fuels. A third major source is from the digestive tracts of ruminants (cattle, sheep, goats) and termites. Claims have sometimes been made that methane release occurs mostly in low-income countries in the tropics—where the ruminant and termite populations and rice production are high. However, recent estimates[4] show that these sources together produce only about 30 to 40% of released methane. Other sources are not centred in the tropics and some, such as landfills and emissions associated with fossil fuels, are actually at a maximum in highly industrialized societies. The principal sink for methane decomposition is oxidation *via* hydroxyl radicals in the troposphere, as discussed earlier (p. 75 ff).

$$CH_4 + \cdot OH \rightarrow \cdot CH_3 + H_2O \tag{8.10}$$
$$\hookrightarrow \text{other reactions}$$

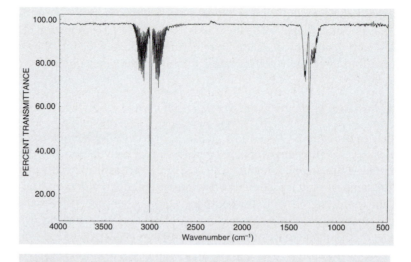

Fig. 8.8 Infrared absorption spectrum of methane. The relation between wave number and wavelength is: wave number $(cm^{-1}) = 10\,000$/wavelength (μm) (© BIO-RAD Laboratories, Sadtler Division, 2000).

Smaller amounts of methane are lost when taken up by soils and by leakage into the stratosphere.

The extent to which oxidation occurs depends on the availability of the hydroxyl radical and this is determined to a large extent by the availability of carbon monoxide which also reacts readily with hydroxyl radicals. Emissions of carbon monoxide are high in industrialized countries; in this way additional atmospheric methane build-up is indirectly caused by excessive use of fossil fuels with consequent increased carbon monoxide release.

Of considerable concern is a possible positive feedback process by which additional enormous quantities of methane could be released to the atmosphere. It is known that reserves of clathrate hydrates such as $CH_4 \cdot 6H_2O$ are trapped in ice crystals in Arctic sediments and soils below the permafrost. The quantities could amount to 10^{17} kg or even more of carbon—much more than the present carbon content of the atmosphere ($\sim 10^{15}$ kg). A somewhat warmer climate could favour the release of a significant portion of these hydrocarbons, greatly multiplying the warming trend.

At present, methane contributes $1.7\,W\,m^{-2}$ to greenhouse warming.

8.3.4 Ozone

Ozone absorbs at between 9 and 10 μm in the IR region and it therefore acts as a highly efficient greenhouse gas. In the troposphere, ozone concentrations are highly variable, in part because it has a short lifetime. Greater production of NO_x by fossil fuel burning and forest and grassland fires has resulted in net lower troposphere ozone concentrations increasing by about 1.6% per year in the northern hemisphere. The decrease in concentration of ozone in the stratosphere, allowing more UV radiation to reach the Earth, also contributes to warming. The overall contribution of ozone is approximately $1.3\,W\,m^{-2}$.

8.3.5 Nitrous oxide

Nitrous oxide absorbs IR radiation in the ranges 3 to 5 μm and 7.5 to 9 μm (again in the infrared window) (Fig. 8.9). The present (1995) concentration is approximately 312 ppbv and it is increasing at a rate of 0.3% per year.

Some nitrous oxide is released from industrial processes such as the production of adipic acid and nitric acid. The major sources, however, are from microbial denitrification, conversion of nitrate to nitrous oxide, in soils, lakes, and oceans. The microbiological reactions describing how nitrous oxide is generated will be discussed in Chapter 15. There is an anthropogenic component to this natural process in that increased application of nitrogenous fertilizer, including animal manure, augments the supply of the nitrate substrate required for denitrification. The amount of nitrous oxide released from soils is much greater where temperature and soil moisture levels are high and where oxygen has been depleted. Additional emissions are produced from urban waste landfill sites, and where there is direct sewage disposal into large water bodies. The influx of this and other types of organic matter leads to emissions from the oceans especially in coastal regions and estuaries. There are no important tropospheric sinks for this gas, so it is lost only by slow leakage into the stratosphere where it undergoes photolytic degradation as described in Chapter 3; it therefore has a substantial tropospheric residence time estimated to be about 120 years. It has approximately the same effect on greenhouse warming as does ozone.

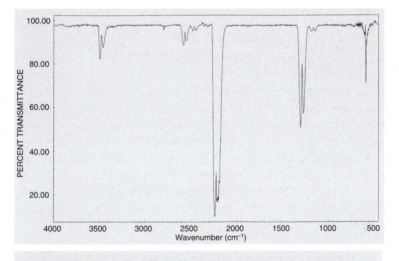

Fig. 8.9 Infrared absorption spectrum of nitrous oxide. The relation between wave number and wavelength is: wave number (cm^{-1}) = 10 000/wavelength (μm) (© BIO-RAD Laboratories, Sadtler Division, 2000).

8.3.6 Chlorinated fluorocarbons (CFCs) and other halogenated gases

In addition to their role as agents for the catalytic decomposition of stratospheric ozone, CFCs are also important greenhouse gases. They absorb in the range 8 to 12 μm with each

CFC having specific absorption bands in this region. For example CFC-11 absorbs at 9.5 and 11.5 μm, CFC-12 at 9.5 and 11.0 μm (see Fig. A.1 in Appendix 2). The recently developed HCFCs also attenuate radiation within the same range, but their residence time in the troposphere is substantially shorter than that of the CFCs. Overall, the present concentration of all CFCs and HCFCs taken together is between 1 and 2 ppbv and, until the time of the Montreal Protocol, had been growing at a rate of nearly 5% per year. The rate of increase of the CFCs has declined by a factor greater than two in the past decade, but HCFC concentrations are rising at a much higher rate.

Three fully fluorinated gases of industrial origin have recently become recognized as potential contributors to greenhouse warming. They are present in trace amounts, but have lifetimes of thousands or tens of thousands of years. Tetrafluoromethane (CF_4) and hexafluoroethane (C_2F_6) both arise during electrolysis of alumina (Al_2O_3) in cryolite (Na_3AlF_6) at carbon electrodes, and release of the gases is estimated to be at about 0.77 and 0.1 kg respectively per tonne of aluminium produced. Together, their atmospheric concentration is approximately 0.08 ppbv. The other gas is sulfur hexafluoride (SF_6), which is formed during magnesium production.

8.3.7 Aerosols

Clouds are the most important atmospheric aerosol in terms of reflecting and absorbing incoming radiation and emitted radiation from the Earth. The cooling effects of clouds on warm days, and their warming effects on cool nights, are phenomena we have all recognized. Other aerosols, too, add to the complexities of the global energy balance situation. In particular, ammonium sulfate and other sulfate-based solid aerosols are becoming increasingly important. The sulfate aerosol derives from natural oceanic sulfide, particularly dimethyl sulfide, emissions as well as from anthropogenic sources of sulfur dioxide. In the northern hemisphere, about 90% results from human activities, while in the south, most has a natural origin. Besides its direct role in backscattering incoming short-wave solar radiation, the presence of sulfate in the aerosol also affects processes of cloud formation. The net result of the direct and indirect processes is complex and varies from region to region but, overall, aerosols contribute to a measure of atmospheric cooling. As noted earlier, sulfate particulates that are periodically injected by volcanoes into the stratosphere also contribute to a cooling of the troposphere.

Biomass aerosols derive from combustion, with the release of fine smoke and soot to the atmosphere. Their extent varies around the globe and from year to year. Recent years, especially 1997–8, have seen increased incidents of widespread biomass burning in countries like Malaysia and Indonesia, as well as in parts of North America. The higher average global temperature in those years may have contributed to drying of forested areas and greater opportunities for fires to ignite and spread. Aerosols of industrial origin are also largely combustion based, and are usually found in the lower (< 2 km) parts of the troposphere. Because they are readily washed out with precipitation, they have small atmospheric residence times, of the order of a few days, so their contribution to greenhouse warming is local and short lived.

8.4 Relative importance of the changes in greenhouse gas concentrations

At the concentrations that are now found in the atmosphere, all of the above species attenuate infrared radiation. In the atmospheric chemistry literature, this absorption of IR radiation is referred to as *radiative forcing*. At least three factors are involved in determining the relative importance of increases in atmospheric concentrations of a gas with respect to radiative forcing.

- The present atmospheric concentration of the species; increasing the concentration of trace components of the atmosphere can have a large effect on radiative forcing. If large concentrations are already present so that essentially complete absorption of particular wavelengths of IR radiation already occur, an additional amount of that species can only increase radiative absorption on the wings of the absorption lines.
- The wavelengths at which the gas molecules absorb; if a species absorbs only in regions of the IR spectrum where absorption is nearly complete due to other species, then an increase in concentration will have only a minimal effect. On the other hand, species that absorb in the 'window' region, have a much greater potential effect on warming.
- The strength of absorption per molecule; a small additional concentration of a strongly absorbing species can have a very large effect.

The three factors relate to immediate effects, resulting from an incremental increase in the amount of the gas being present in the atmosphere. To estimate the long-term effects, the residence time of the gas must be considered as an additional factor.

In order to quantify the relative contribution of the various greenhouse gases to radiative forcing, it has been a goal to take into account the various factors and incorporate them into a single index. In fact, two such indices have been described.

The first index is the relative instantaneous radiative forcing (RIRF). As the name suggests, the RIRF value is a measure of the ability of an incremental addition of a gas, in the present atmosphere, to increase the absorption of infrared radiation. The RIRF value for carbon dioxide is arbitrarily set at 1, and the RIRF for other gases are relative to this value. For instance, the greater than four orders of magnitude ratio of RIRF for CFCs compared with carbon dioxide is due to the ability of CFCs to absorb strongly in the open window portion of the carbon dioxide and water spectra. This explains why an incremental addition of CFCs increases IR absorption to a very much larger extent than does the same mass increment of carbon dioxide.

A second index takes a longer view of the potential for contributing to greenhouse warming. To do this, Lashof and Ahuja[5] have defined an index of global warming potential (GWP) applicable to any gas:

$$\text{GWP} = \frac{\int_0^t a_i(t) C_i(t) \, \mathrm{d}t}{\int_0^t a_C(t) C_C(t) \, \mathrm{d}t} \tag{8.11}$$

where $a_i(t)$ is the instantaneous radiative forcing due to a unit increase in the concentration of the gas i. The ratio of a_i/a_0 is equivalent to the RIRF value; $c_i(t)$ is the fraction of gas i remaining at time t; 0 to t represents the time span over which integration is carried out. The corresponding values with subscript C are for carbon dioxide and are in the denominator.

Defined in this way, the GWP describes the long-term contribution of any gas by comparison to that of carbon dioxide. Any period can be selected for the time over which integration is done. For carbon dioxide itself, both numerator and denominator terms are the same and so the GWP has a value of unity. Other greenhouse gases have GWP values ranging up to more than a thousand (Table 8.3). From the table, for example, we see that addition of a particular amount of nitrous oxide to the atmosphere would contribute 310 times more to greenhouse warming over a 100 year period than would addition of the same amount of carbon dioxide.

In keeping with the principles outlined above, a_i depends on the concentration of the gas (e.g. raising the concentration of carbon dioxide from 360 to 365 ppmv will have a greater effect than raising it from 450 to 455 ppmv). The value of a_i also depends on the absorptivity of the molecule and the wavelength(s) at which absorbance occurs. In particular, gases that absorb at wavelengths within the window region—covering a large portion of the wavelength range from 8 to 12 μm—make an especially significant contribution to additional radiative forcing.

The value of c_i depends on the residence time of the gas and can be calculated if the rate of all processes leading to loss or destruction of the gas are known. Because the GWP includes an accounting of the residence time, gases that rapidly dissipate from the atmosphere have a smaller long-term effect than do those with very long lifetimes. For example CFC-11 and

---

Table 8.3 Relative contributions to global warming

Gas	Residence time/y[a]	Relative instantaneous radiative forcing	Global warming potential[c]
CO_2	50–200[b]	1	1
CH_4	12	43	21[d]
N_2O	120	250	310[d]
CFC-11	60	15000	3400[e]
CFC-12	195	19000	7100[e]
HCFC-22	12	13000	1600[e]

[a] Most of the atmospheric lifetime values are taken from ref. 1.
[b] Reported residence time values for carbon dioxide are highly variable. Differences are associated with the way in which oceanic uptake is measured, particularly whether the surface layer or the entire ocean is considered in the calculation.
[c] GWP values are obtained by integration over a 100 year period.
[d] GWP values from CO_2/Climate Report, Issue 97-1, Environment Canada, Spring 1997.
[e] GWP values from CO_2/Climate Report, Special Issue, Environment Canada, 1993. These values relate to direct effects; interactions of CFCs with ozone in the lower stratosphere may reduce the amount of radiation into the lower atmosphere, contributing to a cooling effect. The GWP values would be correspondingly reduced.

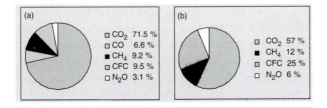

Fig. 8.10 Relative contributions of gases to global warming potential (a) in 1985, and (b) throughout the 1980s. (Reference in Note 5. Reprinted with permission.)

HCFC-22 have similar RIRF values, but the GWP for CFC-11 is more than two times greater than that of HCFC-22 as a consequence of the former compound's longer residence time.

The left-hand sector graph in Fig. 8.10 shows the relative contributions toward the GWP for gases in the atmosphere in 1985, while on the right side is a similar plot showing relative contributions toward greenhouse forcing added during the 1980s. The long-term contribution of persistent compounds like CFCs is one obvious feature of these charts.

While the physics and chemistry of the greenhouse effect are relatively well understood, there is great difficulty in predicting what climatic effects will result from increasing concentrations of the greenhouse gases. That the concentrations will continue to increase is beyond dispute. The difficulties in prediction are associated with the complex interactions between the various environmental processes and the resultant positive and negative feedbacks that result. For example, what happens to the additional carbon dioxide that each year is released into the atmosphere? We have indicated that there are several well known sinks for this gas—the oceans being a major reservoir. But quantitative estimates describing carbon dioxide partitioning in water, the circulation patterns of the oceans, the geochemical and biological processes that take up and release various carbonate species, all require much more detailed knowledge than we presently have available. Adding to this are the equally complex terrestrial processes—both macro- and microbiological—and the interactions between them. A great deal of research effort is presently being devoted to modeling the fate of greenhouse gases. Further discussion of these important issues is beyond the scope of this book.

8.5 Energy resources

The greenhouse gases arise from many sources, but perhaps more than anything else the increasing release of several of the species into the atmosphere is associated with energy production and use. The life of all living species including humans depends on a supply of energy. Energy is involved in providing such basic necessities as the growing and preparing of food, and providing warmth. It also fuels much of modern industry and is an aspect of all forms of transportation.

In considering sources and types of energy, certain ones such as fossil fuels and electricity readily come to mind. But there are many other forms as well, ranging from human and animal energy to energy obtained from the Sun and water, or from nuclear fission. Whatever sources are used, the processes of extracting or manufacturing the energy and then using it all have an effect on the atmospheric environment. Some of the effects relate to smog, acid precipitation precursors, and aerosol production. Here we will focus on effects connected with release of greenhouse gases. One way of classifying the forms of energy is shown in Table 8.4.

The fundamental unit of energy is the joule ($J = N m = kg m^2 s^{-2}$) but it has become customary to use units like barrels of oil, tonnes of coal, or kilowatt hours of electricity that relate to the particular energy form under consideration. Table 8.5 lists equivalents in joules for a variety of frequently encountered units. In this book, the joule is used as a common unit.

Some forms of energy such as fossil fuels are obtained as mineable deposits and these are non-renewable, but others, including the diverse forms of biomass, may be replenished and are available over an essentially limitless period. There are, however, costs (including energy costs) required to sustain the renewability of such resources. For example, to maintain the productivity of a tree farm as an energy plantation, cultivation and fertilizers are needed. Both of these inputs have a significant energy component. We will look at these and other questions related to renewable resources in later sections.

Table 8.4 Sources of usable energy	
Primary sources	
Solar energy	Used directly or converted into electricity *via* photoelectric cells. Also is the driving force for the water cycle, the ultimate energy source creating fossil fuels, and (through differential heating) causing wind and wave action
Lunar energy	The cause of tides which may be converted to electricity
Geo energy, nuclear energy, geothermal energy	Includes nuclear and geothermal energy. May be used as a source of heat which, in turn, is converted to mechanical or electrical energy
Derived sources	
Fossil fuels	Includes coal, petroleum and natural gas from various sources. These are primary combustion sources used as fuel for engines or to generate heat which is often converted to electricity
Biomass	Includes wood, straw, animal dung, sugar cane, corn, waste paper products, etc. Used as fuels, or converted to other fuels or to electricity
Hydro energy, wind energy, wave energy	Through the Sun's heating action on land and water these forms of energy are developed and the power can be used directly, but is most often converted to electricity
Tidal energy	May be used to generate electricity
Electricity	Always a derived form of energy based on primary sources (solar, nuclear) or on other derived sources (fossils fuels, hydro power, etc.)

Table 8.5 Units of energy and equivalents in joules

Energy source	Unit	Abbreviation	Equivalent in joules
Natural gas	Cubic metre	m^3	3.7×10^7
	Cubic foot	ft^3	1×10^6
Petroleum	Barrel	bbl	5.8×10^9
	Tonne	t	3.9×10^{10}
Tar sand oil	Barrel	bbl	6.1×10^9
Shale oil	Tonne	t	4.1×10^{10}
Coal			
Anthracite	Tonne	t or TCE	3.0×10^{10}
Bituminous	Tonne	t or TCE	3.0×10^{10}
Sub-bituminous	Tonne	t or TCE	2.0×10^{10}
Lignite	Tonne	t or TCE	1.5×10^{10}
Charcoal	Tonne	t or TCE	2.8×10^{10}
Biomass (all on a dry weight basis)			
General	Tonne	t	1.5×10^{10}
Miscellaneous farm wastes	Tonne	t	1.4×10^{10}
Animal dung	Tonne	t	1.7×10^{10}
Assorted garbage	Tonne	t	1.2×10^{10}
Wood	Tonne	t	1.5×10^{10}
	Cubic metre	m^3	5×10^9
	Cord	$128\ ft^3$	2×10^{10}
Fission			
Natural	Tonne	t	8×10^{16}
Complete mass $\rightarrow$ energy conversion, $E = mc$	Tonne	t	9×10^{19}
Electricity	Kilowatt hour	kwh	3.6×10^6
	Terawatt year	Twy	3.2×10^{19}
General units	Erg	erg	1×10^{-7}
	Calorie	cal	4.18
	British thermal unit	BTU	1.05×10^3
	(10 BTU)	therm	1.05×10^8
	(10 BTU)	quad	1.05×10^{16}
	(10 BTU)	Q	1.05×10^{21}
	Horsepower hour	hp h	2.7×10^6

The data for individual commodities are obtained from several sources and many are estimates. Values for particular materials vary, especially for highly heterogeneous substances and substances with variable moisture content, such as different forms of biomass.

Table 8.6 Annual commercial energy consumption in the regions of the world

Region	Energy consumption		Population (millions)	Annual commercial energy consumption per capita
	EJ y^{-1}	% of total		GJ pc y^{-1}
Africa	9.8	3	731	13
Asia Pacific	89	27	3300	27
Mid-East	13	4	160	81
USSR (f)	36	11	293	122
Europe	70	21	507	138
Latin America	13	4	488	27
North America	96	29	295	325
World	327	100	4500	67

1996 energy consumption data from *BP Statistical Review of World Energy*, 1997. The British Petroleum Company p.l.c.; 1997.
USSR (f) refers to the countries that made up the former Soviet Union.

The total amount of energy used annually in the world is about 330 EJ (1 EJ $= 10^{18}$ J) but the patterns of commercial energy consumption are highly variable as shown in Table 8.6. Commercial energy is made up of forms that are purchased through regular channels, and excludes energy sources collected and used by individuals.

Many factors contribute to the great variability in energy consumption per capita. Some relate to the size of the country and its effect on transportation needs, others to climate, especially cold climates where heating of buildings is a necessity. Still others relate to the degree of industrialization and the nature of industry. Perhaps most important is the population's expectation with respect to all these issues. It appears certain, however, that energy consumption will increase substantially in coming years, with much of the increase occurring in regions—Asia, Africa, South America—where energy consumption is much lower than the global average. The increase could mean a doubling of world energy consumption over a 40 year time span beginning in 1975.

With this background, we can now look at some individual energy sources and the extent to which they contribute to global release of greenhouse gases. Earlier and later in the book we deal with some of the other environmental impacts associated with energy extraction and use.

8.6 Greenhouse gases associated with use of carbon-based fuels

8.6.1 Coal

The term 'coal' covers a range of sedimentary materials derived from the residues of plant materials that have been buried and subjected to high temperatures and pressures over extended periods of geological time. The degradation processes, which may initially take place under near-surface but anaerobic conditions, first lead to the formation of humic-like material (see Chapter 12) called peat. Further degradation under conditions of high heat and pressure is called coalification and produces coal (Fig. 8.11) with a spectrum of properties depending on the time and intensity of the reactions. Table 8.7 lists some chemical properties of representative American examples of four ranks of coal. The degree of coalification is lowest for lignite and increases from left to right in the table to anthracite.

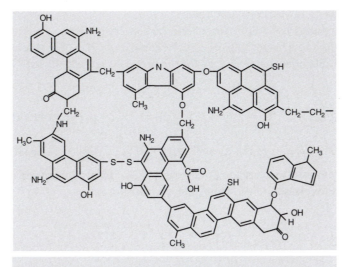

Fig. 8.11 A hypothetical structure of coal. Note the similarities and differences between this structure and that of a hypothetical humate molecule as shown in Fig. 12.3.

Table 8.7 Chemical properties of representative coal types

Coal rank	Lignite	Sub-bituminous	Bituminous	Anthracite
Location	McLean, North Dakota	Sheridan, Wyoming	Muhlenberg, Kentucky	Lackawanna, Pennsylvania
Moisture/%	37	22	9	4
Carbon/%	41	54	65	80
Ash/%	6	4	11	10
Sulfur/%	0.9	0.5	2.8	0.8

US Department of Energy, *Coal Data: Cost and Quality of Fuels*; 1979.
The ranking, from left to right, represents increasing coalification. Ash and sulfur concentrations are independent of the coalification processes.

Increasing coalification results in lower moisture and higher carbon percentages, which factors are reflected in a larger energy content per tonne of coal (Table 8.5). From this perspective, anthracite is most desirable but there is only a limited quantity of this type of coal and bituminous and sub-bituminous resources are most often exploited. The sulfur and ash contents vary depending on source, and are not directly related to rank of coal.

Of the atmospheric pollutants given off during coal combustion, a number have been subjected to stringent controls by various political jurisdictions. We have earlier indicated problems with acid precursors and with ash. As an intrinsic byproduct of carbon combustion, carbon dioxide is released and the release cannot be controlled in any practical sense. The same is true of any carbon-based fuel.

One can compare the combustion of the carbon component of coal (C) with that of natural gas (CH_4) and a heavy oil (C_xH_{2y}, represented as $C_{20}H_{42}$). The combustion reactions and the magnitude of enthalpy change for the reactions are as follows:

$$\text{Coal} \qquad\qquad C + O_2 \rightarrow CO_2 \qquad\qquad \Delta H = -393.5\,\text{kJ} \qquad\qquad (8.12)$$

$$\text{Natural gas} \qquad CH_4 + 2O_2 \rightarrow CO_2 + 2H_2O \qquad \Delta H = -890.3\,\text{kJ} \qquad\qquad (8.13)$$

$$\text{Heavy oil} \quad C_{20}H_{42} + 30\tfrac{1}{2}O_2 \rightarrow 20CO_2 + 21H_2O \quad \Delta H = -13\,315.2\,\text{kJ} \qquad (8.14)$$

During the combustion of carbon, for every GJ ($1\,\text{GJ} = 10^9\,\text{J}$) of heat produced the amount of carbon dioxide released to the atmosphere is therefore

$$\frac{10^9\,\text{J GJ}^{-1}}{393.5 \times 10^3\,\text{J}} \times 44 \times 10^{-3}\,\text{kg} = 112\,\text{kg GJ}^{-1}$$

Corresponding amounts for the two other fuels are $49\,\text{kg GJ}^{-1}$ for natural gas and $66\,\text{kg GJ}^{-1}$ for oil. Of the fossil fuels, then, coal makes the greatest greenhouse gas contribution in terms of release of carbon dioxide per unit of energy produced.

The newer types of combustion processes such as fluidized-bed combustion, as was described in Chapter 6, enhance efficiency of coal combustion and of heat transfer, and therefore serve to maximize energy output for a given amount of fuel.

Other traditional and emerging technologies, developed in order to produce a cleaner and more convenient fuel, convert the coal to gaseous or liquid products. A number of processes are possible as routes to coal gasification (Table 8.8). Several of these occur sequentially in commercial coal gasifiers, some of which have been in use for decades, especially where natural gas has not been available. The product mixtures generally have low heat content and widespread commercial applications await further technological developments.

Similarly, liquefaction of coal is used to produce a substitute for various types of petroleum products. The liquefaction can be done indirectly by first producing gaseous products and following this by a separate process to convert the gases to a liquid form. This procedure has been the basis of the SASOL method, developed and used in South Africa to produce more than 10 million litres per day of synthetic liquid petroleum. The process begins by heating the coal to temperatures between 600 and 800 °C to partially volatilize it, producing a mixture of methane, hydrogen, and carbon. The mixture then moves through a

Table 8.8	Reaction sequences used in coal gasification processs	
Process	**Reactions used**	
Partial gasification	$Coal \xrightarrow{500-700\,°C} C + CH_4 + H_2$	Produces a relatively small amount of high energy content gases
Carbon–oxide	$C + O_2 \rightarrow CO_2$ $2C + O_2 \rightarrow 2CO$	The carbon monoxide is a combustible product
Steam–carbon	$C + H_2O \xrightarrow{heat,air} CO + H_2$	The two product gases are both combustible, but are diluted by nitrogen, giving a low energy content fuel
Catalytic methanation	$3H_2O + CO \rightarrow CH_4 + H_2O$	Catalysts such as nickel oxide can enhance conversion of carbon monoxide into ethane, which has a higher energy content

heated region containing air and steam. The product gas mixture resulting from the complex reactions is a made up of about 10% methane, 20% carbon monoxide, 30% carbon dioxide, and 40% hydrogen. The mixture is treated by a process called the Fischer–Tropsch synthesis where the gas reacts under a pressure of 2×10^6 Pa in the presence of an iron catalyst, releasing heat and producing a liquid product. The hydrocarbon product contains many of the typical low and high molar mass species found in crude oil and may be refined into various commercial products.

Potentially more efficient are direct methods of liquefaction in which the coal is 'cracked' into large subunits that then undergo hydrogenation in a solution or slurry. Several direct approaches have been developed. *Solvent refined* coal is produced by mixing coal with a solvent (actually a product of the process itself), distilling off the solvent and then heating to 450 °C in the presence of hydrogen gas at a pressure of about 1×10^7 Pa. The products include both solid and liquid hydrocarbons. The use of a cobalt molybdenate catalyst can increase the yield of useful liquid products.

While the various conversion methodologies produce fuels that are more convenient to store, transport, and use, and are cleaner, the production processes themselves create negative environmental impacts. The ash and gaseous byproducts like hydrogen sulfide must be disposed of, and large amounts of energy are required to effect the conversion. Typically for every one unit of energy equivalent in the product fuel, 1.5 units of energy input are required. The energy is usually supplied in the form of coal itself so that the total quantity of some waste materials, like ash, are 2.5 times the amount that would result from the direct use of coal for producing the equivalent amount of heat.

8.6.2 Petroleum

Petroleum is the fuel of choice in many situations. Petroleum is relatively easily and safely pumped from oil-bearing reservoirs. By separating the fractions during refining it becomes the source of a range of useful products, all of which are readily transported and stored.

As a fuel, the products are fairly low in sulfur and are suitable for burning to produce heat in domestic or industrial units or for producing mechanical power in internal combustion engines.

Consumption patterns are highly tilted toward high-income nations—about two thirds are used by the wealthiest one quarter of the world's population.

Refined petroleum is essentially a mixture of hydrocarbons so that combustion takes a form such as that shown in eqn 8.15:

$$C_xH_{2y} + \left(x + \tfrac{1}{2}y\right)O_2 \rightarrow xCO_2 + yH_2O \tag{8.15}$$

At present, about 40% of commercial energy consumed throughout the world is produced from crude oil. The use of petroleum-based fuels is therefore a major source of carbon dioxide emissions even though it is more efficient than coal on the basis of heat generated. Furthermore, although oil contains a smaller concentration of sulfur than coal, it does contribute to sulfur dioxide release to the atmosphere. Some non-conventional precursors of oil products—particularly shale oil and tar sands—do have much higher concentrations of the element (the Athabasca tar sand product contains about 4% sulfur) and this must be removed during purification.

While carbon dioxide and sulfur dioxide emissions contribute to *global atmospheric pollution* problems, burning of oil products in the transportation sector is especially important in contributing to *urban atmospheric pollution*. We have already noted some problems associated with carbon monoxide release and with photochemical smog caused, in part, by nitrogen oxides and volatile organic carbon. All these compounds are present in emissions from internal combustion engines, although improved design has significantly lessened the quantities of these gases.

8.6.3 Natural gas

Natural gas is found either in association with or independent of oil deposits. There is a considerable range in the composition of gas at the various locations but most deposits contain the components listed in Table 8.9.

Table 8.9 The principal components of natural gas

Methane (75 to 100%)	Used as an industrial and domestic fuel
Ethane (6 to 10%)	Used as a fuel or as a feedstock for petrochemical plants manufacturing ethylene
Propane and butane (5 to 8%)	Liquefied petroleum gases (LPGs)—used as fuels or as petrochemical feedstocks
Pentane and heavier hydrocarbons (1 to 4%)	Condensate, used as petrochemical feedstock
Nitrogen, carbon dioxide, hydrogen sulfide, helium (variable)	Components other than hydrocarbons

In many ways, natural gas is the most desirable of the fossil fuels. On a mass basis it produces more heat and less carbon dioxide than either coal or petroleum products. In most cases, it has fewer undesirable impurities, such as sulfur, than other fuels and therefore burns more cleanly. Applications range from heat production for electric power generation, domestic and industrial space heating, fuel for transport vehicles, to providing a feedstock in the petrochemical and nitrogen fertilizer industry. Besides being (inevitably) a source of atmospheric carbon dioxide, leakage of natural gas at gas wells, along with losses during processing and transport, may be the source of as much as 20% of the world release of methane—another important greenhouse gas.

Another environmental concern arises from transport of the gaseous products by pipeline where leakage and subsequent explosions may result. The liquid petroleum gases are gases under normal conditions (for butane $T_b = -0.5\,°C$ and for propane $T_b = -42\,°C$ at $P°$) and are carried in pressurized and/or refrigerated containers. In 1979 a train derailment at Mississauga, Ontario exposed the community to a potentially dangerous mix of propane and other chemicals and it was necessary to evacuate more than 200 000 persons from their homes. Nevertheless, the environmental dangers associated with the use of natural gas are somewhat less acute than those for other fossil fuels.

8.6.4 Biomass

Biomass refers to organic material produced as the solid product when photosynthesis occurs in growing plants. As such, it is a form of derived energy with the solar flux being the primary source driving biomass production. We think of biomass as a renewable resource because the average energy of the Sun reaching Earth remains constant, and crops can be grown and harvested year after year. However, solar energy is not the only required input. Water is obviously essential and soil too plays a role as a source of nutrients. To sustain the ability to produce biomass, the chemical and physical integrity of these two resources must be maintained. Besides being renewable, there is another unique and important feature associated with biomass as a fuel source. In principle, it is possible to sequester approximately the same amount of carbon dioxide during the growth phase as that released by combustion.

There is a wide variety of forms of biomass for use in energy production. Wood from trees is the most common form but other crops such as sugar cane or maize (corn) are also grown for the specific purpose of producing a fuel. In some cases biomass energy sources are a byproduct of other agricultural operations. Straw and animal dung are two examples. In the latter case, the byproduct is a tertiary energy form being obtained from animals which feed on plants whose growth depends on the Sun. Although the byproducts are often referred to as waste materials, this is not strictly true as they can serve many other useful purposes—for example as animal feed, soil conditioners, or fertilizers.

The various forms of biomass are converted to energy in two different ways. One is to burn them directly as a fuel source. If the biomass is considered to be a complex form of carbohydrate, the combustion reaction is expressed in simplest form as

$$\{CH_2O\} + O_2 \rightarrow CO_2 + H_2O \qquad \Delta H = -440\,kJ \tag{8.16}$$

As you can see, reaction 8.16 is the reverse of the photosynthesis reaction (8.9). It is for this reason that we can think of biomass growth and use as a fuel as having little or no net impact on atmospheric carbon dioxide levels. Based on reaction 8.16, one tonne of dry biomass would release 1.5×10^{10} J of energy during combustion but actual values can be greater or (usually) less than this due to the fact that biomass is not exclusively carbohydrate. Included in the listing in Table 8.5 are several sources of biomass and their estimated energy conversion factors.

Biomass is used throughout the world as a source of energy for heating, cooking, and other purposes. In many countries it is the principal source but because its use may not be part of the commercial energy network, quantitative details about its production and consumption are difficult to obtain. Even tables of national energy use are usually incomplete with respect to information on biomass. In low-income countries, biomass has been estimated to contribute, on average, about 40% to energy consumption. The figure for higher income countries is 1% and the global average is 14%. This means that, at present throughout the world, about 45 EJ of energy are supplied annually by biomass.

For the future, biomass energy is likely to remain equally important in spite of the fact that obtaining and using it is less convenient than for many of the commercial energy forms. In both high- and low-income countries, there is considerable interest and research directed toward developing sustainable and productive systems of *energy plantations* as sources of domestic and industrial fuel. Clearly, there are environmental considerations with respect to the present situation and the future prospects.

The efficiency of conversion of solar energy into biomass is a simultaneous function of many factors. Of the total solar flux striking the Earth, on average only 46% is available to be absorbed at the Earth's surface. Of this, only about 43% can be used for photosynthesis by the green parts of growing plants. The 43% represents the portion of the solar spectrum between 400 and 700 nm that can be absorbed by chlorophyll in the chloroplasts. This is called photosynthetically active radiation (PAR) and it provides energy for the reaction involving carbon dioxide and water to produce carbohydrate and oxygen.

In some plants the conversion process takes place *via* a reductive pentose phosphate (RPP) cycle that produces, as its primary carboxylation product, a three-carbon acid. Such species are called C3 plants. The rest of the plant kingdom uses mechanisms in addition to the RPP cycle and incorporates carbon dioxide into a four-carbon acid. Naturally, these are termed C4 plants.

Amongst the C3 species are wheat, rice, soya beans, tomatoes, potatoes, and sugar beets. The C4 species include sorghum, maize, sugar cane, and desert grasses—all species that have the potential to produce large yields of biomass. In general C3 species are common in temperate regions while C4 types dominate in the tropics and subtropics, especially in more arid regions.

Differences between C3 and C4 plants are important in terms of photosynthetic and water-use efficiency. The C4 species have higher rates of net photosynthetic production and transpire about 500 mol of water for each mol of carbon dioxide incorporated. In this way they are efficient biomass producers. This is in contrast to C3 species from which 1000 mol or more of water are lost per mol of carbon dioxide fixed. The lower rate of net photosynthesis in C3 plants is due to the fact that, in a warm sunny environment, a part of the photosynthesized material is lost *via* its reoxidation to carbon dioxide. This reverse process does not occur in C4 plants.

Of interest in the present context is the question of whether photosynthetic rates increase due to increasing concentrations of carbon dioxide in the atmosphere. Because C4 plants efficiently carry out photosynthesis, they respond only slightly to carbon dioxide 'fertilization'. However, at least in the short term, C3 plants may grow much faster in an atmosphere containing more carbon dioxide. Laboratory growth experiments under an atmosphere with double the present carbon dioxide concentration have shown that photosynthesis increases substantially, leading to between 20 and 40% greater biomass production. More efficient photosynthesis is largely due to reduced rates of photorespiration. The increased growth rate is one of the sinks counteracting increasing releases of carbon dioxide to the atmosphere.

For any plant, the maximum theoretical efficiency obtainable in the photosynthesis process may be calculated with some precision. Eight quanta of PAR are required to fix one molecule of carbon dioxide. Choosing a mean PAR wavelength of 575 nm, the energy required to fix one mol of carbon dioxide is

$$E = \frac{n N_A h c}{\lambda}$$

$$= \frac{8 \times 6.02 \times 10^{23} \text{ mol}^{-1} \times 6.62 \times 10^{-34} \text{ J s} \times 3.00 \times 10^8 \text{ m s}^{-1}}{575 \times 10^{-9} \text{ m}}$$

$$= 1660 \text{ kJ mol}^{-1}$$

The $\Delta G°$ for the overall photosynthetic reaction is 477 kJ. Therefore photosynthesis could be 28% efficient in terms of PAR or $0.43 \times 28\% = 12\%$ efficient in terms of the portion of the total solar spectrum that is absorbed by the Earth's surface.

While this is a theoretically achievable efficiency, there are practical reasons that lead to much lower rates of productivity than are estimated using the 12% figure. Respiration (chemically, the reverse of the photosynthesis reaction) reduces productivity by 20 to 80% or more. Microbial decomposition of synthesized material is another source of loss. As a result, the seasonal maximum growth rates measured for C4 plants are in the range of 22 g m^{-2} d^{-1} (equivalent to 3.8 GJ ha^{-1} d^{-1}) and for C3 species 13 g m^{-2} d^{-1} (equivalent to 2.2 GJ ha^{-1} d^{-1}). A reasonable average total quantity of energy absorbed at the Earth's surface is 160 GJ ha^{-1} d^{-1}. Therefore, the actual maximum conversion efficiency is about 1.4% and 2.4% for C3 and C4 plants respectively. Even these percentages are high compared with what is attainable in practice using good agronomic practices. Actual averages for well managed production of maize grain plus stalk is 0.6% and for wheat grain plus straw is 0.3%. A large-scale global average efficiency might be close to 0.25%.

It is interesting to calculate how much energy could actually be produced annually in a given area of land. India has a land mass covering about 3 300 000 km^2. Of this, approximately 2 200 000 km^2 are arable or forested. A 0.25% photosynthetic energy conversion efficiency means that

$$0.25/100 \times 160 \text{ GJ ha}^{-1} \text{d}^{-1} \times 365 \text{ d y}^{-1} \times 10^2 \text{ ha km}^{-2} = 1.46 \times 10^4 \text{ GJ km}^{-2} \text{y}^{-1}$$

photosynthetic energy is produced, for a total energy production of $3.2 \cdot 10^{19}$ J $= 32$ EJ which is about 34 GJ per capita in that country. The conclusion arising from this type of calculation is that planting the *entire* arable and forested land mass of India with energy-producing crops could generate a biomass supply sufficient to meet a modest energy demand. It would come nowhere near providing energy at the level presently consumed in high-income countries. And this unlikely scenario leaves no land for food production or other essentials. In other words, biomass alone is not a solution to the world's energy needs.

Having said this, it must be recognized that biomass is now, and will remain, the basic energy source for domestic use by vast numbers of persons—especially those in rural communities in much of Africa, Asia, and Latin America. It has been estimated that 40 to 50% of the world's inhabitants depend entirely or in major part on wood fuel as their source of energy. A frequently cited average annual wood consumption figure (for persons relying on this energy source) is $1.0\,m^3$ per capita per year, representing about 5 GJ of energy.

In the context of the present chapter, there is also the question of carbon dioxide emissions. We have already compared carbon dioxide release from various fossil fuels. For coal, 112 kg CO_2 is released for every GJ of energy produced. The corresponding figure for biomass (as $\{CH_2O\}$) is approximately 100 kg. In that sense, there could be a marginal benefit in using biomass rather than coal. More importantly, the question has been raised as to whether growing crops (often trees) as biomass sources can counteract carbon dioxide increases in the atmosphere by their photosynthetic activity. The answer to this question is complex and depends on the number of years of growth and the end use of the biomass produced. It has been shown[6] that the optimum situation is to grow trees for short periods—usually up to ten years. During this growth period there are high rates of net carbon accumulation, but such rates cannot be sustained as the tree ages. Therefore at this stage the tree should be harvested, dried, and used as a fuel in replacement of fossil fuels. In this way the accumulated and released carbon dioxide would roughly balance, while at the same time producing usable energy.

Table 8.10 Biomass conversion processes and products

Feedstock	Process	Products
Dry biomass—wood, straw, husks, etc.	Gasification	Liquids—methanol Gases—hydrogen, ammonia
	Pyrolysis	Solids—charcoal Liquids Gases
	Fermentation and distillation	Ethanol
Wet biomass—domestic and animal wastes, aquatic plants, etc.	Anaerobic digestion	Methane
Sugars—from juices and hydrolysed cellulose	Fermentation and distillation	Ethanol

Another way of using biomass is to convert it by microbiological and/or chemical processes into other types of fuel, which may then be burned. We will see in Chapter 16 that sewage sludge from the aerator of a secondary waste-water treatment plant is frequently digested anaerobically to produce methane. This kind of process is carried out in large-scale industrial operations as well as in small domestic units, many of which have been installed in low-income countries. Depending on the feedstock and the process, a variety of products or product mixtures are obtained. Table 8.10 lists some such feedstocks, processes, and products. The chemistry and microbiology of some these processes will be discussed in later chapters.

The main points

1 Solar energy in the ultraviolet and visible regions of the electromagnetic spectrum is the primary source of energy available on the Earth's surface. When it is absorbed, it is converted into longer wavelength (infrared) radiation that is emitted outward from the Earth.

2 The chemical composition of the Earth's atmosphere affects the absorption of radiation, both incoming and re-emitted. Through this, it influences the global climate.

3 Water and carbon dioxide absorb large amounts of IR radiation, and contribute to a warmer climate than would otherwise exist on the Earth. Several other trace gases—methane, ozone, nitrous oxide, and CFCs—also absorb IR radiation. All of these are referred to as greenhouse gases.

4 The tropospheric concentration of all (except water) of these gases is increasing. The degree to which this will affect the global climate depends on the extent of this increase as well as on complex positive and negative feedback mechanisms resulting from the chemical changes.

5 Energy production is a major source of greenhouse gases. There are environmental implications of many types associated with each means of energy production.

6 All fossil fuels release carbon dioxide when burned. The ratio of energy produced/carbon dioxide released is in the order: natural gas > petroleum > biomass > coal. Biomass, however, offers the possibility of sequestering atmospheric carbon dioxide during its growth.

Additional reading

1 Houghton, J. T., L. G. Meira Filho, B. A. Callander, N. Harris, A. Kattenberg, and K. Maskell, eds, *Climate Change* 1995: *The Science of Climate Change*, Cambridge University Press, Cambridge; 1996.

2 Miller, D. H., *Energy at the Surface of the Earth*. Academic Press, New York; 1981.

3 Penner, S. S. and L. Icerman, *Energy: Demands, Resources, Impact, Technology and Policy*. Addison-Wesley, Reading; 1981.

Problems

1 Use eqn 8.4 to show why total flux of solar radiant energy is about 10^5 times greater than that from the Earth.

2 The current concentration of carbon dioxide in the atmosphere is 365 ppmv. It was indicated in the text that annual anthropogenic additions to the atmosphere are about 7 Gt (as C) of which about 4 Gt are removed into oceans and the terrestrial environment. Use these numbers to estimate the yearly net increase in atmospheric carbon dioxide mixing ratio in ppmv.

3 Express the amount (65 Mt) of carbon dioxide derived from the Kuwait fires in 1991 as a percentage of the total annual anthropogenic addition of the gas. Note, as indicated in the previous question, that the calculated increase in atmospheric carbon dioxide is only a fraction of what goes into the atmosphere.

4 There has been a steady decrease in the ratio of ^{14}C to ^{12}C in the atmosphere over the past decade. Explain how this is consistent with the view that the well documented increase in atmospheric carbon dioxide concentrations is primarily due to emissions from the combustion of fossil fuels.

5 Estimates (ref. 1) for emissions of methane to the atmosphere are given in the table below and the current atmospheric concentration is 1.77 ppmv. Calculate its residence time.

Sources of atmospheric methane in million tonnes per year	
Wetlands and other natural sources	160
Fossil-fuel-related sources	100
Other anthropogenic sources of biological origin	275

There may be 10^{14} t of methane hydrate ($CH_4 \cdot 6H_2O$) in the permafrost below the ocean floors. If 1% of this were to melt per year, what would be the increased concentration of methane (ppmv y^{-1}) in the atmosphere neglecting any removal processes? What sinks for methane would play a role in reducing this concentration?

6 Recent work has shown that the flux of methane released from fens in the boreal forest area of Saskatchewan, Canada range from 176 to 2250 mmol m^{-2} y^{-1}. Daily fluxes range from 1.08 to 13.8 mmol m^{-2} d^{-1}. The data indicate that there are correlations between methane release and water depth (negative), water flow (negative), temperature (positive), and inorganic phosphorus in the sedimentary interstitial water (positive). Suggest reasons for these correlations. (Rask, H., D. W. Anderson, and J. Schoenau, Methane fluxes from boreal forest wetlands in Saskatchewan, Canada, *Can. J. Soil Sci.*, **76** (1996), 230.

7 The Arrhenius parameters for the reaction

$$N_2O \rightarrow N_2 + O$$

are $A = 7.94 \times 10^{11}$ s^{-1} and $E_a = 250$ kJ mol^{-1}. The reaction is first order. Calculate the rate constant and half-life of nitrous oxide assuming a tropospheric mixing ratio of 310 ppbv N_2O at 20 °C and comment on the environmental significance of these results.

8 Discuss possible effects of the following on greenhouse gas chemistry.

(a) The southern Pacific Ocean is seeded with the algal micronutrients zinc and iron.
(b) CFCs cause further thinning of the ozone layer in the stratosphere.
(c) Urban air pollution leads to increased tropospheric ozone concentrations.
(d) Rice (paddy) is grown, under submerged conditions, on coarse sandy soils, rather than fine clay-rich soils.

9 One 'climate engineering' proposal for reducing the possibilities of global warming is to inject a sulfate aerosol into the stratosphere. Discuss the climatic and other atmospheric implications of this possible human intervention.

10 Tetrafluoromethane and hexafluoroethane produced during aluminium production are potent greenhouse gases, both having GWP values of approximately 8400. Carbon dioxide is also released from the stoichiometric reduction of alumina by carbon. Use the figures given in the text to estimate the relative GWP contribution from these two sources.

Notes

1 The use of the word energy in the context of eqn 8.1 and subsequent ones is common but not quite correct. The unit, watt (W), measures power which is energy per unit time (J s^{-1})

2 Grum, F. and R. J. Becherer, *Optical Radiation Measurements*, Volume 1, *Radiometry*, Academic Press, New York; 1979.

3 The symbol [CH$_2$O] is used here and elsewhere as a representation of the simplest formula for plant biomass, much of which consists of carbohydrate and related materials.

4 *CO$_2$/Climate Report*, Canadian Climate Centre, Atmosphere Environment Service, Downsview, Ontario, Issue 98–1; 1998.

5 Lashof, D. A. and D. R. Ahuja, Relative contributions of greenhouse gas emissions to global warming. *Nature*, **344** (1990), 529–31.

6 Vitousek, P. M., Can planted forests counteract increasing atmospheric carbon dioxide? *J. Environ. Qual.* **20** (1991), 348–54.

The hydrosphere

The shining water that moves in the streams and rivers is not just water
but the blood of our ancestors. If we sell you the land, you must remember
it is sacred and you must teach your children it is sacred, and that each
ghostly reflection in the clear water of the lakes tells of events and memories
in the life of my people. The water's murmur is the voice of my father's father.

Chief Seattle of the Squamish Tribe, 1854

9

The hydrosphere

LIKE air, water is one of the essentials for all forms of life. To many people, the quality of the environment is defined more than anything else by the quality of water they see around them. The water resource—taken as a whole we refer to it as the hydrosphere—is then a second compartment of the environment that must be a subject for study. The hydrosphere occupies 73% of the Earth's surface but water is an important component of the atmosphere and the terrestrial environment as well. The distribution of the Earth's water is shown in Fig. 9.1 and a summary of the global water cycle has been depicted in Fig. 1.4.

Making up some 97% of the total mass, the oceans are quantitatively of primary importance in defining the world's water relations. Besides water, the oceans contain substantial quantities of many dissolved elements. Table 9.1 gives average values for the major constituents that make up sea water.

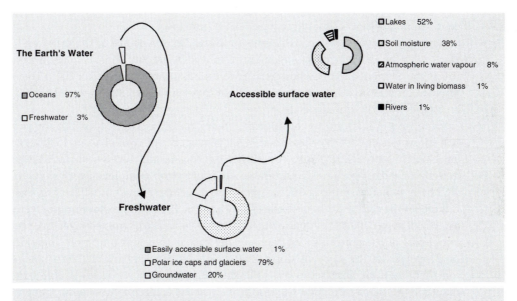

Fig. 9.1 Distribution of global water resources. (G. Lean, D. Hinrichsen, and A. Markham, *Atlas of the Environment*, Prentice Hall, New York; 1990.)

Table 9.1 Composition of sea water—major inorganic chemical constituents

Component	Concentration
Sodium	10 760 mg kg^{-1}
Magnesium	1294
Calcium	413
Potassium	387
Strontium	8
Chloride	19 353
Sulfate	2712
Hydrogen carbonate	142
Bromide	67
Boron	4
Fluoride	1

Data from Martin, D., *Marine Chemistry, Vol. 1 Analytical Methods*, Marcel Dekker, New York; 1968.

The constituents in Table 9.1 are 'conservative' ones with *relatively* constant composition through time and space although many local variations, especially in coastal areas, have been documented. Analysis of trace components indicates that virtually every stable element is present in sea water, in most cases at very low levels. For example the gold concentration has been estimated to be 2×10^{-11} mol L^{-1}.

For conservative elements, residence times in the oceans are of the order of 10^4 to 10^8 years, with the shorter times for elements removed by precipitation as solids. An example of this is the element silicon, whose residence time is approximately 2×10^4 y. The principal removal process involves a biogenic reaction to form the skeletal material of the diatoms, an abundant form of marine algae.

Sometimes it is useful to consider the chemistry of oceans in terms of the vertical distribution of species. The surface microlayer, a layer only micrometres in thickness, is highly enriched in some chemicals as a consequence of the elevated levels of surface-active organic materials which also have considerable ability to form complexes with many metals as well as non-metals and organic compounds. In Chapter 6, we saw that the enrichment influences the composition of the sea-spray aerosol. Below the surface microlayer, down to a depth of about 300 m, is a relatively well mixed volume (the 'mixed layer'). For each element, a plot of concentration vs. depth in the mixed layer has a unique shape depending on properties of the element. Long residence time conservative elements such as rubidium and caesium have relatively uniform concentration vs. depth profiles, influenced mainly by temperature. Major and minor nutrient elements (nitrogen, copper, etc.) are depleted near the biologically active regions where algae proliferate at the ocean surface and their concentrations increase with depth. Non-nutrient elements whose presence in the oceans is influenced by atmospheric sources (lead and vanadium are good examples) show decreasing concentration with depth through the mixed layer. Below the mixed layer is the metalimnion (also called the thermocline), a region of steadily decreasing temperature, extending over several hundred metres. Still deeper is the great mass of ocean water extending to depths of several kilometres (the deepest point in the oceans is over 11 000 m near the island of Guam in the South Pacific) which is only slowly mixed with that on the surface. In calculating the residence times noted above, the total ocean has been taken into account.

Fresh water makes up only about 3% of the total global water resource, yet its importance far outweighs its quantitative contribution. Three quarters of the fresh water is located in polar ice caps and alpine glaciers with 90% of that making up the Antarctic continental ice sheet. Besides being an actual and potential source of high purity water (you have probably heard of proposals to use ships to tow large icebergs from polar regions to areas where there is a shortage of potable water such as the Arabian Gulf—proposals that have never been implemented), the ancient glacial ice of the Antarctic, Greenland and other places serves a useful environmental purpose in that trapped air bubbles and the water itself capture a record of atmospheric conditions through geological time. We have already cited examples of this in considering atmospheric carbon dioxide concentrations and rainfall pH.

Most of the remaining fresh water is found within the ground. Where water is located in soil pores but is subject to periodic evaporation and replacement by air, it is referred to as soil pore water or soil moisture. Where the soil or rock pores are perennially filled—a saturation condition, below the water table—the water is termed groundwater. Some terminology frequently used by Earth scientists is shown in Fig. 9.2. Groundwater makes up about 20% of the world resource of fresh water and is widely used by industry, for irrigation, and for domestic purposes.

Only about 1% of all fresh water is in easily accessible surface forms—rivers, ponds, and lakes. This 0.03% of Earth's total water supply dominates much of our study in environmental chemistry because it is readily available and it is an essential requirement for the

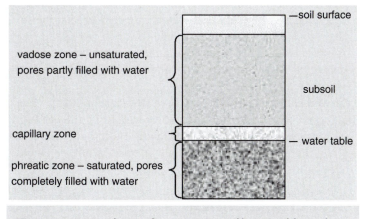

Fig. 9.2 Nomenclature for zones in soil/permeable rock depth profiles.

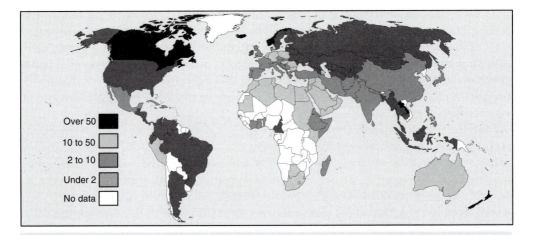

Fig. 9.3 Availability of internal renewable fresh water throughout the world. Map shows internal renewable water resources in thousand cubic metres per year per capita. (Data from Lean, G., D. Hinrichsen, and A. Markham, *Atlas of the Environment*, Prentice Hall, New York; 1990.)

survival and growth of many forms of animal and plant life on the planet. Both quantity and quality of accessible water are of concern.

In terms of quantity, the water resource is unevenly distributed. Figure 9.3 shows water distribution around the globe; quantities are measured as thousands of cubic metres per year per capita ($m^3 y^{-1}$ pc). There is a wide range in water distribution from over 600 000 $m^3 y^{-1}$ being available for each person in Iceland, to 10 000 in Sweden and Malaysia, to 3000 in France and India, to 250 in Israel and Saudi Arabia. Whether the entire population has access to the water is another question. In one assessment, countries which have available less than 2000 $m^3 y^{-1}$ pc of fresh water are considered to be in chronic water deficit.

The quality of water is also highly variable and there are many chemical and microbiological aspects which must be considered in determining its acceptability. This is a subject we will examine in Chapter 16.

A version of the well known water cycle has been described in Fig. 1.4. The cycle includes not only the liquid and solid forms but also water vapour in the atmosphere. The total gaseous mass is not insignificant—equivalent to about 15% of that found in lakes and rivers. As we have seen, the chemistry of rain is strongly affected by anthropogenic processes.

9.1 Physical and chemical properties of water

In order to develop a good understanding of the environmental relations in the hydrosphere, we begin with an examination of some basic principles of aqueous physical chemistry. The properties of water determine the way the Earth is, and the forms of life

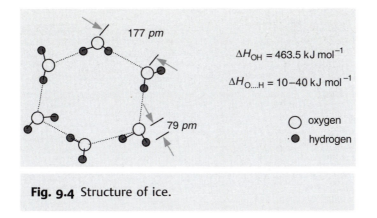

177 pm

$\Delta H_{OH} = 463.5 \text{ kJ mol}^{-1}$

$\Delta H_{O\cdots H} = 10\text{--}40 \text{ kJ mol}^{-1}$

79 pm

○ oxygen
● hydrogen

Fig. 9.4 Structure of ice.

that have developed. Many of these properties are unusual, even unique, in comparison with other compounds related by the position of their atoms on the periodic table. We will look at some of these properties in all three phases of water.

9.1.1 Ice

The structure of ice takes the form of hexagonal puckered rings with bond lengths and bond enthalpies as shown in Fig. 9.4. The density of ice is 0.917 kg L^{-1} so that it floats on the surface of the underlying water—an important property affecting the environment in regions with cold climates. The enthalpy of fusion is 6.02 kJ mol^{-1}, which is higher than that of most other solids. In part, it is because of this high value that temperature fluctuations are reduced in areas adjacent to major water bodies.

9.1.2 Liquid water

When ice melts, only about 12% of the hydrogen bonds are broken, indicating that liquid water at $0\,^{\circ}C$ retains a considerable component of the ice structure. The bonds actually hold the water molecules apart from one another and the limited bond breaking on melting is sufficient to allow the individual molecules to come closer. This is the reason why the density of liquid water at $0\,^{\circ}C$ is greater than that of ice. As water is heated through the liquid temperature range, two competing factors affect its density. One is that there is further breaking of hydrogen bonds (8% additional between 0 and $100\,^{\circ}C$), and this leads to an increase in density. The second is that a higher temperature results in greater kinetic energy of the molecules, causing thermal expansion, and a density decrease. Of these factors the first dominates between 0 and $4\,^{\circ}C$ and the second from 4 to $100\,^{\circ}C$. The greatest density of water is then at approximately $4\,^{\circ}C$.

These density relations lead to a particular pattern of behaviour in lakes and other water bodies that are subject to major seasonal changes in temperature (Fig. 9.5). During the warm summer season, lakes develop a stable structure, with a three-layer profile (epilimnion, metalimnion, hypolimnion) similar to that described above for oceans. Of course the depth of the epilimnion is much less than that in the marine situation. The magnitude

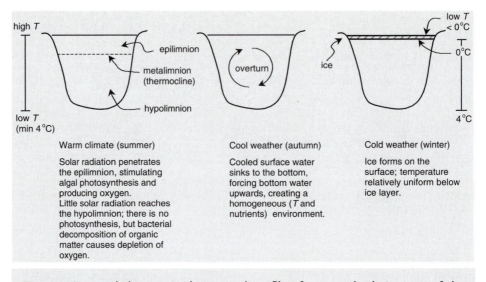

Fig. 9.5 Seasonal changes in the vertical profile of a water body in parts of the world with temperate climate.

of temperature differences and the resulting scale of the layers depend on climatic factors and on the size and depth of the water body. When the warm season ends and a period of prolonged cooling occurs, surface water becomes more dense than the underlying warmer water. This causes it to sink, carrying with it a fresh supply of oxygen and nutrients and lifting the deep water upwards. The mixing process, called overturn, brings about a period of enhanced chemical and microbiological activity. As cold weather continues and a period of sustained sub-freezing temperature sets in, the lake develops a new type of stable (but inverted) structure. In the winter, ice floats on the surface, underlain by water with a temperature at or near $0\,°C$, and the temperature increases downwards to a maximum of $4\,°C$. In the spring, a second period of intense mixing occurs and the cycle repeats in subsequent years.

9.1.3 Hydration

Water has one of the largest dielectric constants of any common liquid, and this determines that it is a good solvent for ionic and polar species. Because of the high value of the dipole moment ($6.1 \times 10^{-30}\,\text{C m}$), the water molecules orient themselves around ions to form hydrated species in solution. The degree of hydration, as represented by the hydration number, depends on several factors, the most important of which is the charge to radius ratio, as is illustrated by the data for alkali metals in Table 9.2. From the table, you can see that there is an interesting inverse relation between ionic radius and hydrated radius. This has implications with respect to ion exchange reactions, involving hydrated ions, that take place on solid surfaces in the water column.

Table 9.2 Charge and radius properties of the alkali metals in aqueous solution

	Li^+	Na^+	K^+	Rb^+	Cs^+
Ionic radius/pm	60	95	133	148	169
Charge density/C pm^{-1}	0.0167	0.0105	0.0075	0.0068	0.0059
Hydrated radius/pm	340	276	232	228	228
Hydration number	23.3	16.6	10.5	10	9.9

9.1.4 Complexation

'Free ions', both cations and anions, are therefore really hydrated ions and these are referred to 'aquo' complexes. The coordinated water is held in position by a combination of electrostatic and covalent forces, depending on the properties of the species with which it is associated.

In most natural water situations, there are many dissolved substances that can act as ligands, displacing the water and forming a new complex with the ion. This implies that the bond between the ion and the different ligand is more stable than the original one involving water. In many cases, the complex involves a central metal ion coordinated to one or more inorganic or organic ligands. Available as potential complexing agents are a whole range of possible substances; examples include sulfate, small organic molecules containing amino groups, or larger organic molecules with several sites for forming a bond. The formation of complexes is a very important feature of aqueous chemistry of metal ions.

Stability of complexes is expressed using *stability constants*, also called *formation constants*. There are two types of such constants, as can be illustrated using a generic example. In the example, M is a metal ion and L is the ligand of interest. For simplicity, the charges on the metal ion and on the ligand (if it has a charge) are omitted. We will assume that it is possible for four ligand molecules to be coordinated with the metal. The complex formation reactions take place in a stepwise fashion, and each step has a corresponding formation constant.

$$M + L \rightleftharpoons ML \qquad K_{f1} = [ML]/[M][L] \tag{9.1}$$

$$ML + L \rightleftharpoons ML_2 \qquad K_{f2} = [ML_2]/[ML][L] \tag{9.2}$$

$$ML_2 + L \rightleftharpoons ML_3 \qquad K_{f3} = [ML_3]/[ML_2][L] \tag{9.3}$$

$$ML_3 + L \rightleftharpoons ML_4 \qquad K_{f4} = [ML_4]/[ML_3][L] \tag{9.4}$$

Each of the K_f values is referred to as a stepwise formation constant. The sum of the four reaction steps describes the total reaction, and the overall formation constant for this process is designated as β_4.

$$M + 4L \rightleftharpoons ML_4 \qquad \beta_4 = [ML_4]/[M][L]^4 \tag{9.5}$$

In the present case, the overall formation constant is related to the stepwise constants:

$$\beta_4 = K_{f1} \times K_{f2} \times K_{f3} \times K_{f4} \tag{9.6}$$

In the general case, where the number of ligands bound to the metal ion is n,

$$\beta_n = K_{f1} \times K_{f2} \times K_{f3} \times \cdots \times K_{fn} \tag{9.7}$$

One further point to reiterate is that the reactions shown imply that the ligand is forming a series of complexes with a *free* or *uncomplexed* metal ion. In fact, as noted above, the so-called uncomplexed metal is invariably present as an aquo complex, and ligand addition to the metal actually involves displacement of a water molecule and replacement by the new ligand.

In the oceans, one ligand obviously present in large concentration is the chloride ion. As a result, some metals exist in the ocean primarily as chloro complexes. Mercury, for example, is found in clean ocean water at a concentration of approximately 5×10^{-12} mol L^{-1}. The complexation reaction between mercury (II) and the chloride ligand is then described as

$$Hg(H_2O)_6^{2+}(aq) + 4Cl^-(aq) \rightleftharpoons Hg(H_2O)_2Cl_4^{2-}(aq) + 4H_2O \tag{9.8}$$

The equilibrium constant for the reaction is 1.3×10^{15} and the tetrachloro species is known to be a principal form of mercury (II) in sea water.

Many other ligands may be present depending on the water under consideration. Soluble organic substances like citrate ion, or macromolecules produced by decomposition of plant and animal tissue, act as coordinating ligands. In other instances, organic substances (such as nitrilotriacetic acid (NTA), which is sometimes used as a water-softening agent in detergents) or inorganic substances (like phosphate, also used in detergents or applied to soil as a fertilizer) of anthropogenic origin form complexes with dissolved substances. The nature and extent of complex formation depends on the properties of the central atom, and the availability and concentration of potential ligands.

9.1.5 Acid–base properties

Water is an amphiprotic substance and it undergoes autoprotolysis to form the hydronium ion and the hydroxyl ion. The equilibrium constant for autoprotolysis (K_w or K_{auto}) depends on temperature:

$$2H_2O \rightleftharpoons H_3O^+(aq) + OH^-(aq) \qquad K_w = 1.01 \times 10^{-14} \text{ at } 25\,^\circ C \tag{9.9}$$

Because the products, H_3O^+ and OH^-, are themselves hydrated, a better representation of H_3O^+ would be $H_3O(H_2O)_3^+$, but we will use the simpler formula. Water also has the ability to accept and donate protons from other substances and therefore acts as a Brönsted base or acid in the presence of proton donors or acceptors. For example, oxalic acid and other carboxylic acids which are degradation products of natural organic matter (NOM) are a source of acidification of water by reactions such as

$$(COOH)_2 + H_2O \rightleftharpoons COOHCOO^-(aq) + H_3O^+(aq) \qquad K_{a1} = 5.6 \times 10^{-2} \tag{9.10}$$

$$COOHCOO^-(aq) + H_2O \rightleftharpoons (COO^-)_2(aq) + H_3O^+(aq) \qquad K_{a2} = 5.2 \times 10^{-5} \tag{9.11}$$

The doubly charged oxalate anion is the principal form of this molecule in most natural water situations.

9.1.6 Redox properties

Water also has redox properties which define upper and lower potential limits for redox reactions of other substances found in aqueous solutions. The oxidation of water leads to evolution of O_2:

$$6H_2O \rightarrow O_2 + 4H_3O^+(aq) + 4e^- \tag{9.12}$$

$$E = 1.23 - 0.0591 \text{ pH at } 25°C \tag{9.13}$$

Under extreme reducing conditions, water decomposes, producing hydrogen:

$$2H_2O + 2e^- \rightarrow H_2 + 2OH^-(aq) \tag{9.14}$$

$$E = -0.0591 \text{ pH at } 25°C \tag{9.15}$$

The linear boundaries (Fig. 9.6) define the area of stability of water as a function of potential (E) and pH. Only species whose potential/pH characteristics locate them within the boundaries of stable water can exist in a stable aqueous system.

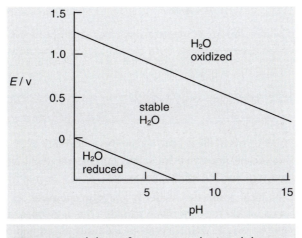

Fig. 9.6 Stability of water as depicted by an E/pH diagram.

9.1.7 Water vapour

The enthalpy of vaporization of water is 40.6 kJ mol^{-1}, which is the highest value for all common liquids. As vaporization occurs, the intact 80% of the hydrogen bonds are broken, so that there is no structure remaining in the vapour state. The large enthalpy value contributes to temperature stability over large water bodies or over land adjacent to these.

Table 9.3 Vapour pressure of water as a function of temperature

$T/°C$	P_{vap} (H_2O)/Pa
0	6.1×10^2
5	8.7×10^2
10	1.2×10^3
15	1.7×10^3
20	2.3×10^3
25	3.2×10^3
30	5.6×10^3

The variation of vapour pressure of water with temperature is shown in Table 9.3. Therefore at 20 °C, 101 kPa pressure, and 100% humidity, the mixing ratio of water in the atmosphere would be

$$\frac{2.3 \text{ kPa}}{101 \text{ kPa}} \times 100\% = 2.3\%$$

The global range of water vapour concentrations averages about 1% but as we showed earlier, actual values are highly variable both temporally and spatially.

The main points

1 Water is essential to all forms of life on the Earth. The availability of an adequate supply of clean fresh water is especially important for supporting development and quality of human life in rural and urban regions throughout the world.

2 In its physical and chemical properties, water is a unique chemical. Its physical properties are a major factor responsible for controlling climate. The acid–base, redox, and solvent properties determine the way in which elements are transported and made available to interact with other components of the living and non-living environment.

Additional reading

1 Berner, E. K. and R. A. Berner, *The Global Water Cycle*. Prentice Hall, Englewood Cliffs, NJ; 1987.

2 Drever, J. I., *The Geochemistry of Natural Waters*, 2nd edn., Prentice Hall, Englewood Cliffs, NJ; 1988.

3 Libes, S. M., *An Introduction to Marine Biogeochemistry*, John Wiley and Sons, New York; 1992.

4 Stumm, W. and J. J. Morgan, *Aquatic Chemistry: Chemical Equilibria and Rates in Natural Waters*. John Wiley and Sons, New York; 1996.

Problems

1 The concentration of gold in the oceans averages approximately $2 \times 10^{-11}\,\text{mol}\,\text{L}^{-1}$. Calculate the total mass in tonnes.

2 The value of the K_w is 0.67×10^{-14} at $20\,°\text{C}$, 1.01×10^{-14} at $25\,°\text{C}$, and 1.45×10^{-14} at $30\,°\text{C}$. Calculate the value of K_w at $0\,°\text{C}$, and determine the pH of pure water at that temperature.

3 In a particular fresh water sample, the concentrations of cations and anions are (in $\mu\text{mol}\,\text{L}^{-1}$):

Na^+	33	Cl^-	120
K^+	4	NO_3^-	13
Mg^{2+}	31	HCO_3^-	270
Ca^{2+}	160	CO_3^{2-}	0.67
		SO_4^{2-}	11

Compare the concentration of total positive and negative charge in the solution. Assume that the difference is due to hydronium or hydroxyl ion, and calculate the pH.

4 The major ions and their concentration ($\text{mmol}\,\text{L}^{-1}$) in sea water are:

Na^+	470	K^+	10
Mg^{2+}	53	Ca^{2+}	10
Cl^-	547	SO_4^{2-}	28
Br^-	1	$HCO_3^- + CO_3^{2-}$	x

Assume that charges of these species balance and calculate the total concentration of negative charge associated with the two carbonate species. With a pH of 8.2, calculate the concentrations of the two individual carbonate species.

5 In the open oceans the concentration of iron is approximately $1 \times 10^{-4}\,\text{ppm}$ in the surface water, and $4 \times 10^{-4}\,\text{ppm}$ in the deep ocean. Corresponding values for concentration of aluminium are $9.7 \times 10^{-4}\,\text{ppm}$ and $5.2 \times 10^{-4}\,\text{ppm}$. Why are the concentrations so low? Why is the ratio surface concentration : deep concentration <1 for iron and >1 for aluminium?

6 Manganese may precipitate as $MnCO_3$ from aqueous solution according to the reaction

$$Mn^{2+}\,(aq) + CO_2\,(aq) + 3H_2O \rightarrow MnCO_3\,(s) + 2H_3O^+\,(aq)$$

The solubility product, K_{sp}, for manganese (II) carbonate is 5.0×10^{-10}. Use this value and the equilibrium constant values for the carbonate system to determine the minimum pH required to precipitate manganese (II) carbonate from a solution which contains

1.0×10^{-3} mol L^{-1} manganese (II) ion. Assume that aqueous carbon dioxide is in equilibrium with atmospheric carbon dioxide.

7 The following are controlling factors for the 'availability' of different elements:

- oxygen availability for iron
- sulfide concentration for zinc
- solution pH for chromium and silicon
- carbonate concentration for calcium
- sorption factors for copper.

Explain the chemical and environmental significance of these factors.

10

Distribution of species in aquatic systems

At the outset of the book, we noted that an accurate description of chemical composition in an environmental compartment requires knowledge of the forms or species of a particular chemical. For the aqueous environment, a classification of this kind is frequently referred to as a description of species distribution or 'speciation'.

The species distribution for an element or compound is calculated in the context of particular environmental conditions, assuming one has available the appropriate analytical data and the required thermodynamic constants. The additional assumptions that are usually made include an idea that distribution is not affected by reaction rate—in other words, the system is at thermodynamic equilibrium. For rigorous calculations, activities rather than concentrations should be used, but this requires knowledge of the total ionic composition of the solution. While recognizing that a simplifying assumption that activity and concentration are equivalent leads to error in the calculation, we choose to use concentrations in most cases in the book. This is done partly to avoid making complex relations appear even more complex. Furthermore, in many situations, the errors generated by neglecting activity coefficients are smaller than those arising from uncertainties in the analytical data available for multi-component environmental materials.

Two calculation situations are encountered. Detailed individual calculations related to specific sets of conditions are important, but also very useful are diagrams which show how species distributions can vary as conditions change. There are many kinds of distribution diagrams. We will look into the construction of several types and we will examine how to interpret these and others. We will do this by way of considering some important environmental examples.

10.1 Single-variable diagrams

10.1.1 Phosphate species

A single variable diagram is a plot of some measure of species concentration versus a particular variable like pH, redox status, or concentration of an important complexing ligand. A well known case, useful for describing the chemistry of species that exhibit acid–base behaviour, is a plot of alpha (α), the fractional concentration, of individual species against pH. A good example of this application is the phosphate system. Phosphorus exists in water almost exclusively as P(V) species, particularly in forms of orthophosphate.

$$H_3PO_4 \rightleftharpoons H_2PO_4^- \rightleftharpoons HPO_4^{2-} \rightleftharpoons PO_4^{3-} \tag{10.1}$$

To calculate this distribution diagram we need only know the values of the acid dissociation constants that are given in Table 10.1. These equilibrium constants apply to the successive dissociation from the acid of the three protons, and apply to an aqueous solution at 25 °C. For example, the first dissociation is approximated by the relation

$$K_{a1} = \frac{[H_2PO_4^-][H_3O^+]}{[H_3PO_4]} \tag{10.2}$$

The fraction, $\alpha_{H_3PO_4}$, of undissociated H_3PO_4 in a solution containing phosphate species is

$$\alpha_{H_3PO_4} = \frac{[H_3PO_4]}{[H_3PO_4] + [H_2PO_4^-] + [HPO_4^{2-}] + [PO_4^{3-}]} = \frac{[H_3PO_4]}{C_p} \tag{10.3}$$

where C_p = the total concentration of all orthophosphate species. For the other phosphate species, similar fractions are given by

$$\alpha_{H_2PO_4^-} = \frac{[H_2PO_4^-]}{C_p} \tag{10.4}$$

$$\alpha_{HPO_4^{2-}} = \frac{[HPO_4^{2-}]}{C_p} \tag{10.5}$$

$$\alpha_{PO_4^{3-}} = \frac{[PO_4^{3-}]}{C_p} \tag{10.6}$$

Table 10.1 Acid dissociation constants for phosphoric acid		
	K_a	pK_a
First dissociation	7.1×10^{-3}	2.15
Second dissociation	6.3×10^{-8}	7.20
Third dissociation	4.2×10^{-13}	12.38

The three dissociation constant expressions can be rearranged to give the concentration of each individual species in terms of $[H_3PO_4]$ and $[H_3O^+]$:

$$[H_2PO_4^-] = \frac{K_{a1} \times [H_3PO_4]}{[H_3O^+]} \tag{10.7}$$

$$[HPO_4^{2-}] = \frac{K_{a1} \times K_{a2} \times [H_3PO_4]}{[H_3O^+]^2} \tag{10.8}$$

$$[PO_4^{3-}] = \frac{K_{a1} \times K_{a2} \times K_{a3} \times [H_3PO_4]}{[H_3O^+]^3} \tag{10.9}$$

$$C_p = [H_3PO_4] + [H_2PO_4^-] + [HPO_4^{2-}] + [PO_4^{3-}]$$
$$= [H_3PO_4] \left(1 + \frac{K_{a1}}{[H_3O^+]} + \frac{K_{a1} \times K_{a2}}{[H_3O^+]^2} + \frac{K_{a1} \times K_{a2} \times K_{a3}}{[H_3O^+]^3} \right) \tag{10.10}$$

From eqn 10.3,

$$\alpha_{H_3PO_4} = \frac{[H_3PO_4]}{[H_3PO_4] \left(1 + K_{a1}/[H_3O^+] + K_{a1} \times K_{a2}/[H_3O^+]^2 + K_{a1} \times K_{a2} \times K_{a3}/[H_3O^+]^3 \right)} \tag{10.11}$$

We then multiply the top and bottom of the right-hand side of the equation by $[H_3O^+]^3$:

$$\alpha_{H_3PO_4} = \frac{[H_3O^+]^3}{[H_3O^+]^3 + [H_3O^+]^2 \times K_{a1} + [H_3O^+] \times K_{a1} \times K_{a2} + K_{a1} \times K_{a2} \times K_{a3}} \tag{10.12}$$

Using similar calculations, we find that

$$\alpha_{H_2PO_4^-} = \frac{[H_3O^+]^2 \times K_{a1}}{[H_3O^+]^3 + [H_3O^+]^2 \times K_{a1} + [H_3O^+] \times K_{a1} \times K_{a2} + K_{a1} \times K_{a2} \times K_{a3}} \tag{10.13}$$

$$\alpha_{HPO_4^{2-}} = \frac{[H_3O^+] \times K_{a1} \times K_{a2}}{[H_3O^+]^3 + [H_3O^+]^2 \times K_{a1} + [H_3O^+] \times K_{a1} \times K_{a2} + K_{a1} \times K_{a2} \times K_{a3}} \tag{10.14}$$

$$\alpha_{PO_4^{3-}} = \frac{K_{a1} \times K_{a2} \times K_{a3}}{[H_3O^+]^3 + [H_3O^+]^2 \times K_{a1} + [H_3O^+] \times K_{a1} \times K_{a2} + K_{a1} \times K_{a2} \times K_{a3}} \tag{10.15}$$

These four α functions are then evaluated and plotted as a function of pH. This is most conveniently done using a computer. Figure 10.1 is the plot for phosphate species distribution.

Using this plot, estimates of the phosphorus species distribution are readily made. For example, in open water of the Kenyan part of Lake Victoria, phosphorus levels are typically about $12\,\mu g\,L^{-1}$ (as P) and the pH of the water is 8.2. Under these conditions, the fractional values of the principal species are given by points along line 'a' on Fig. 10.1 and it is determined graphically that $\alpha_{H_2PO_4^-} = 0.08$ and $\alpha_{HPO_4^{2-}} = 0.92$, giving concentrations $[H_2PO_4^-] = 1\,\mu g\,L^{-1}$ and $[HPO_4^{2-}] = 11\,\mu g\,L^{-1}$. In another setting—a sample of pore water from a forest soil in the Canadian Shield—soluble phosphorus is $62\,\mu g\,L^{-1}$ and the pH 4.3. In this situation (points on line 'b' on Fig. 10.1) the principal species is almost exclusively $[H_2PO_4^-] = 62\,\mu g\,L^{-1}$.

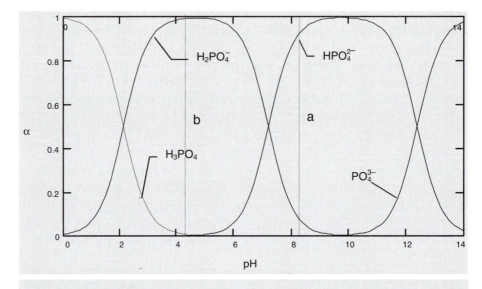

Fig. 10.1 Distribution of phosphorus species expressed as the fraction, α, as a function of aqueous solution pH.

We have mentioned that there are assumptions in these calculations. One assumption is that there are no interactions with other species in the solutions. Furthermore, as noted at the beginning of the chapter, no account was taken of the solution ionic strength. Soil solutions typically have an ionic strength of 0.002 which leads to an activity coefficient of about 0.95 for a singly charged ion. Incorporating the activity coefficient into the calculation would make only a small difference and the correction would be even smaller for the lake water. While assuming zero ionic strength in this situation leads to a small error, the error could be much larger for a medium like sea water which contains a high ionic concentration. Apparent acid dissociation constants (which are affected by large changes in activity coefficient due to the high ionic strength) have been calculated for phosphoric acid in sea water and they are

$$K_{a1} = 2.4 \times 10^{-2}$$
$$K_{a2} = 8.8 \times 10^{-7}$$
$$K_{a3} = 1.4 \times 10^{-9}$$

A substantially different distribution diagram therefore would apply to sea water compared with that calculated above for a 'fresh water' system.

10.1.2 Cadmium complexes with chloride

A second version of a single variable distribution diagram is a plot of the logarithm of the concentration of particular species versus a chosen variable, usually the concentration of an important ligand. By using a log function, a much wider range of concentration can be shown than when the linear α function is used.

As an example, we will consider the distribution of aqueous cadmium chloro complexes as a function of chloride ion concentration. Again we will use concentrations rather than activities in the calculation. We start by assuming that, in the absence of chloride or any other complexing ligand, cadmium exists in aqueous solution as an aquo complex, perhaps $Cd(H_2O)_4^{2+}$. (We will omit the complexed waters and refer to the aquo species as Cd^{2+} in further discussions.) This aquo complex remains without significant deprotonation (see Section 13.1) as long as the pH is less than about 8.5. Chloride ion forms complexes with cadmium in a stepwise fashion, with displacement of a water molecule each time a chloride is added. We apply the general relations for complex formation (eqns 9.1 to 9.7) to the cadmium/chloride situation:

$$Cd^{2+} + Cl^- \rightleftharpoons CdCl^+ \qquad K_{f1} = \frac{[CdCl^+]}{[Cd^{2+}][Cl^-]} = 7.9 \times 10^1 \qquad (10.16)$$

$$CdCl^+ + Cl^- \rightleftharpoons CdCl_2 \qquad K_{f2} = \frac{[CdCl_2]}{[CdCl^+][Cl^-]} = 4.0 \qquad (10.17)$$

$$CdCl_2 + Cl^- \rightleftharpoons CdCl_3^- \qquad K_{f3} = \frac{[CdCl_3^-]}{[CdCl_2][Cl^-]} = 2.0 \qquad (10.18)$$

$$CdCl_3^- + Cl \rightleftharpoons CdCl_4^{2-} \qquad K_{f4} = \frac{[CdCl_4^{2-}]}{[CdCl_3^-][Cl^-]} = 0.6 \qquad (10.19)$$

The reactions may also be described using 'overall' steps and the overall stability constants are symbolized as β_f. It is readily seen that

$$\beta_{fn} = K_{f1} \times K_{f2} \times \cdots \times K_{fn} \qquad (10.20)$$

$$Cd^{2+} + Cl^- \rightleftharpoons CdCl^+ \qquad \beta_{f1} = K_{f1} = 7.9 \times 10^1 \qquad (10.21)$$

$$Cd^{2+} + 2Cl^- \rightleftharpoons CdCl_2 \qquad \beta_{f2} = K_{f1} \times K_{f2} = 3.2 \times 10^2 \qquad (10.22)$$

$$Cd^{2+} + 3Cl^- \rightleftharpoons CdCl_3^- \qquad \beta_{f3} = K_{f1} \times K_{f2} \times K_{f3} = 6.4 \times 10^2 \qquad (10.23)$$

$$Cd^{2+} + 4Cl^- \rightleftharpoons CdCl_4^{2-} \qquad \beta_{f4} = K_{f1} \times K_{f2} \times K_{f3} \times K_{f4} = 3.8 \times 10^2 \qquad (10.24)$$

Depending on the measurement conditions and methodology, note that there can be significantly different values for the stability constants, K_f or β_f in the literature. One reliable compilation is Hogfeldt, E., *Stability Constants of Metal–Ion Complexes*, IUPAC Chemical Data Series, No 21, Pergamon Press, Oxford; 1982.

The total concentration of cadmium in an aqueous solution containing chloride is then

$$C_{Cd} = [Cd^{2+}] + [CdCl^+] + [CdCl_2] + [CdCl_3^-] + [CdCl_4^{2-}] \qquad (10.25)$$

We can derive expressions for the concentration of each of the five cadmium species in the following way. We begin by dividing eqn 10.25 by $[Cd^{2+}]$:

$$\frac{C_{Cd}}{[Cd^{2+}]} = 1 + \frac{[CdCl^+]}{[Cd^{2+}]} + \frac{[CdCl_2]}{[Cd^{2+}]} + \frac{[CdCl_3^-]}{[Cd^{2+}]} + \frac{[CdCl_4^{2-}]}{[Cd^{2+}]} \qquad (10.26)$$

Substituting the expressions for the β functions,

$$\frac{C_{Cd}}{[Cd^{2+}]} = 1 + \beta_{f1}[Cl^-]^1 + \beta_{f2}[Cl^-]^2 + \beta_{f3}[Cl^-]^3 + \beta_{f4}[Cl^-]^4 \qquad (10.27)$$

Rearranging eqn 10.27:

$$[Cd^{2+}] = \frac{C_{Cd}}{1 + \beta_{f1}[Cl^-] + \beta_{f2}[Cl^-]^2 + \beta_{f3}[Cl^-]^3 + \beta_{f4}[Cl^-]^4} \tag{10.28}$$

Similarly, the concentrations of other cadmium chloro species are given by

$$[CdCl^+] = \frac{\beta_{f1}[Cl^-]C_{Cd}}{1 + \beta_{f1}[Cl^-] + \beta_{f2}[Cl^-]^2 + \beta_{f3}[Cl^-]^3 + \beta_{f4}[Cl^-]^4} \tag{10.29}$$

$$[CdCl_2] = \frac{\beta_{f2}[Cl^-]^2C_{Cd}}{1 + \beta_{f1}[Cl^-] + \beta_{f2}[Cl^-]^2 + \beta_{f3}[Cl^-]^3 + \beta_{f4}[Cl^-]^4} \tag{10.30}$$

$$[CdCl_3^-] = \frac{\beta_{f3}[Cl^-]^3C_{Cd}}{1 + \beta_{f1}[Cl^-] + \beta_{f2}[Cl^-]^2 + \beta_{f3}[Cl^-]^3 + \beta_{f4}[Cl^-]^4} \tag{10.31}$$

$$[CdCl_4^{2-}] = \frac{\beta_{f4}[Cl^-]^4C_{Cd}}{1 + \beta_{f1}[Cl^-] + \beta_{f2}[Cl^-]^2 + \beta_{f3}[Cl^-]^3 + \beta_{f4}[Cl^-]^4} \tag{10.32}$$

Again, using an appropriate computer program, the fraction, α, of each cadmium species can be calculated as a function of chloride ion concentration. For the chloride ion, we choose a range of values from 0 to 0.56 M, the latter being the concentration in sea water. The plot then forms another type of distribution diagram and is shown in Fig. 10.2.

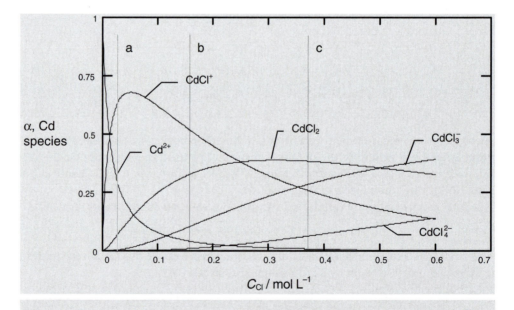

Fig. 10.2 Distribution of cadmium chloro complexes as a function of the concentration of chloride ion in water. The range of chloride concentrations is from zero to 0.56 mol L^{-1}; the latter value is the approximate concentration in sea water.

Figure 10.2 can be useful in several situations. For example, we may use it to describe the changing cadmium species distribution in an estuary. An estuary has been defined[1] as 'a semi-enclosed coastal body of water which has a free connection with the open sea and within which seawater is measurably diluted with fresh water derived from land drainage'. A principal feature of any estuary is a regular variation in dissolved salt concentration as one moves from the inflowing river to the mouth where it opens out into the ocean. In the vertical dimension, there is a steady but sometimes irregular increase in salt concentration with depth. There are also marked seasonal fluctuations in concentration, with reduced levels being characteristic of periods of heavy precipitation or surface runoff. The Chesapeake Bay in Maryland, USA is at the mouth of the Susquehanna River and is the largest estuary on the Atlantic coast of the United States. Figure 10.3 shows the spring-time variation in *salinity* in a vertical section along the axis of Chesapeake Bay.

Salinity is the mass in grams of the solids that can be obtained from 1 kg of sea water after all the carbonate is converted to oxide, the bromine and iodine replaced by chlorine, all organic matter oxidized, and the residue dried at 480 °C to constant weight. The symbol commonly used for salinity is ‰, which is equivalent to per cent, but with a base of one thousand. Sea water has an average salinity of 35‰ (or 3.5%) representing a chloride concentration of approximately 0.56 mol L^{-1}. For the present discussion we will assume a direct linear relation between salinity and chloride molarity.

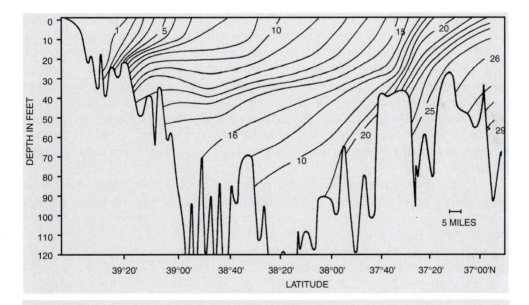

Fig. 10.3 Salinity contours for Chesapeake Bay in spring-time. The left (west) side of the diagram is the inland part of the estuary and salinity approaches that of fresh water. The right (east) side is where the estuary opens out into the Atlantic Ocean. (Schubel, J. R., *The Estuarine Environment*, American Geological Institute; 1971.) Note also the increase in salinity with depth, indicating that the fresh water from the river layers on to the surface of the saline matrix. Reprinted with permission.

We can now estimate the fractional distribution of cadmium species throughout the estuary. At the outlet of the Susquehanna River (Salinity $\simeq 1‰$, $[Cl^-] \simeq 0.02$ mol L^{-1}) as represented by line 'a' on Fig. 10.2, the approximate fractions of the three principal species are $Cd^{2+} = 0.29$, $CdCl^+ = 0.64$, $CdCl_2 = 0.07$. It is interesting that even at very low chloride ion concentrations there is considerable tendency for formation of complexes. About half way (144 km) down the Bay the surface water salinity has increased to 10‰, corresponding to $[Cl^-] = 0.16$ mol L^{-1} (line 'b' on Fig. 10.2). At this point the species have fractions as follows: $Cd^{2+} = 0.04$, $CdCl^+ = 0.52$, $CdCl_2 = 0.33$, $CdCl_3^- = 0.10$, and $CdCl_4^{2-} = 0.01$. Near the mouth of Chesapeake Bay, the salinity of surface water is 23‰ or $[Cl^-] = 0.37$ mol L^{-1}. This corresponds to line 'c' on the distribution diagram. Here the cadmium species fractional distribution is $CdCl^+ = 0.26$, $CdCl_2 = 0.39$, $CdCl_3^- = 0.29$, $CdCl_4^{2-} = 0.06$. It should not be surprising that the cadmium present in estuarine water shows increasing tendency to be complexed with chloride as one moves to higher salinity water. The values we have calculated are fractional results. Using these along with appropriate measurements of total cadmium concentration, we could estimate the amount of each species at any point in the estuary.

10.2 Two-variable diagrams: p*E*/pH diagrams

Both the phosphorus and cadmium distribution diagrams share the obvious limitation that they describe behaviour of a particular chemical in terms of only a single environmental variable—pH or chloride concentration, respectively. Clearly, in complex real systems, there are many variables operating simultaneously. A further step towards accurately describing natural systems in graphical form is then to create a two-variable diagram. In such diagrams, the dominant species are plotted in two dimensions as a function of two independent variables. There are obvious merits in creating diagrams that take into account the simultaneous involvement of two factors, but we shall see that some information (a detailed description of concentrations) is lost in order to depict the results on a flat surface.

A very widely used type of two-variable diagram for describing chemical behaviour in the hydrosphere is the p*E*/pH diagram, also called a Pourbaix diagram. We shall discuss how to construct and interpret such plots, but before doing this it is necessary to introduce the concept of p*E*.

Analogous to pH, the measure of acidity in aqueous solutions, p*E* is defined as the negative logarithm of the electron activity:[2]

$$pE = -\log a_e \qquad (10.33)$$

A large negative value of p*E* indicates a large value for the electron activity in solution which implies that reducing conditions obtain. Conversely, a large positive value of p*E* implies low electron activity in solution and oxidizing conditions. In practice p*E* values in water range from approximately -12 to 25. We shall see the reason for this range shortly.

While the definition of p*E* is simple and understandable to chemists, direct measurements of electron activity are not easily made in the way that measurements of pH are done. To show how p*E* is calculated and measured it is helpful to consider some examples.

Consider the simple half reaction

$$Fe^{3+}(aq) + e^- \rightleftharpoons Fe^{2+}(aq) \tag{10.34}$$

$$K_{eq} = \frac{a_{Fe^{2+}}}{a_{Fe^{3+}} \times a_{e^-}} \tag{10.35}$$

$$\frac{1}{a_{e^-}} = \frac{K_{eq} \times a_{Fe^{3+}}}{a_{Fe^{2+}}} \tag{10.36}$$

Using the definition of pE and taking logs of both sides of eqn 10.36, we have

$$pE = -\log a_{e^-} = \log K_{eq} + \log \frac{a_{Fe^{3+}}}{a_{Fe^{2+}}} \tag{10.37}$$

Since

$$\Delta G^\circ = -2.303\, RT \log K_{eq} \tag{10.38}$$

$$= -nFE^\circ \tag{10.39}$$

(n has the usual electrochemical meaning—i.e. the number of electrons transferred in the half reaction), at 298 K ($R = 8.314\,J\,K^{-1}\,mol^{-1}$ and $F = 96\,485\,C\,mol^{-1}$), we have

$$\log K_{eq} = \frac{nFE^\circ}{2.303\, RT} = \frac{nE^\circ}{0.0591} \tag{10.40}$$

In this case, $n = 1$, so

$$\log K_{eq} = \frac{E^\circ}{0.0591} \tag{10.41}$$

and

$$pE = \frac{E^\circ}{0.0591} + \log \frac{a_{Fe^{3+}}}{a_{Fe^{2+}}} \tag{10.42}$$

Under standard conditions, $a_{Fe^{3+}} = a_{Fe^{2+}} = 1$:

$$\log \frac{a_{Fe^{3+}}}{a_{Fe^{2+}}} = 0 \tag{10.43}$$

and

$$pE = pE^\circ = \frac{E^\circ}{0.0591} \tag{10.44}$$

For non-standard conditions:

$$pE = pE^\circ + \log \frac{a_{Fe^{3+}}}{a_{Fe^{2+}}} \tag{10.45}$$

When the standard pE° value and the actual activities (usually approximated by concentrations) of Fe^{3+} and Fe^{2+} are substituted into this equation, the pE of a particular environmental system can be calculated.

In the general case, for a reaction

$$aA + ne^- \rightleftharpoons bB \qquad (10.46)$$

where A and B are the oxidized and reduced forms of a redox couple, the reaction quotient (Q) is defined as

$$Q = \frac{(a_B)^b}{(a_A)^a} \simeq \frac{[B]^b}{[A]^a}. \qquad (10.47)$$

The reaction quotient takes the form of an equilibrium constant, but uses activities (or, as an approximation, concentrations) that obtain under any conditions, not just those at equilibrium. The general form of eqn 10.45 is then

$$pE = pE^o - \frac{1}{n}\log Q. \qquad (10.48)$$

10.2.1 Methods of calculating pE^o

Values of pE^o for a number of half reactions of environmental interest are given in Appendix 11. When additional values are required, there are several methods that make use of other readily obtainable information. The first method involves using eqn 10.44. Where a tabulated E^o value for a half reaction is available, the pE^o is readily calculated. For

$$Fe^{3+}(aq) + e^- \rightleftharpoons Fe^{2+}(aq) \qquad (10.34)$$
$$E^o = +0.771\,V$$

Therefore

$$pE^o = +\frac{0.771\,V}{0.0591\,V} = 13.0.$$

Being a ratio of two potentials, the pE^o value is a dimensionless number.

A second method for calculating pE^o makes use of the relations in eqns 10.40 and 10.44.

$$\log K_{eq} = \frac{nE^o}{0.0591} = npE^o \qquad (10.49)$$

$$pE^o = \frac{\log K_{eq}}{n}. \qquad (10.50)$$

This relation is applicable when an E^o value is not available, but where the appropriate equilibrium constant is known.

In some cases, several reactions may be combined to produce an overall half reaction. Consider the half reaction for which no tabulated E^o value is easily found:

$$Fe(OH)_3 + 3H_3O^+(aq) + e^- \rightleftharpoons Fe^{2+}(aq) + 6H_2O \qquad (10.51)$$

The reaction is the sum of

$$Fe(OH)_3 \rightleftharpoons Fe^{3+}(aq) + 3OH^-(aq) \qquad (10.51a)$$

$$Fe^{3+}(aq) + e^- \rightleftharpoons Fe^{2+}(aq) \qquad (10.51b)$$

$$3H_3O^+(aq) + 3OH^-(aq) \rightleftharpoons 6H_2O \qquad (10.51c)$$

For reaction 10.51a, $K_a = K_{sp} = 9.1 \times 10^{-39}$ and $\log K_a = -38.0$. For reaction 10.51b, $pE^o_b = \log K_b = 0.771/0.0591$ and $\log K_b = +13.0$. For reaction 10.51c,

$$K_c = \frac{1}{(K_w)^3} = 10^{42}$$

and $\log K_c = +42.0$. For the original, overall reaction

$$\log K_m = \log K_a + \log K_b + \log K_c$$
$$= -38.0 + 13.0 + 42.0$$
$$= +17.0$$

Using equation 10.50 ($n = 1$):

$$pE^o_{overall} = +17.0.$$

There is a third method for calculating pE^o values and this requires combining eqns 10.38 and 10.49:

$$\Delta G^o = -2.303 \, RT \, n \, pE^o \tag{10.52}$$

$$pE^o = \frac{-\Delta G^o}{2.303 \, RT \, n} \tag{10.53}$$

Consider the redox reaction

$$SO_4^{2-}(aq) + 10H_3O^+(aq) + 8e^- \rightleftharpoons H_2S(aq) + 14H_2O \tag{10.54}$$

$$\Delta G^o = \Delta G^o_f(H_2S) + 14\Delta G^o_f(H_2O) - \Delta G^o_f(SO_4^{2-}) - 10\Delta G^o_f(H_3O^+) - 8\Delta G^o_f(e^-)$$

Using thermochemical tables and noting that ΔG^o_f for the aqueous electron is 0, and for the hydronium ion, ΔG^o_f has the same value as for water:

$$\Delta G^o = -27.86 + 14 \times (-237.18) - 10 \times (-237.18) - (-744.60)$$
$$= -231.98 \, kJ$$

$$pE^o = \frac{-(-231.98) \, kJ \times 1000 \, J \, kJ^{-1}}{2.303 \times 8.314 \, J \, mol^{-1} \, K^{-1} \times 298.2 \, K \times 8 \, mol}$$
$$= 5.08$$

This final method for calculating pE^o values is perhaps the most generally useful but, depending on circumstances, any one of the three methods may be employed. Once pE^o values are available, it is possible to determine pE for particular non-standard environmental conditions. A two-part example follows.

10.2.2 Chromium in tannery wastes

Traditional leather tanning processes involve treating the hides with an aqueous solution of chromium. Suppose the waste water from a tannery contains $26 \, mg \, L^{-1}$ chromium, originally in the Cr^{3+} state. As the effluent flows downstream, the dissolved oxygen can oxidize Cr^{3+} to $Cr_2O_7^{2-}$. We will calculate the extent of oxidation for a situation where oxygen in the stream water is in equilibrium with atmospheric oxygen, and has a pH of

6.5 ($a_{H_3O^+} = 10^{-6.5}$). The first step is to calculate the value of pE. We can then use this value to calculate the concentrations of Cr^{3+} and $Cr_2O_7^{2-}$ assuming that the chromium species are also in equilibrium with the system.

The relevant reaction for atmospheric O_2 in equilibrium with water (often called a *well aerated system*) is

$$O_2 \text{ (g)} + 4H_3O^+ \text{ (aq)} + 4e^- \rightleftharpoons 6H_2O \qquad (10.55)$$

$$E^\circ = 1.23\,V$$

$$pE^\circ = \frac{1.23\,V}{0.0591\,V} = 20.8$$

Using eqn 10.48:

$$pE = pE^\circ - \frac{1}{n}\log\frac{1}{P_{O_2}/P^\circ \times (a_{H_3O^+})^4}$$

$$= 20.8 - \frac{1}{4}\log\frac{1}{0.209 \times (10^{-6.5})^4}$$

$$= 14.1$$

Note that in these calculations, the pressure is given as a ratio of P_{O_2}/P° which is numerically identical to pressure in atmospheres. For oxygen, which makes up 20.9% of the atmosphere, $P_{O_2} = 21\,200$ Pa; $P^\circ = 101\,325$ Pa.

For the Cr system:

$$Cr_2O_7^{2-} \text{ (aq)} + 14H_3O^+ \text{ (aq)} + 6e^- \rightleftharpoons 2Cr^{3+} \text{ (aq)} + 17H_2O \qquad (10.56)$$

$$E^\circ = 1.36V \text{ and } pE^\circ = 23.0$$

$$pE = pE^\circ - \frac{1}{6}\log\frac{[Cr^{3+}]^2}{[Cr_2O_7^{2-}](a_{H_3O^+})^{14}}$$

Since the chromium and oxygen systems are in equilibrium, pE is the same for both:

$$14.1 = 23.0 - \frac{1}{6}\log\frac{[Cr^{3+}]^2}{[Cr_2O_7^{2-}](10^{-6.5})^{14}}$$

$$= 23.0 - \frac{1}{6}\log\frac{1}{(10^{-6.5})^{14}} - \frac{1}{6}\log\frac{[Cr^{3+}]^2}{[Cr_2O_7^{2-}]}$$

$$= 7.8 - \frac{1}{6}\log\frac{[Cr^{3+}]^2}{[Cr_2O_7^{2-}]}$$

$$\log\frac{[Cr^{3+}]^2}{[Cr_2O_7^{2-}]} = -37.8$$

$$\frac{[Cr^{3+}]^2}{[Cr_2O_7^{2-}]} = 1.6 \times 10^{-38}$$

This very small ratio indicates that virtually all of the Cr^{3+} would be oxidized to $Cr_2O_7^{2-}$ and this, in fact, has serious environmental consequences. A widely used method of leather tanning involves a two-bath process in which the hides are soaked for several hours in a tank containing chromic acid in order to effect cross-linking between proline and hydroxy-proline residues in protein molecules in the animal skin as a way of toughening the material. The hides are then transferred into another tank containing reducing agents such as sucrose in order to reduce the excess chromium (VI) to chromium (III). The excess is then discharged in the waste water.

The leather industry is important throughout the globe, with India the world's major producer of leather goods. In that country there are large industrial establishments manufacturing shoes, luggage, garments, and so on, but much of the industry is small-scale and scattered in numerous villages in rural areas of the country. The northern part of Tamil Nadu state in South India is the centre of the leather industry.

The quantity of chromium (III) wasted in effluents of a small-scale chrome tannery in India is estimated to be approximately 0.4 kg per 100 kg of raw hides. This is about one half of that added in the first bath. Chromium (III) has low toxicity, but in the hexavalent form it is both toxic and carcinogenic to humans. It is also toxic to plants and large areas of the leather-producing areas of Tamil Nadu are devoid of vegetation. The consequences to humans in that region are not documented. In any case, the release of such large amounts of chromium with its subsequent conversion to the thermodynamically stable $Cr_2O_7^{2-}$ species is a major environmental issue.

10.2.3 pE/pH diagrams

We are now in a position to construct a pE/pH diagram. The diagram will take the form of a two-dimensional plot of pE (ordinate, y axis) versus pH (abscissa, x axis) in which areas in the diagram define the regions where particular species are dominant. In saying a particular species dominates, we must define conditions at the boundary between domains. This requires that we calculate and draw lines on the diagram corresponding to these boundaries.

An important issue, then, is how to define the boundary conditions in different situations. For the case of a reaction between two species where one is dissolved in the water and the other is a gas, the boundary condition is set as $P^o = 101\,325\,Pa = 1\,atm$. This is the minimum pressure required for gas evolution from the aqueous solution. In other words, when P_{gas} is greater than $101\,325\,Pa$, the gas phase itself is said to predominate. As an example, this type of boundary condition is used in the case of the evolution of hydrogen gas from an acidic solution:

$$2H_3O^+ \text{ (aq)} + 2e^- \rightleftharpoons H_2 \text{ (g)} + 2H_2O \tag{10.57}$$

Another type of condition is required for the situation where only soluble species are involved, or where there is a soluble species reacting to form one that is insoluble. Examples of these two cases are

$$Sn^{4+} \text{ (aq)} + 2e^- \rightleftharpoons Sn^{2+} \text{ (aq)} \tag{10.58}$$

and

$$MnO_2 + 4H_3O^+ \text{ (aq)} + 2e^- \rightleftharpoons Mn^{2+} \text{ (aq)} + 6H_2O \tag{10.59}$$

In these cases, we arbitrarily define a concentration (as an estimate of activity) below which a particular species is considered to be soluble. Any concentration may be chosen but it is common to use values between 10^{-5} and $10^{-2}\,mol\,L^{-1}$. Of course, in constructing a diagram one should consistently use the same concentration for all boundaries.

We will illustrate how boundary conditions are chosen and used in the construction of a pE/pH diagram by considering the aqueous sulfur system where the species of interest are SO_4^{2-} (aq), HSO_4^- (aq), S (s), HS^- (aq), and H_2S (aq). Before beginning calculations involving these species, however, there are some preliminary calculations common to all diagrams pertaining to the hydrosphere.

As we had noted in Chapter 10, water itself has a limited range of pE and pH values within which it is stable. Under highly reducing conditions (low pE), water is reduced:

$$2H_2O + 2e^- \rightleftharpoons H_2 \text{ (g)} + 2OH^- \text{ (aq)} \tag{10.60}$$

For this reaction

$$E^o = -0.828\,V$$
$$pE^o = -14.0$$

Equation 10.48 for this reaction is written as

$$pE = pE^o - \frac{1}{2}\log(P_{H_2}/P^o \times (a_{OH^-})^2) \tag{10.61}$$

(note again that here and in subsequent calculations, we will use concentrations rather than activities except for hydroxyl and hydronium ions).

For the boundary involving a gas, we choose the condition that

$$P_{H_2} = P^o = 101\,325\,Pa$$
$$pE = -14.0 - \log(a_{OH^-})$$
$$= -14.0 + pOH$$

Since

$$pH + pOH = 14$$

then

$$pE = -pH$$

This line then defines the boundary for water stability with respect to reduction and is shown as the lower line on Fig. 10.4. Where the pE value is less than the pH value, water is unstable. It is possible also to calculate the same line by taking the reduction reaction to be

$$2H_3O^+ \text{ (aq)} + 2e^- \rightleftharpoons H_2 \text{ (g)} + 2H_2O \tag{10.62}$$

Considering the other extreme, highly oxidizing conditions, water is unstable with respect to O_2 evolution and the reaction is written as

$$6H_2O \rightleftharpoons 4H_3O^+(aq) + O_2 + 4e^- \tag{10.63}$$

For this reaction

$$E^o = 1.229\,V$$
$$pE^o = E^o/0.0591 = 20.80$$

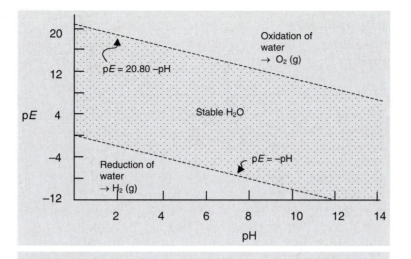

Fig. 10.4 Region of stability and stability boundaries for water on a pE/pH diagram.

The production of oxygen from water is an oxidation reaction. It is important to note, however, that we use the IUPAC convention of defining E^o and E, and pE^o and pE, in terms of the reverse, reduction process:

$$pE = pE^o - \tfrac{1}{4}\log\left(1/P_{O_2} \times (a_{H_3O^+})^4\right) \qquad (10.64)$$

Once again, the boundary condition requires that the pressure of the gas equal atmospheric pressure

$$P_{O_2} = P^o = 101\,325\,\text{Pa}$$
$$pE = 20.80 - \log(1/a_{H_3O^+})$$
$$= 20.80 - pH$$

This, then, defines the upper line on Fig. 10.4 and the region between the two lines is the stability region for water on a pE/pH diagram. Above the top boundary, water is oxidized with the evolution of oxygen, while below the lower boundary it is reduced and releases hydrogen.

10.2.4 The sulfur system

Now we can deal with the sulfur species. To create the pE/pH diagram, we will be defining boundaries between species and superimposing these on the water stability diagram. We may consider the various pairs of species in any order but it is helpful to have an idea of what to expect before beginning calculations—for example we expect that HSO_4^- will be important at low pH and SO_4^{2-} at high pH. Similarly SO_4^{2-} would exist under oxidizing, high pE conditions, while HS^- is a reduced, low pE form. For all soluble species, we will choose 10^{-2} mol L^{-1} as our arbitrary definition for the boundary.

The SO_4^{2-}/HSO_4^- boundary

The equation describing this boundary requires hydronium ion, but there is no oxidation or reduction involved:

$$SO_4^{2-}(aq) + H_3O^+(aq) \rightleftharpoons HSO_4^-(aq) + H_2O \qquad (10.65)$$

To make things easier, we use H^+ as an abbreviation for H_3O^+, the hydronium ion, but of course the result would be the same if the latter species were used in the equations and calculations.

$$\Delta G^o = \Delta G_f^o\ (HSO_4^-) - \Delta G_f^o\ (SO_4^{2-}) - \Delta G_f^o\ (H^+)$$
$$= -755.99 - (-744.60) - 0$$
$$= -11.39\,kJ\ = -11\,390\,J$$

$$\log K = \frac{-\Delta G^o}{2.303\,RT}$$
$$= \frac{+11\,390}{2.303 \times 8.314 \times 298.2}$$
$$= 1.995$$

$$K = \frac{[HSO_4^-]}{[SO_4^{2-}]a_{H^+}}$$

At the boundary, $[HSO_4^-] = [SO_4^{2-}] = 10^{-2}$ M, and

$$K = \frac{1}{a_{H^+}}$$
$$\log K = pH = 1.995$$

Therefore, the boundary between HSO_4^- and SO_4^{2-} is a vertical line at pH 1.995. Below this value, HSO_4^- is the dominant form; above it SO_4^{2-} is most important. See line 'a' on Fig. 10.5.

The HSO_4^-/S^o boundary

$$HSO_4^-\ (aq) + 7H^+\ (aq) + 6e^- \rightleftharpoons S^o\ (s) + 4H_2O \qquad (10.66)$$

$$\Delta G^o = \Delta G_f^o\ (S) + 4\Delta G_f^o\ (H_2O) - \Delta G_f^o\ (HSO_4^-) - 7\Delta G_f^o\ (H^+) - 6\Delta G_f^o\ (e^-)$$
$$= 0 + 4(-237.18) - (-755.99) - 0 - 0$$
$$= -192.73\,kJ = -192\,730\,J$$

$$pE^o = \frac{-\Delta G^o}{2.303\,nRT} = \frac{+192\,730}{2.303 \times 6 \times 8.314 \times 298.2} = 5.626$$

$$pE = pE^o - \frac{1}{6}\log\frac{1}{[HSO_4^-](a_{H^+})^7}$$

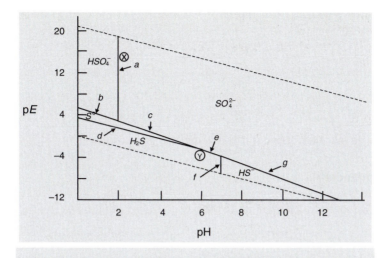

Fig. 10.5 The pE/pH for the aqueous sulfur system.

At the boundary, $[\text{HSO}_4^-] = 10^{-2}\,\text{mol}\,\text{L}^{-1}$:

$$pE = 5.626 - \frac{7}{6}\log\frac{1}{a_{\text{H}^+}} - \frac{1}{6}\log\frac{1}{10^{-2}}$$

$$= 5.626 - \frac{7}{6}\text{pH} - 0.333$$

$$= 5.293 - 1.167\,\text{pH}$$

This is line 'b' on Fig. 10.5; above it is the domain of HSO_4^-; below it is elemental sulfur, S°.

The SO_4^{2-}/S° boundary

$$\text{SO}_4^{2-}\ (\text{aq}) + 8\text{H}^+\ (\text{aq}) + 6\text{e}^- \rightleftharpoons S^\circ\ (\text{s}) + 4\text{H}_2\text{O} \tag{10.67}$$

$$\Delta G^\circ = \Delta G_{\text{f}}^\circ\ (S) + 4\Delta G_{\text{f}}^\circ\ (\text{H}_2\text{O}) - \Delta G_{\text{f}}^\circ\ (\text{SO}_4^{2-}) - 8\Delta G_{\text{f}}^\circ\ (\text{H}^+) - 6\Delta G_{\text{f}}^\circ\ (\text{e}^-)$$

$$= 0 + 4(-237.18) - (-744.60) - 0 - 0$$

$$= -204.12\,\text{kJ} = -204\,120\,\text{J}$$

$$pE^\circ = \frac{\Delta G^\circ}{2.303\,nRT} = \frac{+204\,120}{2.303 \times 6 \times 8.314 \times 298.2}$$

$$= 5.958$$

$$pE = pE^\circ - \frac{1}{6}\log\frac{1}{[\text{SO}_4^{2-}](a_{\text{H}^+})^8}$$

At the boundary, $[SO_4^{2-}] = 10^{-2} \, \text{mol L}^{-1}$:

$$pE = 5.958 - \frac{8}{6} \log \frac{1}{a_{H^+}} - \frac{1}{6} \log \frac{1}{10^{-2}}$$

$$= 5.958 - \frac{8}{6} pH - 0.333$$

$$= 5.625 - 1.333 \, pH$$

This is line 'c' on Fig. 10.5; above it is the domain of SO_4^{2-}, below it S°.

The S°/H_2S boundary

$$S(s) + 2H^+ (aq) + 2e^- \rightleftharpoons H_2S (aq) \tag{10.68}$$

$$\Delta G^\circ = \Delta G_f^\circ (H_2S) - \Delta G_f^\circ (S) - 2\Delta G_f^\circ (H^+) - 2\Delta G_f^\circ (e^-)$$

$$= -27.86 - 0 - 0 - 0$$

$$= -27.86 \, \text{kJ} = -27\,860 \, \text{J}$$

$$pE^\circ = \frac{-\Delta G^\circ}{2.303 \, nRT} = \frac{+27\,860}{2.303 \times 2 \times 8.314 \times 298.2}$$

$$= 2.400$$

$$pE = pE^\circ - \frac{1}{2} \log \frac{[H_2S]}{(a_{H^+})^2}$$

At the boundary, $[H_2S] = 10^{-2} \, \text{mol L}^{-1}$:

$$pE = 2.440 + 1 - pH$$

$$= 3.440 - pH$$

This is line 'd' on Fig. 10.5; above it is the domain of S°, below it H_2S.

The SO_4^{2-}/H_2S boundary

$$SO_4^{2-} (aq) + 10H^+ (aq) + 8e^- \rightleftharpoons H_2S (aq) + 4H_2O \tag{10.69}$$

$$\Delta G^\circ = \Delta G_f^\circ (H_2S) + 4\Delta G_f^\circ (H_2O) - \Delta G_f^\circ (SO_4^{2-}) - 10\Delta G_f^\circ (H^+) - 8\Delta G_f^\circ (e^-)$$

$$= -27.86 + 4(-237.18) - (-744.60) - 0 - 0$$

$$= -231.98 \, \text{kJ} = -231\,980 \, \text{J}$$

$$pE^\circ = \frac{-\Delta G^\circ}{2.303 \, nRT} = \frac{+231\,980}{2.303 \times 8 \times 8.314 \times 298.2}$$

$$= 5.079$$

$$pE = pE^\circ - \frac{1}{8} \log \frac{[H_2S]}{[SO_4^{2-}](a_{H^+})^{10}}$$

At the boundary, $[H_2S] = [SO_4^{2-}] = 10^{-2}\,mol\,L^{-1}$:

$$pE = 5.079 - \frac{10}{8}\log\frac{1}{a_{H^+}}$$
$$= 5.079 - 1.25\,pH$$

This is line 'e' on Fig. 10.5 (a very small segment, difficult to distinguish from line 'd'); above it is the domain of SO_4^{2-}, below it H_2S.

The H₂S/HS⁻ boundary

$$H_2S\ (aq) \rightleftharpoons HS^-\ (aq) + H^+\ (aq) \tag{10.70}$$

As for the HSO_4^-/SO_4^{2-} boundary, this is not a redox reaction and therefore the line will be vertical with the protonated species on the left.

$$\Delta G^o = \Delta G_f^o\ (HS^-) + \Delta G_f^o\ (H^+) - \Delta G_f^o\ (H_2S)$$
$$= 12.08 + 0 - (-27.86)$$
$$= 39.94\,kJ = 39\,940\,J$$

$$\log K = \frac{-\Delta G^o}{2.303\,nRT} = \frac{-39\,940}{2.303 \times 8.314 \times 298.2}$$
$$= -6.995$$
$$K = \frac{[HS^-]a_{H^+}}{[H_2S]}$$

At the boundary, $[HS^-] = [H_2S] = 10^{-2}\,mol\,L^{-1}$:

$$K = a_{H^+}$$
$$\log K = \log[H^+] = -pH = -6.995$$
$$pH = 6.995$$

This is line 'f' on Fig. 10.5; to the left is the domain of H_2S and to the right HS^-.

The SO₄²⁻/HS⁻ boundary

$$SO_4^{2-}\ (aq) + 9H^+\ (aq) + 8e^- \rightleftharpoons HS^-\ (aq) + 4H_2O \tag{10.71}$$

$$\Delta G^o = \Delta G_f^o\ (HS^-) + 4\Delta G_f^o\ (H_2O) - \Delta G_f^o\ (SO_4^{2-}) - 9\Delta G_f^o\ (H^+) - 8\Delta G_f^o\ (e^-)$$
$$= 12.08 + 4(-237.18) - (-744.60) - 0 - 0$$
$$= -192.04\,kJ = -192\,040\,J.$$

$$pE^o = \frac{-\Delta G^o}{2.303\,nRT} = \frac{+192\,040}{2.303 \times 8 \times 8.314 \times 298.2}$$
$$= 4.204.$$
$$pE = pE^o - \frac{1}{8}\log\frac{[HS^-]}{[SO_4^{2-}](a_{H^+})^9}$$

At the boundary, $[HS^-] = [SO_4^{2-}] = 10^{-2}\,mol\,L^{-1}$:

$$pE = 4.202 - \frac{9}{8}\log\frac{1}{a_{H^+}}$$
$$= 4.204 - 1.125\,pH$$

This is line 'g' on Fig. 10.5; above it is the domain of SO_4^{2-}, below it HS^-.

The completed diagram is shown in Fig. 10.5. To interpret the plot, it is helpful to make use of a template (Fig. 10.6) which contains the H_2O stability boundaries and within these, *approximate* pE/pH regions for a variety of environments.

Two examples of particular environmental situations follow. For mine wastes, at a pH of 2.5 and exposed to the atmosphere, so that they are well aerated ($pE \simeq 15$), corresponding to point X on Fig. 10.5, the most important sulfur species in solution would be sulfate. Where the originally mined material was sulfide ore, such as from the copper–nickel ores of Sudbury, Ontario, Canada, sulfur might initially go into solution as sulfide but would eventually (the kinetics are relatively fast) be oxidized to sulfate. For a swamp or paddy (rice) field where soil containing a high content of organic matter (OM) is submerged, the OM acts as a reducing agent and creates low pE conditions. For example, such a soil might have $pH = 6$ and $pE = -3$, corresponding to point Y in the figure. This is near several boundaries, but it would not be surprising to detect the presence of H_2S in the interstitial water of the sediment. Organic matter in soil typically contains about 2% sulfur.

Additional important pE/pH diagrams will be given in the problems section of this chapter and others are discussed at later points in the book.

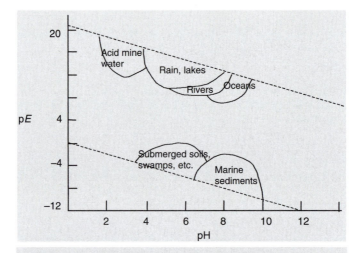

Fig. 10.6 Template for use with pE/pH plots. The indicated areas show typical pE and pH values for commonly encountered aqueous, soil, and sediment environments.

10.3 Measurements of p*E*

In principle it should be simple to measure p*E* in a real environment. In practice, non-equilibrium conditions make stable and meaningful readings exceedingly difficult. The requirements for measurement are an indicator electrode which is inert and develops a potential in response to the ratio of redox couples which are in true equilibrium. A platinum electrode of small size can serve this purpose. In addition, a reference electrode is used so that the indicated potential may be measured against a known value. As with all potential measurements it is essential to ensure that negligible current be drawn during the measurement process. This is readily ensured by using an electronic voltmeter. Figure 10.7 shows an apparatus suitable for determining p*E*.

Suppose the potential in soil pore water is measured in the field and the value is found to be $+713\,mV$ vs. the saturated calomel electrode (SCE). The value of E is recalculated vs. the normal hydrogen electrode (NHE), using the known potential, $+0.242\,V$ of the SCE:

$$E \text{ vs. NHE} = 0.713 + 0.242\,V$$
$$= 0.955\,V$$

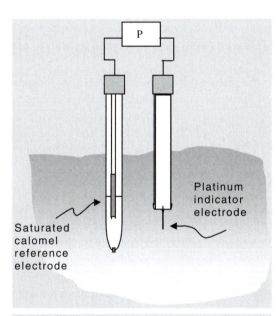

Fig. 10.7 Apparatus for measuring p*E* in water, soil, or sediment, in the laboratory or in the field. The indicator electrode is often a platinum disc or wire; any appropriate reference electrode can be used. P = the potential measuring device, usually an electronic voltmeter.

The value of pE is then readily calculated using eqn 10.44:

$$pE = \frac{0.955}{0.0591} = 16.23$$

A pE of 16.23 indicates an oxidizing regime.

As has been suggested, frequently non-equilibrium conditions obtain in water, soil, and sediments. Furthermore, a typical environmental sample will comprise many species that are part of several different redox couples and that are themselves not in equilibrium. The result will be an unstable, drifting potential. For this and other reasons it is frequently not possible to make accurate and precise pE readings as can be done for pH. It is, however, usually possible to obtain an approximate value and define a regime as generally in the oxidizing or reducing categories. As the template indicates, an intermediate redox status is rare and usually transient as the system moves to a more stable higher or lower pE value.

The main points

1 The environmental behaviour of an element or compound depends on the particular form of the species that is present. In the aqueous environment; the species distribution depends on a number of factors including solution pH, pE, and the nature and availability of complexing ligands.

2 In order to determine the species distribution of a particular substance, it is usual to make an assumption that all forms are in equilibrium with the surroundings. Furthermore, it is frequently assumed that concentration and activity of species are the same. Accurate calculations, however, require activity estimations based on detailed knowledge of the solution composition and ionic strength. The two assumptions and others mean that calculations in actual complex situations are only approximate and can sometimes be erroneus.

3 Single variable distribution diagrams show how the concentrations of different species change as a function of the one defined variable.

4 Two variable diagrams indicate regions on a two-dimensional plot where individual species predominate. No detailed information about concentrations within the domains is given.

5 The pH and pE of the aqueous environment are two key properties which define the nature of chemical species. The pE is a measure of redox status and is high when the water is well aerated and low where oxygen is excluded or has been consumed.

Additional reading

1 Brookins, D. G., *Eh–pH Diagrams for Geochemistry*. Springer, Berlin; 1987.

2 Morel, F. M. M. and J. G. Hering, *Principles and Applications of Aquatic Chemistry*, John Wiley & Sons, New York; 1993.

Problems

1 Consider the case of sulfurous acid (H_2SO_3), which is quantitatively formed when sulfur dioxide gas dissolves in water.

(a) Without calculation, but referring to the appropriate dissociation constants, carefully sketch a single-variable species distribution diagram (α vs. pH) for this acid in water.

(b) Calculate accurately the concentration of all the sulfite species at pH = 7.0.

2 Use standard thermodynamic relations and assumptions to show that

$$n(pE^\circ) = \log K_{eq}$$

3 Consider the equilibrium of an aqueous solution containing permanganate, hydronium ion and solid manganese dioxide as indicated in the following equation:

$$MnO_4^-(aq) + 4H_3O^+(aq) + 3e^- \rightleftharpoons MnO_2(s) + 6H_2O$$

How are the solution pH and pE affected by the individual addition of each of the following:

$$MnO_2, CO_2(g), Fe^{2+}(aq), \text{ and } CaCO_3?$$

4 A potential measurement in a paddy field yields a reading of −278 mV vs. the saturated calomel reference electrode. What is the pE value for this soil?

5 Iron toxicity has occasionally been observed in rice plants grown in lowlands. (submerged soils). Use the simple pE/pH diagram for iron in Fig. 10P.1 to explain how this can happen.

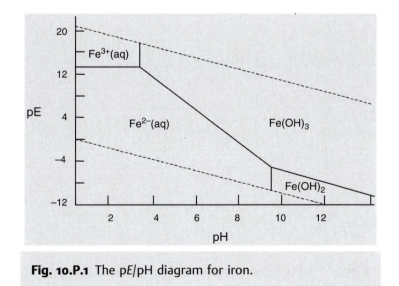

Fig. 10.P.1 The pE/pH diagram for iron.

6 Iron (III) hydroxide can act as an oxidizing agent as indicated by the following half reaction:

$$Fe(OH)_3 + 3H_3O^+(aq) + e^- \rightarrow Fe^{2+}(aq) + 6H_2O$$

(a) Assuming pH = 7, and otherwise standard conditions, can the oxidation of NH_4^+ to NO_3^- be effected by this reaction in the hydrosphere?

(b) Could the reaction bring about the oxidation of HS^- to SO_4^{2-} at a pH of 9 (again, otherwise standard conditions)?

7 Consider the following aqueous uranium species:

$$UO_2^{2+} \rightleftharpoons U^{4+} \rightleftharpoons UOH^{3+}$$

Calculate the UOH^{3+} concentration under typical pE and pH conditions for acid mine drainage waters assuming $C_U = 1 \times 10^{-5}$ mol L^{-1}. ΔG_f^o values/kJ mol^{-1}:

$$
\begin{array}{ll}
UO_2^{2+} & -989.5 \\
UOH^{3+} & -810.0 \\
U^{4+} & -579.3
\end{array}
$$

8 Given below is a partially completed pE/pH diagram for arsenic species (Fig. 10P.2).

(a) Calculate the equation for the line (shown) for the boundary $H_2AsO_4^-/H_3AsO_3$.

(b) Calculate equations for the lines (not shown) for the boundaries $H_3AsO_4/H_2AsO_4^-$ and H_3AsO_4/H_3AsO_3.

Assume a temperature of 25 °C. For boundaries use $C_{species} = 1 \times 10^{-4}$ mol L^{-1}.

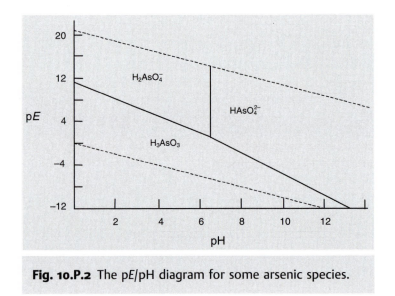

Fig. 10.P.2 The pE/pH diagram for some arsenic species.

(c) Comment on predicted As species in sea water and acid mine drainage; in both situations assume that the solution is well aerated. $\Delta G_f^\ominus$ values/kJ mol^{-1}:

H_3AsO_4	-769.3
$H_2AsO_4^-$	-748.8
H_3AsO_3	-640.0
$H_2AsO_3^-$	-587.7
H_2O	-237.2

Notes

1 Pritchard, D. W., What is an estuary: physical viewpoint. In *Estuaries*, American Association for the Advancement of Science, Publication No 83, Washington, DC; 1967.

2 Both pH and pE are *defined* in terms of activity. In equilibrium derivations in this section, we will employ activities for all species. However, while recognizing the inconsistency, we will revert to use of concentrations for soluble species other than hydronium and hydroxyl ions and the electron in subsequent calculations.

11

Gases in water

GASES (like oxygen) and other volatile compounds (like gasoline) whose vapours may be present at low concentration in the atmosphere, distribute themselves between air and the aqueous part of the environment. A quantitative description of the distribution depends on properties of the substance itself, such as its vapour pressure, solubility, and ability to react with components of the hydrosphere. The distribution also depends on properties of the environment, such as atmospheric conditions, and the water solution composition. One frequently encountered situation is the need to describe the extent to which a gas, present in a steady state mixing ratio in the atmosphere, dissolves in natural waters. In many such cases, an assumption of equilibrium between the two phases is appropriate. A second important type of situation is to describe the volatilization of neutral molecules from water into the atmosphere. If the atmosphere is enclosed and limited in volume, an equilibrium calculation may suffice here as well. More commonly, volatilization takes place into the open atmosphere and equilibrium is never attained. For this situation, the rate of volatilization is markedly influenced by air and water movement and temperature along with the other factors noted above. In this chapter, we will be examining only the equilibrium type of situation.

11.1 Simple gases

In the present context, we define a simple gas as one such as oxygen, that does not react with water during dissolution. At equilibrium, the concentration of a simple gas is adequately described by a Henry's law relationship. Henry's law is based on Raoult's law and is valid at small concentrations. In its fundamental form, it may be written as

$$P_g = KX_l \tag{11.1}$$

where P_g is the partial pressure of the gas in the bulk atmosphere/Pa; K is the Henry's law constant/Pa (K is a function of the temperature, the gas, and the solvent. In the environmental context, the solvent is always water); X_l is the equilibrium mole fraction (a dimensionless number) of the solute in the bulk liquid (aqueous) phase.

For our purpose, a more useful form (and sufficiently accurate at low concentrations in environmental samples) uses a reciprocal version of eqn 11.1 and employs molar units for the aqueous solution concentration:

$$[G]_l = K_H P_g \tag{11.2}$$

where $[G]_l$ is the equilibrium concentration of solute in the bulk liquid (aqueous) phase/ $mol\,L^{-1}$; K_H is the Henry's law constant/$mol\,L^{-1}\,Pa^{-1}$; P_g is, as in eqn 11.1, the partial pressure of the gas in the bulk atmosphere/Pa.

When referring to other data sources where Henry's law is applied to environmental calculations, it is important to be aware of the algebraic form of the equation employed as well as the units used to express concentration in the two phases (see problem 4 in this chapter). These can always be determined by careful examination of the units for the Henry's law constant.

Values of K_H, as defined in eqn 11.2, for various gases are given in Table 11.1.

Table 11.1 Henry's law constants for selected gases dissolved in water at 25 °C

Gas	K_H/mol L^{-1} Pa^{-1}
O_2	1.3×10^{-8}
N_2	6.4×10^{-9}
CH_4	1.3×10^{-8}
CO_2	3.3×10^{-7}
SO_2	1.2×10^{-5}
NH_3	5.7×10^{-4}
Hg	8.6×10^{-7}
CCl_4	3.7×10^{-7}
CH_3COCH_3	3.9×10^{-3}

The equilibrium concentration of oxygen in water at 25 °C is calculated using the value of K_H from this table. Before using eqn 11.2, however, the partial pressure of oxygen in a moist atmosphere must be calculated. To do this, we need to know the mol fraction of oxygen (X_{O_2}) in the atmosphere and the vapour pressure of water (P_{H_2O}) at 25 °C. These values can be obtained from Table 8.1 and Fig. 8.1. The partial pressure of oxygen is then given by:

$$P_{O_2} = (P^o - P_{H_2O})X_{O_2} \tag{11.3}$$

where P^o = atmospheric pressure at 25 °C. P_{H_2O} at 25 °C is 3.2×10^3 Pa.

$$P_{O_2} = (1.01 \times 10^5 - 3.2 \times 10^3)\,Pa \times 0.209$$
$$= 2.04 \times 10^4\,Pa.$$

Using eqn 11.2:

$$[O_2]_{aq} = 1.3 \times 10^{-8}\,mol\,L^{-1}\,Pa^{-1} \times 2.04 \times 10^4\,Pa$$
$$= 2.7 \times 10^{-4}\,mol\,L^{-1}$$
$$= 8.5\,mg\,L^{-1}.$$

The $8.5\,mg\,L^{-1}$ concentration of oxygen in water at $25\,°C$ is important to keep in mind when one is discussing the concept of biological oxygen demand (BOD) (see Chapter 15).

As is well known, the Henry's law constant and therefore the solubility of oxygen (and most other gases) in water increases with decreasing temperature. This implies that oxygen dissolution is an exothermic process. At $5\,°C$ the solubility is approximately $12.4\,mg\,L^{-1}$ while at $30\,°C$ it is $7.5\,mg\,L^{-1}$. The solubility also depends on ionic strength of the solution and is somewhat lower under high salinity (sea water) conditions.

Of course, Henry's law calculations may also be used in a reverse fashion. Consider an enclosed container with 100 mL of an aqueous solution containing 0.5 g of acetone. The concentration is $0.5\,g/(58.1\,g\,mol^{-1} \times 0.1\,L) = 0.0861\ mol\,L^{-1}$; the value of K_H for acetone is $3.9 \times 10^{-3}\,mol\,L^{-1}\,Pa^{-1}$:

$$[Ac]_{aq} = K_H P_{ac} \tag{11.4}$$

The equilibrium vapour pressure of acetone above the solution is then

$$P_{ac} = [Ac]_{aq}/K_H = 0.0861\,mol\,L^{-1}/3.9 \times 10^{-3}\,mol\,L^{-1}\,Pa^{-1}$$
$$= 22\,Pa$$

As we noted above, this 'reverse' type of Henry's law calculation assumes an equilibrium situation and cannot be used where evaporation occurs into the open, constantly replenished atmosphere. There are other features that may also limit the establishment of equilibrium. For some gases such as oxygen, nitrogen, carbon dioxide, methane, and nitrous oxide, diffusion through the liquid film in contact with the surrounding atmosphere is slow and controls the rate of transfer from the atmosphere to water; while diffusion in the gas phase controls the reaction rate in the case of sulfur dioxide, sulfur trioxide, and hydrogen chloride. In all cases, the rate of establishment of equilibrium depends on the degree of contact between the two phases.

11.2 Gases that react with water

Many gases of environmental importance, including sulfur dioxide and ammonia, react with water as they go into solution. The extent of reaction must be taken into account in calculating solubilities. Perhaps the most important environmental example of a reacting gas is carbon dioxide; calculating the total solubility and individual concentrations of all carbonate species requires modification of the simple Henry's law calculation.

When carbon dioxide dissolves in water, to a small extent it reacts to form carbonic acid, H_2CO_3. The extent of this equilibrium is described by $K_r = [H_2CO_3]/[CO_2\,(aq)]$ which has a value of 2×10^{-3}. The small value indicates that most of the undissociated acid actually is present as $CO_2\,(aq)$. For this reason, when we describe the way in which carbon dioxide in water acts as an acid, we usually use a K_{a1} that combines the K_r and the constant (K'_{a1}) for the loss of the first proton of H_2CO_3. The complete picture relating the gaseous and aqueous forms of carbon dioxide then appears as follows:

$$K_H = 3.3 \times 10^{-7} \qquad K_r = 2 \times 10^{-3} \qquad K_{a1}' = 2 \times 10^{-4} \qquad K_{a2} = 4.7 \times 10^{-11}$$

$$CO_2\,(g) \; \rightleftharpoons \; CO_2\,(aq) \; \rightleftharpoons \; H_2CO_3\,(aq) \; \rightleftharpoons \; HCO_3^-\,(aq) \; \rightleftharpoons \; CO_3^{2-}\,(aq)$$

$$K_{a1} = K_{a1}' \times K_r = 4.5 \times 10^{-7}$$

$$(11.5)$$

In the atmosphere where its mixing ratio is 365 ppmv, the partial pressure of carbon dioxide is

$$P_{CO_2} = (1.01 \times 10^5 - 3.2 \times 10^3) \times 365 \times 10^{-6}\,Pa = 35.8\,Pa$$

Applying eqn 11.2, the concentration of $CO_2\,(aq)$ is then calculated to be

$$[G]_1 = [CO_2\,(aq)] = 3.3 \times 10^{-7}\,mol\,L^{-1}\,Pa^{-1} \times 35.8\,Pa = 1.2 \times 10^{-5}\,mol\,L^{-1}$$

The other aqueous carbonate species are determined using the standard acid–base relations. In these and other calculations, the square brackets refer to aqueous molar concentrations, an approximation for activities as described in Chapter 10.

$$[H_2CO_3\,(aq)] = K_r \times [CO_2(aq)] = 2.4 \times 10^{-8}\,mol\,L^{-1} \qquad (11.6)$$

$$K_{a1} = \frac{[HCO_3^-][H_3O^+]}{[CO_2\,(aq)]} = 4.5 \times 10^{-7} \qquad (11.7)$$

If, besides carbon dioxide, there are no additional sources of acid or carbonate species:

$$[HCO_3^-] = [H_3O^+] = ([CO_2(aq)] \times K_{a1})^{\frac{1}{2}} = 2.2 \times 10^{-6}\,mol\,L^{-1} \qquad (11.8)$$

$$[CO_3^{2-}] = \frac{[CO_2(aq)] \times K_{a1} \times K_{a2}}{[H_3O^+]^2} = 5.0 \times 10^{-11}\,mol\,L^{-1} \qquad (11.9)$$

Since $H_2CO_3\,(aq)$ is always present in a low concentration compared to $CO_2\,(aq)$ (0.2% of the latter value as defined by the K_r value) its concentration usually is not calculated. The relative contributions of the HCO_3^- and CO_3^{2-} species depend on the nature of the solution, including its pH. In the present example, the total solubility of all carbonate species is $1.4 \times 10^{-5}\,mol\,L^{-1}$ with $CO_2\,(aq)$ and $HCO_3^-\,(aq)$ being the quantitatively important forms.

Note that the above calculation indicates that the hydronium ion concentration of water in equilibrium with atmospheric carbon dioxide is $2.0 \times 10^{-6}\,mol\,L^{-1}$ corresponding to a pH of 5.7. This is the pH of 'clean' rain or of any other pure water sample in equilibrium with atmospheric carbon dioxide, as has been discussed in Chapter 5.

In many environmental situations, water is exposed to atmospheric carbon dioxide and is also in contact with limestone ($CaCO_3$). If the system is at equilibrium, the concentrations of all important species in the water are calculated as follows:

$$K_H = \frac{[CO_2]}{P_{CO_2}} = 3.3 \times 10^{-7} \qquad (11.10)$$

$$K_{a1} = \frac{[HCO_3^-][H_3O^+]}{[CO_2]} = 4.5 \times 10^{-7} \tag{11.11}$$

$$K_{a2} = \frac{[CO_3^{2-}][H_3O^+]}{[HCO_3^-]} = 4.7 \times 10^{-11} \tag{11.12}$$

$$K_{sp} = [Ca^{2+}][CO_3^{2-}] = 5 \times 10^{-9} \tag{11.13}$$

In any solution, the total ionic positive and negative charge must be equal. In the present case, this charge balance is described by eqn 11.14.

$$2[Ca^{2+}] + [H_3O^+] = [HCO_3^-] + 2[CO_3^{2-}] + [OH^-] \tag{11.14}$$

From Fig. 1.2, in the pH range 6–9:

$$[H_3O^+] \ll [Ca^{2+}]$$
$$[CO_3^{2-}] \quad \text{and} \quad [OH^-] \ll [HCO_3^-].$$

therefore

$$2 \times [Ca^{2+}] = [HCO_3^-] \tag{11.15}$$

From eqns. 11.10, 11.11, and 11.12

$$[CO_3^{2-}] = \frac{K_{a1} \times K_{a2} \times K_H \times P_{CO_2}}{[H_3O^+]^2} \tag{11.16}$$

From eqns. 11.13 and 11.16

$$[Ca^{2+}] = \frac{K_{sp}[H_3O^+]^2}{K_{a1} \times K_{a2} \times K_H \times P_{CO_2}} \tag{11.17}$$

From eqns. 11.10, 11.11, 11.15, and 11.17

$$[HCO_3^-] = \frac{P_{CO_2} K_H K_{a1}}{[H_3O^+]} = \frac{2K_{sp}[H_3O^+]^2}{K_{a1} \times K_{a2} \times K_H \times P_{CO_2}} \tag{11.18}$$

From eqn 11.18

$$[H_3O^+]^3 = \frac{K_H^2 K_{a1}^2 K_{a2} P_{CO_2}^2}{2K_{sp}} \tag{11.19}$$

We showed above that the atmospheric partial pressure of carbon dioxide is 35.8 Pa and the equilibrium aqueous concentration is $1.2 \times 10^{-5} \text{ mol L}^{-1}$. From eqn 11.19

$$[H_3O^+] = \left(\frac{(3.3 \times 10^{-7})^2 \times (4.5 \times 10^{-7})^2 \times 4.7 \times 10^{-11} \times (35.8)^2}{2 \times 5 \times 10^{-9}} \right)^{\frac{1}{3}}$$

$$= 5.1 \times 10^{-9} \text{ mol L}^{-1}$$

$$pH = -\log[H_3O^+] = 8.3$$

From eqn 11.18

$$[HCO_3^-] = \frac{3.3 \times 10^{-7} \times 4.5 \times 10^{-7} \times 35.8}{5.1 \times 10^{-9}} = 1.04 \times 10^{-3} \, mol \, L^{-1}$$

From eqn 11.16

$$[CO_3^{2-}] = \frac{4.5 \times 10^{-7} \times 4.7 \times 10^{-11} \times 3.3 \times 10^{-7} \times 35.8}{(5.1 \times 10^{-9})^2}$$
$$= 9.6 \times 10^{-6} \, mol \, L^{-1}$$

From eqn 11.17

$$[Ca^{2+}] = \frac{5 \times 10^{-9} \times (5.1 \times 10^{-9})^2}{4.5 \times 10^{-7} \times 4.7 \times 10^{-11} \times 3.3 \times 10^{-7} \times 35.8}$$
$$= 5.2 \times 10^{-4} \, mol \, L^{-1}$$

The calculations confirm that the two most important species contributing to mass balance are Ca^{2+} and HCO_3^-. The pH of the solution is 8.30. Undissociated CO_2 and CO_3^{2-} are present only in small concentrations.

Where P_{CO_2} levels differ from the atmospheric values, the solubility of calcium carbonate is correspondingly affected. This situation arises when biological respiration releases large amounts of carbon dioxide (as for example in soil containing a substantial population of microorganisms), or where photosynthesis takes place decreasing the P_{CO_2} (as for example in a water body where algae are actively growing). The plot of Ca^{2+} vs P_{CO_2} given in Fig. 11.1 shows how P_{CO_2} affects the solubility of calcium carbonate.

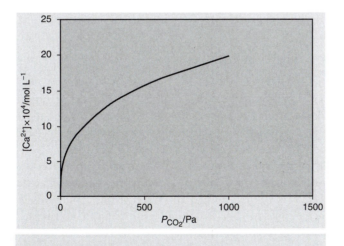

Fig. 11.1 Solubility of calcium carbonate as a function of the partial pressure of carbon dioxide in the atmosphere.

Box 11.1 Lake Nyos

On 21 August, 1986 at 7:30 pm, the community inhabiting a region close to Lake Nyos in the north-western part of Cameroon (Fig. 11B.1) was startled by a series of strange rumbling sounds. Around the same time a white cloud appeared, remaining suspended at the surface of the lake. Without warning, a great plume of water spewed upward from the lake surface. Within seconds the entire population of the vicinity had passed into unconsciousness. Some time later many people regained consciousness but they found that 1700 others, as well as most of the cattle population, had succumbed and died.

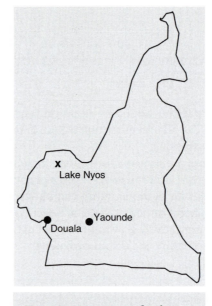

Fig. 11B.1 Location of Lake Nyos (**x**) in Cameroon.

This mysterious and almost unprecedented tragedy has become the subject of intense scientific investigation. Many details related to what happened remain unclear, but it is known that the deaths are associated with a massive release of over 240 000 t of carbon dioxide from the lake.

Lake Nyos lies along a fault line and is of volcanic origin. It has a surface area of 1.48 km^2 and with a shape approximating that of a truncated cone, reaches a depth of up to 210 m (Fig. 11B.2). Vents under the water allow continuous entry of carbon dioxide at a rate estimated to be 10 m^3 min^{-1}. The lake can hold about 1.5 km^3 of the gas in solution so that saturation would be reached in just over 20 years.

Opinion is not unanimous regarding the reasons for the carbon dioxide release. One widely accepted theory suggests that a cold rain, which had fallen for several days prior to the disaster, cooled the surface water increasing its density, causing it to sink and forcing deeper water upwards allowing the carbon dioxide to be released.

The maximum concentration of carbon dioxide in the water as a function of h, the depth in metres, has been calculated to be:[a]

$$[CO_2\,(aq)] = 40 + 2.9\,h\,mmol\,L^{-1}.$$

This is shown by the broken line in Fig. 11B.2.

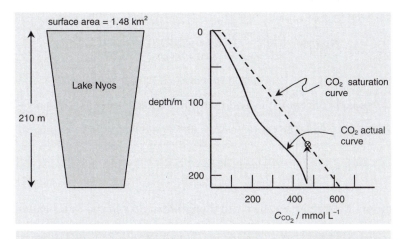

Fig. 11B.2 Cross-section of Lake Nyos shown as a truncated cone with vertical scale compressed. Graph shows the carbon dioxide concentration profile (solid line) and saturation curve (broken line).

Therefore at the original depth of 200 m the concentration limit is 620 mmol L^{-1} and the actual concentration, 475 mmol L^{-1}, is less than this saturation value. The depth at which this latter concentration equals the saturation value is 150 m. Therefore any deep water raised to a level higher than 150 m deep would release carbon dioxide gas. It is believed that the massive overturn of the lake caused an expulsion of the gas along with ejection of water upward and on to the surrounding shore. Soluble iron in the iron (II) form was brought up from the deep reducing regions and oxidized, producing a large (200 m diameter) red spot in the centre of the lake. The red colour was presumably due to precipitated hydrous iron (III) oxide.

$$4Fe(HCO_3)_2\,(aq) + O_2 + 2H_2O \rightarrow 4Fe\,(OH)_3 + 8CO_2 \qquad (11B.1)$$

There is concern that this event could recur and plans have been developed to assist a continuous slow release of carbon dioxide from the deep parts of Lake Nyos.

[a] Nojiri, Y., Gas discharge at Lake Nyos. *Nature*, **346** (1990), 323.

11.2.1 Alkalinity

Arising out of a knowledge of the chemistry of the carbonate species in water is the concept of alkalinity. Alkalinity is a measure of the ability of a water body to neutralize acids and is very important in predicting the extent of acidification of lakes and rivers. A definition of alkalinity takes into account the important proton-accepting components of most natural waters and is given by eqn 11.20:

$$\text{alkalinity} = \underbrace{[OH^-] + [HCO_3^-] + 2[CO_3^{2-}]}_{\text{proton acceptors}} - \underbrace{[H_3O^+]}_{\text{proton donor}}$$

(11.20)

A related but broader concept is the acid neutralizing capacity (ANC):

$$\text{ANC} = \underbrace{[OH^-] + [HCO_3^-] + 2\,[CO_3^{2-}] + [B(OH)_4^-] + [H_3SiO_4^-] + [HPO_4^{2-}] + [HS^-] + [NOM^-] + ...}_{\text{proton acceptors}} - \underbrace{[H_3O^+] - 3\,[Al^{3+}] - ...}_{\text{proton donors}}$$

(11.21)

In this definition, NOM refers to the natural organic matter in the water column.

The boron, silicon, phosphorus, and sulfur species that are able to contribute to acid-neutralizing capacity are ones whose pK values are in the range of the pH of natural water. The usual reactions are:

$$B(OH)_4^-(aq) + H_3O^+(aq) \rightarrow H_3BO_3 + 2H_2O$$

(11.22)

$$H_3SiO_4^-(aq) + H_3O^+(aq) \rightarrow Si(OH)_4 + H_2O$$

(11.23)

$$HPO_4^{2-}(aq) + H_3O^+(aq) \rightarrow H_2PO_4^-(aq) + H_2O$$

(11.24)

$$HS^-(aq) + H_3O^+(aq) \rightarrow H_2S + H_2O$$

(11.25)

The corresponding pK_a values for H_3BO_3, $Si(OH)_4$, $H_2PO_4^-$, and H_2S are 9.14, 9.66, 7.20, and 7.00. While the values are close enough to the pH of various natural waters, the contribution of these species to neutralization is, in almost all situations, minimal because their concentrations are generally too small to have a significant effect. In many water bodies, then, alkalinity is approximately equal to the ANC, which implies that the only proton-accepting species present in significant concentrations are the carbonate species and/or hydroxyl ions.

Typical values of alkalinity span a range from less than 50 to greater than 2000 μmol L^{-1}.

Among the additional species included in an ANC calculation, the natural organic matter (NOM) concentration is highly variable and sometimes makes an important contribution. While the chemical structure of NOM depends on its source and history, a typical value for proton-accepting sites on NOM is 10 mmol g^{-1}. Therefore, a lake containing 10 μg mL^{-1} NOM (and no other proton donors or acceptors) would have an ANC of approximately

$$10\,\frac{\mu g}{mL} \times 10\,\frac{mmol}{g} = 100\,\mu mol\,L^{-1}$$

due to NOM alone. A more detailed picture of the nature of NOM and its ability to act as a proton acceptor is given in Chapter 12.

Alkalinity is readily measured by titration of a water sample with acid. The end point for the carbonate to hydrogen carbonate plus hydroxyl to water part of the titration occurs near pH 8, and such an experiment gives a value for *carbonate alkalinity*. If the titration is continued to pH 4.5, the hydrogen carbonate is further protonated to form aqueous carbon dioxide and titration to this pH yields a value for *total alkalinity*. Actually these experiments measure the concentration of all species which are titrated up to the particular end-point pH values. While the method is usually considered as a means of determining alkalinity, it is really a method for ANC.

Consider the two following hypothetical examples involving water and carbonate species alone. Water sample 1 is water with pH $= 9$, but no carbonate species present. The concentrations of the four species needed to calculate alkalinity are:

$$[H_3O^+] = 10^{-9} \, \text{mol L}^{-1}$$
$$[OH^-] = 10^{-5} \, \text{mol L}^{-1}$$
$$[HCO_3^-] = [CO_3^{2-}] = 0 \, \text{mol L}^{-1}$$

Alkalinity can then be calculated using eqn 11.20:

$$\text{alkalinity} = [OH^-] + [HCO_3^-] + 2[CO_3^{2-}] - [H_3O^+]$$
$$= 10^{-5} + 0 + 0 - 10^{-9}$$
$$= 10^{-5} \, \text{mol L}^{-1}$$
$$= 10 \, \mu\text{mol L}^{-1}$$

Water sample 2 has a slightly lower pH of 8.3 but it contains dissolved $NaHCO_3$ at a concentration of 0.01 mol L^{-1}. By consulting Fig. 1.2, you can see that at pH 8.3, HCO_3^- is the only significant carbonate species present in water. The concentrations of all species contributing to alkalinity are therefore:

$$[H_3O^+] = 10^{-8.3} \, \text{mol L}^{-1}$$
$$[OH^-] = 10^{-5.7} \, \text{mol L}^{-1}$$
$$[HCO_3^-] = 0.01 \, \text{mol L}^{-1}$$

Rearranging and using eqn 11.12:

$$[CO_3^{2-}] = \frac{K_{a2} \times [HCO_3^-]}{[H_3O^+]} = \frac{4.7 \times 10^{-11} \times 0.01}{10^{-8.3}} = 9.4 \times 10^{-5} \, \text{mol L}^{-1}$$

The alkalinity of water sample 2 is then

$$\text{alkalinity} = 10^{-5.7} + 0.01 + 2 \times 9.4 \times 10^{-5} - 10^{-8.3}$$
$$= 0.01 \, \text{mol L}^{-1} = 10\,000 \, \mu\text{mol L}^{-1}$$

Clearly then, while water sample 2 has a slightly lower pH than water sample 1, it has a much higher alkalinity.

These artificial cases help us to see a fundamental difference between the expression of acid-base properties in terms of pH and alkalinity. The pH can be considered to be an *intensity factor* which measures the concentration of alkali or acid immediately available for reaction. In contrast, the alkalinity is a *capacity factor* which is a measure of the ability of a water sample to sustain reaction with added acids or base. Note also that it is possible to have a negative value of alkalinity. Negative values imply that extensive acidification has already occurred so that no significant amount of proton acceptors remain in the water, but rather there is an excess of proton-donating species.

The use of a titration to measure alkalinity has meaning additional to that given by the analytical result. Acidification of a lake in its natural setting is itself analogous to a macroscale titration and lakes are sometimes termed 'well buffered', 'transitional', or 'acidic' (Fig. 11.2) depending on their position on the 'titration curve'. The alkalinity then is a measure of the extent of the well buffered region. When a lake approaches the transitional category, only very little additional acidification is required to drive it past the end point with its precipitous drop in pH. Such lakes are considered to be highly sensitive to acid inputs from natural or anthropogenic sources.

A sensitivity classification (Table 11.2) may be expressed in terms of alkalinity using units of $\mu mol\,L^{-1}$ proton-accepting capacity. A high value of alkalinity means that a large amount of acid would have to be added to bring the lake into the transitional and acidic ranges. An alternative way of reporting alkalinity uses the stoichiometric neutralization reaction between carbonate and protons, and gives values in $mg\,L^{-1}$ $CaCO_3$ or $mg\,L^{-1}$ Ca. The relation between alkalinity expressed as proton accepting capacity and as calcium carbonate (molar mass $= 100\,g\,mol^{-1}$) is based on the following calculation:

$$1\,mg\,CaCO_3 = 1000\,\mu g \text{ which, on a molar basis is}$$
$$1000\,\mu g/100\,g\,mol^{-1} = 10\,\mu mol\,CO_3^{2-}$$

Since each carbonate is capable of neutralizing two hydronium ions, $10\,\mu mol$ of CO_3^{2-} is equivalent to $20\,\mu mol$ proton-accepting capacity. Therefore, the alkalinity of a solution

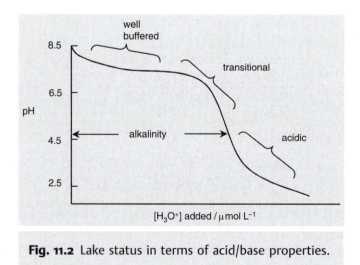

Fig. 11.2 Lake status in terms of acid/base properties.

Table 11.2 Sensitivity classification of lakes expressed by various measures of alkalinity

Sensitivity	Proton acceptors/$\mu mol\,L^{-1}$	$CaCO_3/mg\,L^{-1}$	$Ca^{2+}/mg\,L^{-1}$
High	< 200	< 10	< 4
Moderate	200–400	10–20	4–8
Low	> 400	> 20	> 8

containing $20\,\mu mol\,L^{-1}$ of proton-accepting species could also be reported as $1\,mg\,L^{-1}$ $CaCO_3$.

While the latter two conventions seem to imply that the only significant proton-accepting species in the water body is carbonate, this will not actually be the case in most situations. The HCO_3^- species is usually more important unless the pH is very high.

The following example shows a simple prediction for a highly sensitive isolated lake. Many lakes with the properties noted below have been identified in the Canadian Shield region as well as at other locations around the world.

Alkalinity   = $25\,\mu mol\,L^{-1}$ proton-accepting capacity, $1.25\,mg\,CaCO_3\,L^{-1}$.
Average depth = $20\,m$.
Area   = $37\,ha$ (not needed for calculation).
Rainfall   = $1000\,mm$ mean annual amount; average pH = 4.0.

Consider a square column representing a section of the lake with surface area of $1\,dm^2$ as shown in Fig. 11.3.

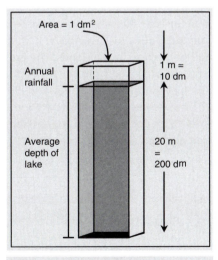

Fig. 11.3 A square column representing a section of a lake.

$$V_{column} = 20\,m \times 10\,dm/m \times 1\,dm^2 = 200\,dm^3 = 200\,L.$$

This volume contains $200\,L \times 25\,\mu mol\,L^{-1} = 5000\,\mu mol$ of proton neutralizer. The volume of rainfall falling on the horizontal surface of this section of the lake during one year would equal the annual depth of rainfall times the surface area:

$$V_{rainfall} = 1000\,mm \times 1\,dm^2 = 10\,L.$$

This volume contains $10\,L \times 10^{-4}\,mol\,L^{-1}\,H_3O^+ = 1000\,\mu mol$ of protons.

Therefore the calculation predicts that the lake would become acidic in about 5 years. In doing this, there are many assumptions, including that the lake is self-contained with neither inlet nor outlet, that there is no interaction with sediment or bedrock, and no losses of acidity through seepage to groundwater. In real situations, drainage could import additional acidity. Neutralization by slow reactions with sediment or bedrock could retard acidification.

Typical values of alkalinity for selected important water bodies are given in Table 11.3.

Table 11.3 Alkalinity values of selected water bodies around the world

Water body	Alkalinity/mg L^{-1} CaCO$_3$
Lake Baikal (Russia)—open lake	104
Lake Ontario (Canada–USA)—open lake	110
Rhine River (Germany–Holland border)	225
Lake Victoria (Uganda)—near Jinja	52
Lake Magadi (Kenya)—surface water	280 000

The highly anomalous value for Lake Magadi requires some comment. This lake is situated in the lowest part of the Rift Valley in East Africa. It is supplied with groundwater which originates from rainfall in the nearby highlands. The groundwater is enriched with sodium, carbonate, and other ions extracted from the bedrock and soil, and these species are further concentrated due to evaporation in the hot arid climate of the region. A solid compound called 'trona', which is essentially sodium carbonate, precipitates out and the *lake* is mostly solid, interspersed with pools of brine and interstitial water. The water has a pH greater than 10 and extremely high alkalinity.

The main points

1 The equilibrium relations between gaseous and water-soluble species of volatile compounds are described by Henry's law. For simple species—that is, those which do not react with water—this law is sufficient to evaluate the equilibrium distribution. For species which take part in reactions in aqueous solution, it is necessary to invoke additional

relations along with Henry's law in order to determine the total concentration and species distribution in the water. Many organic compounds are in the simple species category, while gases with acid–base properties have solubilities that depend on pH.

2 Alkalinity of a water body is a measure of the water's ability to neutralize inputs of acid. Alkalinity is defined in terms of the dissociation products of water and of aqueous carbon dioxide.

3 The acid-neutralizing capacity of water takes into account additional soluble species, including natural organic matter and some inorganic compounds, that can also neutralize acid.

Additional reading

1 Drever, J. I., *The Geochemistry of Natural Waters*, 2nd edn, Prentice Hall, Englewood Cliffs, NJ; 1988.

2 Schwarzenbach, R. P., P. M. Gschwend, and D. M. Imboden, *Environmental Organic Chemistry*. John Wiley and Sons, New York; 1993.

Problems

1 Lake Titicaca is situated at an altitude of 3810 m in the Bolivian Andes. Calculate the solubility of oxygen in the lake at a temperature of 5 °C. (The Henry's law constant at 5 °C is $1.9 \times 10^{-8} \, mol \, L^{-1} \, Pa^{-1}$).

2 At 30 °C, the solubility of oxygen in water is 7.5 mg L^{-1}. Consider a water body at that temperature containing 7.0 mg L^{-1} oxygen. By photosynthesis, 1.5 mg of carbon (as carbon dioxide) is converted to organic biomass ($\{CH_2O\}$) during a single hot day. Is the amount of oxygen produced at the same time sufficient to exceed its aqueous solubility?

3 Thermal power plants, whether powered by fossil fuel or nuclear energy, discharge large quantities of cooling water to a lake or river. Discuss the meaning of *thermal pollution* in the context of this chapter.

4 One of the alternative expressions for Henry's law when applied to oxygen is

$$P_{O_2} = K'_H X_{O_2}$$

where P_{O_2} is the pressure of oxygen in the gas phase and X_{O_2} is its mole fraction in solution. Calculate the value of Henry's law constant, K'_H, at 25 °C in units of MPa.

5 Methane and carbon dioxide are produced under anaerobic conditions in wetlands by fermentation of organic matter approximated by the following equation:

$$2\{CH_2O\} \rightarrow CH_4 + CO_2$$

Calculate the total pressure (P° plus that due to water) at a depth of 5 m. Gas bubbles are evolved at that pressure and remain in contact with water at the sediment surface long enough so that equilibrium is attained. Calculate the methane concentration (mol L^{-1}) in the interstitial water at 25 °C.

6 A sample of water from Lake Huron (one of the Great Lakes in North America) has a pH of 7.34 and an alkalinity of 1.21 mmol L^{-1} as calcium carbonate. Assume equilibrium with the calcium carbonate in the sediment and calculate the concentration of calcium ion (in mg L^{-1}) in the water (25 °C). What assumptions are made in carrying out the calculation?

7 In a soil atmosphere, the mixing ratio of carbon dioxide may be much higher than that in the normal atmosphere due to respiration of microorganisms. If the mixing ratio of carbon dioxide in the air occupying soil pore were 5000 ppmv, calculate the pH of the associated soil solution (25 °C), assuming no other sources of proton donors or acceptors.

8 In Mexico City, the concentration of sulfur dioxide sometimes reaches $200 \,\mu\text{g m}^{-3}$. Calculate the pH of an aqueous aerosol in equilibrium with this gas at 25 °C. As a second case, assume that all of the sulfur dioxide is oxidized in the aerosol by ozone and hydrogen peroxide to produce sulfuric acid. Calculate the pH after such reactions have occurred.

9 Sea water is well buffered at a pH of about 8 and has a total concentration of carbonate species of 2 mmol L^{-1}. Calculate its alkalinity on the basis of this information. Discuss whether the buffering could be due to the carbonate species alone.

12

Organic matter in water

ORGANIC matter (OM), in dissolved or particulate forms, is found in oceans and fresh water of all types. Clear evidence of its presence is the characteristic yellow-brown colour of a bog or swamp and, to a lesser extent, of some lakes and rivers. Even 'clean' water, such as that from a deep lake in a remote part of the world or from the open ocean, contains at least a small fraction of OM—typical concentrations being in the range of 1 to $3\,\text{mg}\,\text{L}^{-1}$.

Most analytical methods for measuring organic matter in water actually determine the carbon content. Carbon, the essential element of organic compounds, is also present in the environment as a component of inorganic species—particularly those in the carbonate family. In measuring the carbon content then, it is necessary to distinguish between organic and inorganic forms. Nevertheless, through reactions such as photosynthesis, respiration, and chemical weathering, the various species can be interconverted. As you might expect, many reactions that bring about synthesis or degradation of organic matter are biological reactions.

Carbon is present in the atmosphere mainly as carbon dioxide, in water as various carbonate species and as carbonate minerals in sedimentary rocks. Smaller, but highly significant amounts of organic carbon are present in all of the compartments of the environment. Estimates of the total amounts of carbon-containing compounds in the atmosphere, in water and on land are given in Table 12.1.

While organic forms of carbon represent only a small proportion of the global carbon reservoir, they are instrumental in many reactions and influence environmental chemistry far beyond their mass contribution. Further discussion of relations between the biological components in the global carbon cycle is given in Chapter 15 and elsewhere throughout the book. The present discussion relates to chemistry of non-living organic species in the water column.

Various ways of classifying organic matter in water help us to understand the relations between the various chemical types of OM and the natural or anthropogenic origins of particular compounds. Where the OM is derived from natural sources, it is often referred to as natural organic matter (NOM). The OM is further divisible into two broad categories—dissolved organic matter (DOM) and particulate organic matter (POM).

Table 12.1 Carbon pools in the global environment

Carbon reservoir	Mass of carbon/Pg
Atmosphere	720
Terrestrial environment	
Plants	830
Soil surface detritus	60
Soil organic matter	1400
Peat	500
Fossil Fuels	5000
Aquatic environment (oceans)	
Living organisms	3
Dissolved organic matter	1000
Dissolved inorganic matter	37 000
Sedimentary carbonate material	20 000 000

Data from Bolin, B., Requirements for a satisfactory model of the global carbon cycle and current status of modeling efforts, in *The Changing Carbon Cycle: A Global Analysis*, eds Trabalka, J. R. and Reichle, D. E., Springer, New York; 1986.
Pg = petagram = 10^{15} g.

12.1 Origins of organic matter in water

The OM that is of natural origin is derived primarily from plant and/or microbial residues. On land, plants grow, shed leaves, and die leaving roots within the upper soil layers and 'litter' on the soil surface. Microorganisms also flourish within the soil. When they die their biomass adds to the soil organic content. Quantitatively smaller, but of considerable biological importance is the contribution from substances secreted by growing macro- and microorganisms. Such molecules, when released in the region directly adjacent to the microbe or plant root (the rhizosphere), play important roles in facilitating or preventing uptake of chemicals by the organisms.

In their original or chemically modified form, the residues of organic matter produced on land are available to be transferred from the soil into the hydrosphere. Transport usually occurs due to rainfall that runs off or percolates through the soil column carrying soluble and particulate OM to streams, lakes, and oceans or into groundwater.

Organic matter is also produced *in situ* in a water body. Once again, wetlands are a prime example; there, the luxuriant growth of vegetation produces a thick mat of roots and aerial material that, upon death, is deposited in the water. The top layer of the swamp sediment consists almost entirely of organic material in various stages of decay. To a smaller degree, aquatic plants as well as animals grow in other water bodies (rivers, lakes, and oceans) and leave their organic remains in the sediment. As is the case with soil, microorganisms inhabit the hydrosphere in great numbers, depending on environmental conditions. Some, like algae, are photosynthetic organisms that require sunlight and oxygen, and therefore grow in the near-surface layers of water. Bacteria and other types of microorganisms live in the sediments. All microorganisms require a supply of nutrients in order to grow and multiply. During their survival period, all microscopic plants release some of their photosynthetic products to the water column, on a global scale it is estimated that around 10% of the photosynthetic activity in water goes to the production of DOM. Microscopic animals also release soluble organic matter from their bodies.

After dying, the organic residues from larger plants and animals, as well as the microbial biomass, become chemically modified by a variety of decomposition and synthesis processes. Many of the reactions to produce altered organic species are facilitated by the presence of living microorganisms in the soil or water. Some of the organic species generated through degradation and synthesis move into the water column in soluble or particulate forms.

Besides the natural sources, there are human inputs that contribute to the organic matter in water. These include large-volume poorly defined wastes, such as domestic sewage or pulp-mill effluent, which are sometimes discharged in treated or untreated form into rivers, lakes, and oceans. Anthropogenic sources also include specific organic compounds—products or byproducts of industrial processes. The range of these is as broad as the range of organic chemistry itself.

The distinction between natural and human sources is not always simple. Falling definitely into the natural category are the organic products that result from leaf decay, while release of nitrilotriacetic acid (NTA) into a waste stream from a detergent-manufacturing facility is an example of an indisputably anthropogenic event. Other cases are less clear cut. Trichloromethane (chloroform) is a chlorinated hydrocarbon that we might assume is produced only *via* industrial processes. However, it has been shown that millions of tonnes of this compound are also formed through natural reactions each year. In fact over 1500 organochlorines have been identified as natural products found in living organisms. While many of these are low molar mass compounds such as mono- and trichloromethane and 2,4,6-trichlorophenol, there are also complex, relatively high molar mass chlorinated organics that occur naturally in the environment. There is therefore a substantial list of chemicals—not just organochlorines—that have both natural as well as anthropogenic origins.

When we consider the chemistry of organic matter found in the hydrosphere, you will see that it is frequently convenient to subdivide the species into two categories based on molecular size.

- Discrete small molecules, like monosaccharides or low molar mass organic acids, whose chemical structure and properties are amenable to individual specific study. Individual polluting species like chemical pesticides fall in this category as well.
- Macromolecules, which are treated as classes in terms of their general structural properties and reactivity. Characterization of macromolecules is often based on an operational definition—that is, a definition arising out of a particular analytical protocol, and not based on fundamental structural properties.

Many of the natural forms of organic matter fall in the macromolecule category, and we will focus on these kinds of compounds in the present chapter.

12.2 Environmental issues related to aqueous organic matter

Organic matter in water is of environmental importance for several reasons. For one thing, particular compounds may be toxic in varying degrees to living organisms, including humans. Polyaromatic hydrocarbons, polychlorinated biphenyls, and dioxins are all well known, and have received considerable publicity and study because of their contribution to real and alleged environmental problems. Residues of pesticides and their metabolic products can also be carried into water. Due to their widespread application in agriculture and forestry, even in urban settings, it is important to understand how they move and react in the environment. Some aspects of the environmental chemistry of biocides will be discussed in Chapter 20.

There are also less direct ways in which organic matter influences environmental processes and merits study. Through its interaction with other organic or inorganic species that are also present in the hydrosphere, in many cases organic material alters the behaviour of these other aquatic chemicals. For example, inorganic tin may undergo alkylation in aquatic environments to form compounds such as monomethyl tin (CH_3Sn^{3+}) and dimethyl tin (($CH_3)_2Sn^{2+}$). The alkylation is biological in that it occurs in the fish gut or *via* microorganisms in the water column and the organotin species produced are more toxic to aquatic biota than are the original inorganic tin compounds. Toxicity becomes greater as the number of organic groups increases in the series $R_nSn^{(4-n)+}$ for $n = 1$ to 3. Toxicity is also inversely related to the lengh of R and is at a maximum where R is a methyl or ethyl group. There are many additional examples of organometallic compounds and metal–organic ligand complexes in environmental situations. Invariably the environmental chemical and toxicological properties of the compounds are different from those of the free metal. In some cases toxicity is enhanced, in other cases it is reduced by reaction with organic matter in the hydrosphere.

A third environmental feature of aqueous organic matter, in particular the bulk residues of plants and animals or some industrial discharges, is that the non-living organic material can be oxidized by oxygen and other oxidizing agents in water. Therefore, when released into a water body the bulk OM degrades, consuming oxygen, and leaving the system in an oxygen-deprived state. The anoxic conditions lead to low pE environment and therefore can change the chemistry of the entire system. The loss of oxygen also creates stress on many aquatic organisms, including fish.

Each of these environmental aspects related to aqueous organic matter will come up in later sections. For now we will look at chemical properties of the broadly defined naturally occurring macromolecules present in soluble, colloidal, and sedimentary forms in the hydrosphere. Related macromolecules are also important in the terrestrial environment.

12.3 Humic material

Humic material (HM) is naturally occurring organic matter of plant or microbial origin. The humic substances are produced and reside in soil and water and form a major component of both the terrestrial (soil organic matter) and aquatic (dissolved organic matter) carbon pools (Table 12.1). In the hydrosphere, HM typically makes up about 50% of the dissolved organic matter (DOM) in surface water,[1] as well as much of the organic sediment. Humic material (also called humate or humus) is subdivided in an operational sense into three categories:

- fulvic acid (FA) is the fraction of humic matter that is soluble in aqueous solutions that span all pH values;
- humic acid (HA) is insoluble under acid conditions (pH 2) but soluble at elevated pH;
- humin (Hu) is insoluble in water at all pH values.

12.3.1 Formation of humic material

Humic substances are formed *via* a complex sequence of only partly understood reactions. Several hypotheses have been proposed to account for their synthesis in nature. According to one set of hypotheses, plant biopolymers are modified through degradation to form the central core of humic substances. These theories propose that labile macromolecules such as carbohydrates and proteins are degraded and lost during microbial attack, while refractory compounds or biopolymers—for example, lignin, paraffinic macromolecules, melanins and cutin—are selectively transformed to produce a high molar mass precursor of humin. Further oxidation of these materials generates increased oxygen functionality and as this process continues, the molecules become small enough and hydrophilic enough to be soluble in alkali. Eventually the molecules become even smaller and sufficiently oxygen-rich to dissolve in both acid and base. The extensive degradation produces structures that retain some original features but have considerable structural dissimilarity from the parent material.

As an alternative, hypotheses based on ideas of condensation polymerization suggest that plant biopolymers are initially degraded to small molecules after which these molecules are repolymerized to form humic substances. It has been proposed that polyphenols synthesized by fungi and other microorganisms, together with those liberated from the oxidative degradation of lignin, undergo oxidative polymerization. A consequence of this scheme is that fulvic acid would be a precursor of humic acid and then of humin (the reverse of the degradative theory). The hypothesis could explain the observed considerable similarity of humic materials formed from a diversity of precursor macromolecules in differing environments.

The two proposed pathways have overlapping features and both routes could contribute to the actual formation of humic material depending on the environment. For instance, the degradative pathway might predominate in wet sediments and in the aquatic environment in general, whereas the conditions in soils that are subject to harsh continental climates could favour a condensation polymerization scheme.

While the detailed features of the pathway are not entirely clear, in either case the formation of humic material does involve degradation of fresh organic matter. When oxidative decomposition processes go to completion the end products are carbon dioxide and water and we describe the reaction in simplified form as

$$\{CH_2O\} + O_2 \rightarrow CO_2 + H_2O \tag{12.1}$$

assuming that the bulk of the OM input is carbohydrate material. Other organic components of plant material also produce carbon dioxide and water as principal products of oxidative decomposition. When in buried material, the final product of decomposition contains carbon in a reduced chemical state—as a fossil fuel of some form. Most of the oxygen in the original biological material has been lost in the process.

The humification reactions can be considered to produce intermediates—the humic material—in these overall degradation reactions. But HM is a stable intermediate, especially under conditions of limited oxygen availability. Carbon-14 dating has indicated a range of age from about 20 years for stream humate, up to 500 or 1000 years for soil humus, and much longer for that associated with buried peat or coal deposits.

As an aside, we may speculate that rates of degradation, specifically oxidation, of soil humus would be affected by the use to which the soil is put. For example, it has been shown that deforestation contributes to an increased carbon dioxide content in the atmosphere not only because of diminished photosynthesis and release by burning the forest slash but also from enhanced carbon dioxide evolution from the exposed and degradable forest humus.

12.3.2 Composition and structure

Analysis of a wide variety of humic substances shows approximate elemental composition ranges (%) as follows:

C	O	H	N	Inorganic elements (ash)
45–60	25–45	4–7	2–5	0.5–5

Because the carbon content is frequently close to 60%, an approximate factor of 1.7 is used as an estimate to convert mass values from organic carbon (OC) to organic matter (OM). This factor is also employed for forms of organic matter other than humic material in both water and soils. For example, if the dissolved organic carbon (DOC) concentration in a lake is measured to be $2.3\,mg\,L^{-1}$, the dissolved organic matter (DOM) content would be estimated as $3.9\,mg\,L^{-1}$. As noted above, it is organic carbon that is usually measured analytically. A brief discussion of parameters used as measures of organic matter in water is given in Chapter 15.

While the elemental concentration ranges indicated above cover most analyses of the three types of humic substances, usually the carbon content of material from a given source increases in the series

$$FA < HA < Hu.$$

Oxygen content of the same set of materials follows the reverse trend.

The environment in which the material is formed also has some influence on the composition, with carbon content being larger in soil humus than in humate from lakes and oceans. The reverse is true for oxygen and nitrogen. In sedimentary material, there is some evidence that the relative proportion of Hu compared to HA and FA increases with depth. Depth is directly related to age of the sediment and so a greater age of the buried material is reflected in a larger carbon content. Such trends are common but have not been observed in all situations.

The presence of oxygen and nitrogen in HM is an indication that certain functional groups are present in the humate molecules. We will now consider the nature of this functionality. The most important oxygen-containing functional groups along with typical ranges of content (mmol functional group per g of humic material) are carboxyl (2–6), phenolic –OH (1–4), alcoholic –OH (1–4), carbonyl, both ketones and quinones (2–6), and methoxyl (0.2–1). It is these groups that, individually or together, enable specific reaction of humic substances with inorganic elements and with other organic molecules in soil–water systems. The functional groups are also a major contributor to the ion-exchange properties of soils and sediments. In addition to being present in the above groups, both oxygen and nitrogen appear as bridging units and in ring structures.

Quantitative structural information is obtained using classical chemical procedures. Spectroscopic methods, including infrared (IR) and carbon-13 nuclear magnetic resonance (^{13}C-NMR) spectroscopy, can be employed to indicate independently the presence of structural features in the material. In addition to confirming the presence of various functional groups, these spectroscopic methods indicate that there is both aliphatic and aromatic character to HM. Figure 12.1 shows IR spectra obtained on fulvic and humic acid from a Kenyan soil and Table 12.2 lists information which can be used to interpret this and other such spectra.

Infrared spectra of many other samples of humic material from around the world have been reported and, in their principal features, are similar to the ones shown here. The common features in the IR bands for various samples of both FA and HA suggest that there are similar structural properties shared by all types of humic materials. The observed bands are consistent with the presence of the expected chemical units, including phenolic and aliphatic -OH groups, carbonyl and carboxyl groups, and there is evidence of aromaticity and double bonds in aliphatic units. Quantitative assessments of the 'concentration' of functional groups is not possible from the IR spectrum.

Carbon-13 NMR spectra for the same Kenyan humate samples are shown in Fig. 12.2.

Nuclear magnetic resonance spectra of humic material characteristically show broad overlapping peaks due to the presence of paramagnetic species such as Fe^{3+} and stable organic radicals from residual soil components in the samples. Nevertheless, useful information is obtained and chemical shift assignments have been made as in Table 12.3.

Although the size of peaks depends on the number of carbon atoms in a particular structural unit, it is not advisable to make quantitative interpretations from ^{13}C-NMR spectra because peak size also depends on the microenvironment of the atoms and on variations in instrumental conditions. In the spectra shown, there is clear evidence of the presence of aliphatic groups (30 ppm), alkoxyl groups (55 ppm), carbohydrate-type functionality (75 ppm), aromatic components (the broad band from about 100–140 ppm), and

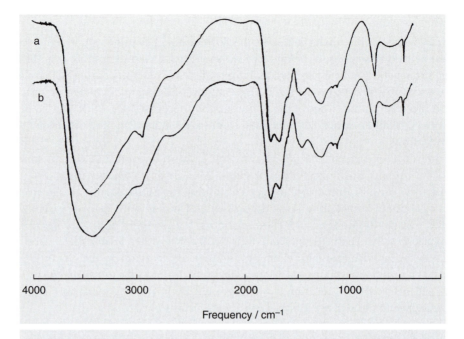

Fig. 12.1 IR spectra of (a) humic acid and (b) fulvic acid from soil obtained near Lake Nakuru in Kenya. (Moturi, M. C. Z., Studies on humic substances extracted from neutral and alkaline soils and sediments from Kenya, MSc Thesis, Queen's University, Kingston, ON; 1991.)

Table 12.2 Some of the prominent bands observed in IR spectra of humic and fulvic acids

Absorption band	Assignment
$3400\ cm^{-1}$	Hydrogen-bonded O–H stretching vibrations from phenolic and aliphatic OH groups
$2920\ cm^{-1}$	Asymmetric stretching vibrations of aliphatic C–H bands in -CH$_3$ and -CH$_2$ units
$2860\ cm^{-1}$	Symmetric C–H stretch of aliphatic bonds in -CH$_3$ and -CH$_2$ units
$1720\ cm^{-1}$	C=O stretching vibrations from -COOH and probably ketonic carbonyls
$1630\ cm^{-1}$	May be due to C=C stretching in aromatic rings, asymmetric stretching of -COO$^-$, hydrogen-bonded C=O, or C=C stretching alkenes conjugated with carbonyl groups or other double bonds
$1540\ cm^{-1}$	Aromatic C=C stretching vibrations or N–H deformations
$1420\ cm^{-1}$	O–H bending vibrations of alcohols, carboxylic acids, and phenols, or C–H deformations of -CH$_2$ and -CH$_3$ aliphatic groups
$1240\ cm^{-1}$	C–O stretching and O–H deformation of -COOH
$1050\ cm^{-1}$	May be due to O–H deformations and C–O stretching in polysaccharides, phenolic and alcoholic groups, or Si–O of silicate impurities
$940\ cm^{-1}$	Aromatic C–H out-of-plane bending vibrations
$800\ cm^{-1}$	Aromatic C–H out-of-plane bending vibrations

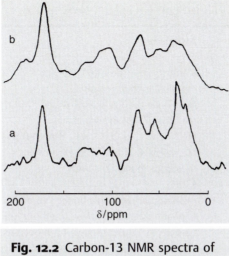

Fig. 12.2 Carbon-13 NMR spectra of (a) humic acid and (b) fulvic acid from soil obtained near Lake Nakuru in Kenya. (Moturi, M. C. Z., ibid, Fig. 12.1.)

carboxylic groups (175 ppm). Other humic material NMR spectra from varied sources around the world show similar features.

Using all these types of information, it becomes possible to develop an idea of the molecular structure of humic substances. Figure 12.3 shows a portion of a hypothetical structure of a generic humate molecule. All of the features—the aliphatic and aromatic character, the varied functionality, the polymeric nature, are represented here.

It is of course essential to realize that there is no such thing as a single specific humate molecule in any of the three classes. Relative proportions of the structural components vary from situation to situation and so does the molar mass. If we isolate a sample of 'pure' fulvic acid, it will have a range of molar mass, as will similar samples of humic acid or humin. The molar mass of polydisperse humic material is difficult to measure, but average values ranging from a few hundred daltons for fulvic acid to hundreds of thousands of daltons for humin have been estimated.

12.3.3 Forms of humic materials

Humic materials are found in the aqueous and terrestrial environments in a variety of forms.

- Free HM consists of soluble or insoluble forms of the material itself.
- Complexed HM is chemically bound to metals, other inorganic species such as phosphate, or organic molecules. The complexed HM is either in solution or in particulate form.
- Surface-bonded HM is chemically bonded to other solids such as clay minerals or iron and aluminium oxides. In this way, the surface of the inorganic material is altered so

Table 12.3 Chemical shift assignments of ^{13}C-NMR spectra of fulvic and humic acids

Chemical shift	Assignment
0–50 ppm	Unsubstituted saturated aliphatic C
10–20 ppm	Terminal methyl groups
15–50 ppm	Methylene groups in alkyl chains
25–50 ppm	Methine groups in alkyl chains
29–33 ppm	Methylene C, α, β, δ, and ϵ from terminal methyl groups
35–50 ppm	Methylene C of branched alkyl chains
41–42 ppm	α-carbons in aliphatic acids
45–46 ppm	R_2NH_3
50–95 ppm	Aliphatic carbon singly bonded to one O or N atom
51–61 ppm	Aliphatic esters and ethers, methoxy, ethoxy
57–65 ppm	Carbons in CH_2OH groups, C_6 in polysaccharides
65–85 ppm	Carbon in CH(OH) groups, ring C of polysaccharides, ether-bonded aliphatic C
90–110 ppm	Carbon singly bonded to O atoms, anomeric C in polysaccharides, acetal or ketal
110–160 ppm	Aromatic and unsaturated C
110–120 ppm	Unsubstituted aromatic C, aryl H
118–122 ppm	Aromatic C ortho to O-substituted aromatic C
120–140 ppm	Unsubstituted and alkyl-substituted aromatic C
140–160 ppm	Aromatic C substituted by O or N, aromatic ether, phenol, aromatic amines
160–230 ppm	Carbonyl, carboxyl, amide, ester C
160–190 ppm	Carboxyl C
190–230 ppm	Carbonyl C

Most of the details are from Hayes, M. H. B., P. MacCarthy, R. L. Malcolm, and R. S. Swift, eds, *Humic Substances II: In Search of Structure*, J. Wiley and Sons, Chichester; 1989.

that its chemical properties are determined largely by the organic coating. The HM can then react in a manner similar to the pure material itself.

12.3.4 Aqueous humic material as a proton acceptor

As their names indicate, the free forms of humic materials are acids. The acidic character is associated largely with the carboxylate and phenolic groups. The former have pK_a values in the range between 2.5 and 5 depending on the proximity of electronegative atoms with respect to the carboxyl groups, while the phenolic hydrogens have pK_a values around 9 or 10. The carboxyls, therefore, are sufficiently strong acids that they remain substantially deprotonated when dissolved in water at a low concentration. When a water body is poorly buffered because it lacks other proton acceptors such as hydrogen carbonate, dissolved humic material will acidify the water so that it has a typical pH of 5.5. to 6.5. In

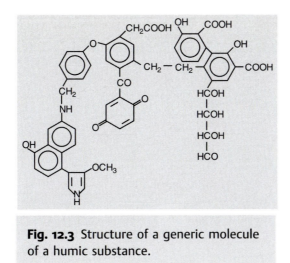

Fig. 12.3 Structure of a generic molecule of a humic substance.

these situations, the humic material itself is negatively charged, and in some cases may be the major source of anionic charge in the dissolved phase.

Consider the following example. A sample of water containing a dissolved humic material concentration of 8.0 mg L^{-1} from the Canadian Shield has measurable concentrations (C_X) of the following ions:

H$^+$	pH = 5.88	Cl$^-$	0.138 mg L^{-1}
NH$_4^+$	3.6 µg L^{-1} (as N)	NO$_3^-$	7.0 µg L^{-1} (as N)
Na$^+$	75.9 µg L^{-1}	HCO$_3^-$	14.4 µg L^{-1} (as C)
K$^+$	50.8 µg L^{-1}	SO$_4^{2-}$	59.4 µg L^{-1} (as S)
Mg^{2+}	0.124 mg L^{-1}		
Ca^{2+}	0.569 mg L^{-1}		

If we carry out a charge balance calculation on the water we find that there is an apparent large excess of positive (cationic) charge. The charge due to each cation and anion is calculated below.

Considering the species indicated, therefore, there is a major charge imbalance as a net positive charge $(44.7 - 9.3) = 35.4$ µmol L^{-1} is indicated. Of course, electroneutrality must obtain in the solution and so a balancing negative charge of 35.4 µmol L^{-1} must exist. To account for this we must include in the calculation the contribution due to dissolved humic material which was present at a concentration of 8.0 mg L^{-1}.

If this material has a carboxyl group concentration of C_{coo^-} mmol g^{-1}, the associated negative charge is

$$8.0 \, \text{mg L}^{-1} \times C_{coo^-} \, \text{mmol g}^{-1} = 8C_{coo^-} \, \mu\text{mol L}^{-1}.$$

	Conc/μmol L^{-1}	Charge/μmol L^{-1}
H$^+$	1.3	1.3
NH$_4^+$	0.2	0.2
Na$^+$	3.3	3.3
K$^+$	1.3	1.3
Mg^{2+}	5.1	10.2
Ca^{2+}	14.2	28.4
$\sum$+ve charge		44.7
Cl$^-$	3.9	3.9
NO$_3^-$	0.5	0.5
HCO$_3^-$	1.2	1.2
SO$_4^{2-}$	1.9	3.7
$\sum$−ve charge		9.3

At pH $= 5.88$, we would expect that close to 100% of the carboxyl groups would be deprotonated. Therefore, assuming that the required negative charge comes exclusively from the organic matter,

$$8\,\text{mg L}^{-1} \times C_{\text{coo}^-}\,\text{mmol g}^{-1} = 35.4\,\mu\text{mol L}^{-1}$$

and

$$C_{\text{coo}^-} \simeq 4.5\,\text{mmol g}^{-1}.$$

The calculated concentration of carboxyl groups in the dissolved organic matter is therefore 4.5 mmol g^{-1}, which is well within the expected range of 2–6 mmol g^{-1}. It should be emphasized that this type of calculation is only approximate as there is the possibility of error associated with each of the analytical measurements which, in sum, leads to considerable uncertainty in the final result.

When we considered the concepts of alkalinity and acid neutralization capacity in Chapter 11, it was noted that the difference between the two was, in some cases, due in large part to organic matter. The previous example illustrates this, as the slightly acidic lake had a very small hydrogen carbonate concentration and a more important proton acceptor was the dissolved humic material. The actual alkalinity titration measurement might account for an indefinite proportion of the dissolved humic matter—indefinite because the pK_a values of HM cover a range from 2.5 to 5 and so only part of these would be protonated during a titration to an end point at pH 4.5, as is specified for the determination of alkalinity. Nevertheless such material is the most important buffering constituent in some slightly acidic lakes with large DOM concentrations.

12.3.5 Humic material as a complexing agent for metal ions

Besides playing a role as a proton acceptor and in contributing to charge balance in aqueous systems, the dissolved humic substances also react with metals in solution through the formation of ionic or covalent bonds. The strength of the metal–organic interaction is measured in terms of a stability constant of the form described in Chapter 10. The actual value depends on the nature and number of binding sites on the HM, on the properties of the metal, and on other environmental factors such as pH and the presence of other competing ligands. We will say more about this in the next chapter.

12.3.6 Reactions between humic material and small organic molecules

Humic substances also are capable of reacting with many specific anthropogenically derived organic compounds found in water and in soil. Organic pesticides are a prime example. Interaction occurs *via* several mechanisms. Van der Waals forces are always a source of weak attractions. Stronger interactions occur due to processes such as the following.

Reaction between positively charged species and negatively charged humate sites (Fig. 12.4)

The example is for atrazine, a widely used herbicide in the triazine class. Under typical environmental conditions (pH < 8) the atrazine molecule is positively charged due to protonation of one of the ring nitrogens and this allows for reaction with the carboxyl groups of the humic material.

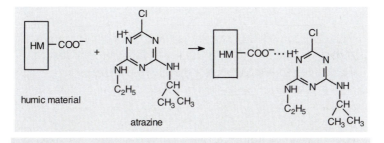

Fig. 12.4 Atrazine bonded to HM by electrostatic forces.

Hydrogen bonding (Fig. 12.5)

A variety of hydrogen-bonding reactions are possible, involving oxygen- and nitrogen-containing functional groups of both the humic material and the organic pesticide molecule. In the example shown, the insecticide carbaryl is hydrogen bonded to HM using both a nitrogen and an oxygen of the carbamate structure.

Salt linkage or ligand exchange (Fig. 12.6)

In this example, assuming a near-neutral pH of 6 to 8, the chlorophenoxy herbicide 2,4-dichlorophenoxyacetic acid (2,4-D), would be present in solution with its carboxyl group deprotonated—as is also the case for the carboxyls of the humic material.

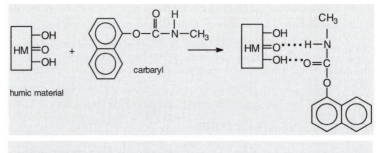

Fig. 12.5 Binding of the pesticide carbaryl to HM through hydrogen bonding.

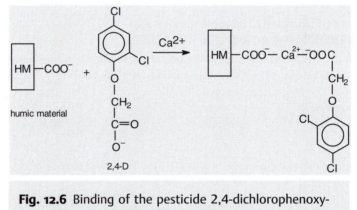

Fig. 12.6 Binding of the pesticide 2,4-dichlorophenoxy-acetic acid to HM through a metal ion bridge.

Transition metals or even common cations such as Ca^{2+} then are able to form a metal bridge or salt linkage so that the herbicide is bound to the soluble or particulate humic material.

Hydrophobic interactions (Fig. 12.7)

For essentially non-polar molecules like dichlorodiphenyltrichloroethane (DDT), the molecules find compatible settings within mostly hydrocarbon regions of humate molecules. The association involves a net energy loss since the small hydrophobic molecule no longer disrupts the ordered structure of the water solvent as it moves from the aqueous phase to the humic substance.

The combination of all these factors contributes to organic solute binding by humic substances. For most small neutral organic molecules present in water, their behaviour is strongly influenced by the presence of humic material in either dissolved or particulate forms. A detailed description of the interactions requires taking into account other environmental factors such as solution pH and ionic strength.

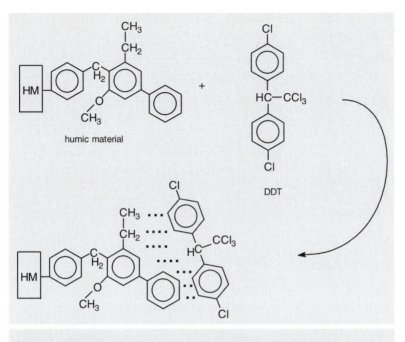

Fig. 12.7 Binding of the insecticide DDT to HM through hydrophobic interactions. An essentially hydrophobic portion of the hypothetical humate molecule is specifically shown in this figure.

As a consequence of the solute-binding processes, organic substances may be brought into the aqueous phase through association with soluble material. If and when the humate becomes insoluble—as could occur when riverine sources encounter sea water in estuaries—the aggregated material may bring down with it the organic solutes. In this case, and also by direct association with solid humic substances, such solutes are 'immobilized'. Therefore, depending on circumstances, humate either enhances solubility or removes from solution substances like pesticides. Furthermore, once the association is established, reactivity of the solute is altered. Examples, such as the hydrolysis of atrazine, have been cited where degradation is enhanced by association with HM. There are also examples in which degradation is inhibited by the same association—for example the base hydrolysis of DDT.

Humic material is therefore found in water in free form or complexed with metal ions or other organic species. The third form in which HM occurs in the environment is in association with other solids in soil/water or sediment/water systems. A particular association is that with clay minerals and it is well known that humic substances are frequently observed as coatings on such mineral surfaces.

The surface bonding occurs through specific covalent interactions between metals (Al^{3+}, Fe^{3+}) on the clay surface or metals (Ca^{2+}, Al^{3+}) in the adjacent solution and the humate

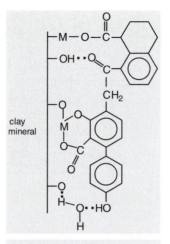

Fig. 12.8 Types of bonding interactions involved in forming the clay–humate complex.

molecules, as well as *via* weaker hydrogen bonds (Fig. 12.8). Because both the clay surface and the humic material have net negative charges, simple electrostatic interactions cannot play a role.

The fact that a portion of the humic material may be associated with the clay mineral fraction in a water/solid system adds a further dimension to the complications involved in describing the chemistry of the humate. As a result of their interaction, the surface properties of both the clay minerals and the humate are altered, thus affecting subsequent interactions with other metals and organics, and this in turn influences the reactivity of the retained species.

The main points

1 Organic matter is present in dissolved and particulate forms in all natural waters. The OM originates from both natural and anthropogenic sources.

2 Specific organic species can be identified, usually in low concentration, in samples, but much of the organic matter is present as poorly defined broad classes of material.

3 Humic substances, natural material of both terrestrial and aquatic origin, make up about half of the organic matter present in many waters.

4 The humic material affects the charge balance and alkalinity of water, and it interacts with metals, small neutral organic molecules, and sediment/soil solids.

Additional reading

1 Aiken, G. R., ed., *Humic Substances in Soils, Sediments and Water: Geochemistry, Isolation and Characteristics*. J. Wiley and Sons, New York; 1985.

2 Rashid, M. A., *Geochemistry of Marine Humic Compounds*. Springer, New York; 1985.

Problems

1 Use diagrams of a type similar of Figs. 12.4 to 12.7 to show the kinds of interactions you might expect between humic material and phenanthrene, trichloroethylene (TCE, $CHCl=CCl_2$), copper (II), and copper (II) complexed with oxalate.

2 Explain how molecular size and oxygen concentration differences could explain the relative solubilities of humic acid and fulvic acid.

3 The ^{13}C-NMR spectra of six humic acid samples from marine and estuarine sediments are shown in Fig. 12P.1 (Hatcher, P. G. and Orem, W. H., Structural interrelationships among humic substances in marine and estuarine sediments. In Sohn, M. L., *Organic Marine Geochemistry*, American Chemical Society, Washington, DC; 1986). The samples are from various locations—the Potomac River, on the coast of the United States; New York Bight, 16 km off the New York coast, on the continental shelf; Walvis Bay, from the deep continental shelf on the Namibian coast; Mangrove Lake, from a marine lake in Bermuda.

We indicated that accurate quantitative analysis of solid samples by NMR is problematic, but it is appropriate to develop semi-quantitative ideas based on the spectra. Which HA samples appear to be lowest in aromatic carbons? Where marine algal material are the principal source of HA, the structure is predominantly aliphatic. Which samples may have the highest aromatic content? Why? The Mangrove Lake surface sediment HA appears to be enriched in carbohydrate material. Why might the concentration decline with depth? To what might the small peak at 50 ppm be due?

4 Water from a lake in an area with limestone bedrock has chemical composition as follows:

Calcium	$95\,mg\,L^{-1}$
Magnesium	$13\,mg\,L^{-1}$
Sodium	$17\,mg\,L^{-1}$
Potassium	$4\,mg\,L^{-1}$
Hydrogen carbonate (as HCO_3^-)	$338\,mg\,L^{-1}$
Sulfate (as SO_4^{2-})	$7\,mg\,L^{-1}$
Chloride	$12\,mg\,L^{-1}$
Fluoride	$0.2\,mg\,L^{-1}$
Nitrate (as NO_3^-)	$3\,mg\,L^{-1}$
pH	6.8

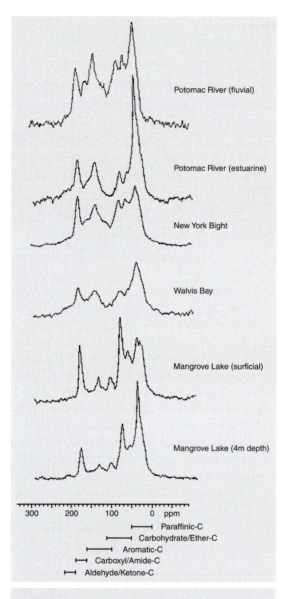

300   200   100   0 ppm

Paraffinic-C
Carbohydrate/Ether-C
Aromatic-C
Carboxyl/Amide-C
Aldehyde/Ketone-C

Fig. 12.P.1 Carbon-13 NMR spectra of six humic acid samples from marine and estuarine sediments.

If the organic matter (HM) content of the water is $12\,mg\,L^{-1}$, show that charge balance is adequately maintained.

5 The chemical nature of the HM in very deep (up to $> 1000\,m$) cores of sediments from the Black Sea have been studied in detail in order to understand the processes that occur over very long periods. Average percentage values of carbon, hydrogen, and oxygen

concentrations in two depth ranges are:

	C	H	O
Shallow	56.3	5.2	31.9
Deep	60.9	5.2	27.9

What do these results suggest about the long-term processes occurring within the organic sediments? (Data summarized from work by Huc, A. Y., B. M. Durand, and J. Monin as reported in ref. 2.)

6 The equivalent weight of HM is sometimes defined as the molar mass per mol of carboxylic plus phenolic groups. For a particular sample of fulvic acid whose molar mass is 6036 daltons, there are 4.2 and 2.2 mmol of carboxylic and phenolic groups respectively per gram of the FA. What is the equivalent weight?

Note

1 As an approximation, the remaining 50% of DOM consisits of low molar mass acids such as oxalic, citric, formic, and acetic acids ($\sim$25%), neutral compounds much of which is carbohydrate material ($\sim$15%), and other species.

Metals in the hydrosphere

OF all the elements on the periodic table, metals make up about 75% and they are found as ions and complex compounds throughout the hydrosphere. The concentration of metal species in various types of water covers a wide range. In the oceans, the concentration of sodium ion is approximately $0.46\,\text{mol L}^{-1}$ (Table 9.1) and magnesium ion is present in sufficiently large amounts that extraction from marine water is a viable source of the element. Other sea water metal concentrations range down to ultratrace levels.

Metals in fresh water are usually present in small concentrations. In water bodies located in regions with carbonate bedrock, calcium ion concentrations are in the mmol L^{-1} range, but most other elements are found at much lower levels. However, there are special situations. Saline lakes such as those in the Rift Valley of East Africa have very large amounts of alkali and alkaline earth elements. The concentration of sodium carbonate in Lake Magadi in Kenya (previously described in Chapter 11, in terms of alkalinity) greatly exceeds the solubility limit and much of the 'lake' is actually solid, with only scattered pools of saturated brine. Pit lakes—abandoned open pit mines that penetrate below the water table—often have elevated levels of metal ions derived from the ore minerals. For example, some pit lakes from the Robinson Mining District in Nevada, USA have copper and zinc concentrations as high as 0.6 and $0.8\,\text{mmol L}^{-1}$ respectively. In submerged soils, the concentration of the redox-sensitive elements iron and manganese may also approach the mmol L^{-1} range in the interstitial water.

Metals in the hydrosphere are of environmental interest and importance because of their interactions with solid phase materials of geological origin and also because of their influence on biological processes. Metals such as potassium and calcium are important nutrients required in substantial amounts by plants, animals, and microorganisms. Other metals such as copper and zinc are also nutrients, but the amounts required by organisms are very small. These metals, if present in excessive amounts can be toxic, so there is a range of concentrations, sometimes narrow, that is suitable for supporting life processes. Still other metals such as cadmium and mercury are not essential nutrients and even very small concentrations can be toxic to many living organisms. In general, a description of the pathways and cycles through which metals in water interact with the associated soils, sediments, and biota, is referred to as the subject of *metal biogeochemistry*.

The discussion in this chapter will centre on developing general ideas which allow us to understand the range of behaviour exhibited by various metals in different environmental situations.

Metals exist in the aqueous environment in a variety of forms—fully protonated or partially deprotonated aquo complexes, as complexes with inorganic ligands such as chloride and carbonate, and as complexes with naturally occurring organic molecules either low molecular weight like citrate or macromolecules such as fulvic acid.[1] In a given situation, the particular *species distribution* of a metal depends on the properties of the metal itself as well as on the availability and nature of potential ligands. Other bulk aqueous solution properties including the ionic strength, pH, and redox status also play a role in defining species distribution. It is helpful to assemble some of the vast collection of knowledge about individual cases into some general principles. To do this we will make use of basic concepts of inorganic chemistry.

13.1 Aquo complexes of metals

Where no other ligand is available to form complexes with a metal in aqueous solution, it exists as an aquo complex, as we showed in Chapters 9 and 10. Depending on circumstances, however, some of the coordinated water molecules may lose a proton. The degree to which deprotonation of aquo complexes occurs is, to a large extent, a property of the metal ion under consideration. The pH of the solution is also important in determining whether protons are lost. This type of reaction is an acid–base reaction and the general equation is

$$M(H_2O)_a^{b+} + H_2O \rightarrow M(H_2O)_{a-1}(OH)^{(b-1)+} + H_3O^+ \tag{13.1}$$

An alternative simplified description of the reaction omits the waters of hydration in the formulae:

$$M^{b+} + 2H_2O \rightarrow MOH^{(b-1)+} + H_3O^+ \tag{13.2}$$

Further deprotonation steps may also occur. Since these reactions ultimately involve a separation of two positive charges, they are favoured in the case of more highly charged metal ions and smaller ions, usually expressed in combination as Z^2/r, where Z is the numerical value of the charge and r is the ionic radius in nm. The inverse relation between pK_{a1} and Z^2/r holds up well for the main group elements but other factors are more important in the case of transition (especially heavy ones) metals. Table 13.1 lists the ratio and the pK_{a1} values for selected metal aquo complexes.

The pK_{a1} is the pH at which the aquo complex is present with half in the fully protonated form and half having lost a single proton. From these data, therefore, it is evident that the waters of hydration surrounding metal ions with a single positive charge exist exclusively in protonated form throughout the entire pH range. Of the +2 ions, deprotonation occurs more readily for smaller species (due to the larger value of Z^2/r). In aqueous solutions with pH > 6.5, $Be(OH)^+$ will be more important than Be^{2+}. The $MgOH^+$ and $CaOH^+$

Table 13.1 Values of Z^2/r (units are nm^{-1}) and pK_{a1} for aquo complexes of selected metal ions

Metal ion	$Z^2 r^{-1}/nm^{-1}$	pK_{a1}	Metal ion	$Z^2 r^{-1}/nm^{-1}$	pK_{a1}
Na$^+$	8.6	14.48	Ni^{2+}	48	9.40
K$^+$	6.6	> 14	Cu^{2+}	46	7.53
Be^{2+}	68	6.50	Zn^{2+}	46	9.60
Mg^{2+}	47	11.42	Cd^{2+}	37	11.70
Mn^{2+}	48	10.70	Hg^{2+}	34	3.70
Fe^{2+}	43	10.1	Al^{3+}	133	5.14
Co^{2+}	45.2	9.6	Fe^{3+}	115	2.19

pK_a values taken from Yatsimirksii, K. B. and V. P. Vasil'ev, *Instability Constants of Complex Compounds*, Pergamon, Elmsford, N Y, 1960.
In each case, the radius used is for the the 6-coordinate high-spin metal aquo complex.

species occur only at very high pH. Deprotonated species begin to assume significance in environmentally common situations for +3 ions including Fe^{3+} and Al^{3+}. A sample calculation illustrates this point.

Iron (III) exists in pure water as an aquo complex (for simplicty, we write its structure without the coordinated waters) and being a +3 ion, there is considerable tendency for deprotonation to occur. In the example we consider the possibility of two deprotonation steps as well as the formation of a bridged dimer. This *polynuclear species* of iron has the structure shown below:

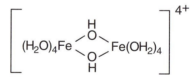

Aluminium and some other metals are also known to form polynuclear species.

Taking into account the four species indicated in the equations below, we will calculate their concentrations in pure water at pH = 7.00. The relevant equilibria are as follows:

$$\text{Fe(OH)}_3 \rightleftharpoons \text{Fe}^{3+}(\text{aq}) + 3\text{OH}^-(\text{aq}) \qquad K_{sp} = 9.1 \times 10^{-39} \qquad (13.3)$$

$$\text{Fe}^{3+}(\text{aq}) + 2\text{H}_2\text{O} \rightleftharpoons \text{FeOH}^{2+}(\text{aq}) + \text{H}_3\text{O}^+(\text{aq}) \qquad K_{a1} = 6.5 \times 10^{-3} \qquad (13.4)$$

$$\text{FeOH}^{2+}(\text{aq}) + 2\text{H}_2\text{O} \rightleftharpoons \text{Fe(OH)}_2^+(\text{aq}) + \text{H}_3\text{O}^+(\text{aq}) \qquad K_{a2} = 3.2 \times 10^{-4} \qquad (13.5)$$

$$2\text{Fe}^{3+}(\text{aq}) + 4\text{H}_2\text{O} \rightleftharpoons \text{Fe}_2(\text{OH})_2^{4+}(\text{aq}) + 2\text{H}_3\text{O}^+(\text{aq}) \qquad K_{ad} = 1.3 \times 10^{-3} \qquad (13.6)$$

Note that the waters of hydration associated with the iron are not shown. Using the first reaction, 13.3:

$$[Fe^{3+}][OH^-]^3 = K_{sp}$$

$$[Fe^{3+}][10^{-7.00}]^3 = 9.1 \times 10^{-39}$$

$$[Fe^{3+}] = 9.1 \times 10^{-18}\,mol\,L^{-1}$$

Using the second reaction, 13.4:

$$\frac{[FeOH^{2+}][H_3O^+]}{[Fe^{3+}]} = 6.5 \times 10^{-3}\,mol\,L^{-1}$$

Using the value of $[Fe^{3+}]$ calculated above:

$$[FeOH^{2+}] = \frac{6.5 \times 10^{-3} \times 9.1 \times 10^{-18}}{1.0 \times 10^{-7}}$$

$$= 5.9 \times 10^{-13}\,mol\,L^{-1}$$

Similar calculations are done to determine the equilibrium concentrations of the two additional species:

$$[Fe(OH)_2^+] = 1.9 \times 10^{-9}\,mol\,L^{-1}$$

$$[Fe_2(OH)_2^{4+}] = 1.1 \times 10^{-23}\,mol\,L^{-1}$$

Therefore, the total concentration of the various aquo species of iron (III) in water at $pH = 7.00$ is $1.9 \times 10^{-9}\,mol\,L^{-1}$ ($= 0.11$ ppb), and the most significant species is $Fe(OH)_2^+$.

Clearly, iron (III) is extremely insoluble in pure water. The solubility is greatly enhanced when ligands are present to form a stable complex. In the hydrosphere these ligands are frequently derived from the natural organic matter (NOM) present in the water. However, in general, simple iron (II) species are more soluble than the corresponding iron (III) species, so that reducing, low pE, conditions lead to increased concentrations of the element in the hydrosphere. That is evident in Fig. 10P.1.

13.2 Classification of metals

The term 'heavy metals' is frequently encountered in the environmental literature and is usually used to designate metals which are the cause of adverse biological reactions. Originally, the expression had a scientifically legitimate origin as it was coined to refer to metals such as lead and mercury. Lead has an atomic mass of $207.2\,g\,mol^{-1}$ and a specific gravity of 11.34; the corresponding values for mercury are $200.59\,g\,mol^{-1}$ and 13.55. They are therefore heavy metals in every sense of the name. The terminology, however, is sometimes applied indiscriminately and it is not unusual to find a list of heavy metals that includes elements like aluminium (atomic mass $26.98\,g\,mol^{-1}$ and specific gravity 2.70). Aside from the semantic problem, there is no chemical basis for deciding which metals should be included in this category.

13.2.1 Traditional classifications of metals

Other categories or classifications of metals have been proposed for general use, not just for environmental purposes. Amongst the useful classifications is one that divides metals into type A, type B and transition metals as first described by Ahrland *et al.* (1958).[2] In this system, type A metal ions are those with inert gas (d^0) electronic configurations. Such ions are characterized by spherical symmetry and low polarizability. The grouping is almost synonymous with the hard sphere metal category of Pearson (1963).[3] Included in this grouping are the environmentally important cations Na^+, K^+, Mg^{2+}, Ca^{2+} and Al^{3+}.

For type A ions, an electrostatic model approximately explains stability of metal–ligand complexes—that is, the stability of these complexes is positively correlated with the ratio Z^2/r (the (charge)^2/radius ratio) for both the metal ion and the ligand species. Therefore the stability of most complexes with the alkaline earth metals is in the order

$$Mg^{2+} < Ca^{2+} < Sr^{2+} < Ba^{2+}$$

Some of the properties of type A metals are

- a preference of the metal for O- or F-containing ligands over sulfur and higher halides—for example, $Al(H_2O)_5(OH)^{2+}$ and $Al(H_2O)_4F_2^+$ are important soluble aluminum species. Weak complexes with oxyanions such as SO_4^{2-}, NO_3^-, and oxygen-containing functional (e.g. $-COOH$, $>C=O$) groups of organic molecules are also found;
- the metal ion may form insoluble OH^-, CO_3^{2-} or PO_4^{3-} compounds—for example, $CaCO_3$ and $AlPO_4$ are important solid forms of these elements;
- complexes with OH^- are more stable than those with HS^- or S^{2-}. Metal sulfide soluble complexes or precipitates are not important;
- complexes with Cl^-, Br^-, and I^- tend to be weak;
- complexes with H_2O are more stable than those with NH_3 or CN^-.

Type B metal ions are ones with nd^{10} and $nd^{10}(n+1)s^2$ type electronic configurations. Such cations have readily distorted electronic distributions; in other words, they exhibit high polarizability. This class is approximately equivalent to the soft sphere metal ion category of the Pearson system. Included in the group are Ag^+ $(Kr(4d^{10}))$, Zn^{2+} $(Ar(3d^{10}))$, and Pb^{2+} $(Xe(4f^{14}5d^{10}6s^2))$.

For type B ions, covalent bonding plays a role in complex formation and therefore an electrostatic model alone is unable to explain stability relations. On the other hand a major factor affecting complex stability is the ability of the metal to accept electrons from the ligand. Therefore high electronegativity (in relative terms) of the metal and low electronegativity of the ligand donor atom would be consistent with high stability of a complex with type B metals.

For example, in the IIB group of elements, the Pauling electronegativities (values in brackets) are in the order

$$Zn(1.6) < Cd(1.7) < Hg(1.9)$$

and complex stability is generally in the same order.

The trend of electronegativity (en) of ligand donor atoms, and stability of their complexes (as measured by the overall stability constant β_f) with type B metal ions, is

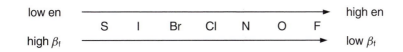

Therefore type B cations exhibit the following complexation properties.

- In general, these metals form more stable complexes than do the type A cations.
- Complex stability with the halides is in the order $I^- > Br^- > Cl^- > F^-$. This is the reverse of the situation observed with type A metals.
- Complexes with ligands containing nitrogen are more stable than those with oxygen—for example, NH_3 is favoured over H_2O, and CN^- is favoured over OH^-.
- Complexes with sulfide or organosulfides are common and stable. Such compounds are frequently insoluble.
- Complexes with carbon, for example, organometallic complexes, are observed. The mercury compounds, CH_3Hg^+ and $(CH_3)_2Hg$, are well known examples.

Transition metal cations are those with nd^x ($0 < x < 10$) electronic configurations. Pearson has described a borderline class which is similar, but not identical to the transition metals. Defined either way, this class exhibits properties that are intermediate between those of the the type A and B classes. Borderline metals are able to form complexes with all types of donor ligands, with the relative importance depending on a number of factors. Second row transition elements usually show more type B character than do those in the first row, and type B character also tends to increase somewhat as one moves from left to right on the periodic table.

Electrostatic factors play a role in determining stability of borderline metals and this is reflected in a general trend (Fig. 13.1, straight line) showing increasing stability for high-spin octahedral complexes with increasing atomic number of the $+2$ ions of the first transition metal series.

The increase in stability can be attributed to a stronger electrostatic effect because of increasing Z^2/r associated with decreasing atomic radius across the series. The other important factor which leads to deviations from the straight line is crystal-field stabilization energy. Placement of electrons in a t_{2g} orbital stabilizes the ion (as for Sc^{2+}, Ti^{2+}, V^{2+}), while electrons in e_g orbitals (as in Cr^{2+}, Mn^{2+}) reduce stability. This argument is a simplification; for example Cu^{2+} complexes tend to have greater stability than do those of Ni^{2+} due to a different structure of the complexes.

It should be noted that the trend presented here assumes a consistent $+2$ oxidation state. For some of these metals, this oxidation state is rarely found in the hydrosphere. Vanadium (II), for example, is virtually unknown in the environment, the common forms being vanadium (V) as the highly deprotonated VO_2^+ cation or the HVO_4^{2-} anion and, under reducing conditions, vanadium (IV) as the VO^{2+} species.

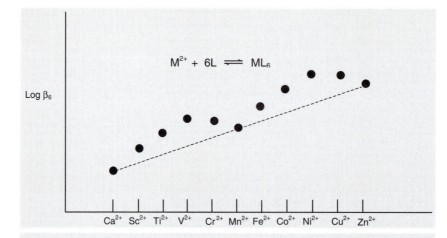

Fig. 13.1 Trend in overall stability constants (β_{f6}) for high-spin octahedral complexes involving +2 ions of transition (borderline) metals. As one moves from calcium to zinc across the periodic table, there is a gradual decrease in radius associated with the *lanthanide contraction*. Therefore, for +2 ions, the value of Z^2/r increases with increasing atomic number in this row of elements.

13.2.2 An environmental classification of metals

A more recent attempt at classifying environmentally important metals has been made by Nieboer and Richardson (1980).[4] Their system builds on the earlier concepts of Ahrland but takes into account bonding due to both covalent and ionic interactions. To do this, a covalent index is plotted versus an ionic index (Fig. 13.2).

The covalent index $X_m^2 r$ (X_m = metal ion electronegativity, r = ionic radius of metal) is a reflection of the ability of the metal to accept electrons from a donor ligand, and becomes the chemical parameter used to differentiate between type A, borderline and type B metals. Values of the index are smallest for type A and largest for type B ions. The ionic index Z^2/r measures the possibility of ionic bond formation and thus more highly charged species tend to be found on the right-hand side of the diagram. These are also the species that tend to act as Brönsted acids, as noted earlier. Overall, considering a natural sample containing a variety of ligands with different donor atoms (for example humic material), the tendency to exist in complexed form might be expected to follow an angled trend as shown in Fig. 13.2.

Some of the functional groups found in naturally occurring ligands and involved in forming metal complexes are listed later in this chapter (Table 13.3).

It is interesting that elements such as potassium and calcium which serve as macro-nutrients for microorganisms, plants, and animals are found in the type A class. When dissolved in water and in their interaction with complexing ligands, they are usually found to be associated with oxygen electron donors. Most biological micronutrients including manganese, copper, and zinc are found in the borderline group. In contrast to the type A

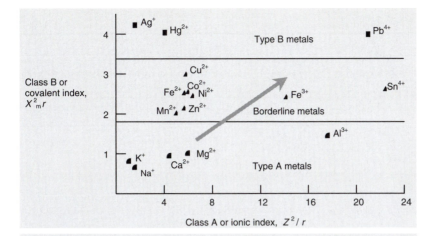

Fig. 13.2 Classification of some metals of environmental importance. (Redrawn from Nieboer and Richardson, 1980.) Subdivisions based on the traditional, Ahrland *et al.*, and Pearson classifications are indicated by the horizontal lines. Stability of complexes increases with increasing ionic and/or covalent index, as shown by the arrow.

metals, the borderline ions form stable complexes with a variety of electron donor atoms including oxygen, nitrogen, and sulfur. The type B metals include several that are known to be toxic to organisms. In general (there are exceptions), toxicity increases in the order type A < borderline < type B metals. While type B ions have a strong affinity for sulfur donor atoms, they also form more stable complexes with oxygen-donating compounds than do borderline and type A metals. The ability to form methylated derivatives that are stable in aqueous solutions is another feature that is characteristic of metals in this class. Methyl derivatives of type A metals decompose in water and the same is true of most borderline metals.

In spite of limitations, the above concepts do help to understand some of the features of natural aqueous ion solution chemistry. Table 13.2 gives the principal inorganic aqueous species of some environmentally important metal ions. In compiling the table, it was assumed that the metals are present within the typical 'normal' concentration range in water and that the water contains carbonate, sulfate, and chloride at levels approximately equal to that found in average river water. Additional species associated with the marine environment are also noted.

Several of the general features of the classifications described above are underlined by the data in the table.

- All the metals form aquo complexes in the aqueous media. Metals such as sodium and potassium have a small value of Z^2/r and the coordinated water molecules remain protonated in all situations. For metals with a larger ratio (e.g Al(III)), deprotonation

Table 13.2 Principal aqueous species of environmentally important metal ions. (Only inorganic species are considered)

	pH = 4		pH = 7		pH = 10	
	oxidizing environment	reducing environment	oxidizing environment	reducing environment	oxidizing environment	reducing environment
Sodium	Na^+	Na^+	Na^+	Na^+	Na^+	Na^+
Potassium	K^+	K^+	K^+	K^+	K^+	K^+
Magnesium	Mg^{2+}	Mg^{2+}	Mg^{2+}, $MgSO_4^0$ (sw)	Mg^{2+}, $MgSO_4^0$ (sw)	Mg^{2+}	Mg^{2+}
Calcium	Ca^{2+}	Ca^{2+}	Ca^{2+}, $CaSO_4^0$ (sw)	Ca^{2+}, $CaSO_4^0$ (sw)	Ca^{2+}	Ca^{2+}
Aluminium	Al^{3+}, $AlOH^{2+}$	Al^{3+}, $AlOH^{2+}$	$Al(OH)_2^+$, $Al(OH)_3^0$, $Al(OH)_4^-$ (sw)	$Al(OH)_2^+$, $Al(OH)_3^0$, $Al(OH)_4^-$ (sw)	$Al(OH)_4^-$	$Al(OH)_4^-$
Vanadium	$H_2VO_4^-$, VO_2	VO^{2+}	$H_2VO_4^-$, HVO_4^{2-}, $V_{10}O_{28}^{6-}$	VO^{2+}	VO_4^{3-}	
Chromium	$HCrO_4^-$	$CrOH^{2+}$	$HCrO_4^-$, CrO_4^{2-}	$CrOH^{2+}$, $Cr(OH)_2^+$	CrO_4^{2-}	$Cr(OH)_4^-$
Manganese	Mn^{2+}	Mn^{2+}	MnO_2^0, $MnCl^+$ (sw)	Mn^{2+}, $MnCl^+$ (sw), $MnSO_4$ (sw)	MnO_2^0	$MnCO_3^0$
Iron	$FeOH^{2+}$, $Fe(OH)_2^+$	Fe^{2+}	$Fe(OH)_3^0$	Fe^{2+}, $FeCO_3^0$	$Fe(OH)_4^-$	$FeOH^+$, $Fe(OH)_2^0$
Cobalt	Co^{2+}	Co^{2+}	Co^{2+}, $CoCO_3$	$CoCO_3$	Co_3O_4	$CoCO_3$
Nickel	Ni^{2+}, $NiSO_4^0$	Ni^{2+}	Ni^{2+}, $NiHCO_3^+$, $NiCl^+$ (sw)	Ni^{2+}, $NiHCO_3^+$, $NiCl^+$ (sw)	$NiOH^+$, $Ni(OH)_2^0$, $NiCO_3$	$NiOH^+$, $Ni(OH)_2^0$, $NiCO_3$
Copper	Cu^{2+}	Cu^{2+}	Cu^{2+}, $CuOH^+$, $CuHCO_3^+$, $CuCl^+$ (sw)	Cu^{2+}, $CuOH^+$, $CuHCO_3^+$, $CuCl^+$ (sw)	$Cu(OH)_2^0$, $Cu(CO_3)_2^{2-}$	$Cu(OH)_2^0$, $Cu(CO_3)_2^{2-}$
Zinc	Zn^{2+}	Zn^{2+}	Zn^{2+}, $Zn(OH)_2^0$, $ZnCl^+$ (sw)	Zn^{2+}, $Zn(OH)_2^0$, $ZnCl^+$ (sw)	$Zn(OH))_2^0$	$Zn(OH))_2^0$
Molybdenum	$HMoO_4^-$		$HMoO_4^-$		$HMoO_4^-$, MoO_4^{2-}	
Lead	Pb^{2+}, $PbSO_4^0$	Pb^{2+}	Pb^{2+}, $PbOH^+$, $PbHCO_3^+$, $PbCl^+$ (sw), $PbSO_4^0$ (sw)	Pb^{2+}, $PbOH^+$, $PbHCO_3^+$, $PbCl^+$ (sw)	$Pb(OH)_2$, $PbCO_3$, $Pb(CO_3)_2^{2-}$	$Pb(OH)_2$, $PbCO_3$, $Pb(CO_3)_2^{2-}$
Mercury	$HgOH^+$, $Hg(OH)_2^0$, $HgCl_2^0$	Hg^0	$Hg(OH)_2^0$, $HgCl_2^0$, $HgCl_4^{2-}$ (sw), $HgCl_3^-$ (sw)	Hg^0	$Hg(OH)_2^0$	Hg^0

Sulfate, chloride, and carbonate in concentrations which approximate those found in average river water are assumed to be present in the water. The metal concentrations are also assumed to be in the range of those found in 'normal' water. Some neutral species such as $Fe(OH)_3$ and MnO_2 are highly insoluble and will be present as colloids even when the metal concentration is very small.

The notation (sw) indicates that this additional species is present in sea water. Note that sea water pH is approximately 8.

Coordinated water molecules are not included in the formulae.

occurs more readily, and with metals in very high oxidation states (e.g. Cr(VI) and Mo(VI)), oxyanions are the principal species.

- Those metals which are subject to redox reactions are present as different species in oxidizing (high pE) compared to reducing (low pE) environments.
- In sea water, the high concentration of chloride and, to a lesser extent, sulfate favours formation of complexes with these ligands in the case of certain metals.
- The tendency for formation of complexes with ligands other than water is generally in the order type B metals (e.g. Pb(II), Hg(II)) > borderline metals (e.g. Mn(II), Zn(II)) > type A metals (e.g. Ca(II), Al(Ill)).

13.2.3 Complexes with humic material

As illustrated in Fig. 13.3 humic material, whether it is dissolved in water or present as part of the solid phase in soils and sediments, has functional groups that are capable of acting as ligands in forming complexes with metals. Some metals such as the alkaline earth elements react by forming relatively weak ionic bonds at the negative sites on the depronated humic molecules. On the other hand, other elements such as Cu(II), Pb(II), and the trivalent metals Al(III) and Fe(III) have large stability constants. Complexation in these cases involves covalent bonding and bidentate chelates are likely important. Figure 13.3 shows some of the functional groups associated with humate molecules.

A possible reaction between lead and a portion of a HM molecule is given in reaction 13.7. The chelate formed with the salicylate functional group is a stable six-membered system. The waters of hydration have been omitted for clarity.

$$\text{Pb}^{2+} + \text{HM} \underset{\text{OH}}{\overset{}{\bigcirc}} \text{COOH} \longrightarrow \text{HM} \bigcirc \underset{\text{O}}{\overset{}{\text{C}=\text{O}}} + 2\text{H}^+ \tag{13.7}$$

The stability constant, K_{f1}, for this reaction has been determined to be approximately 10^6. In a natural situation, the extent of complexation depends on a number of factors, as listed below.

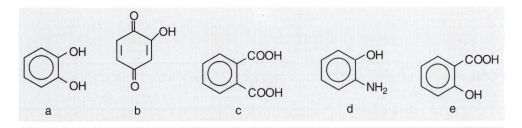

Fig. 13.3 Examples of humic material functional groups available for complexation reactions. Structures c (phthalate) and e (salicylate) are thought to be particularly important players in chelate formation processes.

- The nature of the metal ion. Because bonding may involve both ionic and covalent forces, various properties of the metal affect stability as we have seen earlier. Most important among these are the ionic index—so that trivalent ions like Al^{3+} tend to be strongly bonded and monovalent ions like Na^+ and K^+ have only a weak association—and the ability to form covalent linkages as indicated by the covalent index. The alkaline earth metal ions Ca^{2+} and Mg^{2+} react to a much smaller extent than divalent transition metal ions such as Cu^{2+} or Pb^{2+} due to their inability to bond covalently.

- Ambient solution pH. A low solution pH means competition from hydrogen ions for sites on the functional groups which react with metals, and therefore reduced tendency for complex formation. Conformational changes occurring at low pH may further inhibit reaction with metals.

- Ionic strength. The ability of humic material to react with transition metals is inversely related to solution ionic strength. There are two reasons for this: competition for ligand sites from the cations (especially alkaline earth cations) which contribute to the increased ionic strength, and availability of anions (like Cl^-, SO_4^{2-}, and HCO_3^-) to react with metals, thus inhibiting metal–humate reactions.

- Availability of functional groups. This depends on both the concentration and nature of the humic material. An estimate of the *maximum* complexation capacity may be made by assuming that metals react in a 1:1 ratio with humic functional groups.

Given that humic materials contain a heterogeneous collection of functional groups that vary from source to source, it becomes difficult to estimate the extent of complex formation between particular metals and humic acid in general. Nevertheless, stability constants for such reactions have been determined and are useful in a semiquantitative sense for estimating the degree to which complexes might be expected to form in a given situation. Because protonated functional groups have a wide range of pK_a values, it is best to report the constants as conditional stability constants (K_f') defined at a particular pH. A further refinement would be to establish the ionic strength associated with the defined values. Table 13.3 gives values for (K_f') for soluble fulvic acid and several metals at pH 5.

Consider water of pH 5, containing $85\,\mu g\,L^{-1}$ ($1.44\,\mu mol\,L^{-1}$) of nickel and $8\,mg\,L^{-1}$ of soluble fulvic acid. In order to calculate the concentration of complexed nickel, we must know or estimate the concentration of functional groups capable of binding with nickel. As we saw in Chapter 12, a reasonable estimate for this is around $5\,mmol\,g^{-1}$ of the fulvic acid.

Table 13.3 Conditional stability constants (pH 5.0) for soluble fulvic acids and selected metals

	Mg^{2+}	Ca^{2+}	Mn^{2+}	Co^{2+}	Ni^{2+}	Cu^{2+}	Zn^{2+}	Pb^{2+}
K_f'	1.4×10^2	1.2×10^3	5.0×10^3	1.4×10^4	1.6×10^4	1.0×10^4	4.0×10^3	1.1×10^4

Schnitzer, M. and S. U. Khan, *Soil Organic Matter*, Elsevier, Amsterdam; 1978.

This then means that the concentration of potential binding sites is

$$8 \, \text{mg} \, \text{L}^{-1} \times 5 \, \text{mmol} \, \text{g}^{-1} = 40 \, \mu\text{mol} \, \text{L}^{-1}$$

The $40 \, \mu\text{mol} \, \text{L}^{-1}$ corresponds to an ability to form soluble complexes with about $2.3 \, \text{mg} \, \text{Ni}^{2+} \, \text{L}^{-1}$.

It is essential to realize that determining the concentration of binding sites allows us to estimate only the *potential* for reacting with a metal. The calculated value is called the complexation capacity and is characteristic of a water body at a given time. Typical values for complexation capacity for particular types of water bodies are:

$$\begin{aligned} \text{rivers} \quad & 1-2 \, \mu\text{mol} \, \text{L}^{-1} \\ \text{Lakes} \quad & 2-5 \, \mu\text{mol} \, \text{L}^{-1} \\ \text{Ponds} \quad & 5-15 \, \mu\text{mol} \, \text{L}^{-1} \\ \text{Swamps} \quad & > 15 \, \mu\text{mol} \, \text{L}^{-1} \end{aligned}$$

The complexation capacity used in the above example is clearly characteristic of water containing a large concentration of soluble organic matter.

The *extent* to which reaction actually occurs depends on the other factors as noted previously. Using the conditional stability constant in Table 13.3, we can calculate the percentage of nickel that would be complexed with FA in the present situation.

Without showing charges and form of the species involved, the general reaction of nickel (Ni^{2+}) with fulvic acid (FA) can be represented as

$$Ni + FA \rightarrow NiFA \tag{13.8}$$

For the reaction, the conditional stability constant is

$$K'' = \frac{[NiFA]}{[Ni_u][FA_u]} = 1.6 \times 10^4 \tag{13.9}$$

In this formulation, $[Ni_u]$ refers to the total concentration of all soluble nickel species that are not complexed with fulvic acid and $[FA_u]$ is similarly the concentration all of the fulvic acid not complexed with nickel. The latter concentration is expressed in terms of functional groups available for complexation and we are assuming that no other metals are present to compete for these sites.

Since the ligand is available in large excess, we can make the approximation that $C_{FA} = [FA_u] = 4 \times 10^{-7} \, \text{mol} \, \text{L}^{-1}$. The total concentration of nickel is $1.44 \, \mu\text{mol} \, \text{L}^{-1}$. Let the value of uncomplexed nickel, $[Ni_u] = u$.

$$\frac{(1.44 \times 10^{-6} - u)}{u \times 4.0 \times 10^{-7}} = 1.6 \times 10^4$$

Therefore

$$u = [Ni_u] = 8.8 \times 10^{-7} \, \text{mol} \, \text{L}^{-1}$$

and

$$[NiFA] = 5.5 \times 10^{-7} \, \text{mol} \, \text{L}^{-1}$$

About 40% of the nickel in this water is complexed by the fulvic acid. In a real environmental situation, the calculation would be complicated by the fact that other metals present would compete with nickel for binding sites on the soluble organic matter. We must also restate the caveat that these calculations assume equilibrium conditions.

13.3 Three metals—their behaviour in the hydrosphere

The literature on the environmental properties and behaviour of metals is extensive. There is considerable information about baseline levels in various types of water and many examples exhibiting contamination have been described in detail. In this section, we will focus on how principles outlined above are expressed in three important cases.

13.3.1 Calcium

Calcium, at 3.6%, is the fifth most abundant element in the Earth's crust. Important mineral forms of calcium include limestone ($CaCO_3$) and dolomite ($CaMg(CO_3)_2$) as well as aluminosilicate minerals such as the feldspar anorthite and the clay mineral montmorillonite. Weathering of these and other minerals leads to some dissolution in water. The solubility of the particular mineral and other environmental factors—particularly the concentration of carbonate species in water—determine the final concentration of calcium in a particular water body.

Aside from the metallic form, calcium has a single common oxidation state, $+2$, and so it is not directly influenced by the redox status of the water. It is a type A metal and its behaviour in the hydrosphere is characteristic of that class. It is therefore usually associated with an oxygen-donor ligand environment which means that, in addition to the aquo species, it is able to form complexes with species such as phosphate, carbonate, and sulfate when they are present. Except in soil solutions, phosphate is usually found in such low concentrations, that its reaction with calcium is insignificant. Where carbonate (usually as hydrogen carbonate) and sulfate are in the mmol L^{-1} concentration range, their complexes with calcium are of some importance.

As shown by the stability constant (Table 13.3), calcium interacts to a limited extent with dissolved humic material. Bonding presumably involves the oxygen-containing functional groups. Depending on the concentration of the dissolved HM, in water whose pH is near neutral, a small but significant portion of the calcium may be in the complexed form. Under somewhat acid conditions, hydrogen ions compete effectively for sites on the humate and the calcium reverts to the simple ionic form. Calcium also forms weak complexes with anions of some organic acids—oxalate and citrate being two examples. For example, in a solution containing 2.1 mg L^{-1} Ca^{2+} and 2.3 mg L^{-1} oxalate at pH 6.5, the fraction of calcium in soluble complexed forms is about 0.1%. The stability constants, K_{f1} and K_{f2}, used in making this calculation were 4.6×10^1 and 1.1×10^1. Other metals in water will, of course, compete with calcium for sites on the complexing ligand.

Most suspended and sedimentary mineral material contains calcium at the per cent level. Dissolution of calcium from these solids is enhanced under acidic conditions. Therefore, internal or external processes which produce acidity—acid rain, nitrification, and so

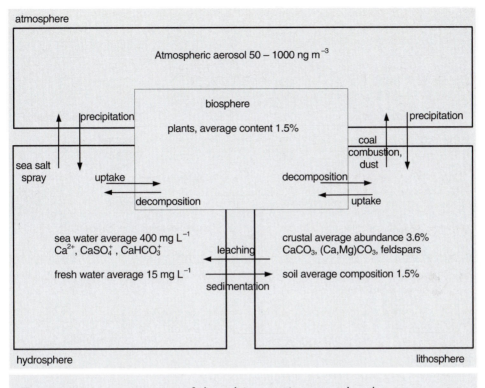

Fig. 13.4 Major components of the calcium environmental cycle.

on—increase the concentration of calcium in the associated water. This is especially true where the calcium is present as carbonate minerals. In soils, calcium is usually the principal ion occupying exchange sites on the organic and mineral material (see Chapter 18) and is an important nutrient for plants. The exchangeable calcium is also readily displaced by acidity in water passing through the soil.

Figure 13.4 summarizes major components of the environmental cycle for calcium.

13.3.2 Copper

Copper is an economically important element that is found in only trace quantities (global average abundance is $63\,\mu g\,g^{-1}$) in the Earth's crust. For both plants and animals it is required as a trace nutrient, but excessive amounts are toxic. Mineral forms include the free metal, a number of silicate and oxide species and mixed copper/iron sulfide minerals such as chalcopyrite ($CuFeS_2$). The principal oxidation state of copper in the hydrosphere is the $+2$ state. Monovalent copper species are also known but they dissociate to form copper (0) and copper (II) in most instances. An exception is in sea water where, under reducing conditions, copper (I) chloro species are stable.

Copper is a borderline metal and therefore has a good ability to form complexes with a variety of ligands. Ligands with nitrogen donor atoms are especially favoured. In natural

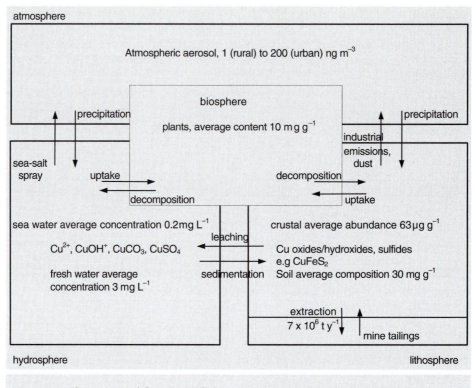

Fig. 13.5 The principal features of the environmental cycle for copper.

fresh water that is in equilibrium with the atmosphere, at low pH the aquo complex of copper is the principal species. In the neutral pH range, partially deprotonated forms become important as does a complex with hydrogen carbonate. At still higher pH values, further deprotonation takes place and a complex involving two carbonate ions is formed (Table 13.2).

Copper forms complexes with dissolved humic material. It is likely that the small fraction of nitrogen in this organic matter plays an important role in binding with copper and creating a stable species. When the humic material is present in soluble form in water, it serves to increase the solubility of copper. Conversely, particulate humic material—either suspended or precipitated—serves to remove soluble copper from the water column.

Significant features of the global cycle of copper are summarized in Fig. 13.5.

13.3.3 Mercury

In the Earth's crust, the concentration of mercury is three orders of magnitude smaller than that of copper, the global average value being $89 \, ng \, g^{-1}$. Where it is found in nature, it is in elemental form or combined with sulfur; the most important mineral is cinnabar, a form of mercury (II) sulfide (α-HgS). In the hydrosphere and associated sediments it can exist as 0, +1, and +2 species depending on redox and other environmental conditions. The mercury (II) forms comprise the most important aqueous species when aerobic conditions obtain.

Mercury aquo complexes readily tend to deprotonate. Therefore, even in moderately acidic conditions at pH 4, the mono and dihydroxy complexes are the two dominant forms if no other complexing agents are present. Mercury (II) is a type B metal. One consequence of this is its strong affinity for sulfur as evidenced in the most common mineral form of the element. The tendency to form complexes with other chelating ligands is also strong. This is clear in Fig. 1.3 where we saw that even when the chloride ion concentration in well water was as low as $9.5 \, \mu g \, mL^{-1}$, the principal mercury (II) species is $HgCl_2$ (aq). Mercury (II) also has a strong affinity for organic ligands in water. In the pore water of forest soils that are rich in organic matter, essentially all of the dissolved mercury exists as complexes with the soluble organic matter. The stability constants for the 1 : 1 complex involving mercury and humate have been estimated to be larger than those of the metals in Table 13.3. Binding may involve the nitrogen and sulfur groups that are present in small amounts in the humate, as well as the more common oxygen-containing functional groups.

Mercury (II) also has the ability to form a bond with carbon in the form of methyl mercury species. Methylation occurs under anaerobic conditions by a pathway which involves transfer of a methyl group from methylcobalamin, a derivative of vitamin B_{12}, to the mercury atom. This reaction requires reducing conditions as the methylcobalamin is formed through the action of methane-generating bacteria. In the anaerobic environment, a competitive reaction occurs in which mercury (II) is reduced to the metallic state, or in the presence of sulfate-containing sediments reacts with sulfide (formed during reduction) to form a very insoluble mercury (II) compound.

An alternative pathway occurs under aerobic conditions inside bacterial cells where the mercury binds with an enzyme, after which a methyl group is transferred to the mercury.

Both pathways are favoured by a moderately low pH and result in the production of monomethyl mercury (II) (CH_3Hg^+). This ion readily forms complexes with a variety of ligands. For example, compounds such as CH_3HgCl, and $(CH_3Hg)_2S$ are formed and are highly stable with respect to breaking the Hg–C bond. To some extent, however, chemical, photochemical and biotic degradation does occur.

Further methylation also takes place to produce dimethyl mercury ($(CH_3)_2Hg$). This neutral compound is highly insoluble and volatile. Once in the atmosphere it eventually photolyses to regenerate ionic mercury (II) and is rained out.

Environmental methylation is not unique to mercury. Methylated derivatives of tin (IV) and lead (IV) have been observed in natural, unpolluted situations; the mechanism involves oxidative methylation of the lead (II) or tin (II) ions by the methyl carbonium ion. This route is in contrast to that for mercury (II) where the transfer occurs *via* a methyl carbanion from methylcobalamin.

Figure 13.6 shows the principal features of the global cycle of mercury. This element is unique amongst metals in that its environmental cycle includes an important contribution from gas-phase species.

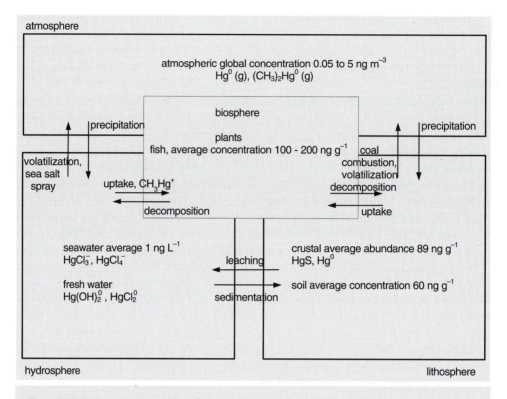

Fig. 13.6 Major components of the environmental cycle of mercury.

Box 13.1 Mercury in the Amazon Basin

Since the late 1970s when major deposits of gold were discovered in the Brazilian Amazon region, there has been a 'gold rush' bringing in persons hoping to make a fortune on finding and recovering gold from the soils and river sands. The operations involve over a million people in an area of about 170 000 km^2 and several hundred tonnes of gold are obtained annually.

The technology used is simple. A common recovery process begins with dredging of the river sand. After sieving, the material is passed through carpeted riffles to retain the heavier particles. In another method, soils or sediment are excavated and any consolidated material is crushed before being subjected to centrifugal separation to produce a concentrate.

The separated concentrates are placed in amalgamation drums where they are mixed with mercury which is able to dissolve gold. The solution, called an amalgam, is very dense and is separated from the concentrate matrix by panning. The mercury is then recovered from the amalgam by 'roasting' it in partially enclosed retorts or even in the open air. The 'pure' gold which is produced and supplied to the market may still contain up to 5% (by mass) mercury and therefore reburning is frequently required.

Without proper recovery, this adds to the atmospheric release in the small towns where gold dealers locate. It has been estimated that each year more than 100 tonnes of mercury are lost to the atmosphere in the recovery and purification operations.

While global background atmospheric concentrations of mercury are typically 0.05 to 5 ng m^{-3}, levels in the Brazilian mining areas have been found to range from 20 to 500 ng m^{-3}. In towns near dealers' shops, atmospheric concentrations reach as high as 3 μg m^{-3}, and inside the shops, levels of up to 300 μg m^{-3} have been measured. The World Health Organization limit for public exposure has been set at 1 μg m^{-3}.

The atmospheric mercury is oxidized to mercury (II) by photochemical reactions involving ozone and water vapour. In its ionic form, mercury (II) is rained out on to land or water, where it takes part in other reactions. In other areas of the Amazon basin remote from mining, where atmospheric deposition is the only source, sediment concentrations, have been positively correlated with organic matter content, reflecting the element's tendency to form complexes with ligands from the humate material. Conditions in the Amazon basin—high temperature, high biological activity, slightly acidic conditions, a plentiful supply of organic matter—favour methylation processes. Methylated compounds are therefore also present.

Additional mercury, perhaps half the amount that is released to the atmosphere, is returned to the river as part of the sediments. The element has low mobility and therefore sediment concentrations near sources of mining are high. Small streams feeding into the Madiera River (a major tributary of the Amazon and a centre of the gold deposits) have shown sedimentary levels up to 20 μg g^{-1}. In the short term the mercury is quite unreactive and remains in the elemental form, sometimes visible in exposed sediments during the dry season. Near dredged areas in the river itself, concentrations are an order of magnitude lower, and only a few km downstream the sediment concentration declines by a further factor of ten. Lateral transport is mainly in the form of suspended material and most movement downstream occurs in 'white water' during the rainy season when the suspended solids load is at a maximum. Most of the mercury is associated with suspended matter and the total concentration in the murky water has been measured to be as large as 13 μg L^{-1}; the dissolved concentration is small. In surface sediments, oxidizing conditions enable slow oxidation of mercury to form soluble mercury (II) chloro species. In these forms, some is accumulated by the plants growing in the nutrient-rich waters. Where sediments contain abundant organic matter methylation can occur, and this form of the element is taken up by fish. Elevated levels of mercury are observed in both plants and fish in rivers near the mining regions.

13.4 Metal complexes with ligands of anthropogenic origin

In addition to the naturally occurring ligands—sulfate, chloride, anions of organic acids, humic material—that are found in water in various environments, there are also compounds resulting from industrial, general urban and agricultural activities which find their way into surface and ground water. Some of these are also able to complex with metals. When an insoluble compound is formed, the effect is to remove the metal from the aqueous phase. If the compound is soluble, the mobility of the metal in the environment is increased. There are a number of situations where enhanced solubility is of major concern.

Some complexing agents that are released as effluents into water bodies include the following:

- ammonia—resulting from the decay of nitrogen-containing organic wastes;
- sulfide, sulfite, and sulfate—discharged from pulp and paper mills, depending on the processes used;
- phosphate—a constituent of some detergents and therefore present in municipal waste water; also released from agricultural run-off where phosphate fertilizers have been used;
- cyanide—used in various industrial processes including extraction of gold from ore minerals;
- EDTA (ethylenediaminetetraacetic acid)—used for industrial cleaning, in the photographic, textile, and paper industries, and in detergents in some countries such as Germany;
- NTA (nitrilotriacetic acid)—a detergent builder employed in some countries such as Canada and several countries in northern Europe.

The behaviour of NTA in water provides an interesting case study of the effect of introducing a human-produced complexing agent into natural water. Where its use is permitted, NTA is present in detergents at a concentration of about 15% by mass in order to act as an efficient binder for calcium and other ions present in water. Sequestering the metals improves the action of the surfactants in the detergent. The structural formula of NTA (molar mass = 191.8 daltons) is

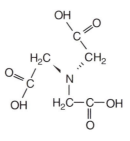

As can be seen from the formula, NTA is a triprotic carboxylic acid; the acid dissociation constants are $pK_{a1} = 1.66$, $pK_{a2} = 2.95$, and $pK_{a3} = 10.28$.

Complexation with metal ions is a simple one-step reaction in which the NTA chelate is coordinated with the metal in a tetrahedral configuration involving the three carboxyl groups and the nitrogen atom. The complex with copper is shown in Fig. 13.7. Table 13.4 gives the stability constants, as log K_f for a number of metals with NTA.

There is concern that NTA that is present in an urban waste water stream will be released into water bodies where it could enhance the solubility of metals, especially the

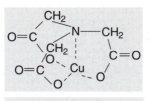

Fig. 13.7 The tetrahedral complex formed between nitrilotriacetic acid (NTA) and copper.

Table 13.4 Log K_f values for the 1:1 complexes between selected metals and NTA

Metal ion	log K_f	Metal ion	log K_f
Mg^{2+}	5.47	Mn^{2+}	7.46
Ca^{2+}	6.39	Cu^{2+}	12.94
Fe^{2+}	8.82	Zn^{2+}	10.66
Fe^{3+}	15.9	Pb^{2+}	11.34

Martell, A. E., the influence of natural and synthetic ligands on the transport and function of metal ions in the environment. *Pure Appl. Chem*, **44** (1975), 81.

borderline and type B metals whose presence at elevated concentrations in drinking water is undesirable. When waste water is subjected to biological treatment processes, such as the activated sludge process (Chapter 16), it has been shown that a considerable amount of NTA is degraded by microbial digestion to produce carbon dioxide, water, inorganic nitrogen, and cellular mass. Breakdown is slower under low temperature conditions. Where the waste water contains relatively high concentrations of metals such as copper, nickel, and lead, these metals form chemically stable complexes with NTA. In contrast to the free ligand, the metal complexes are quite resistant to microbial degradation.

Concentrations of NTA in water just below waste streams have been found to reach several hundered $\mu g \, L^{-1}$, especially in cold climates where biodegradation is incomplete.

Consider the following situation. The total NTA and copper concentrations in waste water effluent (pH = 7.5) are $100 \, \mu g \, L^{-1}$ ($= 5.2 \times 10^{-7} \, mol \, L^{-1}$) and $2.0 \, mg \, L^{-1}$ ($= 3.1 \times 10^{-5} \, mol \, L^{-1}$) respectively. We will determine the species in which these two entities exist in the water, neglecting the presence of other possible complexing agents or metals.

In order to carry out the calculation, we begin with the formation reaction for the complex, using H_3T as an abbreviation for the triprotic NTA.

$$Cu^{2+}(aq) + T^{3-}(aq) \rightarrow CuT^{-}(aq) \qquad \log K_f = 12.94 \qquad (13.10)$$

While this is the reaction corresponding to the definition of formation constant, in the water itself only a fraction of the copper and NTA are in the simple forms shown. We can calculate those two fractions using methods described earlier.

It is important to recall that copper, like any other metal ion, exists in the aqueous solution as a hydrated species, and deprotonation of the coordinated waters can take place. For copper in water, the first (and only significant) deprotonation step has a pK_{a1} value of 7.53 (Table 13.1). For the 4-coordinated aquo complex, the reaction can be written as

$$Cu(H_2O)_4^{2+}(aq) + H_2O \rightleftharpoons Cu(H_2O)_3(OH)^+(aq) + H_3O^+(aq) \tag{13.11}$$

The formula, $Cu(H_2O)_4^{2+}$, can be simplified as Cu^{2+} and $Cu(H_2O)_3(OH)^+$ as $CuOH^+$.

The fraction of species present as the fully protonated Cu^{2+} is calculated using the method described in Chapter 10, Section 10.1. At $pH = 7.5$, the calculation is simple since the copper aquo complex can be thought of as a monoprotic acid.

$$\alpha_{Cu^{2+}} = \frac{Cu^{2+}}{Cu^{2+} + CuOH^+} = \frac{H_3O^+}{H_3O^+ + K_{a1}} = 0.52$$

The NTA exists in four possible forms with varying degrees of deprotonation. The fraction of the T^{3-} form is calculated as follows:

$$\alpha_{T^{3-}} = \frac{T^{3-}}{H_3T + H_2T^- + HT^{2-} + T^{3-}}$$

$$= \frac{K_1 K_2 K_3}{[H_3O^+]^3 + K_1[H_3O^+]^2 + K_1 K_2 [H_3O^+] + K_1 K_2 K_3}$$

$$= 1.66 \times 10^{-3} \tag{13.12}$$

The K_f expression can then be rewritten in the following form:

$$K_f = \frac{[CuT^-]}{[Cu^{2+}][T^{3-}]} = \frac{[CuT^-]}{[Cu_u]\alpha_{Cu^{2+}}[NTA_u]\alpha_{T^{3-}}} \tag{13.13}$$

where $[Cu_u]$ and $[NTA_u]$ refer to the molar concentration of the uncomplexed forms of copper and NTA respectively. The conditional stability constant applies to specific soultion conditions and is defined as

$$K_f' = \frac{[CuT^-]}{[Cu_u][NTA_u]} = K_f \alpha_{Cu^{2+}} \alpha_{T^{3-}} \tag{13.14}$$

In the present case for water at pH 7.5,

$$K_f = 10^{12.94} = 8.7 \times 10^{12}$$

$$K_f' = 8.7 \times 10^{12} \times 0.52 \times 1.66 \times 10^{-3} = 7.5 \times 10^9$$

Since, in the present example, copper is in a 60-fold excess compared to NTA, essentially all of it is in the uncomplexed form and $[Cu_u] = C_{cu}$. Therefore the fraction of the NTA that is complexed with copper is

$$\frac{CuT^-}{[NTA_u]} = K_f' \times C_{Cu} = 7.5 \times 10^9 \times 3.2 \times 10^{-5} = 2.4 \times 10^5$$

The large ratio of NTA complexed with copper (CuT^-) compared with uncomplexed NTA (NTA_u) shows that essentially all of the NTA is in the metal-bound form. Such forms are stable with respect to further biodegradation, keeping the metal in solution and this is one of the concerns regarding the release of this water-softening agent into the environment.

13.5 Suspended matter in the hydrosphere

Thus far in the chapter, we have considered metal ions only when they are present in dissolved forms. In groundwater, filtered through layers of permeable soil material, water is usually clear, indicating that any metals present are in fact dissolved. In other compartments of the hydrosphere, however, there are situations where a considerable fraction of the metal exists as suspended particulate material. You can picture a fast-flowing river carrying an abundant sediment load that gives it a distinct milky appearance—for example the 'White Amazon' mentioned above. A portion of any metal in the river is in the form of dissolved species, but additional metal is also found in the suspended matter. What is dissolved and what is suspended is a complicated matter which will be discussed in Chapter 14. Often the distinction is based on a laboratory procedure where filtration through a membrane with nominal pore size of 0.45 μm is used for the separation.

Suspended material consists of typical soil minerals and organic matter, especially those in the fine (clay-sized) fraction. The elements present in such inorganic and organic structures usually include relatively high concentrations of alkali and alkaline earth metals, aluminium, and iron, along with smaller amounts of other metals depending on the particular materials involved. The sediment therefore carries a considerable fraction of major structural elements in suspended form compared with that dissolved in the water.

On the other hand, some elements of environmental interest, particularly those usually found in trace amounts, are associated with the sediments as adsorbed species on the

Table 13.5 Concentrations and ratios of particulate (p) and soluble (s) cadmium, copper, lead, and zinc in the Yamaska and St Francois rivers

| | Concentration μg L^{-1} | | | | | |
| | Yamaska River | | | St. Francois River | | |
	Particulate (p)	Soluble (s)	Ratio p/s	Particulate (p)	Soluble (s)	Ratio p/s
Cadmium	0.4	<0.1	>4	0.14	<0.2	0.7
Copper	0.9	1.0	0.9	4.1	4.6	0.9
Lead	1.5	1.5	1	3.8	<1	>4
Zinc	4.2	3.2	1.3	12.2	6.7	1.8

Campbell, P. G. C., A. Tessier, and M. Bisson, Anthropogenic influences on the speciation and fluvial transport of trace metals. In *Management and Control of Heavy Metals in the Environment*, Proceedings of an International Conference, London, U.K., Sept. 1979. CEP Consultants, Edinburgh.

surface of the fine particles. The surface-absorbed elements are more available than structural elements; for example, when a sediment-bearing river discharges water into an estuary, most of the adsorbed ions are displaced by sodium ions *via* an ion-exchange reaction and become part of the solution phase.

The ratio of adsorbed to dissolved cations in rivers is quite variable, with values typically ranging from less than 0.1 to much greater than 1. As an example, the mean particulate and soluble concentrations of four metals in two rivers in South-Eastern Quebec Province in Canada are given in Table 13.5.

The main points

1 The simplest dissolved forms of any metal in the hydrosphere are the aquo complexes. Deprotonation of these complexes occurs to some extent, especially in the case of metal species with a large (charge)^2 to radius ratio. Species such as aluminium (III) therefore act as Brönsted acids in aqueous media.

2 In the presence of natural or anthropogenic ligands, metals may combine to form complexes in solution. The tendency to form such complexes depends on the nature and concentration of the potential ligands in the water. Ligands in the hydrosphere are derived from both natural and anthropogenic sources.

3 Tendency to from complexes depends also on the nature of the metal itself. Various classification schemes have been proposed to explain and predict the favoured types of ligands and degree of complex formation. The schemes are usually based on the ability of metals to form ionic and covalent bonds with various types of ligands.

4 Soluble species are not the only forms in which metals exist in the hydrosphere. Depending on the element and other environmental conditions, a significant portion may be present in association with suspended material.

Additional reading

1 Ferguson, J. E., *Inorganic Chemistry and the Earth*. Pergamon Press, Oxford; 1982.

2 Bodek, I., W. J. Lyman, W. F. Reehl, and D. H. Rosenblatt, eds, *Environmental Inorganic Chemistry: Properties, Processes, and Estimation Methods* (Setac special publication), Pergamon Press, New York; 1988.

Problems

1 Plot the distribution (α vs. pH) of the zinc aquo complex and the first four deprotonated species of this complex. The pK_a values are pK_{a1}=9.6, pK_{a2}=7.9, pK_{a3}=11.3, pK_{a4}=12.3. Note the unusual feature that pK_{a2} is smaller than pK_{a1}. Comment on how this affects the distribution.

2 The stepwise formation constants for the complexes $Pb(OH)^+$ (aq) and $Pb(OH)_2$ (aq) from Pb^{2+} (aq) are 2.0×10^6 and 4.0×10^4 respectively. The reactions can be written in simple form as

$$Pb^{2+} \text{ (aq)} + OH^- \text{ (aq)} \rightleftharpoons PbOH^+ \text{ (aq)}$$

and

$$PbOH^+ \text{ (aq)} + OH^- \text{ (aq)} \rightleftharpoons Pb(OH)_2 \text{ (aq)}$$

Calculate the pK_{a1} and pK_{a2} values for deprotonation of the aquo complex of lead (II), and determine the fractional concentration of the two most important species at pH 7.0.

3 Without calculation, sketch a distribution diagram, α vs. pH, for uncomplexed NTA species in water, covering the entire pH range from 0 to 14. What will be the most important species in water with near neutral pH?

4 Examine Table 13.4 which reports data related to the stability of complexes between various metals and NTA. Comment on the relative values in the table in terms of the environmental classification of metals. Predict the approximate pK_f values for nickel (II) and mercury (II) based on that classification.

5 In lake water containing $0.9 \, \text{mmol L}^{-1}$ calcium and $12 \, \mu\text{g L}^{-1}$ fulvic acid, determine the fraction of the fulvic acid that is bound to calcium, assuming that this is the only metal present in a significant concentration. The pH of the water is 5.0.

6 Note the electronic structure of cadmium (II) and indicate the group (type A, borderline, or type B) in which it should be placed. Predict principal inorganic forms of the element in fresh and sea water under both oxidizing and reducing conditions.

7 Aritificial and natural wetlands are frequently used as catchment basins for urban run-off and storm water. The sediments in these basins have some ability to extract and retain soluble metals from the water as it flows through the pond. For lead, cadmium, and zinc, use information in this chapter and elsewhere to predict the affinity of each of these metals for the carbonate, the iron oxide and the organic matter fractions of sedimentary material suspended and settled in the pond.

Notes

1 No distinction is made here or in subsequent discussions between true complexes and ion pairs. The usual definitions are as follows. In a complex, there is a direct covalent bond between the metal and the ligand. In an ion pair, both species are surrounded by a hydration sphere, but they maintain combined existence as a single unit held together by electrostatic forces.

2 Ahrland, S., J. Chatt, and N. R. Davies, The relative affinities of ligand atoms for acceptor molecules and ions. *Quart. Rev. Chem. Soc.*, **12** (1958), 265–76.

3 Pearson, R.G., Hard and soft acids and bases. *J. Am. Chem. Soc.*, **85** (1963), 3533–9.

4 Nieboer, E. and D. H. S. Richardson, The replacement of the nondescript term 'heavy metals' by a biologically and chemically significant classification of metal ions. *Environ. Pollut. (Series B)*, **1** (1980), 3.

14

Environmental chemistry of colloids and surfaces

O<small>N</small> several occasions earlier in the book, we have indicated that it is not easy to assign discussions of particular processes to unique compartments of the environment. The relation between soil and water is a good example that underlines this difficulty. Soil processes determine metal solubility in the associated pore water and, through leaching, this affects the composition of lakes and rivers. The atmosphere can also be involved. There is an intimate connection between atmospheric carbon dioxide levels and concentrations of carbonate species in water; soil minerals and organic matter also take part in the complex interrelations.

In this chapter, connections between the hydrosphere and the terrestrial environment are especially clear and obvious. The subject matter is concerned with the behaviour of the solid phase, in particular the finely divided solid material when it is in contact with water. The principles discussed are important in terms of suspended solids and sediments in water bodies. They are also important for understanding the relation between soil and its pore water.

Consider the following situation. There is a need to know the concentration of phosphorus in a body of water—say a small lake. A sample, or more likely many samples, are obtained using an appropriate sampling protocol. In the laboratory, analysis for phosphorus is then to be carried out. Several well established standard methods for phosphorus analysis are available. Therefore, while this may seem to be a simple process there are, in fact, enormous complications. Of these, one of the most important is the issue concerning which forms of phosphorus should be considered to be *in* the water. On the one hand there are species, most commonly forms of inorganic orthophosphate as well as organic phosphorus compounds, which unquestionably are dissolved constituents of the aqueous system. At the other extreme, there is inorganic and organic sedimentary material temporarily suspended in the water at the time of sampling and this solid phase material too contains phosphorus in a variety of forms. Between the obviously soluble and clearly insoluble components is a whole range of material having different densities and particle sizes. The question regarding analysis for phosphorus concentration in a water sample then requires decisions (among others) about which size fractions should be considered 'phosphorus *in* the water'.

Figure 14.1 illustrates size-range distributions of some components of a natural aqueous system. An arbitrary but reasonable classification of soluble, colloidal and precipitated forms based on particle size diameters is also given. In this classification, the diameter of colloidal sized particles spans the range from 10 nm to 10 μm.

With regard to the phosphorus analysis problem there are two connected questions related to the sizes described in the figure. One is the *practical experimental issue* of how to separate the desired size fractions. Recent developments in centrifugal and filtration technologies have provided a variety of options in this area but for many routine analytical purposes there is a consensus that filtration through a 0.45 μm controlled pore size filter provides an appropriate division between 'soluble' and 'insoluble' fractions (Fig. 14.1 shows that 0.45 μm is near the centre of the colloidal region.) For more detailed studies, other divisions could be more appropriate.

The other question raises the issue of the *environmental significance* of this experimentally defined size separation. To refer again to the phosphorus example, we answer the question by first considering the purpose of determining the concentration of phosphorus in a water body. Probably the most likely reason is because that element is one of the major nutrients supporting growth of algae and aquatic plants, and where this growth is excessive, eutrophication occurs. Amongst the essential nutrients, phosphorus is most frequently the limiting factor and it therefore controls the rate of eutrophication. Of the phosphorus

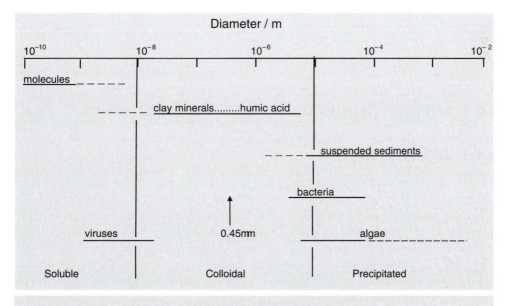

Fig. 14.1 Size classification of materials in the hydrosphere.

species, soluble orthophosphate is the primary nutrient source of the element. But over a longer time-span, other phosphorus-containing species too may release soluble orthophosphate to growing organisms. Included in this category could be easily decomposable non-living biomass and colloidal-sized inorganic matter that supports adsorbed phosphorus on its surface. Much less likely to be available for plant or algal growth would be phosphorus that is present as a structural component of primary minerals like apatite ($Ca_5(PO_4)_3(F,OH)$). At least some of the components containing *biologically available* phosphorus appear in the fractions of soluble and colloidal-sized material which passes through the 0.45 μm filter. Most of the structural phosphorus is in the larger-sized suspended residual portion. Therefore the experimental filtration through a 0.45 μm device provides an approximate but far from perfect separation of biologically accessible forms from resistant forms.

The example of phosphorus is a good illustration of the need to know more about the forms of this element or any other one that is associated with solid-phase material suspended in the hydrosphere. Because much of the suspended material is in the colloidal size range, we will focus on the chemistry of naturally occurring colloids.

Specific surface area

A characteristic of very small particles is their enormous specific surface area, which is usually expressed using units of $m^2 g^{-1}$. Figure 14.2 illustrates the origin of the large surface area of small sized materials. Although the calculation is based on a simplified and artificial situation, it gives values that approach the low end of specific surface areas of some naturally occurring colloids. Real colloids would, of course, have highly irregular geometries and this adds to the surface area.

14.1 Surface properties of colloidal materials

14.1.1 Surface charge

As a consequence of their extremely large surface areas, colloids exhibit unique properties that depend on the particular chemical nature of the colloid. A common and environmentally important feature is that the colloid surface is capable of adsorbing molecules or ions from the surrounding solution. The adsorption then temporarily (if reversible) or permanently (if irreversible) serves to remove such species from solution. In a lake or other water body, adsorption brings about a decrease in the concentration of the soluble species, while in water percolating through the soil, adsorption on to the solid phase attenuates the movement of particular solutes.

There are several types of adsorption phenomena. One is due to electrostatic attraction to a charged surface. Some types of colloids, for example the clay minerals, have a surface

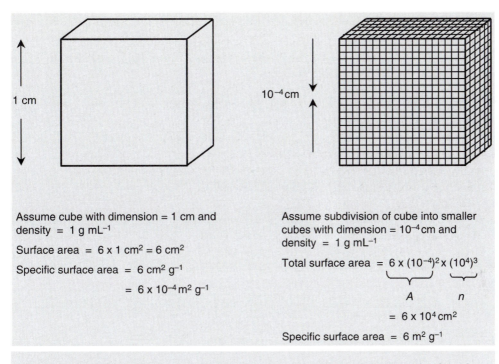

Assume cube with dimension = 1 cm and density = $1 \, g \, mL^{-1}$

Surface area = $6 \times 1 \, cm^2 = 6 \, cm^2$

Specific surface area = $6 \, cm^2 \, g^{-1}$

$= 6 \times 10^{-4} \, m^2 \, g^{-1}$

Assume subdivision of cube into smaller cubes with dimension = 10^{-4} cm and density = $1 \, g \, mL^{-1}$

Total surface area = $6 \times \underbrace{(10^{-4})^2}_{A} \times \underbrace{(10^4)^3}_{n}$

$= 6 \times 10^4 \, cm^2$

Specific surface area = $6 \, m^2 \, g^{-1}$

Fig. 14.2 Specific surface area of colloidal-size cubes. Typical measured values for natural materials are: kaolinite, $5{-}20 \, m^2 \, g^{-1}$; montmorillonite, $700{-}800 \, m^2 \, g^{-1}$; and fulvic and humic acids, $700{-}10\ 000 \, m^2 \, g^{-1}$.

charge whose sign is fixed (negative) and relatively constant in magnitude. In other cases, the nature of the charge is affected in a major way by properties of the surrounding solution. Whether the surfaces of metal oxides, such as iron and aluminium oxides, are protonated or deprotonated depends on the pH of the associated water (Fig. 14.3).

Humic material also has a variably charged surface in that deprotonation of carboxyl groups results in a negative charge while protonation of amino groups generates a positive charge. The position of the protonation/deprotonation equilibrium is a unique property of each material—not only of its chemical composition, but also of the way in which it has been formed.

To measure the pH value at which surface positive and negative charges associated with protonation/deprotonation equilibria are just balanced, the pH of a series of samples is measured after the material has been equilibrated in the presence of different added amounts of acid and base (Fig. 14.4). The set of measurements is repeated in solutions containing various concentrations of non-reactive (indifferent) electrolyte. Ideally, the plot gives a point where all the curves intersect and the pH value at this point is called the zero point of charge (zpc), point of zero charge (pzc), or pH zero (pH_0). We will use the latter term. Some pH_0 values of common environmental colloids are given in Table 14.1.

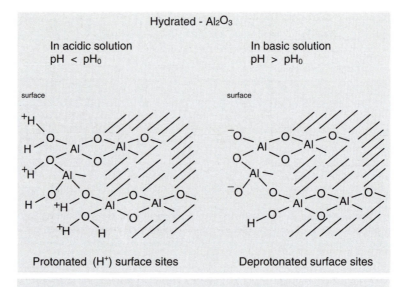

Fig. 14.3 Surface charge of hydrated alumina in environments of varying acidity.

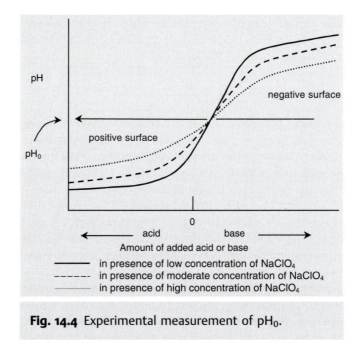

Fig. 14.4 Experimental measurement of pH_0.

Table 14.1 pH_0 values of various natural colloids

Colloid	pH_0	
SiO_2	2.0	
MnO_2	2–4.5	Actual value depends on mode of formation and age of precipitate
Fe_2O_3 hydrated	6.5–9	
goethite	7.5	
haematite	8.5	
Al_2O_3 hydrated	5–9	
Humic material	4–5	
Bacteria	2–3	

In particular situations where the pH of the ambient solution is less than the colloid pH_0, the surface becomes protonated and the colloid therefore attains a net positive surface charge. In this state it is able to attract electrostatically and adsorb negative species (anions). Conversely, where solution pH is greater than the colloid pH_0, the colloid surface is negative and has an electrostatic affinity for positive species (cations). In many environmental circumstances the ambient pH is not far from 7, and the majority of materials listed in the table have a negatively charged surface. In some common situations, however, the hydrated iron and aluminium oxides develop a positive surface.

14.1.2 Electrostatic adsorption

Because many environmental colloids have a negatively charged surface, electrostatic attraction for cations is an important phenomenon and this is usually described in terms of an ion-exchange process. An example showing exchange of adsorbed sodium by potassium ion from the solution is given in reaction 14.1.

$$\text{colloid}^- \cdot Na^+ + K^+(aq) \rightleftharpoons \text{colloid}^- \cdot K^+ + Na^+(aq) \qquad (14.1)$$

The equilibrium position of the reaction depends on the nature of the colloid, and also depends on the nature (principally, charge density) and concentration of the aqueous species adjacent to the colloidal particle. In soils and sediments, the relative importance of cations occupying exchange sites is frequently $Ca^{2+} > Mg^{2+} > K^+ > Na^+$. In low pH environments these adsorbed ions may be partially displaced by Al^{3+} and/or H_3O^+ because these two species become important solution constituents under highly acidic conditions.

A measure of the number of negative exchange sites (which is the same as the amount of positive charge that can be accommodated) on a given amount of material is termed the cation exchange capacity (CEC), the units of which are centimoles of positive charge on the surface per kilogram of solid (cmol (+) kg^{-1}). We will say more about this in a later section (pp. 386–8).

14.1.3 Specific adsorption

The adsorption phenomena that are based on electrostatics can be considered to be physical processes where charge density on both the colloid and solution species determines the extent of adsorption. In contrast, other types of adsorption involve the forming of specific covalent chemical bonds between the solution species and the surface atoms of the colloid. The adsorption of fatty acids by a hydrated iron oxide surface is an example of adsorption involving covalent bond formation (reaction 14.2):

For this type of reaction, the equilibrium may lie far to the right, and while adsorption is favoured under particular pH conditions, it is not simply dependent on whether there is a positive or negative charge on the surface of the solid. Such chemical adsorption processes are, to a significant extent, irreversible and are frequently called *specific adsorption*. Here the adsorption depends on a degree of chemical, as opposed to electrostatic, affinity between the colloid and the adsorbed species.

Metal ions are also capable of forming specific bonds with oxide surfaces, as illustrated in the example with zinc (reaction 14.3):

When specific adsorption occurs, the nature of the surface is altered and so the extent of protonation or deprotonation as measured by the pH_0 also changes. Covalent binding of a cation to the surface shifts the colloid pH_0 to a lower value, while binding of an anionic species produces an upward shift.

14.1.4 The electrical double layer

The charge properties of a colloid surface are often described in terms of an *electrical double layer*. Double layer theory has been developed with varying degrees of sophistication and a simple version is illustrated in Fig. 14.5.

As we have shown, a colloid has charge associated with the surface species, such as protonated or deprotonated functional groups, or other charged atoms in contact with the solution. The charge serves to attract oppositely charged *counterions* that are present in the

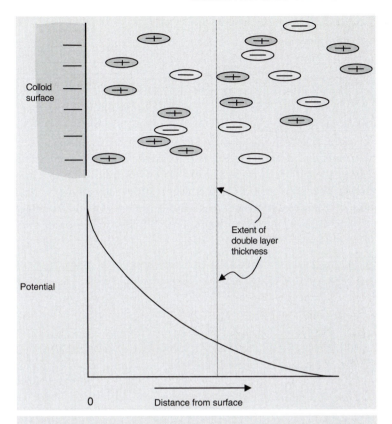

Fig. 14.5 The electrical double layer. In this case, the surface of the colloid has a net negative charge and this attracts positive counterions to the region adjacent to the surface. The thickness of the layer is that distance from the surface where the charge is reduced to $1/e = 0.37$ times its original value.

surrounding solution and these form a 'layer' adjacent to the colloid surface. One should not think of this layer as a clearly defined set of ions that are exclusively of the opposite charge to that of the surface. Instead, the picture is of a preponderance of the counterions near the surface with the proportions gradually reverting to the normal solution charge balance as one moves out into the bulk solution. As this happens, the potential, which is a maximum at the surface, decreases to zero. The 'thickness' of the counterion layer is defined as the distance at which the potential has decreased to $1/e$ of its value at the surface.

A colloidal system, consisting of uniformly charged particles, is stable because the small charged particles repel each other. As they move about in the liquid medium due to Brownian motion, they are unable to come sufficiently close in order to overcome the repulsive forces of the surface charge and aggregate into larger, settleable units.

There are several mechanisms by which colloids are destabilized. One mode of destabilization results from the presence of a high concentration of electrolyte in the water. This provides a source of counterions which then effectively reduce the thickness of the double layer. The availability of many counterions ensures that they are present in large concentration near the colloid surface so that the surface potential falls more rapidly to zero (Fig. 14.6). A well known situation where these processes apply occurs when a river discharges into an ocean estuary. Sedimentation is frequently observed in the estuary due to destabilization of the river colloids as they encounter the high-salt sea water.

A second mechanism of destabilization is associated with specific binding of an ion of opposite charge to the colloid surface. For example aluminium, as Al^{3+}, may covalently bind to a negative clay surface. In sufficient concentration, the surface charge is reduced to near zero and repulsive interactions are partially eliminated. We shall see later that this is one means of destabilizing and precipitating out colloids during water and waste-water treatment.

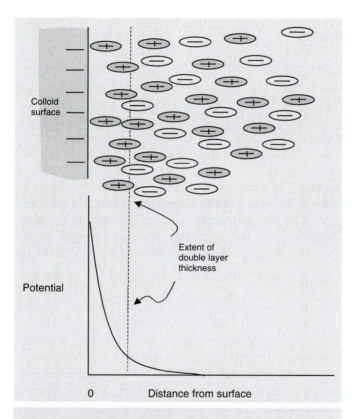

Fig. 14.6 The compression of the electrical double layer surrounding a colloid particle by a high salt concentration of ions enables neutralization of the surface charge over a small distance. Compare with Fig. 14.5.

14.2 Quantitative descriptions of adsorption—I

To describe specific adsorption quantitatively, a variety of mathematical descriptions have been developed for particular cases. We will examine two of the relations that are fundamental and are frequently used.

14.2.1 The Langmuir relation

The Langmuir relation assumes a surface with a specific number of sites each of which is capable of reacting with and binding to a solution molecule. All of the sites are considered to be equivalent and when all are occupied, no further adsorption can occur—in other words adsorption is limited to monolayer coverage (Fig. 14.7). Another way of saying this is that the quantity adsorbed (C_s) reaches the maximum quantity adsorbable (C_{sm}) when all sites are occupied (see eqn 14.4).

A mathematical expression of the Langmuir relation is

$$\frac{C_s{'}}{C_{aq}} = \frac{bC_{sm}}{1 + C_{aq}b} \tag{14.4}$$

where C_s = quantity adsorbed by the suspended solid, soil, or sediment/mol g^{-1}; C_{aq} = equilibrium aqueous solution concentration/mol L^{-1}; b = binding constant/L mol^{-1}— depends on the physical and chemical nature of the solid material; C_{sm} = maximum quantity adsorbable/mol g^{-1}—depends on the nature of the solid as well as the concentration of surface sites. (Note that any unit of mass or quantity may be chosen for use in the C_s and C_{aq} terms as long as they are the same for both the solid and the solution.) A plot of the above relation is given in Fig. 14.8.

The fraction of active sites occupied by adsorbate under specific conditions is given the symbol θ.

$$\theta = \frac{C_s}{C_{sm}} = \frac{bC_{aq}}{1 + bC_{aq}} \tag{14.5}$$

Further algebraic manipulations provide another useful form of the relation:

$$C_s = \frac{bC_{aq}C_{sm}}{1 + bC_{aq}} \tag{14.6}$$

$$\frac{1}{C_s} = \frac{1 + bC_{aq}}{bC_{aq}C_{sm}} \tag{14.7}$$

$$= \frac{1}{bC_{aq}C_{sm}} + \frac{bC_{aq}}{bC_{aq}C_{sm}} \tag{14.8}$$

$$\frac{1}{C_s} = \frac{1}{C_{sm}} + \frac{1}{bC_{aq}C_{sm}} \tag{14.9}$$

When plotted, eqn 14.9 produces a linear plot that may be used to obtain values of C_{sm} and b (Fig. 14.9).

The following example is an application that assumes Langmuir behaviour for the adsorption process. Phosphate adsorption by sedimentary material has often been

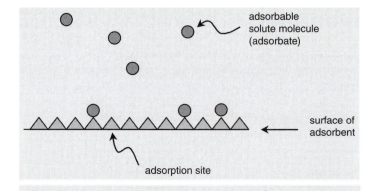

Fig. 14.7 The Langmuir adsorption process. The diagram shows that there are a limited number of sites on the surface, where adsorption can occur. Once these are completely occupied with adsorbate, no further sorption can occur.

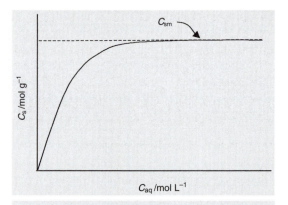

Fig. 14.8 A graphical representation of the Langmuir relation showing that adsorption continues to a maximum concentration (C_{sm}) on the solid surface. C_s and C_{aq} are the equilibrium concentrations of adsorbate on the solid and in solution, respectively.

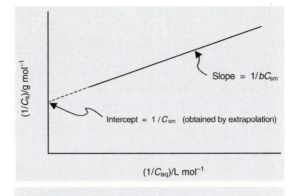

Fig. 14.9 Linearized form of the Langmuir equation. The values of C_{sm} and b may be determined from the values of the slope and intercept.

Table 14.2   Equilibrium values for phosphorus in water and sediment

	Phosphorus concentration	
	In water ($=C_{aq}$) mol L^{-1} (ng mL^{-1})	In sediment ($=C_s$) mol g^{-1} ($\mu g\ g^{-1}$)
Sample 1	4.0×10^{-7} (12)	2.0×10^{-7} (6.2)
Sample 2	1.3×10^{-7} (4)	1.0×10^{-7} (3.1)

described in terms of the Langmuir relation. In the laboratory, two portions of a single sedimentary material are equilibrated with different concentrations of orthophosphate in water and the equilibrium concentrations in the solution and on the solid are determined. The equilibrium values are shown in Table 14.2.

Using eqn 14.9 for subsample 1:

$$\frac{1}{2 \times 10^{-7}} = \frac{1}{C_{sm}} + \frac{1}{4 \times 10^{-7}bC_{sm}}$$

$$\frac{1}{2 \times 10^{-7}} = \frac{4 \times 10^{-7}b + 1}{4 \times 10^{-7}bC_{sm}}$$

$$2bC_{sm} = 4 \times 10^{-7}b + 1$$

$$5 \times 10^{6}C_{sm}b = b + 2.5 \times 10^{6} \qquad (i)$$

and for subsample 2

$$\frac{1}{1 \times 10^{-7}} = \frac{1}{C_{sm}} + \frac{1}{1.3 \times 10^{-7}bC_{sm}}$$

$$\frac{1}{1 \times 10^{-7}} = \frac{1.3 \times 10^{-7}b + 1}{1.3 \times 10^{-7}bC_{sm}}$$

$$1.3bC_{sm} = 1.3 \times 10^{-7}b + 1$$

$$1 \times 10^{7}C_{sm}b = b + 7.7 \times 10^{6} \tag{ii}$$

Multiplying eqn (i) by 2 and subtracting eqn (ii) gives

$$0 = b - 2.7 \times 10^{6}$$

$$b = 2.7 \times 10^{6}\,\text{L mol}^{-1}$$

By substituting this value into eqn 14.8 we obtain a value for C_{sm}:

$$10^{7} \times 2.7 \times 10^{6}C_{sm} = (2.7 \times 10^{6}) + (7.7 \times 10^{6}) = 10.4 \times 10^{6}$$

$$C_{sm} = 3.8 \times 10^{-7}\text{mol g}^{-1}(12.8\,\mu\text{g g}^{-1})$$

This is a measure of the maximum concentration of phosphate that could be adsorbed on the surface of the sedimentary material. While the Langmuir relation allows for predictions of this sort, it gives no indication of the processes and mechanisms of the adsorption.

In the general case, Table 14.2 can be written in the form of Table 14.3. The values of b and C_{sm} can then be calculated from the equations

$$b = \frac{(C_s C'_{aq} - C'_s C_{aq})}{C_{aq}C'_{aq}(C'_s - C_s)} \tag{14.10}$$

$$C_{sm} = \left(\frac{C_s C'_{aq}}{bC_{aq}} + C_s\right) \tag{14.11}$$

Table 14.3 Values for concentration of dissolved species in water (C_{aq}) and in sediment (C_s)—the general case

	Concentration	
	Water	Sediment
Sample 1	C_{aq}	C_s
Sample 2	C'_{aq}	C'_s

14.3 Phosphorus environmental chemistry

We began the chapter by noting that phosphorus exists in many forms in water. This is an appropriate place for us to look at its environmental chemistry in more detail. The global phosphorus cycle is described in Fig. 14.10. There are no common gaseous forms of the element and it is found in the atmosphere only in association with dust particles. In water, phosphorus derived from a variety of inorganic and organic sources hydrolyses to release orthophosphate in various degrees of protonation depending on the aqueous pH. Terrestrial phosphorus is composed of a number of specific minerals including apatite, $Ca_5(PO_4)_3(F,Cl,OH)$, and vivianite, $Fe_3(PO_4)_2 \cdot 8H_2O$. It is also present in all soils where it cycles through plants as they grow and decay.

In water, phosphorus solubility is controlled by the availability of iron and aluminium under acid conditions and calcium under alkaline conditions; each of these metals forms insoluble phosphates (Fig. 14.11). Where the pH is on the slightly acid site, phosphorus has its maximum solubility, and the predominant aqueous species under these conditions is $H_2PO_4^-$.

The environmental significance of phosphorus arises out of its role as a major nutrient for both plants and microorganisms. In considering substances which pollute the

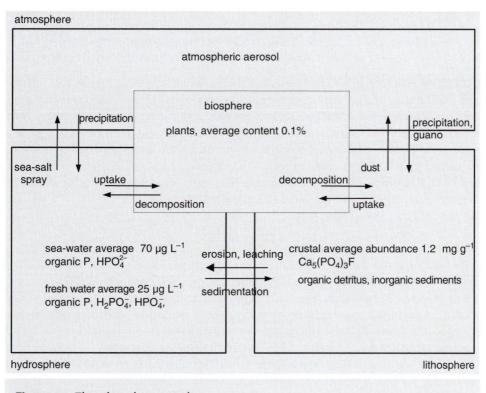

Fig. 14.10 The phosphorus cycle.

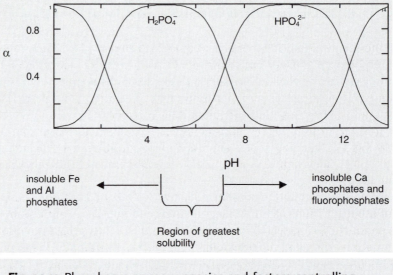

Fig. 14.11 Phosphorus aqueous species and factors controlling solubility in water and the associated suspended material, sediment, or soil.

hydrosphere, one category is the nutrient elements—especially those required in relatively large amounts by micro- and macroorganisms—nitrogen, phosphorus, and potassium. These are considered to be pollutants because excessive concentrations of them in water stimulate the growth of aquatic microorganisms, principally algae. While these organisms are growing, photosynthesis occurs and the organisms serve as oxygen-generating sources (reaction 14.12, left to right). However, when organisms die and decay, their decomposition leads to consumption of oxygen and release of carbon dioxide, creating an anoxic and mildly acidic environment (reaction 14.12, right to left):

$$CO_2 + H_2O \rightleftharpoons \{CH_2O\} + O_2 \tag{14.12}$$

When nutrient-rich conditions lead to excessive biological productivity, the water body is referred to as being *eutrophic*. On the other hand, an *oligotrophic* water body is one where the growth of aquatic organisms, especially algae, occurs only to a limited extent. When a formerly oligotrophic lake rapidly (typically over a period of several years) becomes eutrophic, the situation is usually considered to be undesirable as it inhibits the development and growth of higher forms of life, notably desirable fish species.

We should not assume that intense algal growth is undesirable *per se*. In fact, algae are a food source for fish and when present in moderate amounts, may support a thriving population. In order to ensure an adequate supply of algae, inland fish producers in China, India, and elsewhere actually add nutrients (in the form of inorganic or organic fertilizers) to the water to stimulate microbiological activity. In these instances the decomposition aspect of the algal life cycle takes place in the digestive system of the fish and therefore does not contribute to eutrophication.

Then, too, we should remember that eutrophication is a natural process that occurs over geological time as deep, clear well aerated water bodies are gradually supplied with nutrients and decomposable organic material, and fill with sediment from erosion of the surrounding land surfaces. It is the human-induced premature and often rapid *cultural* eutrophication that we frequently consider to be undesirable.

Of the nutrient elements that contribute to eutrophication, most commonly the limiting one is phosphorus, and low-phosphorus water bodies, when supplied with excessive amounts of the element, tend to become eutrophic. There are some less common situations where nitrogen or other elements serve this function.

The sources of phosphorus found in lakes and rivers may be divided into two categories—point sources, which are well defined discharges from factories or outlets of municipal sewer systems, and diffuse sources, which include run-off from rural and urban landscapes.

Except in specific circumstances, the major type of point source of phosphorus discharge is municipal waste water. Within the municipal discharges, there are two principal sources of phosphorus. The first is from commercial soaps and detergents. These have traditionally contained condensed polyphosphates to react with calcium ion, thus preventing it from causing surface-active ingredients to be removed from solution by precipitation. Condensed polyphosphates include the following:

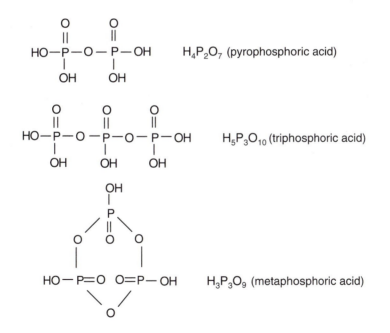

with the triphosphoric acid in a trisodium form being most commonly used in detergent formulations. Typically, detergents contain from 5 to 50% P_2O_5 by weight. The polyphosphates rapidly hydrolyse in aqueous solutions to release orthophosphate species (eqn 14.13):

$$HO-\underset{\underset{O-}{|}}{\overset{\overset{O}{\|}}{P}}-O-\underset{\underset{O-}{|}}{\overset{\overset{O}{\|}}{P}}-O-\underset{\underset{O-}{|}}{\overset{\overset{O}{\|}}{P}}-OH + 2H_2O \longrightarrow 3HO-\underset{\underset{O-}{|}}{\overset{\overset{O}{\|}}{P}}-OH \tag{14.13}$$

The actual orthophosphate product of the reaction adjusts to be in the acid–base form consistent with the pH of the receiving water (Fig. 14.11). Recent legislation in many jurisdictions has limited the use of phosphates in detergents but its replacement by alternatives is not without environment consequences. Nitrilotriacetic acid (NTA), whose chemistry was described in Chapter 13, is one of the alternative sequestering agents.

The second source of phosphorus in municipal waste water is human wastes. In a chemical sense, human faeces is made up of about 25% organic matter (the remainder is mostly water) and it contains about 0.03% nitrogen (largely of bacterial origin) and 0.005% phosphorus. The corresponding figures for urine average 0.5% organic carbon, 1.0% nitrogen, and 0.03% phosphorus. Given that an 'average person' generates approximately 100 g and 1200 g of these wastes respectively each day, a small urban area of 100 000 persons would produce 3600 kg of phosphorus daily. If we assume that water use per capita is 0.5 m^3 (a figure valid for some high-income countries which also have a comprehensive sewage system) the concentration of phosphorus derived from human wastes in sewage influent would be approximately 7 mg L^{-1}. Actual values in many cities are around 10 mg L^{-1}, but this value fluctuates widely depending on season, time of day, and so on. The difference between the two values is due to phosphorus from other sources. If municipal waste water remains untreated, all the phosphorus would be discharged to the receiving water and this can be a major contributor to eutrophication. Where treatment is done, it ranges from use of settling ponds or lagoons, where the natural biological processes cause uptake and removal of limited amounts of pollutants, to primary, secondary, or tertiary treatment plants using a variety of physical, chemical, and/or biological processes. In all cases there is some discharge of phosphorus that inevitably finds its way to streams, rivers, ponds, and lakes. To some degree, the extent of phosphorus release can be controlled by upgrading the treatment process.

Much more difficult to control are the many diffuse sources of phosphorus. These include all forms of urban, agricultural, and forest run-off and leachates. The element is present in low concentrations as soluble organic and inorganic forms and also associated with clay-sized material that is carried into the hydrosphere by erosion.

14.3.1 The Bay of Quinite

There are many water bodies, large and small, throughout the world where phosphorus-controlled eutrophication is evident. One example is the Bay of Quinte (BOQ) (Fig. 14.12) at the north-eastern end of Lake Ontario in Canada.

During the 1960s and 1970s, the eutrophic status of this portion of the lake was evidenced by dense algal blooms that blocked light from reaching the rooted aquatic plants. This

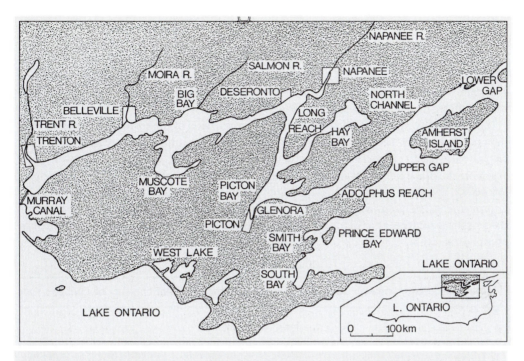

Fig. 14.12 The Bay of Quinte (BOQ), shown here as the narrow Z-shaped water body that extends eastward from Trenton, is situated on the north side of Lake Ontario. While it is part of the lake, a large peninsula isolates it from the main water body, and there is only limited water exchange with the open lake. The BOQ is a long ($\sim$120 km) serpentine stretch of water, bordered by several towns and small cities as well as agricultural and recreational properties. In different ways, all of these contribute to phosphorus loading in the Bay.

resulted in reduced growth of plants which, in turn, contributed to shifts and instability in the fish communities. Drinking water was also affected by taste and odour problems.

The Bay is surrounded by several small cities as well as a considerable extent of agricultural and recreational land. Over half of the phosphorus is from point sources, principally from five waste-water treatment plants located around the Bay. Diffuse sources, including run-off from agricultural and forested land and from urban areas, make up the rest of the annual loading of the element. Since 1978 considerable effort has gone into controlling release from both types of source. Provincial regulations strictly limited the amount of phosphorus allowed in detergents and other cleaning agents. Improvements were made in waste-water treatment plants in accordance with concentration targets of $0.5\,\mu\text{g mL}^{-1}$ (as P) set for phosphorus in the effluent. Encouragement and incentives were provided to improve farm practices in a way that would reduce phosphorus release through run-off and leaching. These included the adoption of conservation tillage in crop production, care in manure storage, improvements in household sewage systems, and minimizing livestock access to pasture adjacent to the shoreline.

The combined result of these measures has led to a marked improvement in water quality in the BOQ, as shown in Table 14.4. Phytoplankton volume and chlorophyll a mass concentrations are both indicators or microbial growth, which in turn depend on the availability of nutrient phosphorus. The vertical light extinction index, ε_{PAR}, is a measure of the absorbance of photosynthetically available radiation (PAR) by turbidity in the water per vertical metre depth; larger values signify greater turbidity.

These data may be compared with a general classification of the trophic status of lakes as shown in Table 14.5. The classification includes lakes that support little biological activity (*ultraoligotrophic*) to ones where productivity of organisms is excessively high (*hypertrophic*). Based on the phosphorus data, one can see that in the years since control measures were instituted, the BOQ has moved from the eutrophic to the mesotrophic category. Major improvements (reductions) in microbial biomass production are also indicated, although classification according to this criterion indicates categories different from that defined by phosphorus levels.

Table 14.4 Water quality data related to eutrophication, for the Bay of Quinte

	Mean values during			
	1971–77	1978–83	1984–89	1991–94
Total P/μg L^{-1}	52	41	38	30
Phytoplankton/mm L^{-1}	8.6	6.1	7.1	6.1
Chlorophyll a/μg L^{-1}	28	23	20	14
Vertical light extinction, ε_{PAR}/m^{-1}	1.43	1.32	1.27	1.09

Table 14.5 Trophic status of lakes,[a] including corresponding Bay of Quinte values before (1971–7) and after (1991–4) phosphorus controls were introduced.

	Total P/μg L^{-1}	Chlorophyll a/μg L^{-1}
Ultraoligotrophic	<4	<1.0
Oligotrophic	4–10	1.0–2.5
Mesotrophic	10–35	2.5–8
Bay of Quinte, 1991–4	30	14
Eutrophic	35–100	8–25
Bay of Quinte, 1971–7	52	28
Hypertrophic	>100	>25

[a]Rast, W. and M. Holland, Eutrophication of lakes and reservoirs: a framework for making management decisions. *Ambio*, **17** (1988), 2–12.

The BOQ trends therefore make clear that, in the years after the implementation of controls (1977), there has been significant improvement in water quality—a conclusion supported by all four types of measurements.

It is now recognized that further improvements may have to take into account sedimentary release of accumulated phosphorus. In the sediment, the element is found as a component of some primary minerals, occluded within the amorphous hydrous oxides of iron and aluminium, adsorbed on the surface of minerals such as calcite and quartz and associated with the organic component of the sediment. In the BOQ about 50 to 75% of the sedimentary phosphorus is associated with calcium carbonate and 20 to 40% is in the organic fraction. Some of the latter category, especially that found in living biomass, appears to be active in terms of refluxing or releasing phosphorus to more available or soluble forms. Endogenously produced phosphatase enzymes allow certain bacteria to mineralize organic phosphorus, releasing soluble orthophosphate into the water column. It is estimated that the extent of reflux is equivalent to the release of approximately 36 kg P d^{-1} in early summer and 72 kg P d^{-1} in late summer when decomposition of accumulated biomass is at a maximum. These amounts are several times larger than the 19 kg P d^{-1} which are now estimated to be released from waste-water treatment plants serving the urban areas along the Bay.

If sedimentary reflux maintains elevated phosphorus levels in the Bay, the introduction of even tighter regulations on phosphorus discharge might have little immediate effect on water quality. Consideration of other strategies that would affect sedimentary release has been given; these include enhancing the flow of water through the Bay in order to 'flush out' the system, dredging portions of highly contaminated sediment, or containing the surface phosphorus by covering the sediment with uncontaminated material. However, each of these possibilities has major practical limitations and is unlikely to be used on a large scale.

In other lakes around the world as well, sedimentary reflux has been shown to be a principal means by which eutrophic conditions persist when phosphorus input levels are reduced.

14.4 Quantitative descriptions of adsorption—II

14.4.1 The Freundlich relation

A second relation used to describe adsorption on environmental colloids is the empirical Freundlich equation:

$$C_s = K_F C_{aq}^{n_F} \qquad (14.14)$$

where C_s = quantity adsorbed per unit mass/mol g^{-1}; C_{aq} = equilibrium solution concentration/mol L^{-1}; and K_F/L g^{-1} and n_F (a dimensionless number) are empirical Freundlich constants (n_F is usually < 1). (Again, any unit of mass or quantity may be chosen for use in the C_s and C_{aq} terms as long as the same unit is chosen for both the solid and the solution.)

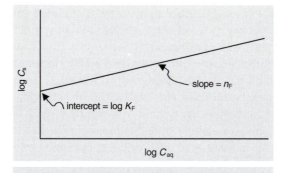

Fig. 14.13 A linearized Freundlich isotherm plot. Values of the constants, K_F and n_F, are determined from the intercept and the slope of the plot.

To linearize the equation, logarithms are taken:

$$\log C_s = \log K_F + n_F \log C_{aq} \tag{14.15}$$

and plots of the type shown in Fig. 14.13 are then made.

The Freundlich relation differs from that of Langmuir in that not all sites on the surface are considered equal but rather that adsorption becomes progressively more difficult as more and more adsorbate accumulates. Furthermore, it is assumed that once the surface is covered, additional adsorbed species can still be accommodated. In other words, no maximum (monolayer) adsorption is predicted by this relation.

The equation does not imply any particular mechanism of adsorption; it is purely empirical, and has been found to be most satisfactory in the low concentration range.

Values of K_F and n_F have been determined for some particular cases. For a forest soil, it has been reported[1] that $K_F = 0.0324 \, \text{L g}^{-1}$ and $n_F = 0.82$ for cadmium. From this we may calculate the equilibrium concentration of adsorbed cadmium in the presence of a $4.0 \, \mu\text{g L}^{-1}$ solution concentration.

$$C_s = K_F C_{aq}^{n_F} \tag{14.14}$$
$$= 0.0324 \, \text{L g}^{-1} \times (4.0 \, \mu\text{g L}^{-1})^{0.82}$$
$$= 0.10 \, \mu\text{g g}^{-1}$$

This is the concentration of adsorbed cadmium in equilibrium with a $4.0 \, \mu\text{g L}^{-1}$ solution of the element.

A slightly different problem is to calculate the concentration of cadmium that would be adsorbed by 10 g of soil under equilibrium conditions *starting with* 1 L of a $4.0 \, \mu\text{g L}^{-1}$ solution of the metal ion. At equilibrium,

$$C_{aq} = 4.0 - C_s$$
$$C_s = 0.0324(4.0 - C_s)^{0.82}$$

Solving the last equation directly is difficult, but it may be done simply enough using an iterative procedure. A starting point is to use the adsorbed concentration calculated above, that is $C_s = 0.10\,\mu g\,g^{-1}$. For this, the first estimate of total mass adsorbed on 10 g of soil would be 1.0 μg and this is subtracted from the 4.0 μg present in the original 1 L of solution:

$$C_{s1} = 0.0324(4.0 - 1.0)^{0.82}$$
$$= 0.080\,\mu g\ g^{-1}(\text{total mass in }10\,g\text{ of soil} = 0.80\,\mu g)$$

Repeating the substitution:

$$C_{s2} = 0.0324(4.0 - 0.8)^{0.82}$$
$$= 0.084\,\mu g\ g^{-1}(\text{total mass in }10\,g\text{ of soil} = 0.84\,\mu g)$$

Again,

$$C_{s3} = 0.083\,\mu g\ g^{-1}$$

and

$$C_{s4} = 0.083\,\mu g\ g^{-1}$$

We now have the correct value. The concentration of cadmium adsorbed on the soil is $0.083\,\mu g\,g^{-1}$ and that remaining in the equilibrium solution is $3.2\,ng\,mL^{-1}$.

14.5 Partitioning of small organic solutes between water and soil or sediment

14.5.1 The distribution coefficient K_d

Like other substances, small organic molecules in water distribute themselves between the aqueous phase and the solid, particulate materials suspended as colloids or in the sediments. A Freundlich relation often provides a satisfactory description of the equilibrium distribution especially when, as is frequently the case, the organic solute is present in very small concentrations. Under these conditions n_F in the expression $C_s = K_F C_{aq}^{n_F}$ can be adequately approximated to have a value of 1. This implies that all sites responsible for retention of the solute molecule are similar and that the initial substance retained on the surface neither enhances nor inhibits subsequent retention. The modified equation may then be written as

$$C_s = K_d C_{aq} \tag{14.16}$$

Based on this relation, K_d, given by eqn 14.16, is the simplified distribution coefficient describing the partitioning of the solute between the solid phase and water:

$$K_d = \frac{C_s}{C_{aq}} \tag{14.17}$$

The value of K_d depends on the organic solute itself, on the chemical and physical nature of the solid phase, and on other environmental properties such temperature and solution ionic strength. The infinite range of these combined properties makes it impossible to tabulate values of K_d. However, two related 'reportable' parameters have been developed for neutral organics in particulate/aqueous systems. These are the octanol/water partition coefficient (K_{OW}) and the organic matter/water partition coefficient (K_{OM}) that are used to assess the ability of a particulate surface to take up (sorb—we define this below) species from the surrounding solution. Environmental engineers and others make use of K_{OW} and K_{OM} data to describe and predict movement of organic contaminants such as pesticides in soil/water and sediment/water systems. We will discuss the chemical basis and some applications of the K_{OW} and K_{OM} parameters and then show how they may be used to estimate the value of K_d.

14.5.2 Sorption of organic species by soils

In the environment, the surfaces taking part in partitioning of solutes include those of inorganic species such as the hydrous oxides and clay minerals which have hydroxyl groups extending into the aqueous solution from the mineral face. A combination of van der Waals, induced dipole:dipole, dipole:dipole, and hydrogen bonding forces (these are listed in increasing order of strength) are responsible for the binding of species to the surface. Where specific binding occurs, retention of small organic molecules by mineral surfaces has been recognized as being important but in most cases, the sorption is very weak. This is because there is a strong attraction between the mineral and water, and retention of a solute must simultaneously involve displacement of water molecules from the surface.

The more active surfaces with respect to neutral organics are those of the natural organic macromolecules present in suspended or sedimentary solids. We have seen in Chapter 12 that the organic matter includes humic material derived from plant or microbial sources. This material has polar properties because of its oxygen-containing functionality. However, also important in the present context, humic matter has major hydrocarbon regions located within the molecule where non-polar solutes encounter little competition from surface-bound water molecules. Thus, small hydrophobic solutes can effectively 'dissolve' within the interior non-aqueous medium. In this sense the retentive forces go beyond being surface forces and the process is more akin to absorption than to adsorption. It is for this reason that the generic term *sorption* is especially appropriate here.

It is worth reminding ourselves too that humic material itself may be bound to mineral phases where it forms a coating. Because the mineral surface is thus blocked from reacting directly with other solutes, this accentuates the importance of the organic matter as a sorbent (Fig. 12.8). The HM is retained on the surface by hydrogen bonds between surface groups such as -SiOH and electronegative atoms in the otherwise hydrophobic organic molecule. Other weaker van der Waals forces make a smaller contribution to the energy of binding.

As a consequence, in sediments and soils an organic matter content as low as 1% can dominate the sorptive properties with respect to small neutral organic species so that the contribution from the mineral fraction is negligible.

Putting these ideas together, in the general case, the equilibrium concentration of a substrate in the soil or sediment, C_S, is given by the relation

$$C_S = f_{OM} \times C_{OM} + f_{MM} \times C_{MM} \tag{14.18}$$

where C_{OM} and C_{MM} are the concentrations of the solute in the organic matter and mineral matter respectively; f_{OM} and f_{MM} are the fractions of these two phases in the whole soil or sediment. In many situations involving neutral organics, C_{MM} is so small that the relation reduces to

$$C_S = f_{OM} \times C_{OM} \tag{14.19}$$

14.5.3 The octanol–water partition coefficient, K_{OW}

Now we can move to look at the distribution coefficients that are used to describe the relationships between dissolved and sorbed species. Octanol (actually n-octanol, $CH_3(CH_2)_7OH$) is an amphiphilic solvent—by this we mean that it has both hydrophilic and hydrophobic character. As such, water has a substantial solubility in the solvent (molar ratio 0.25) while the solubility of n-octanol in water is much less (molar ratio is 8×10^{-5}). The amphiphilic character gives n-octanol a solvating ability not unlike that of humic acid and other naturally occurring organic colloids (Fig. 14.14) in that it has some ability to associate with both polar and non-polar compounds.

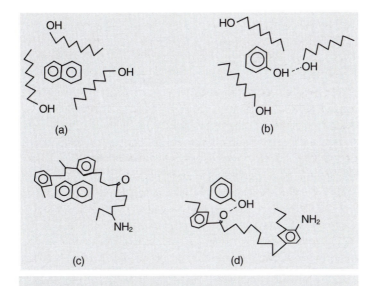

Fig. 14.14 Representation of the ability of n-octanol to solubilize both non-polar (a, naphthalene) and polar (b, phenol) molecules. Similar representation (c and d) showing how a portion of a humate molecule can interact with the same species. Hydrophobic portions of the octanol and HM tend to associate with hydrophobic organic solutes, while hydrogen bonding and other interactions favour reaction with polar groups that may be present on these solutes.

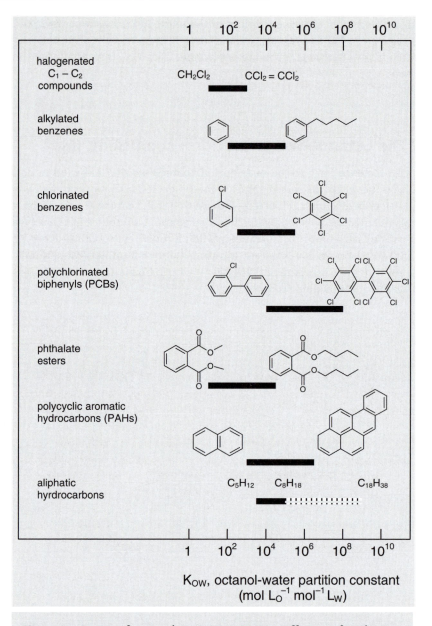

Fig. 14.15 Range of octanol–water partition coefficients for classes of organic compounds. Redrawn from Schwarzenbach, R.P., P.M. Gschwend and D.M. Imboden, *Environmental Organic Chemistry*, John Wiley and Sons, Inc., New York; 1993.

The octanol/water partition coefficient, K_{OW}, then provides a relatively convenient means of predicting the partitioning behaviour of a hydrophobic organic substance between being dissolved in water and being sorbed by solid organic matter associated with the water. The K_{OW} is defined as

$$K_{OW} = \frac{C_O}{C_{aq}} \qquad (14.20)$$

where C_O is the equilibrium molar solubility of the solute in octanol and C_{aq} the corresponding solubility in water. Figure 14.15 shows ranges of K_{OW} values for various classes of (usually synthetic) organic compounds. The range covers more than ten orders of magnitude and several features are qualitatively predictable. Small K_{OW} values—favouring the aqueous phase—are characteristic of low molar mass and oxygen-containing species, while hydrocarbons or compounds with large carbon:oxygen ratios, especially high molar mass ones, have large K_{OW} values and tend to be associated with the suspended or precipitated organic matter in the environment.

Box 14.1 Bioconcentration factors

Bioconcentration is a widely recognized phenomenon whereby organisms—everything from microorganisms to plants to polar bears and whales—accumulate particular contaminant chemicals in their tissues. Uptake occurs *via* a variety of mechanisms depending on the organism and the chemical. In the aqueous environment, filter feeders like molluscs extract nutrients and, with the nutrients, other chemical entities which are not metabolized and accumulate in the soft flesh or hard shells. Microorganisms, aquatic plants, fish, and other animals living in water are also capable of accumulating soluble metals and organic chemicals. One concern is that the assimilated chemicals can be further accumulated and concentrated by other species that use the original organism as a food source. Eventually, at higher levels in the food chain, the amount of accumulated chemical may be large enough to be toxic.

Many chlorinated compounds are chemically and biologically stable and are relatively lipophilic. For these reasons, they are susceptible to accumulation in the fatty tissue of organisms. In the Firth of Clyde off western Scotland, 1989 measurements[a] of various chlorinated compounds including polychlorinated biphenyls (PCBs), and the insecticides dieldrin and DDT showed that concentrations in the soft tissue of mussels were several orders of magnitude higher than concentrations in the ocean water. For example, the aqueous concentration of DDT was as low as $1\ ng\ L^{-1}$ while that in the mussels reached $300\ \mu g\ kg^{-1}$. The ratio of these two numbers, using a common mass basis, is approximately 300 000 and is referred to as the bioconcentration factor (BCF).

There is usually a relationship between the octanol/water coefficient and the value of the BCF. Once again, the relation is based on the idea that octanol has an ability to dissolve lipophilic compounds in a way that can be related to some types of biological tissue. An example is shown in a study of various organochlorine contaminants in water, fish, and seal in Lake Baikal in eastern Siberia.[b]

Persistent compounds such as DDT and polychlorinated biphenlys are transported over long distances *via* the atmosphere. As a result, regions where such compounds have never been used exhibit significant levels in water, and because they are hydrophobic the compounds bioaccumulate to higher levels in the fatty tissue of animals living in the water. The BCF values reported here are ratios of the concentrations (mg kg^{-1}) of a given compound in Baikal seal blubber to the concentration (mg L^{-1}) in the ambient water. A reasonably good correlation was observed between tabulated log K_{OW} values and log BCF. Figure 14B.1 shows the correlation observed for a wide range of organochlorines with respect to accumulation in the Baikal seal. Such correlations serve a predictive function in that biouptake may be estimated for other halogenated compounds for which K_{OW} values are available. However, there are limitations to this kind of exercise. You can see from the figure that the BCF of some compounds is not well described by the correlation. For 4,4'-DDE, a BCF about 10^7 is predicted, while the experimental value is closer to 10^9. As a general approximation, bioconcentration of small organic molecules becomes important when the BCF value is 10 000 or above.

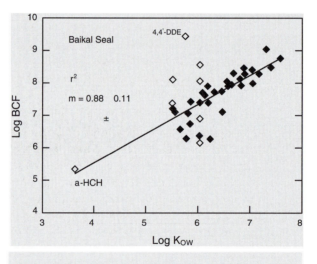

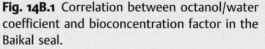

Fig. 14B.1 Correlation between octanol/water coefficient and bioconcentration factor in the Baikal seal.

[a] Porteous, A., *Dictionary of Environmental Science and Technology*, John Wiley and Sons, Chichester; 1992.
[b] Kuklick, J. R., T. F. Bidleman, L. L. McConnell, M. D. Walla, and G. P. Ivanov, Organochlorines in the water and biota and Lake Baikal, Siberia. *Environ. Sci. Technol.* **28** 1994, 31–37.

Octanol/water partition coefficients may be estimated by calculation and are also experimentally measurable in laboratory experiments. Tabulated values for many compounds are available.[2] Using K_{OW} in a relative sense enables us to make qualitative predictions and comparisons between different compounds.

14.5.4 The organic matter–water partition coefficient, K_{OM}

To describe the behaviour of organic compounds in natural water/particulate systems, two additional partition coefficients, the K_{OM} and K_{OC}, have been defined in the following ways.

$$K_{OM} = \frac{C_{OM}}{C_{aq}} \tag{14.21}$$

and

$$K_{OC} = \frac{C_{OC}}{C_{aq}} \tag{14.22}$$

In both these expressions, the numerator is the concentration of the organic compound of interest in the solid NOM component of a soil or sediment. Typical units of C_{OM} could be mol of compound per kg of soil or sediment organic matter and for C_{OC}, mol per kg of organic carbon. In both cases, the denominator is the concentration of the compound, in units such as mol L^{-1}, in the equilibrated aqueous solution. Because the carbon content of NOM is typically about 60%, values of K_{OC} will be approximately 1.7 times greater than those of K_{OM}. Clearly, the two coefficients describe the same partitioning process. In this chapter, we will make use of the K_{OM} value, while the K_{OC} will be used in other examples in Chapter 20.

We have already noted that a simplified version of the Freundlich equation leads to a relation describing the distribution between solid material and the associated water

$$K_d = \frac{C_s}{C_{aq}} \tag{14.17}$$

where C_s is the concentration (mol kg^{-1}) in the total solid and C_{aq} the concentration (mol L^{-1}) in the aqueous solution. For cases where the organic component of the solid phase is the only significant contributor to uptake of the solute, $C_s = f_{OM} \times C_{OM}$, and Equation 14.17 becomes

$$K_d = \frac{C_{OM}}{C_{aq}} \times f_{OM} \tag{14.23}$$

Therefore

$$K_d = f_{OM} \times K_{OM} \tag{14.24}$$

And, from eqn 14.17

$$C_s = f_{OM} \times K_{OM} \times C_{aq} \tag{14.25}$$

It has been found that K_{OM} values for particular chemicals are relatively constant (± 0.3 log K_{OM} units) over a considerable variety of sediments and soils. This confirms that there is a large degree of similarity in the nature of organic matter–substantially humic material–present in many environmental situations.

An example will illustrate some useful features of these concepts. The compound *p*-dichlorobenzene has been found to have $K_{OM} \simeq 630$. For a soil containing 1.6% organic matter, the distribution coefficient would be

$$K_d = 1.6/100 \times 630 = 630 \times 0.016$$
$$= 10$$

In a simple way, this means that if 100 L of water containing 5 ppm dichlorobenzene is spilled on a $0.2\,m^3$ volume of the dry soil, it would distribute itself at equilibrium according to the following calculation.

$$\text{Mass of dichlorobenzene} = 5\,mg\,L^{-1} \times 100\,L$$
$$= 500\,mg$$
$$\text{Mass of soil} \qquad = 0.2\,m^3 \times 1.2\,t\,m^{-3}$$
$$= 240\,kg$$

(A typical soil bulk density is $1.2\,t\,m^{-3}$; see Chapter 18.) If x is the mass, in mg, of dichlorobenzene in the soil after equilibration, then

$$K_d = \frac{C_s}{C_{aq}} = \frac{x/240}{(500-x)/100} = 10$$
$$x = 480$$

Therefore, 480 mg are sorbed to the soil matrix and the concentration is $480/240\,mg\,kg^{-1} = 2.0$ ppm. In the water, 20 mg remains, giving a concentration of $20/100\,mg\,L^{-1} = 0.20$ ppm.

On a different soil with 2.5% organic matter content, the value of K_d would be 16 and the corresponding concentrations are 2.0 ppm in the soil and 0.13 ppm in the water.

We had noted above, when discussing the K_{OW} function, that octanol in some sense has properties that make it a good surrogate for the humic material of water/particulate systems. We could therefore expect there to be a relation between K_{OW} and K_{OM}. Logarithmic relations have been established experimentally[3] and take the form of

$$\log K_{OM} = a \log K_{OW} + b \qquad (14.26)$$

It is perhaps surprising that for a fairly wide range of chemical classes, a single version of eqn 14.26 shows a fairly good correlation between the two parameters. Karickhoff[3] used the equation with coefficients as shown:

$$\log K_{OM} = 0.82 \log K_{OW} + 0.14 \qquad (14.27)$$

for a variety of aromatic hydrocarbons, chlorinated hydrocarbons, chloro-S-triazines, and phenylureas. Overall the correlation coefficient, r^2, was 0.93. Even better correlations can be established when an individual equation is used for each particular class of compounds.

14.6 Colloidal material in the natural environment

There is a wide variety of colloidal components found in different environments. Some colloids have wide-ranging distribution throughout the world. Others are formed under specific conditions and their presence is clear evidence of those conditions.

14.6.1 Colloids formed under specific environmental conditions

Under an oxidizing (high pE) regime, iron and manganese (soluble under reducing conditions as iron (II) and manganese (II) species) are oxidized and deposited in forms of colloidal $Fe_2O_3 \cdot xH_2O$ and $MnO_2 \cdot xH_2O$. In the oceans, this phenomenon is associated with hydrothermal vents that release effluent rich in reduced iron and manganese. When these species encounter an oxygen-rich environment, the insoluble minerals are formed as colloids that eventually precipitate and are incorporated into the sediment. In lakes and rivers, smaller amounts of reduced iron are derived from various sources such as effluent from some mining or mineral processing operations. In an oxidizing aqueous environment, the colloidal hydrous oxides are again formed.

Under reducing conditions, especially in a marine environment, a process occurs that is essentially the reverse of the one described above. Such environments include deep water sediments where the only source of oxygen is the water immediately adjacent to the sediment surface. If the sediment contains reducing materials such as organic matter, the oxygen is used up and anoxic conditions result. The insoluble hydrous oxides of iron (III) and manganese (IV) are then reduced to more soluble iron (II) and manganese (II) species. At the same time sulfate, which is abundant in sea water and present in smaller amounts in fresh water, is reduced to sulfide. As a consequence of these simultaneous processes, colloidal iron and manganese sulfides are precipitated from solution and are therefore found in anoxic sediments. The reduction reactions are of microbiological origin and will be discussed in the next chapter.

When waters containing dissolved organic matter encounter an acidic environment, acid-insoluble humic acid is precipitated from solution as a colloidal deposit. Sediments deposited in other situations contain organic matter of highly variable amount and composition. In lakes and coastal regions of oceans, sediment may contain from 1% to greater than 20% organic matter, much of which is carried in as particulate material from terrestrial sources. Smaller amounts are found buried in deep ocean sediments. Organic compounds are also produced in the water. The growth of both microorganisms and larger organisms depends on nutrients and energy from dissolved constituents and from sunlight. After their life cycle is complete, the residual organic matter forms part of the particulate fraction, remaining suspended or sinking to the lake or ocean floor. While passing through the water column and while in the sediments, decomposition and resynthesis takes place so that other compounds are formed including the aqueous equivalent of terrestrial humic material. The origin of humic substances found in sediments can often be determined, since (for example) terrestrial material contains a higher proportion of aromatic subunits, is more oxidized, and therefore has a greater concentration of acidic functional groups than does humic material produced *in situ* in the aqueous (particularly marine) environment.

Calcium carbonate is a frequently observed colloid formed in an alkaline environment—even one that has only a temporary existence. The following example illustrates a mechanism by which such material is formed.

Consider a shallow eutrophic water body that has pH$= 7.20$ and HCO$_3^- = 1.06 \times 10^{-3}$ mol L^{-1} and Ca$^{2+} = 1.5 \times 10^{-3}$ mol L^{-1}. These properties are characteristic of lakes located in a region where the bedrock and sediments contain limestone. Under these conditions,

$$[CO_3^{2-}] = \frac{K_{a_2} \times [HCO_3^-]}{[H_3O^+]}$$

$$= \frac{4.7 \times 10^{-11} \times 1.06 \times 10^{-3}}{6.3 \times 10^{-8}}$$

$$= 7.9 \times 10^{-7} \text{mol L}^{-1}$$

The reaction quotient,

$$Q_{sp} = [Ca^{2+}][CO_3^{2-}] = 1.5 \times 10^{-3} \times 7.9 \times 10^{-7}$$
$$= 1.2 \times 10^{-9} < 5 \times 10^{-9} = K_{sp}$$

Because the reaction quotient is smaller than the solubility product, this indicates that $CaCO_3$ is sufficiently soluble so that no precipitate forms.

Production of biomass in eutrophic lakes (expressed in terms of carbon) on days with intense sunlight is frequently in the range 1000 to 2000 mg C m^{-2} d^{-1}. Consider that in the present case the production rate is 1700 mg C m^{-2} d^{-1}. We will assume that this production is mostly in the top 50 cm of the water column, where the sunlight is most intense.

At pH 7.20, most of the carbonate is in the hydrogen carbonate form and algal photosynthesis can be described by the following reaction:

$$HCO_3^- (aq) + H_2O \xrightarrow{h\nu} \{CH_2O\} + OH^- (aq) + O_2 \qquad (14.28)$$

During the period of photosynthesis, $1700 \times 10^{-3}/12$ mol of hydrogen carbonate are consumed and the same quantity of hydroxide ion is produced. This takes place in a volume of 1×0.5 m$^3 = 500$ L. The molar concentration changes are therefore 2.8×10^{-4} mol L^{-1}.

To determine how this affects the water chemistry, we should calculate the concentrations of other species before and after the production of biomass.

1. Before biomass has been produced:

$$[H_3O^+] = 6.3 \times 10^{-8} \text{ mol L}^{-1}$$
$$[HCO_3^-] = 1.06 \times 10^{-3} \text{ mol L}^{-1}$$
$$[CO_2] = \frac{[H_3O^+][HCO_3^-]}{K_{a1}} = \frac{6.3 \times 10^{-8} \times 1.06 \times 10^{-3}}{4.45 \times 10^{-7}}$$
$$= 1.5 \times 10^{-4} \text{ mol L}^{-1}$$
$$[CO_3^{2-}] = 7.9 \times 10^{-7} \text{ mol L}^{-1}$$

2. After biomass has been generated by photosynthesis, 2.8×10^{-4} mol L^{-1} of HCO_3^- has been used up, but the 2.8×10^{-4} mol L^{-1} of hydroxide which is produced is sufficient to react with all of the aqueous carbon dioxide, producing an equivalent quantity of hydrogen carbonate. The additional 1.3×10^{-4} mol L^{-1} of hydroxide converts the same amount of hydrogen carbonate to carbonate. The net result of these reactions is

$$[CO_3^{2-}] = 1.3 \times 10^{-4} \text{ mol L}^{-1}$$
$$[HCO_3^-] = (1.06 \times 10^{-3}) - (2.8 \times 10^{-4}) + (1.5 \times 10^{-4}) - (1.3 \times 10^{-4})$$
$$= 8.0 \times 10^{-4} \text{ mol L}^{-1}$$
$$[H_3O^+] = \frac{[HCO_3^-]K_{a2}}{[CO_3^{2-}]} = \frac{8 \times 10^{-4} \times 4.7 \times 10^{-11}}{1.3 \times 10^{-4}}$$
$$= 2.9 \times 10^{-10} \text{ mol L}^{-1}$$

This corresponds to a pH of 9.54. Under the new conditions,

$$Q_{sp} = 1.5 \times 10^{-3} \times 1.3 \times 10^{-4}$$
$$= 2.0 \times 10^{-7} > 5 \times 10^{-9} = K_{sp}$$

Because the solubility product of calcium carbonate has now been exceeded, insoluble calcium carbonate is precipitated out of the water. It is through this biogeochemical process that large amounts of colloidal calcite are produced in lakes and ultimately these become massive deposits of sedimentary limestone.

This phenomenon is associated with biomass synthesis in water during the daytime. Besides the alterations in carbonate chemistry, the pH rises substantially. At night, there is no photosynthesis and the pH declines, but dissolution of calcium carbonate under these conditions is slow, so that most of it remains in suspended or precipitate form.

Calcium carbonate, silica, and smaller amounts of other minerals also arise from deposition of the remains (*detritus*) of plankton and, to a smaller extent, the bones of fish and the shells of crustaceans.

14.6.2 Clay minerals

The above four natural colloid types—hydrated iron and manganese oxides, sulfide minerals, organic material, and carbonate minerals—have formed and continue to form in particular environments but there is a fifth category of colloids that is distributed extensively throughout the globe. This category comprises the clay minerals which are a suite of aluminosilicate minerals with a layered lattice structure. All are products of physicochemical weathering of primary minerals; they are usually of terrestrial origin and have been carried to the hydrosphere as eroded material in run-off and by winds or glaciers.

The clay minerals (also called phyllosilicates) have, as a common structural feature, SiO_4 tetrahedra linked together in a planar structure by three of the oxygen atoms. The 'sheet' thus formed is then joined *via* the additional oxygen to octahedral units of aluminium surrounded by six oxygens or hydroxyl groups. The tetrahedral and octahedral layers together form what is referred to as a 1:1 layer clay mineral structure. This is characteristic of several particular clay minerals, including kaolinite (Fig. 14.16).

Kaolinite and other clay minerals that are products of weathering of primary minerals are frequently encountered in water, sediment, and soil in the colloidal size fraction. Kaolinite is derived from orthoclase feldspar in the following reaction—a very slow process that only occurs over a geological time-span.

$$2KAlSi_3O_8 \text{ (s)} + 2H_3O^+ \text{ (aq)} + 7H_2O \rightarrow Al_2Si_2O_5(OH)_4 \text{ (s)} + 4H_4SiO_4 \text{ (aq)} + 2K^+ \text{ (aq)}$$
$$\text{orthoclase} \qquad\qquad\qquad\qquad\qquad \text{kaolinite}$$

$$(14.29)$$

The H_4SiO_4 (aq) and K^+ (aq) are removed in solution during the weathering process.

The electrostatic adsorption capability for kaolinite arises from its negative charge, which is a consequence of two factors:

- On the edges of units, broken bonds leave oxygen atoms with excess negative charge, the magnitude of which is a function of the number of such exposed atoms, which in turn depends on the particle size of the clay.
- The other, usually less important, source is from dissociation of -OH groups in the octahedral layer. This is a pH-dependent phenomenon.

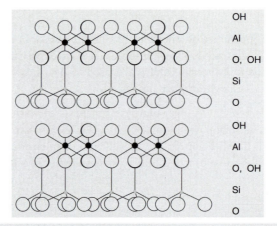

Fig. 14.16 Crystal structure of kaolinite. Each sheet is made up of silicon-oxygen tetrahedra attached to one side of an aluminum-oxygen octahedral sheet to form a 1:1 layer. The octahedra are arranged in pairs and the sheet is referred to as a dioctahedral type. The distance between two layers in kaolinite is about 0.7 nm.

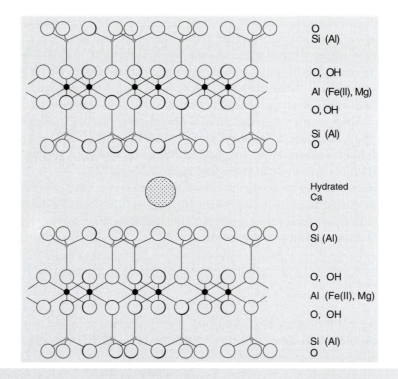

Fig. 14.17 Crystal structure of montmorillonite. The 2:1 sheets consist of a dioctahedral sheet sandwiched between two tetrahedral sheets. There is a considerable degree of substitution of +2 ions for Al^{3+} in the octahedral sheet, giving a substantial net negative charge to the clay. The large distance between individual layers (~1.4 nm) allows hydrated cations such as calcium to readily exchange.

The negative charges generated in these two ways are balanced by cations held at the surface in the adjacent ambient solution, and the total cation exchange capacity (CEC) for kaolinite is in the range 3 to 15 cmol $(+)$ kg^{-1}.

A second clay mineral is montmorillonite (also called smectite) which is one of the class of 2:1 phyllosilicates. The class is so named because the layers consist of an aluminium octahedral sheet sandwiched between two silica tetrahedral sheets. A negative charge is developed on montmorillonite for the two reasons noted above, but there is a third overriding factor as well: a large amount of negative charge is due to isomorphous substitution of aluminium for silicon atoms in the tetrahedral layers, which occurs to a limited extent, and of iron (II) or magnesium for aluminium in the octahedral layer, which occurs to a considerable extent. The net result of these substitutions of $+3$ ions for $+4$ ions and of $+2$ ions for $+3$ ions is a reduction in positive charge—in other words, excess negative charge. Because much of this negative charge is located in the interior of the 2:1 layer, the cations which balance it are only loosely held at the surface and therefore are readily exchangeable.

The CEC of montmorillonite is very large, ranging between 80 and 150 cmol $(+)$ kg^{-1} (Fig. 14.17). A summary of structures and properties of these and other clay minerals is given in Fig. 14.18. CEC values: kaolinite 3–15, halloysite 5–10, montmorillonite 80–150, vermiculite 100–150, chlorite 10–40; all values in cmol $(+)$ kg^{-1}.

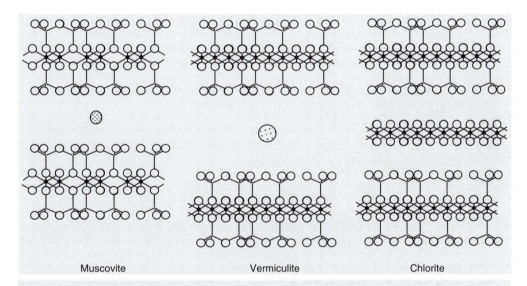

Muscovite   Vermiculite   Chlorite

Fig. 14.18 Crystal structure of three clay minerals. *Muscovite* is similar to montmorillonite, but there is more substitution of Al^{3+} for Si^{4+} in the tetrahedral sheets and much less substitution in the octahedral sheet. As a result, interlayer cations such as potassium are tightly held, the interlayer spacing is smaller ($\sim$1.0 nm) and the clay does not readily swell on wetting. *Vermiculite* is also similar to montmorillonite. The octahedral sheets are both di- and trioctahedral types. The interlayer spacing is quite large ($\sim$1.3 nm) and hydrated cations can readily exchange. *Chlorite* is a 2:1:1 clay mineral with a trioctahedral layer positioned between two typical 2:1 layers. Spacing between adjacent 2:1 layers is $\sim$1.4 nm.

Table 14.6 Cation exchange capacities (CEC) of various environmental materials found in the colloidal size fraction

Material	CEC range (average)/cmol $(+)$ kg^{-1}
Kaolinite	3–15 (8)
Montmorillonite	80–150 (100)
Chlorite	10–40 (25)
Hydrous iron and aluminium oxides	~4
Feldspar	1–2 (2)
Quartz	1–2 (2)
Organic matter	150–500 (200)

Most values from ref. 1 of the Additional reading list in Chapter 17.

The cation-exchange capacities of various environmental colloids, including clay minerals, are given in Table 14.6. The bracketed numbers are 'averages' that may be used in making rough estimates of cation-exchange capacities for particular samples.

Note that the CEC values in the table apply to solid material in the colloidal size fraction—larger sized materials make a negligible contribution to the exchange properties.

Consider sedimentary material containing 8% organic matter and 41% clay minerals. Of the latter, 70% is kaolinite and 30% is chlorite. The CEC may be estimated to be

$$\underset{\text{organic matter}}{0.08 \times 200} + \underset{\text{kaolinite}}{0.41 \times 0.70 \times 5} + \underset{\text{chlorite}}{0.41 \times 0.30 \times 25} = 21 \, \text{cmol} \, (+) \, \text{kg}^{-1}$$

This approximate calculation assumes that each sediment component contributes independently to the cation-exchange properties, whereas it is known that interactions between the components may alter these properties.

The main points

1 In the hydrosphere, the colloidal size range comprises material that falls between soluble and precipitated fractions. It includes a variety of minerals and organic matter.

2 The small colloidal particles have a very large specific surface area. The surface holds an electrical charge, usually negative, which is balanced by ions of opposite charge from the surrounding solution. This leads to the formation of an electrical double layer that serves to prevent particles from coming together, growing, and forming a precipitate.

3 Due to the surface charge and also because of specific chemical bond formation, ions and small molecules may be held (adsorbed) on the surface of the colloids.

4 There are many quantitative descriptions of the adsorption process; two of the most common are the Langmuir and Freundlich descriptions.

5 Simplified versions of the Freundlich relation are used to describe the retention of small organic molecules by colloids in soils and sediments. Arising out of these relations, a number of equilibrium distribution constants have been defined, including K_d, K_{OW}, K_{OM}, and K_{OC}.

6 Colloidal material found in water is derived from terrestrial sources or is generated *in situ*. A variety of colloid compositions is observed depending on the environment in which it has been formed.

7 The clay minerals are a class of colloidal silicate minerals found suspended in water and in soils and sediments. An important property of clay minerals is their ability to act as cation exchangers and this is quantified by measurements of their cation-exchange capacity.

Additional reading

1 Greenland, D. J. and M. H. B. Hayes, *The Chemistry of Soil Constituents*, John Wiley and Sons, Chichester; 1978.

2 Lyman, W. J., W. F. Reehl, and D. H. Rosenblatt, *Handbook of Chemical Property Estimation Methods*, American Chemical Society. Washington DC; 1990.

3 Stumm, W., *Chemistry of the Solid–Water Interface*, John Wiley and Sons, New York; 1992.

Problems

1 Show that a cation exchange capacity (CEC) value in units of cmol $(+)$ kg^{-1} is numerically the same as one using units of meq 100 g^{-1}. An explanation is provided in Chapter 18.

2 A sediment sample containing 47% clay and 3.2% organic matter was found to have a cation-exchange capacity of 10.5 cmol $(+)$ kg^{-1}. Assuming that the organic matter has a CEC of 210 cmol $(+)$ kg^{-1}, show that the clay could be kaolinite but not montmorillonite.

3 A particular lake sediment sample is made up of

> 23% organic matter
> 77% mineral fraction
> 42% clay minerals
> 24% silt
> 11% sand

The clay minerals are 90% kaolinite and 10% halloysite. Calculate an approximate cation-exchange capacity (cmol $(+)$ kg^{-1}).

4 Amongst the forms of phosphorus found in sediments are

- organic phosphorus
- apatite (($Ca_5(PO_4)_3$(F,Cl,OH) is one representation for this mineral)
- adsorbed on hydrated iron (III) oxides
- associated with clay minerals.

How would you expect the solubility of each of these forms to be affected by the ambient pE of the water?

5 (a) Use Fig. 10P.1 to suggest how phosphorus release could occur from buried iron-rich sediments.

(b) In laboratory experiments, it has been shown that phosphorus is released from sediment under dark conditions[4], but is accumulated in the sediment when they are exposed to light. Suggest an explanation.

6 The behaviour of lead in soil/water systems has been described in terms of Langmuir adsorption. From laboratory experiments using 'pure' solids, the following values for Langmuir parameters have been estimated.

For clay, $\quad b = 3200\,\text{L}\,\text{mol}^{-1}$
$\quad C_{sm} = 4 \times 10^{-5}\,\text{mol}\,\text{g}^{-1}$.
For organic matter, $\quad b = 10\,000\,\text{L}\,\text{mol}^{-1}$
$\quad C_{sm} = 4 \times 10^{-4}\,\text{mol}\,\text{g}^{-1}$.

Use these values to calculate the concentration (in $\mu\text{g}\,\text{L}^{-1}$ or ppb) of water-soluble lead in equilibrium with clay containing 5 ppm adsorbed lead, and with organic matter also containing 5 ppm adsorbed lead. Both soil concentrations are those found at equilibrium. Use the information from this question to predict the relative mobility of lead (released from gasoline combustion) deposited on clay-rich vs. organic-rich roadside soils.

7 The K_{OW} function is frequently used to describe the movement of pesticides in the surface soil pore water, but it is used less frequently to describe the behaviour of an organic contaminant in groundwater. Explain.

8 The correlations between K_{OW} and K_{OC} imply that organic carbon in all soils has similar properties. Other relations that have been suggested include correlations between K_{OC} and the ratio $(N+O)/C$ of soil organic matter and between K_{OC} and percentage aromaticity in the organic fraction of the soil. Suggest situations where these correlations might be appropriate.

Notes

1 Sidle R. C. and L. T. Kardos, Adsorption of copper, zinc and cadmium by a forest soil. *J. Environ. Qual.* **6** (1977), 313–7.

2 Tewari, Y. B., M. M. Miller, S. P. Wasik, and D. E. Martire, Aqueous solubility and octanol/water partition coefficient of organic compounds at 25.0 °C. *J. Chem. Eng. Data*, **27** (1982), 451–4.

3 Karickhoff, S. W., Semi-empirical estimation of sorption of hydrophobic pollutants on natural sediments and soils. *Chemosphere*, **10** (1981), 833–46.

4 Moore, P. A. Jr, K. R. Reddy, and D. A. Graetz, *Phosphorus geochemistry in the sediment–water column of a hypereutrophic lake*. *J. Environ. Qual.*, **20** (1991), 869–75.

15

Microbiological processes

Up to now, it is only in passing that we have made a distinction between *abiotic* and *biotic* environmental processes. Abiotic reactions are ones that occur through purely physical and/or chemical means. Such reactions would take place even in a completely sterile environment, if one existed. Biotic reactions, on the other hand, have a biological component. Photosynthesis, by which inorganic carbonate species are converted to organic carbon forms in green plants or other organisms, is a typical example of an environmental biotic reaction. In many cases, the biological component involves *microorganisms*; biota classified as microorganisms are very small, often having dimensions in the μm size range, and are frequently invisible to the naked eye. They are found in great numbers almost everywhere, in water, in soil, even in the air.

Microorganisms play an essential role in facilitating many chemical reactions that occur in the natural environment. A case in point is a process through which ammonium ion is converted to nitrate in water or soil:

$$NH_4^+(aq) + 2O_2 + H_2O \rightarrow NO_3^-(aq) + 2H_3O^+(aq) \tag{15.1}$$

This important oxidation reaction is called nitrification. Although nitrification can be described in chemical terms using a straightforward equation, it is not a single-step reaction and it does not occur to a significant extent as an abiotic chemical process. The reaction has a ΔG° value of -266.5 kJ, indicating that it is a highly favoured process in terms of thermodynamics. None the less, as is well known, an aqueous solution containing ammonium ions is kinetically stable in the presence of air and can be kept without oxidizing appreciably for years. This is not true in natural water or in soil where oxidation of ammonium occurs at a measurable rate due to the mediation of microorganisms. By this we mean that certain classes or specific species of microorganisms are the location of metabolic processes that result in the net chemical transformation shown in the equation. Often, particular enzymes within the microorganism are involved in catalysing the reaction.

In the case of nitrification, it is *Nitrosomonas* sp. and *Nitrobacter* sp. bacteria that make use of the ammonium ion as a substrate for their metabolism. In an aerobic environment in the presence of these bacteria, ammonium is readily oxidized to nitrate, usually in a matter of days.

For this reaction and in many other situations, environmental chemistry is actually environmental microbiological chemistry. The situations include degradation, synthesis, and other chemical transformations. The chemical species involved as reactants include a wide variety of naturally occurring carbon compounds as well as species containing nutrient elements like nitrogen, phosphorus, and sulfur, and even metals like iron or manganese.

In the present part of our survey of environmental chemistry we will introduce some of the fundamental terminology of microbiology and consider overall aspects of a number of important environmental reactions in which microorganisms play a major role. We will not be examining details of the metabolic transformations that occur *within* the organisms themselves.

15.1 Classification of microorganisms

In water, free-floating organisms are termed plankton; these are further subdivided into plant and animal components – phytoplankton and zooplankton respectively. Micro-organisms that exist in or on the sediment at the bottom of a water body form part of the benthos. The soil environment too is home to communities of microorganisms.

There are several systems used for classifying aquatic and soil microorganisms. Each system is useful in being able to describe particular features relevant to studies in environmental chemistry. We begin by looking at definitions within some of the systems.

15.1.1 Classification based on phylum of microbe

Bacteria are probably the most abundant type of microorganism found in both aqueous and terrestrial environments. Bacteria exist in a variety of sizes with diameters ranging from 0.2 to 50 μm, but most have dimensions less than 5 μm. Being of colloidal size, they have a large surface : volume ratio. Exposed to the surrounding solution, on their surface are deprotonated acidic groups and thus bacteria carry a negative charge under common environmental conditions. There are many species of bacteria, taking forms ranging from spherical through ellipsoidal to rod-like. They are found in a variety of environments—ones that are oxygen-rich (oxic, aerobic) as well as those that are devoid of oxygen (anoxic, anaerobic). As nutrients, bacteria require a variety of chemical species including the common ones such as nitrogen, phosphorus, and potassium. As carbon sources required for their own growth, bacteria most commonly modify and incorporate preformed molecules from other organic materials.

The population (nature and number) of bacteria in a given environment depends on many factors including temperature, pH, electrolyte concentration, nutrient supply, and, in the case of soils, moisture content. In sediments and soils, bacterial count is positively correlated with organic matter content and can range from several million to several hundred million bacteria per gram. However, because of their very small size, on a mass basis even the larger number corresponds to less than 0.1% of the total mass of the solid material. In the water column itself bacterial numbers range from 5×10^4 to $5 \times 10^7 \, \mathrm{mL}^{-1}$ depending on the location, season, position in the water column, and the trophic status of the water.

Fungi are a ubiquitous and diverse category of microorganism, commonly thought to be associated almost exclusively with the terrestrial environment. This is because they tend to favour an aerobic, usually somewhat dry, situation. However, fungi also inhabit most oxygenated parts of pools, lakes, rivers, and the oceans. They are found in a wide range of forms and sizes including multicellular species (e.g. mushrooms) readily visible without magnification. Fungi always require presynthesized compounds as carbon sources; in using these molecules to grow, they break down the original substance and incorporate the transformed molecules into their own structure. Fungi therefore play a key role in the degradation of litter in soil—an essential link in the chain of reactions resulting in formation of humic material. Microorganisms that live on decaying organic matter are called saprophytes. In contrast to bacteria, most of which usually prefer neutral to slightly alkaline environments, fungi are better adapted to acidic conditions.

The number concentration of fungi in soil is usually several orders of magnitude smaller than that of bacteria, but being much larger in size, on a mass basis they may be the most important type of microorganism. Many fungal species have a filamentous structure, so they too have a very large surface : volume ratio in spite of their large overall dimensions.

Actinomycetes are a class of unicellular organisms. At one stage of their development they take the form of fine, branching filaments similar to the fungi. In a later stage, they develop into a population of bacteria-like species. They are found in both aquatic and terrestrial situations; in the latter case they prefer aerobic, relatively dry conditions. Actinomycetes are more abundant in warm climates and are particularly numerous in tropical grasslands where they may make up about 30% of the total microbial population. In cooler, wetter areas, they contribute much less to the overall numbers. They are sensitive to acidity but are able to tolerate higher salt concentrations than many bacteria. Like fungi, actinomycetes are frequently involved in degradation processes in both water and soil, but they also are producers of natural antibiotics that control the populations of coexisting microorganisms.

Algae are chlorophyll-containing organisms that range in size from microscopic individuals to large plant-like structures. They are abundant in the upper epilimnion of eutrophic water bodies as well as in the moist surface layer of soil. While fungi and actinomycetes are essentially degraders of pre-existing organic structures, the algae act as producers of new organic matter. They therefore play an important role in converting inorganic carbonate into organic forms. In this way they are an essential link in the global carbon cycle. Another important role is as agents of non-symbiotic nitrogen fixation.

As they convert inorganic carbon into organic compounds, other inorganic nutrients and even non-nutrient species are assimilated in order to build up the algal molecular structure. Because algae accumulate non-essential elements, they are sometimes used as integrating environmental monitors. By comparing the elemental content of algae growing in clean and polluted water, unusually high levels of metals like lead, cadmium, or mercury can be detected. Where excessive values are observed, it indicates that there have been high average concentrations of the contaminant in the ambient water throughout the growing season.

Algae can cope with great temperature extremes and are present in lakes and rivers in all parts of the Earth; they have also been found in polar ice as well as in hot springs.

Protozoa are the simplest form of animal life; they are commonly 5–50 μm in size and are found in both water and in the ground. In soil, they live in thin films of water on the surface of particles. Besides having a requirement for an adequate supply of moisture, they prefer a habitat that is warm (18 to 30 °C), well oxygenated, and with pH between 6 and 8, although they can exist throughout the range 3 to 10. Being consumers (sometimes referred to as grazers or predators) of bacteria and other microorganisms, protozoa regulate the total microbial population.

15.1.2 Classification based on ecological characteristics

A mixed population of microorganisms is present in any environment. Some species are active, while others are dormant. As we have indicated earlier, the numbers of different species at any time depend on many environmental conditions. In a stable situation such as a mature forest, the microbial population is also stable, in balance with the surroundings and with numbers determined by the available food supply and other properties of the local environment. The typical microorganisms that are indigenous in a particular area like this are called the *autochthonous microorganisms*.

If the environmental situation is changed, such as by an influx of a fresh nutrient supply, particular types of microorganisms that can take advantage of the new situation may proliferate, at least until the situation returns to stability. The highly active, but fluctuating population is called a *zymogenous* or *allochthonous population*.

An example of the latter situation is one that occurs after a heavy rainfall when a small nutrient-poor water body is inundated with a quantity of nitrate dissolved in run-off from adjacent fields. In the new situation, algae that can make use of the nitrate are able to flourish. However, the growth of algal numbers is temporary and when the nitrate has dissipated, the baseline conditions resume, with more stable chemistry and the autochthonous microbial population.

15.1.3 Classification by carbon source

Microbial structures consist of organic molecules—carbohydrates, proteins, lipids, and so on—of which carbon makes up about half of the molecular mass. Microorganisms can be further classified on the basis of the source of carbon that is required in order for them to function and build up their own species.

Autotrophs are microorganisms that are capable of growing in a completely inorganic medium using carbonate species as their sole source of carbon. Earlier, we referred to autotrophs as producers or synthesizers. The carbonate may be in the form of atmospheric carbon dioxide or aqueous carbon dioxide, hydrogen carbonate, or carbonate. Green plants (of course these are *macro*organisms), most algae, and some bacteria are all classified as autotrophs. As subdivisions within this category, there are *photoautotrophs*—algae are a well known example—that use solar radiation as an energy source for their synthetic processes. On the other hand, *chemoautotrophs* derive energy from chemical oxidation reactions. Most bacteria are chemoautotrophs. *Nitrosomonas*, for example, is the bacterium that facilitates the first step of nitrification by deriving energy through oxidizing ammonium to nitrite:

$$2NH_4^+ \text{ (aq)} + 3O_2 + 2H_2O \ \rightarrow \ 2NO_2^- \text{ (aq)} + 4H_3O^+ \text{ (aq)} \qquad (15.2)$$

Heterotrophs are microorganisms that make use of presynthesized organic compounds as a carbon source. These organisms are degraders or consumers of other living matter—plant and animal residues in water and soil and many organic waste materials. The heterotroph category includes fungi, actinomycetes, protozoa and most bacteria. Specific microbes are also important as degraders of synthetic organic molecules such as pesticides. For example, the degradation of the carbamate pesticide isopropyl-*N*-phenylcarbamate, IPC, is facilitated by a number of bacterial species such as *Arthrobacter* spp.

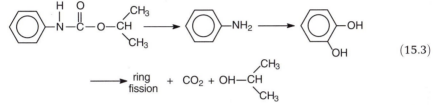

$$(15.3)$$

When degradation of organic molecules proceeds to inorganic products such as carbon dioxide, ammonia, and sulfate, the process is referred to as ultimate degradation or mineralization.

15.1.4 Classification by source of electron acceptor

This classification applies to microorganisms that can facilitate redox reactions. The degradation of organic matter in the sediment of a lake, in a forest soil, in a landfill for solid waste, and in a waste-water treatment plant all takes place *via* various sequences of oxidation steps.

Aerobes are a large category of microorganisms that use molecular oxygen, either in gaseous form or dissolved in water, as the electron acceptor for their oxidation reactions. In the process, the oxygen is reduced and hydronium ions are consumed:

$$O_2 + 4H_3O^+ \text{ (aq)} + 4e^- \ \rightarrow \ 6H_2O \qquad (15.4)$$

In order for aerobic species to be active, the environment must be in contact with a plentiful supply of oxygen-containing air.

When air transfer is limited and the oxygen becomes depleted, the strictly aerobic organisms are no longer able to function. However, oxidation reactions can continue through the activity of *anaerobic* microorganisms. Anaerobes have the capability of effecting oxidation without molecular oxygen. Instead, they use other electron-poor species, for example sulfate, to accept electrons from the reduced substrate. The electron-acceptor species is converted into a more reduced form:

$$SO_4^{2-} \text{ (aq)} + 9H_3O^+ \text{ (aq)} + 8e^- \ \rightarrow \ HS^- \text{ (aq)} + 13H_2O \qquad (15.5)$$

Bacteria make up the largest number of anaerobes.

Within the anaerobic category are *obligate anaerobes* for which oxygen is toxic, so they can only function in the absence of oxygen. The toxicity of oxygen is due to the absence of cytochromes and catalase, allowing for the accumulation of toxic hydrogen peroxide. There are also *facultative anaerobes* which are organisms that are able to make use of either oxygen or other electron acceptors for oxidation.

15.1.5 Classification by temperature preference

All microorganisms follow an activity vs. temperature relationship of the form shown in Fig. 15.1.

As illustrated in this plot, microbial growth rate increases with increasing temperature according to the Arrhenius relation, $\ln k = -E_a/RT + \text{constant}$, where k is the reaction rate constant. However, above a particular maximum temperature, enzymes within the microorganism become denatured and growth declines precipitously. The position of the maximum along the x axis depends on the species of microorganism and they may be defined according to the value of T_{optimum}.

- *Psychrophiles*: $T_{\text{optimum}} < 20\,°C$ (uncommon but found in polar environments);
- *Mesophiles*: T_{optimum} 20–45 °C (most common);
- *Thermophiles*: $T_{\text{optimum}} > 45\,°C$ (found in tropical environments and associated with situations like compost production).

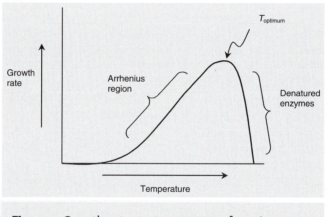

Fig. 15.1 Growth rate vs. temperature for micro-organisms.

15.1.6 Classification by morphology

For purposes of identification and characterization, the shape, form, and colour of microorganisms are important characteristics noted by microbiologists. Some of the many shapes observed, and the nomenclature associated with these, are:

- rod shape: bacilli
- spherical shape: cocci
- spiral shape: spirilla.

Many other morphological categories exist and some species change shape or exhibit a variety of forms.

15.2 Microbiological processes—the carbon cycle

Many of the natural environmental cycles operating in the Earth's environment have key links that are controlled in large part by microorganisms. When human interventions occur, the microbial population adjusts to the new environment, and the chemical reactions within that part of the cycle are affected by the changes in the situation. This point in our study of environmental chemistry is a good place to summarize broad chemical features in some of these important cycles. We will include major abiotic as well as biotic processes, but discussion will centre on the main biological processes that occur in water or on land. In the discussions, only the general types of reactions are considered; there are always a vast number of specific detailed reactions that are not included.

The carbon cycle (Fig. 15.2) is one of the great cycles that determines the forms of life and their interactions within the global environment. Important organic and inorganic carbon-containing species exist in water, on land, and in the atmosphere. Every organic compound and all the inorganic carbonate species fit into this general picture in some way—through synthesis, transformation, and/or decomposition.

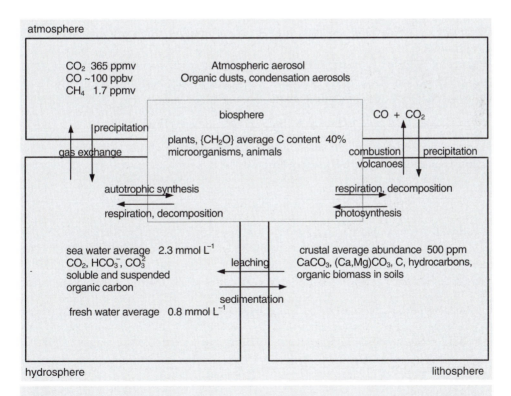

Fig. 15.2 The carbon cycle—a summary of principal processes in the environment.

15.2.1 Carbon species in water and on land

In the terrestrial environment, there are three major reservoirs of carbon. On a mass basis, the largest of these is in the carbonate rocks, limestone ($CaCO_3$), and dolomite ($(Ca,Mg)CO_3$). The second major carbon-containing deposit is made up of material buried in the Earth in the form of fossil fuels of all three phases—solid coal, liquid petroleum, and natural gas. The third large terrestrial reservoir of carbon is organic material in various stages of chemical modification, on or in the soil. Clearly, the fossil fuels and soil organic matter have a biological origin. The carbonate minerals too are at least partly of biological origin.

In the previous chapter we showed, by calculation, how the daily photosynthetic cycle can be an agent through which organic carbon molecules are formed in a water body. At the same time, an increase in pH occurs and calcium carbonate can be precipitated from solution. In the oceans, these kinds of processes operate on an immense scale. Carbonate species are present in all natural water arising from atmospheric carbon dioxide, remineralized organic matter, and dissolved carbonate minerals. The oceans are an especially large reservoir of carbon compounds and play a major role in establishing and maintaining the global carbon balance. In the surface of oceans, through energy supplied by solar illumination, photoautotrophic microorganisms synthesize particulate organic matter using inorganic carbonate species from both water and the atmosphere as their carbon source. Simultaneously, additional carbonate is microbially incorporated to form the hard structures of some oceanic plankton. One net result of carbon assimilation is that surface waters develop a slightly higher pH and smaller alkalinity than water below the thermocline.

In thermodynamic terms, the surface of the oceans is thus supersaturated with respect to the precipitation of calcium carbonate. The reaction that best describes features controlling solubility of this compound is

$$CaCO_3 + CO_2 \text{ (aq)} + H_2O \rightleftharpoons Ca^{2+} \text{ (aq)} + 2HCO_3^- \text{ (aq)} \tag{15.6}$$

The size of the equilibrium constant, a modified solubility product constant, $K'_{sp}(1)$, for reaction 15.6, is calculated in the following way:

$$K'_{sp}(1) = \frac{K_{a1} \times K_{sp}}{K_{a2}} = \frac{1.0 \times 10^{-6} \times 5.9 \times 10^{-7}}{7.8 \times 10^{-10}} = 7.6 \times 10^{-4}$$

The values for the constants (K_{a1}, K_{a2}, K_{sp}) are ones for reactions in surface sea water;[1] because they depend on ionic composition of the water, they are different from the 'fresh water' values given in the Appendix.

For a situation near the ocean surface, typical values of pH, hydrogen carbonate, carbon dioxide, and calcium ion concentration are 8.2, $2.1 \times 10^{-3} \text{ mol L}^{-1}$, $2.5 \times 10^{-5} \text{ mol L}^{-1}$, and $1.1 \times 10^{-2} \text{ mol L}^{-1}$ respectively. The reaction quotient ($Q'_{sp}(1)$) related to these conditions in the present example is therefore:

$$Q'_{sp}(1) = \frac{[HCO_3^-]^2[Ca^{2+}]}{[CO_2]} = \frac{(2.1 \times 10^{-3})^2 \times 1.1 \times 10^{-2}}{2.5 \times 10^{-5}} = 2.0 \times 10^{-3}$$

The ratio of reaction quotient to solubility product constant is a measure of the degree of saturation and a value of the ratio greater than unity indicates a condition of supersaturation. In this case the ratio,

$$\frac{Q'_{sp}(1)}{K'_{sp}(1)} = \frac{2.0 \times 10^{-3}}{7.6 \times 10^{-4}} = 2.6$$

indicating that the surface water sample is highly supersaturated with respect to calcium carbonate and a colloidal precipitate is formed.

The microorganisms and carbonate shells sink into deeper parts of the oceans where the solar flux has declined to negligible values. The net reaction is biological decomposition of organic matter, thus converting the organic carbon forms back into soluble mineral species. Carbon dioxide is released and its hydrolysis causes a drop in pH of about half a unit through the thermocline. A recalculation of the solubility product constant for eqn 15.6 requires using a K_{sp} for calcium carbonate that applies to pressures found at a depth of 2500 m. The value of the K_{sp} under these conditions is 1.3×10^{-6}, and the modified solubility product for reaction $(K'_{sp}(2))$ 15.6 is

$$K'_{sp}(2) = \frac{K_{a1} \times K_{sp}}{K_{a2}} = \frac{1.0 \times 10^{-6} \times 1.3 \times 10^{-6}}{7.8 \times 10^{-10}} = 1.7 \times 10^{-3}$$

Note that we have used the original acid dissociation constant values. A more refined calculation would require that these also be modified to account for the conditions at depth.

The reaction quotient is calculated in the new situation using concentrations of the relevant species in that part of the ocean.

$$Q'_{sp}(2) = \frac{[HCO_3^-]^2[Ca^{2+}]}{[CO_2]} = \frac{(2.4 \times 10^{-3})^2 \times 1.1 \times 10^{-2}}{4.5 \times 10^{-5}} = 1.4 \times 10^{-3}$$

In the deep water situation, the ratio of the reaction quotient to the solubility product$(Q'_{sp}(2)/K'_{sp}(2))$ has become 0.82. Being less than unity it indicates an undersaturated situation. On the basis of these changed thermodynamic conditions, one would predict spontaneous dissolution of the calcium carbonate microbe shells. The fact that calcium carbonate remains in ocean sediments indicates that the dissolution reactions are kinetically hindered. Little dissolution occurs in the time before the sediments are buried, and massive calcium carbonate deposits accumulate in this way.

15.2.2 Biomass degradation

Another process in the carbon cycle which is controlled almost exclusively by microbial activity is decomposition of biomass. In this context, the term biomass refers to dead plant and microbial material—as, for example, leaf, twig, and branch litter in a forest; root and straw residues in a cultivated field; or dead aquatic plants and microorganisms in a lake or swamp. The biomass is to a large extent carbohydrate—in particular cellulose, and cellulose-related compounds. Decomposition is a process of oxidation—with exceptions that will come up later—and the half reaction is expressed in simple form as

$$\{CH_2O\} + 5H_2O \rightarrow CO_2 \text{ (g)} + 4H_3O^+ \text{ (aq)} + 4e^- \tag{15.7}$$

When the oxidative degradation takes place in water whose pH is near 7, we describe the redox potentials in terms of a pE^o (w) (the concept of pE^o (w) is described in Box 15.1). Values of pE^o always refer to reduction reactions. In the present case therefore, the pE^o (w) is for reaction 15.7 when it is read from right to left and it has a value of -8.20. For the oxidation (as written, left to right) reaction, the pE value is $+8.20$. In order for oxidation to occur, the half reaction must be coupled to an appropriate reduction half reaction—that is, an electron-accepting process. By appropriate, we mean that a naturally occurring oxidizing agent must be available in sufficiently high concentration, and the pE for the overall process must be positive. We will consider a number of these oxidizing agents (electron acceptors) in order of decreasing pE^o (w) values.

Box 15.1 pE^o (w) values

The term pE^o has been defined and described in Chapter 10. The superscript ^{o} indicates a pE under standard conditions, including the condition that activity of hydrogen ion is 1, equivalent to a pH in the system of 0. Clearly, in real environmental circumstances, a pH of 0 is highly unusual and it is more likely that the value be near neutrality. We therefore define pE^o (w) as a measure of pE^o at pH $= 7$ rather than pH $= 0$.

Consider the 4-electron process described by the reaction

$$O_2 + 4H_3O^+ \text{ (aq)} + 4e^- \rightarrow 6H_2O \qquad pE^o = 20.80 \qquad (15b.1)$$

$$pE = pE^o - \tfrac{1}{4}\log\frac{1}{P_{O_2}(a_{H_3O^+})^4}$$

The usual standard conditions include $P_{O_2} = 101.3$ kPa $= 1$ atm. At the boundary,

$$pE = pE^o - \tfrac{1}{4}\log\frac{1}{(a_{H_3O^+})^4}$$

If we define $pE = pE^o$ (w) when the pH is 7, then

$$pE = pE^o \text{ (w)} = 20.80 - \tfrac{1}{4}\log\frac{1}{(10^{-7})^4}$$

$$= 20.80 - 7 = 13.80$$

For a general reaction

$$Ox + n_1H_3O^+ + n_2e^- \rightarrow Red \qquad (15b.2)$$

$$pE = pE^o - \frac{1}{n_2}\log\frac{a_{Red}}{a_{Ox} \times (a_{H_3O^+})^{n_1}}$$

Applying standard conditions for all but H_3O^+:

$$pE = pE^o - \frac{1}{n_2}\log\frac{1}{(a_{H_3O^+})^{n_1}}$$

and

$$pE = pE^o(w) = pE^o - 7(n_1/n_2) \qquad (15b.3)$$

15.2.3 Oxygen

When molecular oxygen acts as an oxidizing agent, it is reduced as shown by the half reaction 15.8:

$$O_2 + 4H_3O^+ \text{ (aq)} + 4e^- \rightarrow 6H_2O \qquad pE^o(w) = +13.80 \qquad (15.8)$$

The sum of reactions 15.7 and 15.8 gives the overall reaction for oxidation of biomass by oxygen:

$$\{CH_2O\} + O_2 \rightarrow CO_2 + H_2O \qquad (15.9)$$

and the pE^o (w) is the sum of the pE^o (w) for the oxygen half reaction and the negative value of pE^o (w) for the carbohydrate half reaction:

$$pE^o \ (w)_1 = pE^o \ (w)_{O_2} - pE^o \ (w)_{CH_2O}$$
$$= 13.80 - (-8.20)$$
$$= +22.00$$

The Gibb's free energy for the oxidation at 298 K is a 4-electron process and

$$\Delta G^o(w)_1 = -2.303 \, nRT \, pE^o(w)_1 = -500 \, kJ$$

The reaction as written allows for the oxidation of 1 mol of $\{CH_2O\}$

$$\Delta G^o(w, 1 \, mol)_1 = -500 \, kJ$$

Reaction 15.9 is therefore highly favoured thermodynamically and as a consequence, oxidation of organic matter by oxygen is predicted to occur under aerobic conditions. In an abiotic situation, however, the reaction is very slow, but degradation does proceed at a reasonable rate when microorganisms capable of facilitating the redox process are present. Appropriate microorganisms include a wide range of heterotrophic species of bacteria, fungi, protozoa, and actinomycetes that usually are abundant in well oxygenated water, sediment, and soil.

We have used cellulose, represented as $\{CH_2O\}$, as the model compound but biomass is complex material made up of a wide range of components. Some are more readily degraded than cellulose, many are more resistant to oxidation. To some degree, however, decomposition occurs by reactions similar to reaction 15.9 with oxygen being the prime electron acceptor for microbial oxidative degradation.

15.2.4 Nitrate

Where oxygen is absent, it is still possible for organic matter to be degraded *via* other oxidation reactions. Nitrate can serve as an electron acceptor, being reduced to nitrite, ammonium ion, nitrous oxide, or dinitrogen gas depending on environmental circumstances. The reduction reaction producing dinitrogen gas is important as a link in the global nitrogen cycle and the equation for it is

$$2NO_3^- \text{ (aq)} + 12H_3O^+ \text{ (aq)} + 10e^- \rightarrow N_2 + 18H_2O \qquad pE^o(w) = 12.65 \qquad (15.10)$$

When coupled with oxidation of carbohydrate (reaction 15.7), the overall reaction is

$$4NO_3^- \text{ (aq)} + 5\{CH_2O\} + 4H_3O^+ \text{ (aq)} \rightarrow 2N_2 + 5CO_2 + 11H_2O \tag{15.11}$$

The pE° (w) is calculated as before:

$$\begin{aligned} pE^\circ(w)_2 &= pE^\circ(w)_{NO_3^-} + pE^\circ(w)_{CH_2O} \\ &= 12.65 - (-8.20) \\ &= 20.85 \end{aligned}$$

The Gibb's free energy for the reaction as written (a 20-electron process) is

$$\Delta G^\circ(w)_2 = -2.303 \, nRT \, pE^\circ \,(w)_2 = -2380 \, kJ$$

For the reaction written in terms of 1 mol of $\{CH_2O\}$,

$$\Delta G^\circ(w, 1 \, mol)_2 = -480 \, kJ$$

The negative value of -480 kJ indicates that nitrate is capable of oxidizing organic matter in the absence of oxygen. In Section 15.3, we will show that this reaction—called denitrification—is an important link in the nitrogen cycle. To proceed at a significant rate, it requires a large supply of organic matter and the presence of denitrifying organisms. These conditions apply in stagnant waters supplied with an abundance of decaying biomass. Zones of the ocean where upwelling of the deep waters occurs also produce abundant organic-rich sediments. Parts of the Atlantic and Pacific Oceans adjacent to the west coasts of Africa and South America are such regions.

15.2.5 Sulfate

Where oxygen and nitrate are both absent and sulfate is present, the latter species may serve as an electron acceptor for the oxidation of organic matter:

$$SO_4^{2-} \text{ (aq)} + 9H_3O^+ \text{ (aq)} + 8e^- \rightarrow HS^- \text{ (aq)} + 13H_2O \qquad pE^\circ \text{ (w)} = -3.68 \tag{15.12}$$

The overall reaction is obtained by combining reactions 15.7 and 15.12:

$$2\{CH_2O\} + H_3O^+ \text{ (aq)} + SO_4^{2-} \text{ (aq)} \rightarrow HS^- \text{ (aq)} + 2CO_2 + 3H_2O \tag{15.13}$$

The pE° (w) and Gibb's free energy are calculated as before:

$$\begin{aligned} pE^\circ \,(w)_3 &= pE^\circ \,(w)_{SO_4^{2-}} - pE^\circ \,(w)_{CH_2O} \\ &= -3.68 - (-8.20) \\ &= +4.52 \\ \Delta G^\circ \,(w, 1 \, mol)_3 &= -103 \, kJ \end{aligned}$$

Again, the ΔG° calculated here is for one mole of $\{CH_2O\}$. The negative value indicates that oxidation by sulfate is also thermodynamically favourable. This reaction is important, for example, in oxygen-depleted organic-rich marine sediments where there is an abundance of sulfate available in the sea water. Bacteria of the species *Desulfovibrio* are important mediators for this process.

15.2.6 No oxidants present

In an environment in which no oxygen, nitrate, or sulfate is present, organic matter may still undergo oxidation by means of anaerobic reactions leading to a variety of products. The ultimate products of anaerobic decomposition include methane and are *via* internal redox processes, or redox disproportion reactions. The overall reduction half reaction is

$$CO_2 + 8H_3O^+ \text{ (aq)} + 8e^- \rightarrow CH_4 + 10H_2O \qquad pE^\circ \text{ (w)} = -4.13 \qquad (15.14)$$

Combining 15.14 with the half reaction for the oxidation of carbohydrate (15.7) gives the following overall redox process:

$$2\{CH_2O\} \rightarrow CH_4 + CO_2 \qquad (15.15)$$

The pE° (w) and Gibb's free energy are

$$pE^\circ \text{ (w)} = pE^\circ \text{ (w)}_{CO_2} - pE^\circ \text{ (w)}_{CH_2O}$$
$$= -4.13 - (-8.20)$$
$$= +4.07$$
$$\Delta G^\circ \text{ (w, 1 mol)}_4 = -93 \text{ kJ}$$

The smaller negative free energy value indicates that reaction 15.15 is, thermodynamically, the least favoured of the four that we have described. Nevertheless, it is a spontaneous process and results in the anaerobic decomposition of organic matter. Therefore, decomposition of organic matter can and does occur in environmental situations even in the absence of any other oxidizing agents. Species of actinomycetes are the predominant anaerobes responsible for anaerobic biodegradation. The organisms required for the production of methane are called methanogens. This gas, sometimes called 'marsh gas', is frequently observed in swamps or other wetlands. We have seen that methane release from wetlands contributes almost half to the atmospheric input of this important greenhouse gas.

Anaerobic decomposition of this type is sometimes referred to as a 'fermentation reaction'. Fermentation to produce other products can also occur, In some of these reactions carboxylic acids and hydrogen gas are produced as, for example, according to eqn 15.16:

$$3\{CH_2O\} + 4H_2O \rightarrow CH_3COO^- \text{ (aq)} + HCO_3^- \text{ (aq)} + 2H_3O^+ \text{ (aq)} + 2H_2 \qquad (15.16)$$

Reactions like 15.16 represent a less complete decomposition of the original organic matter than when carbon dioxide and methane are formed.

In summary, there are at least four mechanisms by which biomass is oxidized. For any particular situation, the mechanism that operates depends on the availability of oxidants, on the value of ΔG (the above four reactions were given in order of decreasing negative value of ΔG°(w, 1 mol)) and on the presence of appropriate microorganisms to facilitate the actual reaction. Assuming that the last factor is not limiting, a reaction sequence as in Fig. 15.3 would be expected if each oxidant were to react one after the other. The bracketed

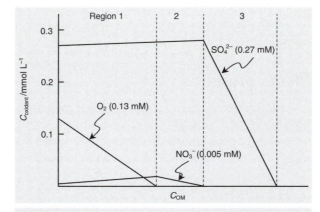

Fig. 15.3 Reaction sequence for oxidation of organic matter in a lake or river. The bracketed values are the original concentrations of each oxidant. Data are taken from samples from the Niagara River. The x axis is not to scale.

concentrations are actual data taken from a Niagara River water sample. In the sequence, oxygen is the primary oxidant followed by nitrate and then sulfate. During the time when oxygen is the oxidant, nitrogen in the biomass—often at a concentration of around 1%—is converted into ammonium ion and then oxidized to produce nitrate, additional to that originally present in the water. Similarly, there may be a small increase in sulfate concentrations during the initial two stages of organic matter oxidation.

The reactions we have discussed occur in water as well as in soils. With respect to soils, there is considerable interest in the redox processes that occur after they have been flooded. Flooded soils are a seasonal feature in many parts of the world and are associated with the cultivation of rice—a practice that makes use of some 170 million ha of agricultural land in tropical countries worldwide.

In newly submerged soils, and also in the case of ocean and fresh water sediments, oxygen is the primary oxidant. The rate at which it is used up depends on surface aeration and water movement, and on production and use of oxygen by plants and microorganisms. After oxygen is depleted, there is a succession of reactions similar to that described above. Table 15.1 reports data describing a microbiological and chemical sequence that occurs after a soil is submerged. The observed data are consistent with the sequence.

When the dry soil is submerged with oxygen-saturated water, it has an initial pE in the aerobic range. The microbial population is small, but aerobic species multiply rapidly in the newly moist environment. The initially high oxygen concentrations decline rapidly as oxygen is consumed by the processes of decomposing biomass that produce carbon dioxide. In a matter of a few days, the oxygen is depleted. Nitrate and sulfate then become the active electron acceptors until they too are depleted. The pE drops dramatically and the activity of aerobes declines while anaerobic microorganisms become dominant. Eventually,

Table 15.1 Chemical and microbiological changes occurring over a 23 day period after a soil is submerged

Microbial population/10^6 g				C_{gas}/mL 100 g^{-1} soil			
Time/days	pE	Aerobes	Anaerobes	O_2	CH_4	H_2	CO_2
0	7.6	34	22	3.2	ND	ND	83
1	3.7	220		0.3	ND	ND	10
2	−0.84	110	33	ND	ND	0.2	172
4.5	−3.9	55	50	ND	0.3	ND	
6				ND	2.2	3.6	280
8	−4.2	53	170	ND	14.7	2.1	
10				ND	21.4	ND	226
13	−4.2	62	130	ND			
23					60.3	3.2	

ND, not detected. From Takai, Y., T. Koyama, and T. Kamura, *Soil and Plant Food*, 2 (1956), 63–6 as reported in Alexander, M., *Introduction to Soil Microbiology*, John Wiley and Sons, New York; 1961.

measurable quantities of methane and some hydrogen are produced—evidence that fermentation reactions have become the means of biomass decay.

The flooded soil case is an example where a substantial area of soil becomes reduced. On a much smaller scale too, it is possible to have reducing conditions in an otherwise aerobic situation. Where soil aggregates are several mm or larger in dimensions, the gas diffusion rate may be insufficient to maintain a significant presence of oxygen in the interior regions of the particle. In this microenvironment, the local pE value then actually reflects reducing conditions.

The reactions that we have been discussing are the dominant ones controlling the redox status in most water or water/soil environments and this helps to explain why pE values tend to be found in one of two general regions. The sets of high and low pE environments shown in Fig. 15.4 are evidence of this phenomenon, one that is sometimes referred to as *redox buffering*. Where oxygen is present, the pH is controlled by the oxygen/water half reactions (region 'a' on Fig. 15.4) but when it is used up, the pE drops substantially to the region 'b' where it is determined by the next available redox couple. This is usually the sulfate/hydrogen sulfide couple as, in most cases in the hydrosphere, nitrate is present in relatively small concentrations. After sulfate is exhausted, a further small decline in pE occurs while the system moves into the fermentation region.

Figure 15.5 shows 'idealized' oxygen and temperature profiles of water (not sediment) in two lakes—one oligotrophic and one eutrophic—that are subject to the seasonal variations characteristic of temperate climate regions. The three profiles for each lake correspond to the situations in winter, spring/autumn, and summer.

In this idealized situation, the temperature profiles in the two lakes are shown to be identical and follow the seasonal pattern explained in Chapter 9. Note the well mixed,

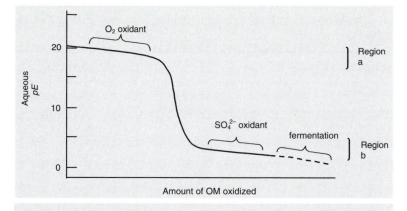

Fig. 15.4 Redox buffering regions. The pE value is maintained at a relatively constant value by oxidant species in the regions shown.

uniform situation in spring and autumn, and the presence of a thermocline in summer. In the oligotrophic lake, the oxygen concentration tracks, in an inverse sense, the temperature profile. This is because of the greater solubility of oxygen in cold compared to warm water.

The picture in a eutrophic lake is influenced by the same principles, but an additional factor is the presence of oxidizable biomass in the deeper regions of the water body. In winter, this leads to a stable situation with a steady decline in oxygen concentration from lake surface to sediment surface. The depletion is due to the low level activities of degradative heterotrophic microorgnisms that consume oxygen as they degrade biomass in the sediment. In this season, oxygen is replenished only by diffusion from surface water that is in contact with the atmosphere. That contact may be made negligible by the ice cover. In spring and fall, the rapid mixing due to overturn provides a relatively uniform oxygen concentration profile as the element is distributed by both diffusion and convection. The summer picture is quite different. The major decline below the thermocline is due to a stable, warm environment where oxygen has been depleted by microbial degradation yet not replenished due to the inability of water in the hypolimnion to mix with the oxygen-rich water of the epilimnion.

Degradation of non-living biomass is not the only microbial process controlling dissolved oxygen content of water bodies. Living microorganisms, especially algae also affect oxygen levels. In Chapter 11, we calculated the concentration of molecular oxygen at 25 °C in a well aerated water body and found it to be close to 8 mg L^{-1}. The growth of algae creates a daily cycle around this value (Figure 15.6). The cycle is readily explained in terms of the reversible chemical reactions corresponding to photosynthesis and respiration:

$$CO_2 + H_2O \underset{\text{respiration}}{\overset{\text{photosynthesis}}{\rightleftharpoons}} \{CH_2O\} + O_2 \tag{15.17}$$

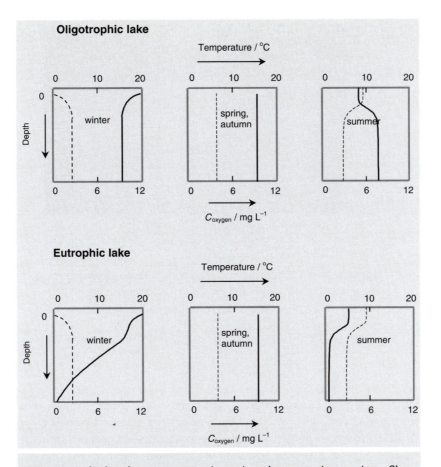

Fig. 15.5 Idealized temperature (-----) and oxygen (———) profiles in oligotrophic and eutrophic lakes, as seasons change in the temperate regions of the world.
(Redrawn from Wetzel, R. G., *Limnology*, Saunders, Philadelphia; 1975.)

As the figure shows, the algal growth cycle creates enhanced aqueous oxygen concentrations in daytime and diminished concentrations at night. Further consequences are changes in pH and possible precipitation of calcium carbonate as described earlier.

Various compounds found in water are substrates that can be degraded and contribute to oxygen depletion. Usually the organic components of non-living biomass—dead algae and plants, and organic wastes from municipal or industrial sources—are the major electron donors. However, inorganic species including ammonium, sulfide, and reduced forms of iron, chromium, manganese, and other metals may also use up oxygen.

Any material in a water body that can react with dissolved oxygen contributes to the biological (or biochemical) oxygen demand (BOD). High concentrations of these oxidizable materials lead to anoxic conditions—stagnant water that does not support higher life forms

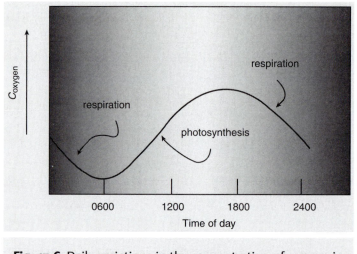

Fig. 15.6 Daily variations in the concentration of oxygen in water containing algae. Actual values of concentration depend on algal population and temperature of the water, among other factors.

such as fish. Therefore BOD is one measure of natural water quality. Also for effluents, the BOD is measured in order to assess the extent to which they could promote anoxic conditions when released into a water body. Analysis of BOD is done by adding appropriate heterotrophic microorganisms to a diluted effluent or water sample, saturating it with air, incubating for 5 days, and then determining the remaining oxygen. The concentration of oxygen used up (mg L^{-1}) during the experiment is the BOD of the diluted sample. On this basis water quality may be described in terms of its BOD values as

- very clean: < 1 mg L^{-1} O$_2$
- fairly clean: 1–3
- doubtful purity: 3–5
- contaminated: > 5.

Many effluents have BOD values much greater than the maximum solubility (8 mg L^{-1}) of oxygen in water; for example, municipal sewage effluent having BOD of 50 mg L^{-1} means that this material diluted 1:10 uses up 5 mg L^{-1} of O$_2$ during the test.

Besides BOD, there are other parameters that are sometimes used to evaluate the organic matter content in a water body or effluent. The chemical oxygen demand (COD) measures substances that can be oxidized by a chemical oxidant, acidified potassium dichromate, when it is refluxed with a sample at the boiling temperature. In most cases, the major contributor to COD is diverse types of soluble and particulate organic matter and a simple representation of the oxidation is

$$3\{CH_2O\} + 16H_3O^+ \text{ (aq)} + 2Cr_2O_7^{2-} \text{ (aq)} \longrightarrow 4Cr^{3+} \text{ (aq)} + 3CO_2 + 27H_2O \qquad (15.18)$$

Suppose a 100.0 mL sample of effluent from a pulp and paper mill is taken for measurement of the COD. The sample is digested in an excess of acidified dichromate solution and by back titration, it is found that 4.64×10^{-4} mol of dichromate have been consumed in the chemical oxidation. According to reaction 15.18, this is equivalent to $3/2 \times 4.64 \times 10^{-4} = 6.96 \times 10^{-4}$ mol of $\{CH_2O\}$. Therefore the concentration of $\{CH_2O\}$ in the original solution is

$$6.96 \times 10^{-4}\,mol \times \frac{1000\,mL\,L^{-1}}{100\,mL} \times 1000\,mmol\,mol^{-1} = 6.96\,mmol\,L^{-1}$$

To convert this value into a measure of COD, we conceptually make use of eqn 15.9,

$$\{CH_2O\} + O_2 \; \rightarrow \; CO_2 + H_2O \tag{15.9}$$

which indicates the molar equivalence of $\{CH_2O\}$ and O_2. Therefore the potential for consumption of oxygen by this effluent is also $6.96\,mmol\,L^{-1}$, and the COD is

$$6.96\,mmol\,L^{-1} \times 32\,g\,mol^{-1} = 220\,mg\,O_2\,L^{-1}$$

A third measure of organic matter concentration in water is the total organic carbon (TOC). Usually TOC is determined using instruments that combust the sample producing carbon dioxide. The amount of carbon dioxide produced is measured using IR spectroscopy.

Continuing with our example, the COD test had measured that $6.96\,mmol\,L^{-1}\{CH_2O\}$ were present in the effluent. The carbon concentration of the sample is then the TOC and equals

$$6.96\,mmol\,L^{-1} \times 12\,g\,mol^{-1} = 84\,mg\,L^{-1} \text{ or } 84\,ppm$$

Further subdivisions of carbon in water have been made to include particulate organic carbon (POC) and dissolved organic carbon (DOC).

We must remember again the assumptions made in these calculations:

- organic matter is represented by the simple formula $\{CH_2O\}$;
- in comparing COD and TOC, we have neglected oxidation of inorganic components of water such as ammonia;
- all of the organic matter is oxidized in the chemical digestion.

Because these assumptions are never totally true, the equivalencies we have calculated will, at best, be approximate.

Since resistant compounds are oxidized under the harsh conditions of this experiment to a greater extent than in a BOD experiment, usually COD > BOD often by a factor of around two or more.

15.3 Microbiological processes—the nitrogen cycle

A second great cycle that affects life in all its varieties on Earth is the nitrogen cycle. We have considered the atmospheric forms of nitrogen and we saw that various nitrogen-containing compounds play a central role in tropospheric gas-phase chemistry. They are important constituents of precipitation, they affect global climate, and influence the stratospheric reactions that determine the ozone synthesis–decomposition balance.

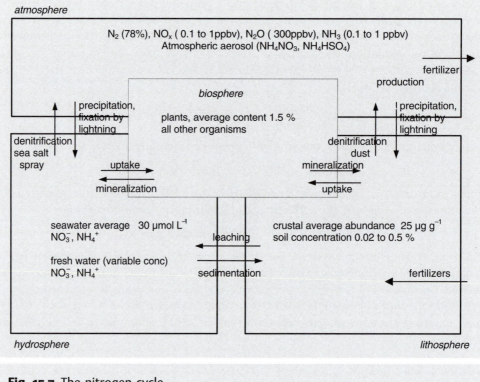

Fig. 15.7 The nitrogen cycle.

In the hydrosphere and on land, nitrogen in various forms is an essential nutrient for plants and animals. But excessive concentrations of inorganic species can contribute to eutrophication and are toxic to some organisms. Here, we will look at some of the interactions between species in the hydrosphere and lithosphere. Again, emphasis will be on chemical processes that are mediated by microorganisms.

A pE/pH diagram for aqueous inorganic nitrogen species is shown in Fig. 15.8.

Under aerobic conditions, nitrate is the stable species in both water and soil, but a low pE state leads to reduction from nitrate through nitrite to ammonia in its protonated and unprotonated forms. Because of the redox buffering effect described above, intermediate pE values are uncommon in water and nitrite is usually a transient species measured only in small concentrations.

The types of processes that interrelate species in the atmospheric/aqueous/terrestrial nitrogen cycle are shown in Fig. 15.7. A number of atmospheric nitrogen compounds whose environmental behaviour had been discussed in Chapter 4 and 8 are included in this diagram.

15.3.1 Nitrogen fixation

Nitrogen fixation reactions are those through which the gaseous dinitrogen molecule is converted into one of the 'fixed' forms of nitrogen associated with the aqueous or

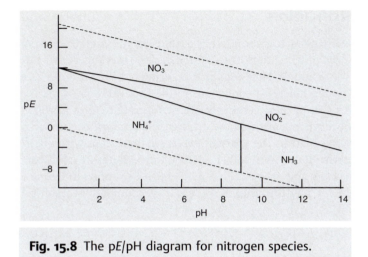

Fig. 15.8 The p*E*/pH diagram for nitrogen species.

terrestrial environments—particularly nitrate or ammonium ions. The conversion requires breaking the strong $N \equiv N$ bond (the bond energy for the $N \equiv N$ triple bond is 945 kJ mol^{-1}); therefore a large energy input is required.

Sources of energy in the atmosphere are lightning discharges, cosmic rays, and meteor trails. An estimated 5×10^6 t y^{-1} of nitrogen is fixed *via* these natural phenomena. Industrial production of ammonia using the well known Haber process accounts for another 6×10^7 t y^{-1} of nitrogen fixation:

$$N_2 + 3H_2 \rightarrow 2NH_3 \qquad (15.19)$$

In the Haber process, air is the nitrogen source and methane in natural gas is the source of hydrogen. The reaction is carried out at about 3×10^4 kPa pressure and 500 °C in the presence of a nickel catalyst. Much of the ammonia produced in this way is used in the manufacture of urea ($(NH_2)_2CO$), most of which is applied to soil as a fertilizer.

Biological nitrogen fixation that is mediated by terrestrial microorganisms can involve organisms like *Rhizobium* sp. that live in a symbiotic relation with nodules on the roots of particular species of plants such as legumes and alders. These organisms are capable of catalysing the conversion of atmospheric nitrogen into forms usable by plants. Likewise there are at least 15 species of non-symbiotic, free-living microorganisms such as *Spirillum lipoferum* that are found in particular environments and are able to fix as much as 100 kg ha^{-1} y^{-1} of nitrogen. In total it is estimated that 1.5×10^8 t y^{-1} nitrogen is biologically fixed in soil.

Biological fixation can also occur through the agency of marine microorganisms, including a variety of species of blue-green algae, *Azotobacter*, and *Clostridium*; the amount has been estimated to be 1×10^7 t y^{-1}.

In total, about 2.3×10^8 t y^{-1} of nitrogen is transferred from the atmosphere to the other compartments of the environment through fixation reactions; about 25% of this is a consequence of human activity.

15.3.2 Denitrification

The ultimate balancing factor leading to the return of nitrogen from land and water into the atmosphere is denitrification. The denitrification reactions occur most commonly in stagnant fresh water and in deep, organic-rich waters of the sea. They also occur to a significant extent under anaerobic conditions in the soil. There are several types of denitrification reactions. One of these is the reduction of nitrate to form nitrogen gas, a reaction we described above when showing how nitrate is able to act as electron acceptor for the oxidation of organic matter. The process involves several steps with nitrite ion and nitric oxide being intermediates. A number of facultative heterotrophic bacteria including species of *Pseudomonas* and *Achromobacter* mediate these processes. Because the organisms are heterotrophic, a plentiful supply of easily decomposable organic matter favours denitrification:

$$4NO_3^-\ (aq) + 5\{CH_2O\} + 4H_3O^+\ (aq) \ \rightarrow \ 2N_2 + 5CO_2 + 11H_2O \qquad (15.11)$$

A second gaseous product that is produced when denitrification occurs under conditions where small amounts of oxygen are present is nitrous oxide (reaction 15.20):

$$2NO_3^-\ (aq) + 2\{CH_2O\} + 2H_3O^+\ (aq) \ \rightarrow \ N_2O + 2CO_2 + 5H_2O \qquad (15.20)$$

The increases in atmospheric nitrous oxide mixing ratios which have been observed in recent decades are due, at least in part, to enhanced rates of denitrification resulting from greater rates of application of nitrogenous fertilizers worldwide.

A third type of denitrification leads to release of ammonium ion, which is then available to be used for the synthesis of cell protein in the microorganisms themselves. When the product is incorporated into the cell structure, the reaction is referred to as *assimilatory denitrification*. However, under alkaline conditions, the ammonia remains in a deprotonated form and is released to the atmosphere. Like the reactions to produce nitrogen and nitrous oxide gases, this is an example of *dissimilatory denitrification*. In the absence of dissolved oxygen, as in waterlogged soils, *Bacterium denitrificans* is one species that mediates the reaction to produce ammonium/ammonia.

$$NO_3^-\ (aq) + 2\{CH_2O\} + 2H_3O^+\ (aq) \ \rightarrow \ NH_4^+\ (aq) + 2CO_2 + 3H_2O \qquad (15.21)$$

15.3.3 Combustion

Fossil fuels and combustible biomass contain nitrogen in varying amounts depending on the nature of the fuel. When these materials are burned, nitrogen is returned to the atmosphere mostly in the form of NO_x compounds. Additional NO_x is generated even in the absence of nitrogen in the fuel, during combustion from combination of nitrogen and oxygen in the air. The extent of NO_x production by this latter reaction is related positively to combustion temperature.

15.3.4 Ammonification

Organic matter contains nitrogen in amounts varying from less than one to several per cent depending on the organism and the type of tissue. Most of the nitrogen is in the form of amino acids in protein. When organic matter decomposes in water and soil, the nitrogen is first released in a reduced form as ammonium ions or ammonia, depending on the ambient pH. The conversion of nitrogen from organic to inorganic forms is a type of mineralization called *ammonification*. Carbon–nitrogen bonds are relatively reactive and ammonification is therefore a rapid reaction:

$$2CH_3NHCOOH + 3O_2 + 2H_3O^+ \text{ (aq)} \rightarrow 2NH_4^+ \text{ (aq)} + 4CO_2 + 4H_2O \qquad (15.22)$$

15.3.5 Nitrification

Nitrification is the microbiological reaction sequence we noted at the beginning of the chapter. Ammonium ion, present in water or in soil as a result of ammonification or added in the form of an ammonium-containing fertilizer (ammonium sulfate, ammonium nitrate, urea, and ammonia itself), is subject to oxidation in an aerobic environment. The optimum environmental pH for nitrification is between 6.5 and 8, and the reaction rate decreases significantly when the pH falls below 6. The reaction takes place in two steps collectively called nitrification. (Note that there are several types of denitrification, only one of which (reaction 15.21) is essentially the reverse of nitrification.) Both steps (reactions 15.2 and 15.23) of the nitrification process are mediated by autotrophic bacteria:

$$2NH_4^+ \text{ (aq)} + 3O_2 + 2H_2O \xrightarrow{\text{Nitrosomonas}} 2NO_2^- \text{ (aq)} + 4H_3O^+ \text{ (aq)} \qquad (15.2)$$

$$2NO_2^- \text{ (aq)} + O_2 + H_2O \xrightarrow{\text{Nitrobacter}} 2NO_3^- \text{ (aq)} \qquad (15.23)$$

The overall reaction (which we have already seen) is

$$NH_4^+ \text{ (aq)} + 2O_2 + H_2O \rightarrow NO_3^- \text{ (aq)} + 2H_3O^+ \text{ (aq)} \qquad (15.1)$$

In another sense, the overall reaction is an oversimplification of what actually occurs. Along with oxidation of reduced nitrogen, some of the element is assimilated into the bacterial protoplasm to form cells with an empirical formula of approximately $C_5H_7O_2N$. Assimilation, however, accounts for only about 0.2% of the original nitrogen and therefore most of the nitrate produced is available for other purposes.

A consequence of the first step in nitrification is the concomitant release of hydronium ions, causing the local environment to be acidified. Suppose we are dealing with water containing alkalinity measured as 90 mg L^{-1} $CaCO_3$, and with pH 6.8. At this pH, all of the alkalinity may be assumed to be due exclusively to the HCO_3^- ion, and therefore its molar concentration would be 1.8×10^{-3} mol L^{-1}. If 6.0 mg L^{-1} (as N) of $NH_4^+ (= 4.3 \times 10^{-4}$ mol $L^{-1})$ in the water is nitrified, then 8.6×10^{-4} mol L^{-1} of hydronium ion is produced. Before nitrification occurs,

$$K_{a1} = \frac{[a_{H_3O^+}][HCO_3^-]}{[CO_2]} = \frac{10^{-6.8} \times 1.8 \times 10^{-3}}{[CO_2]} = 4.5 \times 10^{-7}$$

Under these conditions, the equilibrium solubility of carbon dioxide is 6.3×10^{-4} mol L^{-1}.

After nitrification, assuming a closed system, some of the hydrogen carbonate is converted to aqueous carbon dioxide. The final concentrations are then

$$[HCO_3^-] = (1.8 \times 10^{-3}) - (8.6 \times 10^{-4}) = 9 \times 10^{-4} \, mol \, L^{-1}$$

$$[CO_2] \, (aq) = (6.3 \times 10^{-4}) + (8.6 \times 10^{-4}) \, mol \, L^{-1} = 15 \times 10^{-4} \, mol \, L^{-1}$$

Under the new conditions, the pH of the water is calculated from

$$K_{a1} = \frac{[a_{H_3O^+}][HCO_3^-]}{[CO_2]} = \frac{10^{-pH} \times 9 \times 10^{-4}}{15 \times 10^{-4}} = 4.5 \times 10^{-7}$$

Therefore, the pH within this closed system has been reduced to 6.12. Taking into account carbon dioxide exchange with the atmosphere would alter the calculations.

15.3.6 Uptake (assimilation)

After carbon, oxygen, and hydrogen, nitrogen is quantitatively the most important element required by plants and microorganisms for growth in water and soil. For many agricultural crops, the concentration of nitrogen in plant tissue averages about 1.5% by mass. Both ammonium and nitrate serve as a nutrient source. Ammonium is the preferred form but nitrate is the stable and common species in well aerated environments. The first step of nitrate assimilation is ion exchange at the root or microbe surface and this is essentially an acid-neutralizing process. This is because the anion exchanged from the cell is usually an anion of a weak acid such as carbonate (reaction 15.24). Following release into the solution adjacent to the microbe or plant root cells, the carbonate is able to act as a proton acceptor (reaction 15.25):

$$Root \, CO_3^{2-} + 2NO_3^- \, (aq) \; \rightarrow \; CO_3^{2-} \, (aq) + Root \, (NO_3^-)_2 \qquad (15.24)$$

$$CO_3^{2-} \, (aq) + 2H_3O^+ \, (aq) \; \rightarrow \; CO_2 \, (aq) + 3H_2O \qquad (15.25)$$

In a closed system, then, acidity generated by nitrification is at least partially neutralized by assimilation. Therefore nitrate uptake is a means of biologically immobilizing nitrogen species and, at the same time, neutralizing acidity.

15.3.7 Abiotic ion exchange and adsorption

We have seen that most sediment and soil colloid surfaces are negatively charged, giving them the ability to act as cation exchangers. The positive ammonium ion can therefore be immobilized geochemically through retention by clays, organic matter, and other soil materials.

In contrast, nitrate, the more important inorganic nitrogen species under aerobic conditions, is an anion and does not interact with soil materials in this way. Furthermore, compared to phosphate and other potential ligands, nitrate usually is able to form only weak complexes with most metals. Therefore, there is little tendency to take part in specific adsorption reactions with components of the solid phase. For these reasons, nitrate is a geochemically mobile species.

15.3.8 Leaching

The net result of the biological and geochemical process is that ammonium ion is immobilized by both biological and geochemical processes, whereas nitrate mobility is only subject to biological control. Nitrate is therefore readily leached through soil into surface or groundwater under a number of particular environmental conditions:

- where there are few or no living plants, such as in a fallow field or a clear-cut forest;
- in situations where plants are dormant, as in winter time. Storage of high-nitrogen-content animal wastes outdoors in winter is a particularly serious problem. The nitrogen present in large concentration may readily be leached as nitrate from the biologically dormant manure, through soil and into water;
- in high intensity agriculture where nitrate fertilizer exceeding plant requirements is applied.

There are some negative environmental consequences that arise due to excess nitrogen being leached into a water body.

Being an essential nutrient for aquatic microorganisms and plants, excess nitrate may be a cause of eutrophication in water bodies. As we have noted, however, most often phosphorus, not nitrogen, is the limiting nutrient in lakes and ponds.

Where nitrate is present in water from acid precipitation or as a result of nitrification of reduced nitrogen compounds, the associated acidity may overwhelm the alkalinity of the water body. Figure 15.9 plots pH and alkalinity in the surface water of Little Turkey Lake, a small lake about 65 km north of Sault Ste Marie, Ontario. The data cover the period between 19 March and 17 May in 1982, a time that includes the Spring thaw in early April. During this period, there is an increased flux of acidified water that reduces the background alkalinity and the pH of lakes and rivers in the catchment area. Recovery is due to mixing of surface and less-affected deep waters, and to slow neutralization reactions.

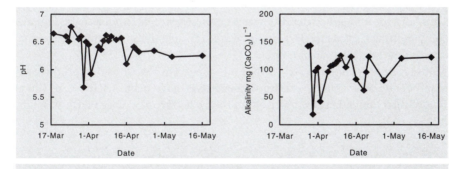

Fig 15.9 Changes in pH and alkalinity (μeq L^{-1}) at a depth of 1 m in Little Turkey Lake over the period between 19 March and 17 May 1982. Note the significant excursions to pH 5.5–6 and the greatly reduced alkalinity.
(From Kwain, W. and Kelso, J. R. M., Risk to salmonids of water quality in the Turkey Lakes Watershed as determined by bioassay. *Can. J. Fish. Aqat. Sci.*, **45** suppl. 1 (1998), 127–35.)

A third problem relates to the toxicity of nitrate to humans and other mammals. In drinking water, maximum allowable concentrations have been set, usually between 10 and 50 mg L^{-1}, by many jurisdictions. Actually nitrate itself is not toxic; rather, nitrite is a highly toxic species and is produced through reduction of nitrate by the bacteria *Escherichia coli* in the mammalian intestinal tract. The nitrite so produced reacts with haemoglobin, causing severe oxygen deprivation, especially in children. Alternatively, it may react with secondary amines and amides to form carcinogenic N-nitrosamines.

15.4 Microbiological processes—the sulfur cycle

Sulfur chemistry has a major influence on processes in all the compartments of the Earth's environment. In the atmosphere, oxidation reactions convert lower oxidation state species into sulfate, sulfate aerosols act as condensation nuclei for cloud formation, and sulfuric acid is one of the principal acidifying components in precipitation. In the hydrosphere and in soil, sulfur is present in many inorganic and organic forms exhibiting oxidation states from -2 to $+6$. The most important reduced and oxidized mineral forms of the element are sulfides, including pyrite (FeS_2), and sulfates, including gypsum ($CaSO_4 \cdot H_2O$). Minor amounts of sulfur are found as 'impurities' in many rocks.

Sulfate concentrations in the ocean are 28 mmol L^{-1}. The concentration in fresh water is much less (0.12 mmol L^{-1}), but it is none the less one of the principal ionic species in lakes and rivers. In reducing environments, species containing sulfur in the -2 state are formed. Like nitrogen, sulfur is an essential nutrient for microorganisms and plants, but the amounts required are approximately one order of magnitude smaller and it is rarely the nutrient that limits biological growth. Nevertheless, the element plays an important role in a number of aquatic and terrestrial environmental processes.

The pE/pH diagram for sulfur was constructed earlier (Fig. 10.5) and shows how the principal inorganic forms of this element in the hydrosphere are sulfate, under aerobic conditions, and protonated or deprotonated hydrogen sulfide, under acidic and alkaline anaerobic conditions, respectively. Elemental sulfur is usually a rare and transient species in aquatic and terrestrial environment.

The sulfur cycle is depicted in Fig. 15.10. Some aspects of reactions that produce gaseous species and of the atmospheric reactions themselves have been discussed in Chapter 5. We will now briefly consider three portions of the cycle that take place in water, sediments, and soils.

15.4.1 Sulfur release during organic matter decomposition

In organic matter, sulfur exists as both carbon-bonded and oxygen-bonded forms:

C-bonded	R-CH$_2$SH	hydrosulfide	in amino acids and other compounds
	R$_3$C-S-S-CR$_3$	disulfide	
O-bonded	$-\overset{\mid}{\underset{\mid}{S}}=O$	sulfoxide	
	R-OSO$_2$OH	sulfonic acid	

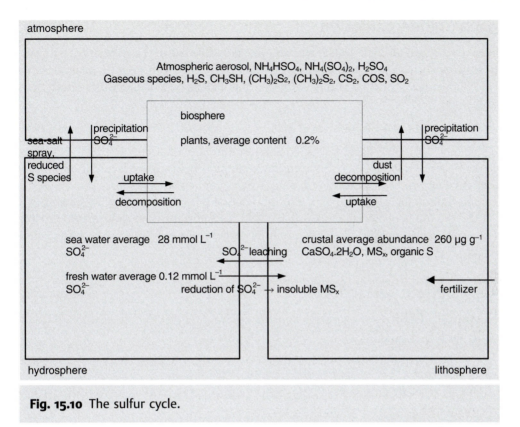

Fig. 15.10 The sulfur cycle.

When organic matter undergoes microbial decomposition, the sulfur-containing groups within the organic compounds are simultaneously transformed. The carbon-bonded organic sulfur is present as a component of protein in the form of the amino acid cysteine and it is mineralized by a process (not a redox reaction) as shown in reaction 15.26. The enzyme responsible for the reaction is associated with a variety of bacterial species.

$$\text{HS--CH}_2 - \overset{\overset{\displaystyle COOH}{|}}{\underset{\underset{\displaystyle NH_2}{|}}{CH}} + H_2O \xrightarrow[\text{desulfhydrase}]{\text{cysteine}} CH_3CO-COOH + HS^- \ (aq) + NH_4^+ \ (aq) \tag{15.26}$$

The pK_{a1} of H_2S is approximately 7. Therefore, in any environment whose pH is below neutrality, a large proportion of the sulfide is present as the undissociated molecule, giving rise to the strong odours characteristic of this gas. Hydrogen sulfide is also a toxic species.

For oxygen-bonded organic sulfur, microbial mineralization involves hydrolytic splitting of the S–O bond *via* sulfatase enzymes:

$$ROSO_2O^- \ (aq) + 2H_2O \ \rightarrow \ ROH + SO_4^{2-} \ (aq) + H_3O^+ \ (aq) \tag{15.27}$$

Under environmental conditions sulfate is not protonated, as the pK_a value is 1.92.

15.4.2 Sulfide oxidation

Under aerobic conditions sulfide is unstable and is easily oxidized *via* a variety of pathways. The sulfide may have been formed during decomposition of organic matter (reaction 15.26) or may have been present as sulfide minerals that had deposited at an earlier stage in the sediment or soil.

$$HS^- (aq) + 2O_2 + H_2O \rightarrow SO_4^{2-} (aq) + H_3O^+ (aq) \qquad (15.28)$$

The chemoautotrophic bacteria responsible for sulfide oxidation include the common species *Thiobacillus thiooxidans*. The bacteria are found in most oxygen-containing waters, sediments, and soils and multiply rapidly when supplied with a source of sulfide. The reaction simultaneously produces hydronium ions and is thus an acidifying process.

These and other related microbial reactions are important in marine coastal environments and within deposits of sulfur-containing wastes such as sulfide mine tailings. Some specific cases will be discussed later (pp. 401–2; pp. 415–17).

15.4.3 Sulfate reduction

In an organic-rich, reducing (low pE) aqueous environment, sulfate is readily reduced to species in the -2 or, less commonly, 0 oxidation states:

$$SO_4^{2-} (aq) + 2\{CH_2O\} + H_3O^+(aq) \rightarrow HS^- (aq) + 2CO_2 + 3H_2O \qquad (15.13)$$

$$2SO_4^{2-} (aq) + 3\{CH_2O\} + 4H_3O^+(aq) \rightarrow 2S^o + 3CO_2 + 9H_2O \qquad (15.29)$$

Reaction 15.13 is the one we considered above to describe the oxidation of organic matter. It leads either to gaseous emissions of hydrogen sulfide (at low pH) or to production of aqueous soluble and insoluble sulfides. Many marine, estuarine, and fresh water sediments as well as waterlogged soils contain sulfate-reducing bacteria. The most common of these is *Desulfovibrio desulfuricans* which is an obligate anaerobe that grows at pH values above 5.5. The sulfide species generated (H_2S or HS^-) are toxic to aquatic life; they may also react with metals present in the sediment/soil, such as iron (II), to produce insoluble sulfides that are components of marine shales.

Dimethyl sulfide can also be a product of reduction of sulfate in low pE environments. Production of this and other reduced sulfur gases is carried out by marine plankton and provides an important route for natural release of sulfur compounds into the atmosphere. Dimethyl sulfide is a particularly important reduced sulfur species and is a key link in the global sulfur cycle.

The main points

1 Microorganisms are small species of flora and fauna that populate all compartments of the environment. They are categorized in various ways, including by phylum, environmental preference, carbon source, electron acceptor type, morphology, or temperature requirements.

2 Many environmental chemical processes in water and on land are facilitated through the activities of microorganisms. Microorganisms play key roles in parts of the global carbon cycle, including synthesis and degradation of many biomolecules. They mediate processes within all the nutrient cycles—the nitrogen cycle being one of overriding importance.

3 Microorganism activities are not restricted to synthesis and degradation of organic compounds; they are involved in mediating transformations of many inorganic species of both non-metals and metals. Sulfur is also an element whose aquatic and terrestrial behaviour is regulated, in large part, by microbial processes.

Additional reading

1 Alexander, M., *Introduction to Soil Microbiology*, 2nd edn, John Wiley and Sons, New York; 1977.

2 Gaudy, A. F. Jr and E. T. Gaudy, *Microbiology for Environmental Scientists and Engineers*, McGraw-Hill, New York; 1980.

3 Rheinheimer, G., *Aquatic Microbiology*, 3rd edn, John Wiley and Sons, Chichester; 1985.

Problems

1 Given that Henry's law constant, K_H, for oxygen in water at 25 °C is $1.3 \times 10^{-8}\,\text{mol L}^{-1}\,\text{Pa}^{-1}$, explain why a biological oxygen demand (BOD) value of $> 5\,\text{mg L}^{-1}$ is indicative of contaminated water.

2 Consider dissolved organic matter to have the generic formula $\{CH_2O\}$. For a $1\,\text{mg L}^{-1}$ (as C) aqueous solution of organic matter, calculate the mg of dissolved oxygen in the same volume required to oxidize it completely. Use this calculation to establish a relation between COD and DOC. Repeat the calculation using the generic formula for dissolved humic material (Fig. 12.3). Assume reaction of only the carbon and hydrogen in the humic material.

3 For an organic-rich soil, under controlled conditions in the laboratory at temperatures maintained in the mesophilic ranges, 20 to 30 °C, the rate of carbon dioxide production commonly ranges from 5 to $50\,\text{mg CO}_2$ per kg soil per day. Using these data estimate the number of kg of carbon dioxide released from a 1 ha field under the same temperature conditions.

4 'Mineralization' refers to the process by which organic forms of an element are broken down and converted to inorganic species. In environmental situations this is most often a microbiological process. For nitrogen, indicate what forms of the element might be present in water or soil as reactants and products of the mineralization process.

5 Use the thermochemical tables to calculate $pE°$ values for the reductive half reactions involving nitrate (NO_3^- (aq)) in three different cases—when the product is NH_4^+ (aq), N_2O (g), and N_2 (g). Then calculate the $pE°$ (w) values for the same three reactions. What is the environmental significance of these results?

6 Evolution of ammonia is one process that can result in the transfer of nitrogen from an aqueous system to the atmosphere. Discuss environmental conditions that would favour such a process. Would you expect ammonia evolution to be significant (a) in an acid bog? (b) in the oceans?

7 For water with pH 6.7 and alkalinity of $110\,mg\,CaCO_3\,L^{-1}$, calculate the maximum concentration $(mg\,N\,L^{-1})$ of NH_4^+ that could be nitrified without the pH of the water falling below 6. Assume a closed system.

8 Why are reduced gaseous species of sulfur emitted from rice (paddy) fields, swamps, and the near shore borders of lakes, but not from open lakes—even though reduced sulfur compounds are found in the sediments in all these situations?

9 Would you expect gaseous sulfur emissions to be greater from the soil of a tropical savannah or of a tropical rainforest? What gaseous species are likely to be released?

10 Using data from the thermochemical tables, calculate the pE /pH equation for the SO_4^{2-}/ SO_3^{2-} boundary. Plot it on the sulfur diagram. What is the significance of this plot in terms of the stability of the sulfite species in the hydrosphere?

11 The pE of a groundwater sample is -1.2 and the pH is 8.83. The concentrations of SO_4^{2-} and HS^- are 2.29 and 0.003 mM respectively. Is the system at equilibrium?

Note

1 Stumm, W. and Morgan, J. J., *Aquatic Chemistry*, John Wiley and Sons, New York; 1981.

Water pollution and waste-water treatment chemistry

16

AT the outset of this book, we emphasized that our focus would be on the composition of the natural environment, the processes that take place within it, and the kinds of changes that come about as a result of human activities. With regard to many issues, we have discussed each of these aspects, including ways in which humans influence composition and processes in the environment. In these discussions, we have occasionally used the word 'pollution' but have not spent time to consider what that word really means. At this point, in the context of water chemistry, we have an appropriate place to consider this concept.

Pollution may be defined in a variety of ways. In terms of the hydrosphere, one extreme definition would be to say that unless water is 100% H_2O (i.e. unless it is pure in the literal chemical sense) it is, at least to some extent, polluted. Obviously such a definition is unrealistic and of limited utility. Every water sample in contact with the atmosphere contains dissolved gases including oxygen and carbon dioxide and every water sample in contact with sediment and rock contains other dissolved constituents such as species of silicon and calcium. In most cases we would not think of such waters as being polluted.

As an indication of concentrations of dissolved species in clean natural water we may examine analyses of supposedly pristine waters. For example, in the copious literature on marine chemistry we can find data on the composition of 'open ocean' water. The term 'open ocean' implies that the water has not been subject to significant additional human-derived inputs of chemicals and represents the true composition of 'natural' sea water. Of course, in the limit, no such water exists. Table 16.1 reports analytical results for some trace elements in one well characterized open ocean sample in the North Atlantic south-east of Bermuda.

As for sea water, there are also data available for 'pure' fresh waters from various locations, often places in far northern or southern latitudes or from glacial sources at high altitude—sources relatively remote from human habitations and activity. Table 16.2 reports data for some anions and cations in snow samples from Terra Nova Bay in Antarctica.

Table 16.1 Composition of some minor and trace elements in a sample of open ocean sea water

Element	Concentration / μg L^{-1} $\pm$ 95% confidence limit
Arsenic	1.65 ± 0.19
Cadmium	0.029 ± 0.004
Chromium	0.175 ± 0.010
Cobalt	0.004 ± 0.001
Copper	0.109 ± 0.011
Iron	0.224 ± 0.034
Lead	0.039 ± 0.006
Manganese	0.022 ± 0.007
Molybdenum	11.5 ± 1.9
Nickel	0.257 ± 0.027
Selenium (IV)	0.024 ± 0.004
Uranium	3.00 ± 0.15
Zinc	0.178 ± 0.025

Sample is NRC Standard Reference Material NASS-2 from a depth of 1300 m at a location in the North Atlantic Ocean, SE of Bermuda. The relatively large concentration for molybdenum is not unusual, but is typical of values in oceans throughout the world.

Table 16.2 Concentration of some anions and cations in Antarctic snow

	Concentration / μg L^{-1}	
	Minimum	Maximum
Chloride	25	40100
Nitrate	8.6	354
Sulfate	10.6	4020
Bromide	0.8	49.4
Phosphate	1.8	49
Fluoride	0.1	0.2
Sodium	15	17050
Potassium	3.1	740
Magnesium	2.7	1450
Calcium	12.6	1010
Ammonium	2.4	46.5

From Udisti, R., S. Bellandi, and G. Piccardi, Analysis of snow from Antarctica: a critical approach to ion chromatography methods, *Fresenius J. Analyt. Chem.*, **349** (1994), 289–93.

Clearly, in the natural environment there is no such thing as pure water in the chemically rigorous definition.

Just as total purity of water is impossible, so also is complete removal of all potential contaminants in a water supply or waste stream. While 'zero discharge' is a commendable goal, it is an ultimately unattainable one.

A second possibility is to define water pollution as any concentration of chemical (or microorganism) in water above the 'natural' level—in other words, added concentrations due to anthropogenic inputs. Compared to our first definition, this one is somewhat more realistic, yet it too presents difficulties. For one thing, it is often impossible to distinguish between human and other environmental factors. Furthermore, certain human inputs added deliberately or inadvertently may result in enhanced water quality. Adding calcium carbonate to an acid lake will certainly increase the calcium ion concentration above its previous 'natural' level but at the same time, it should improve the quality of water in terms of many environmental criteria.

This brings us to a third possible definition of water pollution that is more utilitarian. One version of the definition is that a pollutant[1] is 'a substance or effect which adversely alters the environment by changing the growth rate of species, interferes with the food chain, is toxic, or interferes with health, comfort, amenities, or property values of people'. The definition implies a need to set standards or guidelines in order to indicate that water, whose chemical properties exceed the limits of the standards, may cause a particular environmental alteration or interference.

While we will make use of this last definition, we should recognize that it too has limitations. In order to define criteria, information is required concerning toxicity and other factors. Experimental work is carried out to establish standards and guidelines but there are always assumptions and incomplete data, and conclusions are partly subjective and often treat risk on a statistical basis.

Amongst the criteria, guidelines have been established for the following:

- physical properties of temperature, colour, odour, and turbidity;
- for general chemical classes of chemical properties such as pH, total dissolved solids (TSS), salinity, hardness, biological oxygen demand (BOD), detergents, and petroleum residues;
- for specific elements, complex ions, and organic compounds; for radiological properties, that is levels of radioactivity due to particular isotopes; and
- for microbiological properties, that is, counts of specific organisms and groups of organisms.

Clearly, decisions regarding the particular criteria will depend on the end use of the water. For drinking water, rather stringent requirements will be in order, involving requirements related to many of the classes above, and including properties related to aesthetics such as colour, odour, and taste. For industrial and irrigation purposes, quality demands may be less severe. The most important properties of water to be used for irrigation are the total concentration of soluble salts, the molar ratio of sodium to calcium and magnesium in the water, the concentration of potentially toxic elements especially boron, and the carbonate species concentration.

Box 16.1 details some properties set out for irrigation waters used in India. (These standards were, in part, based on ones established in the United States.) The criteria involve four major inorganic chemical parameters, but there are also guidelines for nature and type of pathogens and potentially toxic chemicals.

Box 16.1 Irrigation water standards for India

Total soluble salts (salinity) are assessed by conductivity:

Conductivity / dS m^{-1} (25 °C)	Quality
< 0.25	Excellent
0.25–0.75	Good
0.75–2.25	Doubtful
> 2.25	Unsuitable

The 'sodium hazard' is defined in terms of the sodium adsorption ratio $C_{Na}/(C_{Ca}+C_{Mg})$, and is a measure of the potential of the water to saturate the exchange sites of the soil with sodium:

Sodium adsorption ratio (SAR)	Quality
< 10	Excellent
10–18	Good
18–26	Doubtful
> 26	Unsuitable

Boron is essential for normal crop growth, but is toxic when present in excessive concentrations. The sensitivity of crops is variable:

Boron concentration/mg L^{-1}			
Sensitive	Semi-tolerant	Tolerant	Quality
< 0.33	< 0.67	< 1.0	Excellent
0.33–0.67	0.67–1.3	1.0–2.0	Good
0.67–1.0	1.3–2.0	2.0–3.0	Doubtful
1.0–1.3	2.0–2.5	3.0–3.8	Unsuitable
> 13	> 2.5	> 3.8	Very toxic

Alkalinity of the water may contribute to harmful shifts in pH of the soil to the alkaline range:

Carbonate alkalinity / meq L^{-1}	Quality
< 1.25	Safe
1.25–2.5	Marginal
> 2.5	Unsuitable

Waste water that is recovered from domestic or industrial sources and is returned to the natural environment is subject to other requirements. In general, if the water is to be discharged into the hydrosphere, it should not contain dangerous levels of toxic chemicals or organisms, it should not supply excessive quantities of readily oxidizable (usually organic) compounds and it should not be a source of nutrients which would support microbial growth.

However, for waste water that is to be used for irrigation of productive lands, it is not necessary to remove nutrients and the benign dissolved and suspended organic matter. There is a wide range of types of industrial waste water and treatment protocols that depend on what, if any, contaminants have been introduced during the industrial processes.

Discussion of aspects of the chemistry of many potential or actual contaminants in water has been or will be presented in other parts of this book. In the present section we will examine principles of chemical methods used to treat waste water of various sorts before it is returned to the environment.

16.1 Community waste-water treatment chemistry

Community waste water (sewage) combines domestic wastes with industrial and other effluents. In most cases, the waste water, whether or not it undergoes treatment, is discharged into a natural water body such as a river, lake, or the ocean. Increasingly, industrial effluents are under strict regulation to limit or eliminate discharge of known contaminants to the public system. The major components of community waste water then derive from domestic and commercial sources and are made up of human wastes, solid and dissolved forms of food wastes, soaps and detergents, and soil residues. Typical measured properties of untreated sewage include the following:

Biological oxygen demand (BOD)	250 mg L^{-1}
Chemical oxygen demand (COD)	500 mg L^{-1}
Total solids (TS)	720 mg L^{-1}
Suspended solids (SS)	220 mg L^{-1}
Total phosphorus (TP)	8 mg L^{-1}
Total nitrogen (TN)	40 mg L^{-1}
pH	6.8

During the day there are fluctuations in both the flow and contaminant concentration with higher values being characteristic of the morning and evening and lowest values occurring at night.

Of the above properties, BOD, SS, and TP are of greatest concern as they may all, in different ways, upset the normal balance of aquatic life. The natural levels of dissolved oxygen in water are sufficient to oxidize small amounts of animal and vegetative wastes *via* aerobic microbial reactions. In the process, the organic wastes are converted into simple organic and inorganic compounds, and oxygen into carbon dioxide. The carbon dioxide, under the influence of light, takes part in photosynthesis and oxygen is returned to the water. This cycle of self-purification is broken by the presence of excessive amounts of degradable amounts organic matter (high BOD) producing anoxic conditions, by turbidity (high SS) that inhibits photosynthesis, and by unusually large concentrations of nutrients (most often, high TP) that stimulate plant and algal growth. For this reason, there may be government regulations to require that the effluent of a treatment facility meet specified criteria—such as BOD of $15 \, mL^{-1}$, SS of $15 \, mg \, L^{-1}$, and TP of $1 \, mg \, L^{-1}$. In order to accomplish this, a variety of treatment procedures have been developed and applied on a scale where up to $10\,000 \, m^3$ of water a day may be processed. Waste-water treatment plants are one of the most common of all chemical plants in the world. There are many processes and designs, but some general features are frequently incorporated in any system. There may be physical, chemical, and biological processes operating sequentially or simultaneously.

Waste-water treatment plants are often described as offering primary, secondary, and tertiary treatment. The primary processes are mainly physical ones. The waste stream is first subjected to pulverization to reduce large-sized solid material to smaller dimensions. It then passes through a screen and the water containing small particles flows into a clarifier where separation occurs by flotation of oil, grease, and foams and simultaneous sedimentation of heavier grit—sand, silt, and other solids. The settling process may be assisted by the addition of chemical coagulants such as alum. More will be said about this below.

The secondary processes are biological, where conditions are adjusted so that aerobic microorganisms are able to thrive. A widely used method of secondary treatment is known as the activated sludge process (ASP). To create an aerobic environment, oxygen is provided by pumping coarse or fine air bubbles through an aeration tank or lagoon, or by vigorously mixing the surface of the waste water in order to increase the area of contact with the atmosphere. A further requirement is that the population of growing microorganisms be augmented by recycling some of the previously activated biological mass. This biological material is called sewage sludge. Under these conditions, heterotrophic aerobic bacteria and protozoa grow and respire; in the process the organic matter of the raw sewage serves as a carbon source. The carbon is partially incorporated into the microbial biomass and the other part oxidized to produce carbon dioxide. At the same time a portion of the nutrients, phosphorus and nitrogen, are removed from solution as they too are taken up by the microbes. Metals that are present in waste water interact with cellular material by passive adsorption through complex formation with extracellular functional groups. Some additional portion of the metals is retained by physical entrapment, and some is incorporated in the cellular structures. Further removal of all these elements and of other contaminants

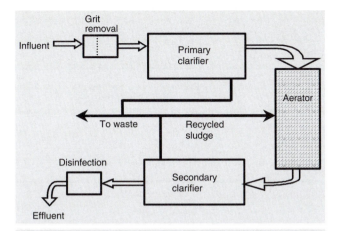

Fig. 16.1 Schematic diagram of an activated sludge waste-water treatment facility. The open arrows show the path of the waste-water stream, while the filled arrows follow the separated sludge.

is accomplished by use of coagulants. After an appropriate residence time—of the order of 4 to 12 hours—the mixed liquor from the aeration tank flows into the secondary clarifier where the biological floc is allowed to settle. Where a chemical coagulant, such as alum or ferric chloride, is added prior to the secondary clarifier, the chemical floc settles along with biological material. A schematic diagram of a typical ASP plant is shown in Fig. 16.1.

An alternative to the aeration tank for stimulating microbial growth makes use of a trickling filter whereby the waste stream is allowed to drain slowly over a large surface-area medium—again in order to provide efficient contact between air and water.

A variety of other processes, used individually or in series, make up a tertiary system. These include filtration through a microscreen or sand bed, precipitation after chemical additions, adsorption on to granular activated charcoal (GAC), ion exchange, reverse osmosis, and disinfection by chlorination or ozonation. Some of these technologies are especially important for the treatment of specific kinds of industrial waste water. The processes are chosen depending on the chemical nature of the waste water steam and the quality requirements of the effluent water.

Typical values indicating the effectiveness of various treatment types (not including the tertiary methods) are given in Table 16.3. The three chemical quality parameters—biological oxygen demand (BOD), suspended solids (SS), and total phosphorus (TP)—are those most commonly monitored and regulated in a community waste-water treatment facility.

16.1.1 Chemical coagulants

Chemical coagulant additions are frequently applied as a component of various levels of waste-water treatment. Coagulants serve two principal functions—to assist in the coagulation and flocculation processes, in order to maximize removal of very small solid particles

Table 16.3 Typical concentrations of waste-water contaminants during the waste-water treatment process (BOD=biological oxygen demand, SS = suspended solids, TP = total phosphorus

	Raw influent	Primary	Secondary (biological)	Chemically asssisted secondary
BOD / mg L^{-1}	250	175	15	10
SS / mg L^{-1}	220	60	15	10
TP / mg L^{-1}	8	7	6	0.1–1

of various compositions, and to react with and remove potential pollutant chemical species such as phosphorus in the water.

To understand how coagulants assist the removal of turbidity, we will once again make use of Stokes' law, this time to estimate settling rates in an aqueous medium. In this application, the Stokes–Cunningham slip correction factor is not required, and density and viscosity of water are used in place of those for air:

$$v_t = \frac{(\rho_p - \rho_a)g\,d_p^2}{18\mu} \tag{16.1}$$

where v_t = terminal velocity of particles / m s^{-1}; ρ_p = density of particle / kg m^{-3}; ρ_a = density of water = 1.0×10^3 kg m^{-3} at P^o and 20 °C; $g = 9.8$ m s^{-2}; d_p = particle diameter / m; and μ = viscosity of water = 1.0×10^{-3} kg m^{-1} s^{-1} at P^o and 20 °C.

If we apply the Stokes relation to a spherical particle of sand, 0.1 mm (10^{-4} m) in diameter, assuming particle density of 2.65×10^3 kg m^{-3}, then

$$v_t = \frac{(2650 - 1000)\,9.8\,(10^{-4})^2}{18 \times 10^{-3}}$$

$$= 9 \times 10^{-3}\,\text{m s}^{-1} = 9\,\text{mm s}^{-1}$$

Where the settling tank has a depth of 4 m, sand particles with these dimensions will settle in $4/(9 \times 10^{-3}) = 440$ s = about 7.4 min. A typical clarifier has been designed so that water takes an average of 1.5 h to pass through it (this is called the detention time[2]). Clearly, sand with particle diameters of 0.1 mm will have sufficient time to settle completely in a tank of the specified size.

We can repeat the calculation for a spherical particle of clay whose diameter is 0.002 mm:

$$v_t = \frac{(2650 - 1000)\,9.8\,(2 \times 10^{-6})^2}{18 \times 10^{-3}}$$

$$= 3.6 \times 10^{-6}\,\text{m s}^{-1}$$

Within 1.5 h, clay particles of this size would therefore have settled to a depth of only

$$3.6 \times 10^{-6} \, \mathrm{m \, s^{-1}} \times 1.5 \times 3600 \, \mathrm{s} = 0.02 \, \mathrm{m} = 2 \, \mathrm{cm}$$

This calculation then shows that, for a settling tank with a detention time of 1.5 h, the 0.002 mm diameter particles will not settle to the bottom before moving out in the waste stream.

Turbidity in waste water is due to a mixed collection of solids having a range of particle sizes, shapes, and densities. Much of the suspended material will, however, consist of colloidal material that settles too slowly to be removed within the detention time in the clarifier. One of the functions of coagulants is to assist and speed up the settling process.

Three coagulants are frequently employed in waste-water treatment. These are alum ($Al_2(SO_4)_3$), obtained as a concentrated solution containing about 5% aluminium or as a solid hydrate (with 14 or 18 waters of hydration), ferric chloride ($FeCl_3$, usually a byproduct of steel manufacturing and obtained in a solution form), and hydrated lime ($Ca(OH)_2$ in a solid form).

Alum and ferric chloride both are sources of trivalent metal cations and their behaviour in waste-water treatment is somewhat similar. We will consider the case of alum. When it is added to any type of water containing sufficient alkalinity, the hydrated aluminium ion undergoes a series of hydrolysis reactions as indicated in the following reactions.

$$Al(H_2O)_6^{3+}(aq) + H_2O \rightarrow Al(H_2O)_5(OH)^{2+}(aq) + H_3O^+(aq) \quad (16.2)$$

$$Al(H_2O)_5(OH)^{2+}(aq) + H_2O \rightarrow Al(H_2O)_4(OH)_2^+(aq) + H_3O^+(aq) \quad (16.3)$$

$$Al(H_2O)_4(OH)_2^+(aq) + H_2O \rightarrow Al(H_2O)_3(OH)_3(s) + H_3O^+(aq) \quad (16.4)$$

$$Al(H_2O)_3(OH)_3(s) + H_2O \rightarrow Al(H_2O)_2(OH)_4^-(aq) + H_3O^+(aq) \quad (16.5)$$

A diagram showing solubility of aluminium species vs. pH is given in Fig. 16.2.

The hydrolysis reactions involve successive deprotonation of the waters of hydration surrounding the central metal ion, and the extent of the reactions depends on solution conditions, particularly the availability of Brönsted bases to act as proton acceptors for the released hydronium ion. As we have seen earlier, a measure of the capacity of water to accept protons is the alkalinity and almost always, community waste water has a relatively high alkalinity–typically in the range from 100 to 300 mg L^{-1} as $CaCO_3$. The pH of such water is usually between 6.5 and 7.5. These properties of the waste water ensure that the equilibrium positions of reactions 16.2 to 16.5 are such that any added aluminium is largely converted to insoluble $Al(OH)_3$. Since the alkalinity is mostly due to hydrogen carbonate ion, the reaction of aluminium in waste water may be approximated by

$$Al(H_2O)_6^{3+}(aq) + 3HCO_3^-(aq) \rightarrow Al(OH)_3(s) + 3CO_2(g) + 6H_2O \quad (16.6)$$

The above chemical processes occur in waste water and play a role in removal of small particles that cause turbidity. The aluminium hydroxide precipitate takes the form of a flocculated material having a very large surface area. Depending on conditions, the precipitate has a pH_o value that is as high as 9. It therefore carries a net positive charge which

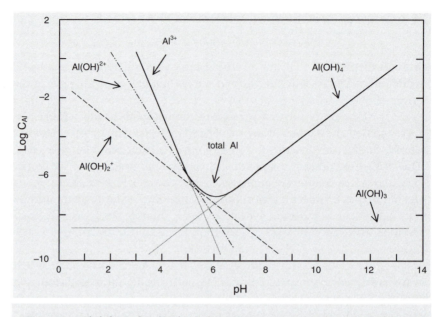

Fig. 16.2 Solubility of individual aluminium species and total aluminium solubility in water as a function of pH.

acts partially to neutralize the negative charges of the suspended matter—the clays, particulate organic matter, and bacteria—all of which are characterized by having much lower pH_o values. Neutralization of the double layer charge allows the colloids to approach each other and come together into larger aggregates that readily settle in the clarifier. A physical sweeping is also responsible for removal of turbidity, especially when a high concentration of coagulant is added to the waste stream. In this 'sweep floc process', the voluminous inorganic floc produced by reaction 16.6 physically entraps particles of clay and organic material as it settles, aiding in the clarification of the mixed liquor.

The second function served by the alum coagulant is for removal of phosphorus from the waste-water stream. Phosphorus is of particular concern because of its role in promoting eutrophication, as shown Chapter 14. In many jurisdictions there are strict limits set for maximum allowable phosphorus discharge; a $1 \, mg \, L^{-1}$ concentration in treated effluent is frequently specified. The biological processes described above serve to remove a portion of the phosphorus—typically about 20%—by incorporation within the insoluble biomass that is subsequently removed in the organic sludge. However, because the influent phosphorus concentration is of the order of 3 to $8 \, mg \, L^{-1}$, the treated water still contains phosphorus in excess of the allowed, $1 \, mg \, L^{-1}$ or lower, concentration. An advanced biological process, which may effect a much more efficient removal of phosphorus, will be discussed briefly later in this chapter.

As well as being an effective coagulant for removal of the turbidity associated with various colloids, alum is able to react with phosphate in order to precipitate out the highly insoluble aluminium phosphate. Usually the reaction is shown simply as

$$Al^{3+} \text{ (aq)} + PO_4^{3-} \text{ (aq)} \rightarrow AlPO_4 \text{ (s)} \tag{16.7}$$

While the reaction is described by the neatly balanced equation, what actually happens is almost certainly a much more complex process. When aluminium ion is added directly to a low alkalinity waste stream containing phosphorus, some direct reaction of the type shown here may occur. However, in the presence of the large excess of high-alkalinity water, aluminium hydrolyses sufficiently rapidly that it is already converted to an insoluble hydrous oxide form before it is able to react with phosphate. Nevertheless, the solid still is able to remove phosphorus by way of specific adsorption on the active surface of the freshly precipitated floc; a possible reaction involves displacement of hydroxide ions by the partially protonated phosphate species:

$$Al(OH)_3 + HPO_4^{2-} \text{ (aq)} + H_2O \rightarrow AlOH(HPO_4) \cdot H_2O \text{ (s)} + 2OH^- \text{ (aq)} \tag{16.8}$$

Following the surface reaction, phosphate slowly migrates to the interior parts of the aluminium colloid, leaving fresh surface to continue to adsorb more phosphate. As these processes continue, the gel-like hydrous aluminium oxide gradually converts more completely into aluminium hydroxyphosphate or aluminium phosphate, although complete conversion may not be effected during the lifetime of the precipitate in a sewage-treatment plant. That lifetime, incidentally, is not just the 4 to 12 hours residence time in the aerator and clarifier, but much longer—perhaps 10 days—during which time most of the sludge is repeatedly recycled into the aerator from the clarifier. A smaller proportion of the sludge, perhaps 20%, is 'wasted' and collected for further treatment (see section below). Recall that the principal purpose of the recycle is to supply to the fresh sewage a substantial quantity of microorganisms that will multiply and grow in the presence of the influx of nutrients in the influent. However, we can now see that a second useful function of the recycled sludge is that the precipitated alum is still capable of reacting with phosphorus by reaction 16.8, thus removing it from solution. In summary, the fresh alum added to a waste water stream reacts in part with soluble phosphorus during or immediately after its hydrolysis; the 'used' hydrolytic precipitate of alum also acts to remove phosphorus even in the absence of new additions of coagulant.

Calculations of the amount of alum required in any given situation are usually based on the stoichiometry of reaction 16.7, assuming that in the end, the precipitate contains a ratio of aluminium to phosphorous that is near 1 : 1. To ensure better nutrient removal, an excess of coagulant is often specified. Because of the large quantities of alum required to treat waste water from a major community, it is advantageous to devote some effort to ensure optimum efficiency in its use. This is achieved by thorough mixing at the time of addition to maximize contact between aluminium and phosphorus during precipitation and equally good mixing during the time that the precipitate is present along with fresh influent.

It is possible to use iron (III) chloride or calcium hydroxide as coagulants in place of alum. The iron (III) species behave in a manner similar to that described for aluminium (III)—hydrolysis, reaction with phosphate, simultaneous removal of suspended materials—but the best conditions for it to be used require a somewhat more acidic waste water medium. A disadvantage of iron (III) chloride is that its solution is acidic and oxidizing, which means that it can be highly corrosive to pumps and other metallic parts of the delivery system.

Calcium hydroxide requires that the sewage-mixed liquor pH be raised to at least 9.0 in order to ensure complete precipitation. Under these conditions, a discrete compound with formula $Ca_5OH(PO_4)_3$ has been identified and so the phosphorus removal reaction may be written as

$$5Ca(OH)_2 \text{ (aq)} + 3HPO_4^{2-} \text{ (aq)} \rightarrow Ca_5OH(PO_4)_3 \text{ (s)} + 6OH^- \text{ (aq)} + 3H_2O \qquad (16.9)$$

Because of the need to carry out precipitation in an alkaline solution, after settling and removal of the sludge but before discharge of the purified water, an additional pH adjustment is required to shift the pH downwards to a more acceptable value, usually between pH 6.5 and 7.5.

Being the limiting nutrient as far as fresh water eutrophication is concerned, phosphorus is usually the element of greatest concern in sewage treatment. However, nitrogen is also very important because of its contribution to eutrophication processes and because, in the ammonium ion form, it can react with and reduce the concentration of dissolved oxygen in receiving waters. Therefore there is increasing concern to develop methods to bring about its removal.

Unlike phosphorus, nitrogen has no easily produced insoluble forms and it therefore cannot be removed by a simple chemical precipitation procedure. When the nitrogen is present in the form of ammonium ion, there are relatively simple chemical procedures that can be employed for its removal. One method is to raise the pH of the water with calcium hydroxide (which precipitates out the phosphorus as described above) and then to pass it through a stripping tower where the water is aerated to strip out the gaseous ammonia:

$$NH_4^+ \text{ (aq)} + OH^- \text{ (aq)} \rightarrow NH_3 \text{ (g)} + H_2O \qquad (16.10)$$

An alternative approach that does not involve pH modification is to remove ammonium ions from the neutral solution by ion exchange using the natural ion exchanger clinoptilolite or synthetic exchangers.

More complete removal of ammonium nitrogen is effected by subsequent chlorination, to form mono-, di-, and trichloramines. Hypochlorous acid is the effective chlorinating species under treatment conditions.

$$NH_4^+ \text{ (aq)} + HOCl \text{ (aq)} \rightarrow NH_2Cl \text{ (aq)} + H_3O^+ \text{ (aq)} \qquad (16.11)$$

$$NH_2Cl \text{ (aq)} + HOCl \text{ (aq)} \rightarrow NHCl_2 \text{ (aq)} + H_2O \qquad (16.12)$$

$$HOCl \text{ (aq)} + NHCl_2 \text{ (aq)} \rightarrow NCl_3 \text{ (aq)} + H_2O \qquad (16.13)$$

In the presence of carbon adsorption filters the chloroamines undergo a heterogenous surface reaction that produces nitrogen gas as one of the products.

16.2 Biological processes for removal of phosphorus and nitrogen from waste water

Biological treatment processes for removal of both phosphorus and nitrogen have been suggested as alternative methods to the chemical treatment systems thus far proposed. The facility required for these processes must be designed specifically to remove one or both of the two nutrients. Of the proven biological phosphorus-removal technologies, a number include steps in which the waste water passes through a series of reactors where the sequence of environments goes from anaerobic to anoxic to aerobic (Fig. 16.3). In this context, an anaerobic environment is defined as one that contains no molecular oxygen (O_2), while an anoxic environment is devoid of oxygen in both free (O_2) and combined (NO_3^-, SO_4^{2-}, etc.) forms. The purpose of the succession of zones is to allow for the development (growth) of microorganisms with specific traits. Bacteria of the genus *Acinetobacter* are particularly known for their ability to assimilate phosphorus during their growth. The biological phosphorus-removal process (Fig. 16.3) begins with waste water moving into basins where anaerobic and anoxic conditions are maintained. In the basins, the normal population of facultative decomposers causes the release of acetate and other fermentation products from soluble organic matter in the waste stream, by way of fermentation reactions such as those described in the previous chapter. The fermentation products are substrates that are favoured by *Acinetobacter* and other phosphorus-storing organisms, and they stimulate growth of these species compared with other microorganisms in the general

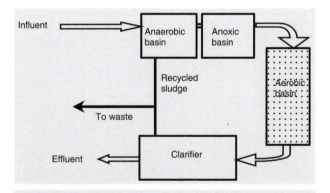

Fig. 16.3 Design for advanced biological phosphorus removal. Note the sequence of anaerobic, anoxic, and aerobic treatments before settling of the phosphorus-enriched sludge in the clarifier. Primary and tertiary treatments can be additional components to this system. The open arrows show the path of the waste-water stream, while the filled arrows follow the separated sludge.

population. Thus the anaerobic conditions provide an environment that results in population selection and development of the phosphorus-storing species. Without the oxygen-free phase, such organisms would be present in only very small amounts in the activated sludge. Upon entry into the aerobic zone (equivalent to an aerator in a conventional ASP), the selected microorganisms efficiently take up the soluble phosphorus from the waste-water stream. The phosphorus-enriched sludge is removed and treated in the usual way.

Nitrogen can be removed from waste water using a 'nitrification–denititrification process', in which the two types of nitrogen reactions occur in series. In the first step, the environment is aerobic, favouring microbial nitrification to convert aqueous ammonium ion to nitrate:

$$NH_4^+ \text{ (aq)} + 2O_2 \text{ (aq)} + H_2O \xrightarrow[\text{Nitrobacter}]{\text{Nitrosomonas}} NO_3^- \text{ (aq)} + 2H_3O^+ \text{ (aq)} \tag{16.14}$$

The second step requires anaerobic conditions. Denitrification occurs in the presence of denitrifying bacteria such as *Pseudomonas*, *Micrococcus*, *Serratia*, and *Achromobacter*. These bacteria mediate reduction processes in which sugars and carbohydrate and other organic compounds in the waste water serve as electron donors and the soluble nitrate is converted into nitrogen gas (reaction 16.15). In many cases, smaller amounts of ammonium ion and volatile ammonia are produced as by products:

$$4NO_3^- \text{ (aq)} + 5\{CH_2O\} \xrightarrow{\text{Denitrifying bacteria}} 2N_2 \text{ (g)} + 4HCO_3^- \text{ (aq)} + CO_2 \text{ (aq)} + 3H_2O \tag{16.15}$$

In reaction 16.15, the electron donor is represented by the generic organic form $\{CH_2O\}$. While waste water biomass can act as the reducing agent for denitrification, in practice methanol is often added as a supplement, because it ensures more rapid and complete reaction under anaerobic conditions.

A simple version of a biological treatment system based on the nitrification–denitrification process is shown in Fig. 16.4.

Although the two biological processes appear not to be compatible based on the sequence of environments required, it is possible that a combined biological/chemical system for sequential nitrogen and phosphorus removal can be achieved by recycle of the waste stream. For example, in the phosphorus-removal system shown in Fig. 16.3, recyling the effluent from the aerator into the anaerobic basin creates a system that allows for nitrification/denitrification to take place. However, removal of the two nutrients using these and other biological waste-water treatment plants requires very careful design and control of the operating conditions. Even in a biological system, it is frequently necessary to add alum (or another coagulant) at an appropriate location within the plant, in order to enhance phosphorus and solids removal.

One of the major drawbacks to the biological systems is the requirement for several large tanks where the various environments and processes are established. There are therefore substantial construction and operational costs as well as the cost of land for the complete facility.

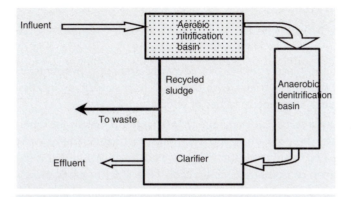

Fig. 16.4 A schematic diagram of a biological nitrification–denitrification process for nitrogen removal from waste water. Note that in some ways, this system reverses the processes of the biological phosphorus-removal system. The open arrows show the path of the waste-water stream, while the filled arrows follow the separated sludge.

16.3 Anaerobic digestion of sludge

Wherever biological processes are a part of a waste-water treatment system, a sludge is produced, and its fate must be considered. The sludge is a mixture of bacterial cells formed in the biological reaction tanks, undigested organic matter, and where chemical coagulants have been used, the inorganic floc. Small concentrations of metals and inorganic species are also present. When removed from the clarifier, the solids content is only 0.1% or less, the rest being water. An effective treatment system subjects the slurry to anaerobic digestion in an enclosed reactor. The complex microbial reactions that occur under these conditions may be summarized by the reaction we described for the fermentation process in the previous chapter:

$$2\{CH_2O\} \rightarrow CH_4 \text{ (g)} + CO_2 \text{ (g)} \tag{16.16}$$

In order to maintain its temperature near the ideal of 35 °C for mesophillic bacteria, part of the methane that is produced can be used as fuel to heat the reaction vessel. This ensures that the fermentation proceeds rapidly. In the early stages, the mixture is stirred, but as decomposition nears completion, stirring is stopped, allowing the solids to settle. During digestion, most of the pathogenic organisms are killed, and the product has little objectionable odour. The process causes the sludge to become more concentrated, so that it finally contains about 5% solids. It may be further dewatered on open drying beds or by using a large centrifuge. We will discuss some uses and problems associated with uses of this material in Chapter 19.

The main points

1 A pollutant is a substance or effect that adversely alters the environment by changing the growth rate of species, interferes with the food chain, is toxic, or interferes with health, comfort, amenities, or property values of people.

2 Pollution of natural water is largely due to discharges of wastes into the water body. In order to minimize this problem, waste water must be treated. The type of process required depends on the end use of the water following treatment. Water that is ultimately going to be used for human consumption will probably have strict regulations, while water which might only be used for agricultural or industrial use may have fewer, but different, restrictions.

3 Waste-water treatment systems are commonly referred to as primary, secondary, or tertiary, depending on the type of processes employed during the treatment.

4 Phosphorus and nitrogen are the two most important chemical constituents which need to be removed from urban waste water in order to lessen the impact of the effluent on the recieveing water body and the surrounding ecosystem. Chemical and biological processes for both phosphorus and nitrogen removal have been developed.

Additional reading

1 Barnes, D. and P. J. Bliss, *Biological Control of Nitrogen in Wastewater Treatment*, E & F.N. Spon, London; 1983.

2 Bowker, R. P. G. and H. D. Stensel, *Phosphorus Removal from Wastewater*, Noyes Data Corp., New Jersey; 1991.

3 Vesilind, P. A., J. J. Pierce, and R. Weiner, *Environmental Engineering*, Butterworth, Boston; 1988. pp. 545.

4 Welch E. B., *Ecological Effects of Wastewater, Applied Limnology and Pollutant Effects*, 2nd edn, Chapman and Hall, London; 1992.

Problems

1 Estimate the total organic carbon (TOC) concentration of waste water whose chemical oxygen demand (COD) is $500 \, mg \, L^{-1}$ (O_2). What fraction of the total (dissolved and particulate) solids content of $720 \, mg \, L^{-1}$ is then made up of organic material? Assume the organic fraction can be represented as [CH_2O]. Of what might the remaining solids consist?

2 An influent waste-water stream contains $330 \, mg \, L^{-1}$ organic matter (both suspended and soluble) and $27 \, mg \, L^{-1}$ ammonium ion (as N). Calculate the total BOD. What assumptions is it necessary to make?

3 A waste water contains $7.2 \, mg \, L^{-1}$ phosphorus and is treated with $15 \, mg \, L^{-1}$ aluminium in the form of an alum solution. Assume that phosphorus is precipitated out by reaction 16.7 and that any excess added aluminium forms $Al(OH)_3$. Calculate the mass of inorganic sludge produced in one day in a plant treating $20\,000 \, m^3$ of waste water.

4 After waste-water treatment in the activated sludge process, nitrogen is mainly in the form of ammonia and ammonium ion. Plot the fraction of nitrogen that is in the ammonia form (and therefore strippable by air purging) as a function of pH (at 25 °C) over the pH range from 6 to 10.

5 A settling tank treating $5.5 \times 10^6 \, L$ of water per day has dimensions as follows: length, $12.2 \, m$; width, $7.0 \, m$; depth, $3.5 \, m$. Calculate the detention time for water in the tank. Calculate the minimum particle size (expressed as diameter of spheres) that could settle in the tank.

6 Use reaction 16.9 to calculate the daily volume of calcium hydroxide solution of concentration $6.4 \, g$ calcium L^{-1} required to treat pH-adjusted waste water containing $6.1 \, mg \, L^{-1}$ phosphorus. The treatment plant processes $27\,000 \, m^3$ of water each day. Assume a safety factor (fractional excess) of 2.

7 Nitrification of ammonium ion is one of the steps during biological nitrogen removal processes. In waste water, whose pH and alkalinity are 7.2 and $156 \, mg \, L^{-1}$ (as $CaCO_3$), a concentration of $7.8 \, mg \, L^{-1}$ (as N) ammonium ion is present before the process begins. Calculate the pH and alkalinity after nitrification has gone to completion, assuming this to be the only reaction that affects the pH.

8 A waste-water treatment plant produces sludge containing $1800 \, kg$ of dry organic solids each day. Assuming the generic formula $\{CH_2O\}$ for the solids and complete anaerobic digestion by reaction 16.16, calculate the fuel value of the generated methane in joules, barrels of oil, and kilowatt hours.

Notes

1 The words pollutant and contaminant are used interchangeably in this book. Some persons make a distinction by defining a contaminant as any substance present in the environments above a 'natural' level whether or not it causes a detrimental effect, whereas a pollutant is a contaminant that causes adverse effects. (Porteous, A., Dictionary of Environmental Science and Technology, John Wiley & Sons, Chichester; 1992.)

2 The detention time is the time that waste water spends in a particular part of the plant. For any tank, it is readily calculated as follows:

$$\text{detention time (time units)} = \text{capacity (volume units)} /$$
$$\text{flow rate (volume units / time units)}.$$

The terrestrial environment

How little I know of this world
Deeds of men, cities, rivers,
Mountains, arid wastes,
Unknown creatures, unacquainted trees!
The great Earth teems
And I know merely a niche.

Rabindranath Tagore, 1913

17

The terrestrial environment

THE terrestrial environment is composed of rocks and soil and the living matter associated with these. Rocks and soil together are referred to as the lithosphere, and it is this compartment of the environment which is of concern to us in the present section of the book. The land area makes up 29% of the total area of the Earth's surface; this area is subdivided into the categories shown in Table 17.1.

The reactivity of solid materials that make up the lithosphere is, in large part, dependent on the particle size of these materials. Soils—finely divided materials with a relatively large surface area exposed to air or water—react much more readily with environmental agents than do massive rocks. This is one reason why, in the short term, chemistry of the terrestrial environment must be concerned mostly with soils. Furthermore, soils (defined in a broad sense) cover approximately 80% of the Earth's land mass and the next largest fraction is made up of snow and ice in Arctic and Antarctic regions. Exposed rocks make up only about 5% of the terrestrial surface.

Over the long period of the history of science, there have been many reasons for studying soils, but two stand out as being of direct and practical importance to humans and other

Table 17.1 The Earth's terrestrial environment		
	Area/10^6km^2	Area/%
Total land area	148	100
Ice-covered land	17.2	12
Arable land	14.8	10
Pasture and meadow	31.5	21
Forest	40.9	28
Other	43.6	29

Most of the data are from the *FAO Production Yearbook*, Vol. 39, Food and Agriculture Organization, Rome; 1986.
The 'Other' category includes mountainous land, deserts, and some land which is potentially available for pasture or direct food production.

living species. One of these is that soils are the principal plant growth medium and form the basis of agriculture and forestry. Especially over the past century, the subject of soil science has developed around these interests. Soil chemists have been particularly concerned with nutrient cycles and the relations between elements in the soil and their uptake by plants. They are also interested in other agronomic factors such as the interconnections between composition of soil materials, particle size, soil texture, and the resultant soil physical properties. Soil science of this type—related to plant production—is a highly developed science with a body of knowledge that has increased in volume and sophistication over more than a century.

Another reason for studying soils is more recent and relates to the fact that soils play a major role as an environmental agent. Key links in the global carbon, nitrogen, phosphorus, and sulfur cycles, as well as in many others, involve soil-chemical processes. Organic matter decomposition, nitrification, denitrification, phosphorus fixation, and sulfide oxidation are just a few of these processes. There are two broad environmental implications related to such reactions. On the one hand, soil-chemical processes affect the nature and amount of elements that are released to the hydrosphere and atmosphere. In turn, soils are the locus of inputs from other compartments of the environment and are themselves affected by processes there. For example, rainfall chemical composition is altered when rain percolates through the soil, perhaps draining into rivers and lakes, or maybe reaching the water table and becoming part of the groundwater reservoir. Through the interactions, soil properties are also altered by their encounter with the rain.

Many soil reactions involved in the global element cycles have been occurring over long periods of geological time, although human activities in recent years have perturbed some of them to a significant extent. There are other specific types of chemical reactions that only recently have come to be played out in the soil environment. A good example is related to the application of pesticides in agriculture. Pesticides are used to control insects, weeds, or pathogenic microorganisms on growing crops. These chemicals degrade over time and their movement and rate of degradation are, in part, determined through their interactions with the soil. Another example is the disposal of waste materials—municipal garbage, mine tailings, sewage sludge, sometimes even known toxic materials—in the soil environment. In other words, soils are an important environmental agent and a study of the environmental properties of soils is important along with studies of their agronomic properties.

17.1 Soil formation

17.1.1 Soil mineral matter

A starting point for examining properties of soils is to consider the natural processes by which they are formed from exposed rocks on the Earth's surface. These processes have been going on throughout the history of the Earth and continue to operate in the present time.

Table 17.2 lists elements in order of abundance in the Earth's crust—the crust being defined as the approximately 32 km thick layer on the surface of the planet. Exposed soil

Table 17.2 Percentage abundance of elements in the Earth's crust (note that Earth Scientists often report elemental analyses in terms of oxides)

Element	Abundance %	Oxide	Abundance %
O	47.4		
Si	27.7	SiO_2	58.2
Al	8.20	Al_2O_3	15.4
Fe	4.10	Fe_2O_3	7.2
Ca	4.1	CaO	5.1
Na	2.83	Na_2O	3.8
K	2.59	K_2O	3.1
Mg	2.30	MgO	3.5

Besides silicon, the only non-metal present in significant concentrations in many rocks is oxygen. Therefore the sum of concentrations of all metal oxides should be close to 100%.

material is a thin surface layer covering part of the crust; the great mass of crustal material consists of the underlying igneous and metamorphic rocks.

The complex processes by which crustal rocks are transformed into soils are collectively known as weathering and we can consider these processes under two general headings.

17.1.2 Physical weathering

Physical weathering results in the breakdown of massive rock materials into smaller aggregates, eventually fine enough and sufficiently well developed that they are considered to be soils.[1] This is brought about in several ways.

- Freeze–thaw in temperate regions leads to enhancement of fracturing along the natural fracture planes of rocks. Further enhancement occurs when water enters into cracks and expands (by 9%) upon freezing.

- Fire causes expansion of rocks, but due to their low thermal conductivity the surface expands much more rapidly than the rock below. This causes strains that are released by fracturing. The same effect occurs to a lesser extent due to daily temperature variations.

- Deposition of salts in already fractured rocks may occur. If these salts have a thermal expansion coefficient higher than the surrounding rock, diurnal and seasonal temperature changes can cause pressure-induced breakage. Similarly some minerals deposited in cracks, especially clays like montmorillonite, undergo major expansion on hydration. This can also result in cleavage of massive material.

- Once finely divided material has formed on a rock surface it is subject to transport by wind, water, or ice. The pressure release resulting from erosion causes expansion in the vertical plane at right angles to the normal horizontal fracture planes, and this results in further cleavage.

- Abrasion due to erosion by wind or water, but particularly under the influence of glaciation, further subdivides rocks into smaller aggregates.
- Where plants are growing, penetration of roots into crevices generates pressure sufficient to cause breakage.

The net effect of these and other processes is that massive rocks are broken down into material with smaller particle size and correspondingly larger specific surface area. We have seen that particle size is very important in terms of chemical behaviour of solids in any environment.

17.1.3 Chemical weathering

Simultaneously with the above and other *physical processes*, a wide range of *chemical reactions* also takes place. Some chemical reactions are associated with the activity of micro- and macroorganisms, while others are purely abiotic.

Hydrolysis is a general term for processes in which water is an essential reactant. Various types of hydrolysis reactions play a major role in the weathering of rocks and minerals. A typical example related to a specific igneous mineral, orthoclase feldspar, shows the production of the secondary clay mineral kaolinite:

$$2KAlSi_3O_8(s) + 2H_3O^+ (aq) + 7H_2O \rightarrow Al_2Si_2O_5(OH)_4 (s) + 4H_4SiO_4 (aq) + 2K^+ (aq)$$
$$\text{orthoclase} \qquad\qquad\qquad \text{kaolinite} \qquad \text{silicic acid}$$

$$(17.1)$$

In the reaction, silicon is released as silicic acid from the soil at the same time as the clay mineral is produced. Further hydrolysis causes additional desilication, with the end result being the production of gibbsite:

$$Al_2Si_2O_5(OH)_4 (s) + 5H_2O (aq) \rightarrow 2Al(OH)_3 + 2H_4SiO_4 (aq) \qquad (17.2)$$
$$\text{kaolinite} \qquad\qquad\qquad \text{gibbsite} \quad \text{silicic acid}$$

This particular weathering sequence is especially important in the humid tropics due to the abundant rainfall and elevated temperatures. At pH values between 2 and 9, silicic acid ($pK_{a1} = 9.7$) remains in the fully protonated form and has an aqueous solubility of approximately 150 mg L^{-1}. Its prolonged solubilization and the accompanying mineral transformations lead to formation of red soils depleted in silica but rich in kaolinite and hydrated aluminium (and also iron) oxides. Depending on their specific properties, such soils are referred to as Laterites, Latosols, or Oxisols.

We can be more general, and consider primary aluminosilicate minerals as a group. They are weathered to produce any one or more of the clay minerals as summarized in the reaction below:

$$\text{Aluminosilicate (s)} + H_3O^+ (aq) + H_2O \rightarrow \text{clay mineral (s)} + H_4SiO_4 (aq) + \text{cation (aq)}$$
$$(17.3)$$

Both the specific and the general reaction show that water and hydronium ions are the agents of weathering. There are several natural sources of hydronium ion, including carbon

Table 17.3   Abrasion pH of minerals

Mineral	pH
Olivine	10–11
Augite	10
Oligoclase	9
Orthoclase	8
Quartz	6–7
Kaolinite	4–7

dioxide released in the soil by microbial respiration, and low molar mass acids that are decomposition products resulting from the decay of organic matter in the soil. In some situations, the natural sources are augmented by anthropogenically derived acids, most especially nitric acid from fertilizers, and sulfuric and nitric acids that are present in rainfall in certain regions of the Earth.

Because hydrolysis involves uptake of hydronium ion and release of alkali and alkaline earth cations by the mineral, it is not surprising that the pH (called the abrasion pH) of a slurry of secondary weathering products tends to be somewhat lower than that measured on the corresponding finely divided primary minerals (Table 17.3). For the unweathered primary minerals, pH is largely controlled by the small solubility of the metal cations in the ambient solution. During weathering, these cations are removed and the slightly acidic properties of the clay mineral products tend to control the acid–base properties of the slurry.

In many cases, chelation contributes substantially to chemical weathering. Iron and aluminium are two major elements whose solubility in uncomplexed forms is extremely small (see the example in Chapter 13). Yet evidence of substantial dissolution of these elements has been observed in many soils. This has been shown to be due to the formation of soluble organic complexes. For example, in forested areas of the temperate zone, translocation of iron and aluminium from upper to lower layers of the soil is associated with solubilization by chelation with ligands derived from soil organic matter. Both calculations and measurements on soil solutions show that more than 90% of the soluble iron and aluminium is present in the form of organic complexes. The ligands include inorganic species, but more important are ones derived from biological sources often associated with decomposition of dead plant and microbial material. The anions of acids like citric acid complex strongly with iron and aluminium and, in addition, are sources of hydronium ions. Humic and fulvic acid are also important multidentate ligands capable of complexing with metals and accelerating chemical weathering.

Consider a rock exposed to the ambient atmosphere. Table 17.4 lists concentration ratios calculated as the concentration of particular elements in the altered material on the weathered rock surface divided by the concentration of the same element in a freshly cut interior portion. Where these ratios differ greatly from one, it indicates a major

Table 17.4 Concentration ratios for Hawaiian basalt that had erupted in 1907 (exposed rock was free of any lichen, while part of the same deposit was covered with the lichen *Stereocaulon volcani*)

		Exposed rock	Lichen-covered rock
Weather rind thickness/mm		<0.002	0.142
Mass concentration ratio, weathered/fresh	Fe	1.21	6.36
	Al	0.47	0.58
	Si	1.20	0.21
	Ti	0.97	0.27
	Ca	1.24	0.004

From Jackson, T. A. and W. D. Keller, A comparative study of the role of lichens and 'inorganic' processes in the chemical weathering of recent Hawaiian lava flows, *Amer. J. Sci.*, **269** (1970), 446–66.

accumulation (large ratio) or loss (small ratio) of the element during weathering. For bare, exposed rock the weathering occurs to minimal thickness, and the chemical changes are relatively small—due largely to hydrolysis as indicated above. For rock covered with lichen, weathering occurs to a much greater depth and there is loss of most of the calcium and an accumulation of iron in the weathered layer. Lichens are microorganisms that have both algal and fungal components. The fungus attaches to the rock or soil surface and extracts nutrients that are used by the algal component. The algae are then able to carry out photosynthesis and produce carbohydrates and other organic molecules, some of which have chelating properties. The data also show that the chemical changes brought about by hydrolysis reactions and chelating reactions may be quite different in nature as well as in quantity.

Oxidation/reduction is a third important chemical weathering process. Oxidation occurs when primary minerals containing oxidizable elements in low oxidation states are exposed to the atmosphere. The resulting increase in oxidation state disturbs the charge balance of the mineral, and loss or gain of other elements in the compound may occur to maintain neutrality. Oxidation of iron in the mica biotite produces the 2 : 1 layer clay mineral vermiculite. Accompanying the oxidation of iron (II) to iron (III), potassium is lost. An idealized conversion is shown:

$$K_2(Mg, Fe(II))_6(AlSi_3O_{10})_2(OH)_4 \rightarrow Mg_{0.84}(Mg_{5.05}, Fe(III)_{0.9})(Si_{2.74}Al_{1.26}O_{10})_2(OH)_4 \quad (17.4)$$
$$\text{biotite} \qquad\qquad\qquad\qquad \text{vermiculite}$$

Other iron-containing minerals provide us with a further example. The redox behaviour of hydrous iron oxide is responsible for many changes in mineral chemistry. Under oxidizing conditions the stable form of iron oxide is Fe_2O_3 in the form of haematite or FeOOH, hydrated forms known as goethite or limonite. These minerals are highly insoluble, but under reducing conditions they may dissolve as iron (II) species, only to be redeposited in the same or some other location under an oxidizing regime. When considering pE/pH diagrams, we observed that this behaviour is possible.

In most discussions about redox chemistry in soils, the focus is centred on iron because of its abundance (Table 17.2) in the Earth's crust. However, the chemistry of many other elements (e.g. manganese, arsenic, chromium) has important redox components too. Even if a particular element is not directly subject to oxidation or reduction, its environmental behaviour can be affected indirectly by changes in the form of a major element such as iron. When the iron is present as an amorphous form of hydrous iron (III) oxide, other metals as well as non-metals are adsorbed or coprecipitated as impurities in the solid. If conditions change, and the iron is reduced to a soluble form, the coprecipitated elements are released and can simultaneously go into solution.

Hydration reactions contribute to physical weathering as shown above but also bring about chemical alteration of certain minerals. The hydration reactions between haematite and goethite and between gypsum and anhydrite are cases in point.

$$\underset{\text{haematite}}{Fe_2O_3} + H_2O \rightarrow \underset{\text{goethite}}{2FeOOH} \tag{17.5}$$

$$\underset{\text{anhydrite}}{CaSO_4} + 2H_2O \rightarrow \underset{\text{gypsum}}{CaSO_4 \cdot 2H_2O} \tag{17.6}$$

Ion-exchange reactions alter the nature of 'available' elements at the surface exchange sites of soil colloids. Actual structural changes can also be associated with ion exchange. These are particularly important with respect to clay minerals, where replacement of one interlayer ion by another alters the interlayer spacing and therefore the chemical and physical properties of the clay. The 2 : 1 clay mineral illite is very similar to montmorillonite, except that much of the isomorphous substitution is due to Al^{3+} replacing Si^{4+} in the tetrahedral layer. Furthermore, the interlayer cation, Ca^{2+}, is replaced by K^+, which is bonded strongly with the adjacent tetrahedral layer. Such chemical changes result in a different type of clay with significantly altered properties—having much reduced CEC and resistance to physical expansion associated with wetting.

All of the physical and chemical weathering processes we have discussed can occur simultaneously or in overlapping sequences and as they take place, they influence one another. The net result of the weathering of rocks over extended periods is the production of a finely divided material that may be classified as a soil. The processes of soil formation and development are continuous and ongoing. The 'end product' is not a stable material but is itself subject to further change, due either to natural or anthropogenic factors. The kinds of changes that occur in the present time are a subject of great interest to those chemists who study the terrestrial environment.

17.1.4 Soil organic matter

Beginning with rocks alone, weathering produces an essentially inorganic mineral soil and this mineral matter does in fact usually make up the larger proportion of actual soils. Soils are, however, not purely inorganic materials. A second component is organic matter, which often contributes from < 1 to 5 weight % to the soil mass. Much higher concentrations of organic matter occur in important types of soils such as those derived from peat and the surface layers of forest soils. In contrast, desert soils are almost purely inorganic. In any case, even when the proportion of organic matter is small, it plays a disproportionately significant role in many soil physical and chemical reactions.

The primary sources of organic matter are plant tissue (roots of growing and dead plants) and litter, such as leaves and branches that have fallen on the surface. All these components are found in various stages of decay ranging from fresh material to a product that is highly decomposed by microbial and chemical processes, creating a structure and appearance that is vastly different from that of the parent material. The biomass of soil microorganisms—bacteria, fungi, actinomycetes, and protozoa—is itself an important contributor to the organic fraction of soil and typically makes up between 0.05 and 0.5% by mass of the top 15 cm of soil in a natural setting. Other small soil animals, principally earthworms, contribute typically one tenth of this fraction to the biomass. Earthworms are only a minor component of the biomass, but they are important because of their ability to enhance aeration and water movement, and to translocate organic matter within the surface soil.

A typical breakdown of the composition of fresh plant residues is summarized in Fig. 17.1 The carbohydrate in most plant tissues is largely in the form of cellulose, hemicellulose, and, to a lesser extent, starch. All are polymers of (different forms of) glucose.

Decomposition is a complex chemical and microbiological process. Considering all the classes of chemical compounds in plant residues, decomposition occurs at varying

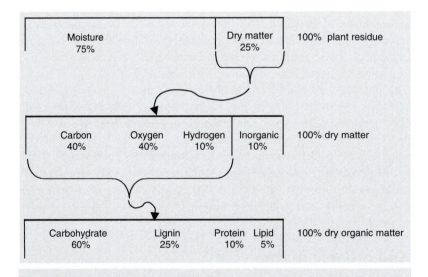

Fig. 17.1 Composition of fresh plant materials. The figure shows components of fresh material, of dry matter, and of the organic fraction.

rates, depending mostly on the suitability of each compound as a food source for microorganisms.

Sugars, starches, simple proteins   rapidly decomposed
Complex proteins
Hemicellulose
Cellulose
Lignins, lipids   slowly decomposed

Heterotrophic organisms incorporate carbon compounds into their own biomass. Degradation and resynthesis produce new soluble or insoluble organic compounds. Inorganic elements from the plant residues are mineralized as species such as NH_4^+, NO_3^-, $H_2PO_4^-$, SO_4^{2-}, Ca^{2+}, and K^+ and/or immobilized in the microbial structure. Simultaneously some of the original organic carbon is evolved by respiration as CO_2. The microbial processes of decomposition are represented in the following sequence which relates to a situation where a soil is supplied with a fresh input of undecomposed organic matter. Examples of this would include a forest soil receiving leaf fall during the autumn season and a field where a 'green manure' crop is grown and then ploughed under.

Soil   Stable microbial population

↓

Soil + fresh organic matter   Decomposition as zymogenous
                                      microorganisms multiply; simultaneous
                                      degradation; synthesis; CO_2 evolution

↓

Soil + microbial biomass +   Development of OM substrate deficiency;
   partially decomposed OM   reduced microbial activity due to death
                                    of microorganisms

↓

Humus   Relatively stable, chemically and
                                    microbiologically

The product resulting from these complex processes is soil humic material, called humus, whose chemistry we examined earlier. This is a relatively stable material, but over years is itself slowly decomposed to produce carbon dioxide as a final product. In a tropical region, where deforestation is brought about prior to use of the land for agriculture, loss of organic matter may be particularly rapid for the first few years—anywhere from 20 to 60% of the original amount may be lost each year. Once it reaches a 'normal' value, further losses occur at a much slower rate.

While we have considered the processes of genesis of inorganic and organic components of soil in separate categories, it is important to realize that there are interactions between the components which means that individual properties are not necessarily additive. For example, iron (III) and aluminium (III) oxide minerals in soils are able specifically to adsorb, and therefore immobilize, some metals that are present in the soil solution. However, there are situations where soil humic material is strongly associated with the same oxides. Coatings of humic substances on the oxide surfaces inhibit the specific adsorption reactions, and mobility is therefore greater than in soils containing uncoated minerals. Therefore, a detailed knowledge of the amount and mineralogy of oxide minerals in soils is alone insufficient to predict the degree of metal retention by the soil. We will come across other cases where interactions produce anomalous effects of this sort.

The complex combinations of various physical and chemical (abiotic and biotic) processes acting on parent rocks and organic residues produce a finely divided material that we call a soil. Reflecting the origin and nature of the parent materials and the geological environment within which they were weathered and transported, soils have a wide range of properties. The variation is evident on both the horizontal and vertical scale, and over time there are continuing changes as well. Common features of all soils include the following:

● soil is a finely divided, heterogeneous, porous material made up of mineral and/or organic matter;

● pore spaces are filled with air and/or water, depending on moisture conditions.

Figure 17.2 shows the principal features common to most soils.

Soils are therefore a three-phase mixture and a complete description of their role in environmental reactions and cycles requires a consideration of the interactions involving all three phases.

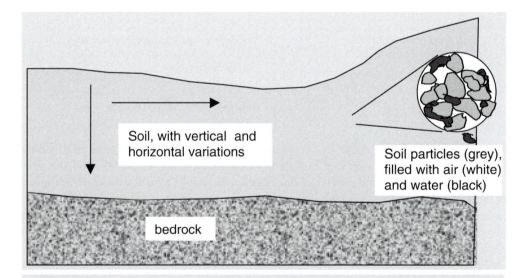

Soil, with vertical and horizontal variations

Soil particles (grey), filled with air (white) and water (black)

bedrock

Fig. 17.2 Soil, the surface layer of much of the terrestrial environment. As a three-phase mixture, it consists of finely divided organic and inorganic particles and pore spaces filled with water and/or air. The soil is highly heterogeneous in both the vertical and horizontal dimensions.

The main points

1 Soils cover a large fraction of the Earth's terrestrial environment. They provide a supporting medium for many forms of life and are the basis of agriculture and forestry.

2 Soils are also an important environmental agent, acting as a 'filter' for aqueous and solid inputs including rain, municipal wastes, pesticides, and other chemicals.

3 Soil is formed over periods of geological time by a combination of physical and chemical (including biotic) processes. The processes act on consolidated rocks to produce the mineral component and on plant and animal material to produce the organic portion of the soil. The product of these weathering processes is a finely divided material that contains air and/or water in the pore spaces. The nature and composition of any soil continues to change and can be affected significantly by inputs from human activities.

Additional reading

1 Birkeland, P. W., Pedology, Weathering and Geomorphological Research, Oxford University Press, NewYork; 1974.

2 Brady, N. C. and Weil, R.R., The Nature and Properties of Soils, 12th edn. Prentice Hall, Upper Saddle, River, N.J.; 1999.

3 McBride, M. B., Environmental Chemistry of Soils, Oxford University Press, New York; 1994.

Problems

1 Consider the terrestrial portions of the global nitrogen cycle (Fig. 15.7). What human activities have perturbed aspects of this cycle, and in what ways would other processes have to adjust to keep the cycle in balance?

2 The weathering of some clay minerals containing potassium is affected by growing plants. Explain why this might be so.

3 The surface soil in the humid tropics is often depleted in silica and enriched in iron and aluminium oxides. In contrast, the surface mineral layer of a forest soil in a temperate region may be devoid of significant iron and aluminium minerals, and have a high concentration of silica. Suggest an explanation.

Note

1 Soil can be defined as the layer of unconsolidated particles that are derived from weathered rock and organic material and that contains water and/or air in the void space. Soil covers the upper surface of much of the Earth and supports plant life. This is not the only definition of the term 'soil' and an excellent discussion of different concepts is given in Chapter 1 of ref. 2.

18

Soil properties

EVERYONE recognizes the importance of soil for the sustenance of human and other living communities on the Earth. Soils are the fundamental resource supporting agriculture and forestry, as well as contributing to the aesthetics of a green planet. They are also a base from which minerals are extracted and on to which solid wastes are disposed. In addition, soils act as a medium and filter for the collection and movement of water. And, critically for life as we know it, by supporting plant growth soil is a major determinant of atmospheric composition and therefore of the Earth's climate. For these and many other reasons, it is of greatest importance to maintain this essential resource. In this chapter, we will spend time examining soil properties in the environmental context.

As we did in the case of soil genesis it is helpful to consider the properties of soil in terms of both their physical and chemical characteristics. We will come to realize that there is a close and overlapping relation between these two aspects.

18.1 Physical properties

18.1.1 Particle size

Particle size is a primary physical property and there are several classification schemes used to define it. Figure 18.1 gives one such scheme—that of the International Society of Soil Science.

By this convention, soil is arbitrarily defined as material with particle size less than 2.0 mm. In fact many soil analysis protocols begin with a sieving (2.0 mm) step to separate soil from non-soil. Within the material defined as soil, there are three primary size categories—in order of decreasing particle size, sand, silt, and clay. Keep in mind our earlier classification of colloids as particles smaller than 10 μm in size. Soil particles in the clay and small silt fractions are therefore in the colloid category.

Sand is 'light', easily worked, has good drainage (but poor water retention) and is readily aerated. Chemically, the most important components of sand are usually primary minerals such as quartz and feldspars; these are relatively inert and poor sources of nutrients. The clay-sized fraction is at the small particle size end of the range. Soils rich in clay are 'heavy', difficult to work, and have poor drainage and aeration. Clay-sized material may be composed of some combination of the clay minerals themselves, organic matter, and of

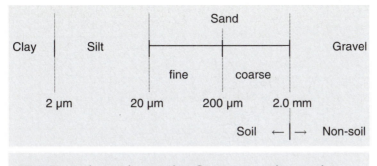

Fig. 18.1 Soil particle size classification according to the International Society of Soil Science.

finely divided primary minerals and hydrous iron and aluminium oxides.[1] All of these fine materials have large surface areas and take part in ion exchange and/or adsorption reactions. Therefore the clay-sized particles are able to interact with and retain nutrients; in this way, they may be productive plant growth media.

18.1.2 Texture

Soil texture is a collective term that defines a real soil by the proportion of different particle size components. Texture nomenclature is based on triangular diagrams such as the one shown in Fig. 18.2.

To use the diagram, consider a case where a soil contains 35% clay, 30% silt, and 35% sand-sized material. Beginning at 35 on the clay axis, a line is drawn parallel to the sand axis. Similarly a line at 30 on the silt axis is drawn parallel to the clay axis. The point of intersection is in the region known as clay loam and the soil is so named.

Soils that are desirable from an agricultural perspective often fall in the middle region of the diagram. Such soils have the beneficial physical properties of lighter soils and tend to be quite easy to work, but also have moderate moisture-retaining ability, and chemical reactivity due to the contribution of the clay-sized materials. Note that the region named 'clay' is large and includes soils that may contain even less than 50% clay. This is because finely divided material has a dominating effect on the behaviour of many soils.

18.1.3 Density

The density of soil reflects that of the minerals or organic components that make up its composition. Density of individual particles—called *particle density*—is considerably less than 1 g mL^{-1} for organic matter, and greater than 5 g mL^{-1} for some metal oxides or 7 g mL^{-1} for less-common minerals such as metal sulfides. Many widely distributed soil minerals including quartz, the feldspars, and clay minerals, have densities that fall inside the approximate range from 2.5 to 2.8 g mL^{-1} and this may be taken as a good estimate of particle density of a number of mineral soils. In the field, the *bulk density* of the soil, which includes the pore spaces between particles, is smaller than this range of values. In the case of mineral soils situated within a depth of 1 m or so from the surface, the bulk density is often

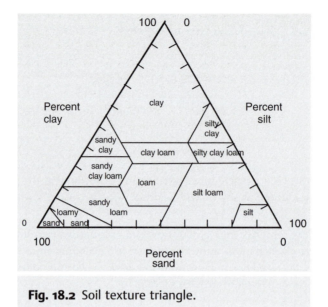

Fig. 18.2 Soil texture triangle.

between 1.2 and 1.8 g mL^{-1} for materials containing a good proportion of sand. For soils with higher clay content, the density falls in a slightly lower range from 1.0 to 1.6 g mL^{-1}. When both particle and bulk densities are known, the pore space of a soil may easily be calculated.

$$\text{Pore space (\%)} = 100 - \frac{\text{bulk density}}{\text{particle density}} \times 100.$$

For silt loam with particle density = 2.65 and bulk density = 1.50:

$$\text{Pore space} = 100 - \frac{1.50}{2.65} \times 100 = 43\%.$$

Sands have pore space values from 35 to 50%, fine textured soils from 40 to 60%, and soils very high in organic matter may have greater than 60% pore space. Subsoils—soils at depth—tend to be more highly compacted than surface soils and usually have low organic matter contents so that the pore space is often only about 25% of the total volume in the field.

18.1.4 Structure

A fourth physical property is structure. Structure is a term used to describe the way in which individual particles are aggregated together to form larger units. Organic matter acts as a cementing agent and plays a key role in developing soil structure. Other chemical entities and the nature of the parent material are also determining factors. Structure is a field characteristic and is usually destroyed when a soil sample is collected and brought to a laboratory for analysis. An abbreviated definition of common structures is given in Table 18.1.

Table 18.1 A partial listing of soil structure types	
Structureless	No observable aggregation or no definite orderly arrangement around natural lines of weakness
Granular	Individual particles held together to form crumb-like aggregates
Block-like	Soil particles arranged around a point and bounded by flat or rounded surfaces—includes blocky to granular-spheroidal forms
Plate-like	Soil particles arranged around a horizontal plane and bounded by flat horizontal surfaces
Prism-like	Soil particles arranged around a vertical axis and bounded by relatively flat vertical surfaces

18.1.5 Permeability

Permeability, also called hydraulic conductivity, is a measure of the ability of soil to conduct water. For example, when it rains heavily, a permeable soil transports the water rapidly downwards. The rate of lateral movement of groundwater in the subsoil is also determined by permeability. Many soils have downward permeabilities in the range 1 to $5 \, \text{cm} \, \text{h}^{-1}$; rates smaller than $0.5 \, \text{cm} \, \text{h}^{-1}$ are very low, while rates greater than $15 \, \text{cm} \, \text{h}^{-1}$ are high.

For soils with little or no structure, permeability is a function largely of soil texture. The coarse sands of the Arabian peninsula have very large permeabilities, while some fine-grained alluvial soils in Iraq have permeabilities in the very low categories.

Structure is also a major determinant affecting permeability, especially in the case of clays. Soils with well developed structure are more permeable than structureless ones. The volume of space created by structural boundaries may be small compared with the total pore volume of the soil, but it can be the major route for water transport. The greater volume of pore space between the fine-grained particles *within* structures then plays only a minor role in movement of water. Water also tends to move *via* cracks created by drying, by earthworms, or by roots of plants.

Permeability tends to decrease with depth in a soil, but the changes are often irregular and discontinuous. Hardpans are regions of impermeable layers in the soil profile. In agricultural soils, a *hardpan* or *ploughpan* is frequently observed just below the depth to which the plough or cultivator extends, and is due to the effect, over many years, of heavy machinery or animals passing over the same area of land. It has been observed that soils with a normal bulk density of 1.2 to $1.5 \, \text{g} \, \text{mL}^{-1}$ may be compacted to densities greater than $2 \, \text{g} \, \text{mL}^{-1}$ in this way. Occasional deep ploughing or, alternatively, the use of no-till technologies are ways employed to prevent the problem of interrupted water percolation in such situations.

There are many ways by which the physical properties of soils influence their environmental behaviour. Permeability is probably the most significant single factor in this regard. Soils with low permeability can become waterlogged and this creates a potentially reducing environment. Likewise, permeability affects transport of chemicals through the soil. Any complete description of movement of chemicals in the soil requires knowledge of both the distribution of the chemical between soil and water, as described in Chapter 14, and the soil hydrological properties. The latter properties are determined by physical attributes of the soil.

18.2 Chemical properties

18.2.1 Total elements

A logical starting point for examining chemical properties of soils is to look at a total analysis of the elements. Ranges of values for the major and some minor elements present in the mineral portion of soils are included in Table 18.2. The chemical composition is determined by the nature of the starting materials from which the soil was formed and by the processes that it has undergone over time.

For many purposes, compositional data are not particularly useful because they do not indicate whether the elements are found as components of the mineral lattice or are associated with surface adsorption phenomena. In the former case, especially for silicate minerals, such elements would be 'insoluble' except over geological time, and therefore would not play a significant role with respect to plant growth or in terms of most environmental processes. Those elements that occupy ion-exchange sites on soil particles or those that are weakly adsorbed are, however, much more available for chemical and biological activity. In fact, the term 'available' and 'extractable' are widely used by Earth Scientists and there are operational definitions and analytical procedures for determining these fractions of the elements. We will look at this later.

The organic portion of soil is intimately mixed into and, in some cases, chemically associated with the mineral soil. Although there are exceptions, a general trend whereby

---

Table 18.2 Ranges of values for major and minor element composition of the mineral component of soils

Major elements/%		Minor elements/mg kg^{-1}	
Si	30–45	Zn	10–250
Al	2.4–7.4	Cu	5–15
Fe	1.2–4.3	Ni	20–30
Ti	0.3–0.7	Mn	$\sim$400
Ca	0.01–3.9	Co	1–20
Mg	0.01–1.6	Cr	10–50
K	0.2–2.5	Pb	1–50
Na	tr–1.5	As	1–20

A more complete list of minor elements in whole soil is given in Table 18.8.
tr = trace.

Table 18.3 The range of organic matter content in soils

Temperate agricultural soils	1–5%
Tropical agricultural soils	0.1–2%
Forest soils (surface horizons)	>10%
Peat soils	>20%

the organic content decreases with depth is usually observed. Table 18.3 reports values for organic *matter* (OM) in a variety of near-surface soil situations. Analytical methods usually give a measure of organic *carbon* (OC) content. Based on a widely held assumption that soil organic matter is about 60% carbon, a factor of 1.7 may be used to convert organic carbon to organic matter values.

Soil organic matter is broadly classified into humic material (HM) and non-humic material. When we considered some characteristics of HM in Chapter 12 we noted there that the HM is derived from either terrestrial or aquatic sources and that there were some general differences between the two types. Terrestrial HM is derived primarily from plant residues—fallen leaves and branches from trees in a forest, dead grass and other meadow plants after a dry season or winter, or crop residues left from harvest and perhaps ploughed into the surface layers of the soil. The partially decomposed and resynthesized matter is relatively stable and contributes to good soil structure and to the cation-exchange capacity of the soil.

The non-humic organic fraction of the soil is made up of many compounds, among them the complex polysaccharides, cellulose, hemicellulose, and pectin. It also includes smaller carbohydrate molecules that have been released as these organic polymers decompose. Because the simple monosaccharides are readily soluble and also serve as a favoured food source for microflora and fauna, their concentration in soil is not great. High molar mass polysaccharides, on the other hand, are relatively stable. Like the HM macromolecules they act as cementing agents, contributing to soil structure by binding individual particles together into larger units. Proteins and amino acids are also an excellent food source for soil microorganisms, but small amounts can be measured at any time in the soil. Lipids are a minor component, but are relatively resistant and long-lived in the soil environment.

Much more resistant to decomposition are the lignins, complex phenolic polymers that are responsible for the toughness of plant parts. Lignins are a major component of the woody components of plants. Tannins are also plant-derived polyphenols that are, in some cases, highly resistant to degradation. For this reason lignins and tannins, like HM, can make up a significant portion of the organic matter especially in forest soils.

Over time, some of the non-humic material is incorporated into the HM fraction and thus, in some ways the non-humic compounds have a transient existence.

18.2.2 Available elements

As noted above, the available element is the portion of the element that can take part in a range of chemical and biological reactions. To some extent, the terms *available* and *extractable* element are synonymous. To measure extractable elements, a soil is shaken in an aqueous solution containing chemicals chosen to displace that portion of the element that is supposed to be readily available for uptake by growing plants. Many extractants and conditions have been recommended for this purpose, depending on the nature of the soil and other environmental circumstances. Most methods make use of the law of mass action in order to displace the cations that are present on the exchange sites (ammonium acetate method) and/or involve chelating agents that combine with the immobilized element to assist its dissolution (DTPA (diethylenetriaminepentaacetic acid) method). The latter class of extractants tends to dissolve a larger fraction of the total metal (Table 18.4).

Table 18.4 Percentage of total metal extracted from soil using two extractants

Extractant	Co	Ni	Cu	Pb	Cr
NH$_4$OAc (pH7)	0.36	0.86	1.1	0.51	0.37
DTPA	1.3	2.7	5.0	8.7	0.11

From McLeod, S. E. and G. W. vanLoon, A study of elemental contamination in Orchard Park, Kingston, Ontario. *Ontario Geography*, **17** (1981), 91–104.

18.2.3 Cation-exchange capacity

In Chapter 14, we showed that components of sediments and soils have the ability to adsorb positive ions electrostatically on to their surface. The clay minerals and organic matter are particularly important in this regard. A quantitative assessment of the ability to interact with cations is the cation-exchange capacity (CEC)—one of the most commonly measured properties of soils. The following example illustrates a calculation of cation-exchange capacity in a soil. The example makes use of one[2] of the many standard methods for determination of exchange cation concentration. It involves extraction by aqueous ammonium chloride at a concentration of 1 mol L^{-1} and with pH adjusted to 4.5. Suppose 1.00 g of soil is extracted with 100 mL of the ammonium chloride solution in order to displace the exchange cations from the soil. After filtering, the dissolved calcium, magnesium, potassium, and sodium are determined by flame atomic absorption spectroscopy. The hydronium ion is determined by titration.

The concentrations of the principal cations found in the extractant solution are as follows:

$$Ca^{2+} \quad 30.3 \, \mu g \, mL^{-1}$$
$$Mg^{2+} \quad 3.2 \, \mu g \, mL^{-1}$$
$$K^+ \quad 2.2 \, \mu g \, mL^{-1}$$
$$Na^+ \quad \text{not detected}$$
$$H_3O^+ \quad 2.60 \, mmol \, L^{-1}.$$

To determine the cation exchange capacity, it is necessary to calculate the total positive charge associated with these ions. In most cases, no other cation would be present in significant amounts. All the cations in solution were originally present on exchange sites and therefore their positive charge must be equivalent to the number of negative sites in 1 g of soil.

For calcium, the positive charge associated with the exchange complex is determined as follows:

The 'concentration' of positive charge due to this divalent cation is equal to

$$2 \times 30.3\,\mu g\,Ca\,mL^{-1}/40.1\,g\,mol^{-1}\,Ca = 1.51\,\mu mol\,mL^{-1}$$

This concentration can be multiplied by 100 mL to give the total positive charge due to exchangeable calcium that had been extracted from 1 g of soil. The final result is, by convention, expressed in terms of cmol (10^{-2} mol) of positive charge per kg (cmol$(+)$kg^{-1}) of soil.

$$1.51\,\mu mol\,(+)\,mL^{-1} \times 100\,mL \times \frac{1000\,g}{kg\,soil} \times \frac{10^{-4}\,cmol}{\mu mol} = 15.1\,cmol\,(+)\,kg^{-1}\,soil$$

Note that this standard unit of cmol$(+)$kg^{-1} has the same value as the traditional and still widely used unit of meq $(100\,g)^{-1}$. A similar calculation can be done for the other elements that were extracted from the exchange complex, and the corresponding values are

$$for\ Mg, \quad 2.6\,cmol\,(+)\,kg^{-1}\,soil$$

$$for\ K, \quad 0.6\,cmol\,(+)\,kg^{-1}\,soil$$

For H_3O^+ the charge is given by

$$2.6 \times 10^{-3}\,\frac{mol\,L^{-1}}{1\,g\,soil} \times 0.100\,L \times \frac{1000\,g}{kg} \times \frac{10^2\,cmol}{mol} = 26.0\,cmol\,(+)\,kg^{-1}\,soil$$

The cation-exchange capacity (CEC) is then the sum of the individual values:

$$CEC = (15.1 + 2.6 + 0.6 + 26.0)\,cmol\,(+)\,kg^{-1}$$

$$= 44.3\,cmol\,(+)\,kg^{-1}\,soil$$

The actual cations that occupy exchange sites depend on the nature of the soil particles as well as on other environmental circumstances. The two alkaline earth elements and two alkali metals used in this example are almost always the most important exchange cations, and for many soils the quantitative order of importance is $Ca^{2+} > Mg^{2+} > K^+ > Na^+$. Only in particular circumstances do other metal cations contribute significantly to the CEC value. Hydronium ion makes a variable contribution to the CEC. Under acid conditions, whether natural or anthropogenic, a large proportion of the cation exchange sites may be occupied by hydronium ions, while in neutral and alkaline soils their contribution is negligible. Where a considerable fraction of the exchange sites is taken up with hydronium ion, the nutrient-supplying capacity is diminished. Furthermore, such soils have reduced capacity to neutralize additional acidity. A measure of proportion of metals (compared with hydronium ions) on exchange sites is the base saturation, defined as

$$Base\ saturation = \frac{number\ of\ exchange\ sites\ occupied\ by\ Ca + Mg + K + Na}{total\ number\ of\ exchange\ sites} \times 100\%$$

In many instances, there is a fairly close positive correlation between the base saturation and the pH of the soil. In the present example,

$$\text{Base saturation} = \frac{(15.1 + 2.6 + 0.6)}{44.3} \times 100$$
$$= 41\%$$

A wide range of CEC values has been observed. Sandy soils, low in organic matter, have very small CEC values, often less than $1 \, \text{cmol} \, (+) \, \text{kg}^{-1}$, while soils high in certain clays and/or organic matter may have a CEC of more than $100 \, \text{cmol} \, (+) \, \text{kg}^{-1}$. The two factors that dominate the size of the CEC are the nature and content of clay minerals and the content and degree of decomposition of the organic matter. This is clear from the Table 14.5 showing CEC values for individual constituents of soil and sediment. Some typical values of whole soil cation-exchange capacity are given in Table 18.5.

Finally, it is important to be aware that there is a very extensive literature on methods for determining CEC, and their significance. One should not embark on a study that interprets and compares CEC values without becoming familiar with some of this literature.

Table 18.5 Cation exchange capacity values/cmol $(+)$ kg^{-1} for a variety of selected surface soils

Kentville sandy loam (Nova Scotia, Canada)	10	Sassafras sand (New Jersey, USA)	2
St Quintin peat (Quebec, Canada)	155	Lipa clay loam (Luzon, Philippines)	36
Darlington clay loam (Manitoba, Canada)	50	Nabha silt loam (Punjab, India)	9.8
Baker Lake sand (NWT, Canada)	18	Gezirach clay (Sudan)	52

18.2.4 Soils of variable charge

What has been described in the previous section implies that the CEC (or the surface charge) of a soil is a constant factor. In some cases this is true—particularly for soils of temperate regions, which tend to contain clay minerals such as montmorillonite. Most of the charge on these clays is fixed charge—that is, the charge is independent of the environment in which the clay is found. Tropical soils however, frequently contain other minerals in the clay-sized fraction, particularly the hydrous oxides of iron and aluminium and these minerals are characterized by having a pH-dependent variable surface charge. In Chapter 14, we discussed the basis of this variable charge phenomenon.

The cation-exchange capacity of variable charge soils depends, in the first instance, on the pH of the surrounding environment. When the ambient pH is lower than pH_0 of the soil component, then that material is protonated, causing its net surface charge to be positive. By this means, the soil develops a certain amount of anion-exchange capacity.

Conversely, when the ambient pH is greater than pH_0, the net surface charge of the variable exchange material becomes negative and contributes to the cation-exchange capacity of the soil. There are many important implications of this. It means that natural and anthropogenic factors (for example acid rain, or nitrification of ammonium ion, both of which add acid to the soil environment) are able to alter the exchange properties of variable charge soils in the field. It also means that measurements of exchange capacity should be carried out under pH conditions that are similar to those of the intrinsic soil environment. For example, determining the cation-exchange capacity of a variable charge soil whose field pH is 5.6 using pH 7.0 aqueous ammonium acetate as extractant would enhance the surface negative charge, resulting in an overestimate of the CEC value.

The pH is not the only environmental factor affecting the charge properties of variable charge soils. The presence of other species that are specifically adsorbed to the soil minerals can substantially change the surface charge. Perhaps the most important example is phosphate, which is strongly retained by iron and aluminium-rich soils. If the phosphate displaces water molecules from the hydrous oxides, it serves to increase the negative charge of the mineral (Fig. 18.3), giving it a higher pH_0 value. Addition of phosphate may, in fact, be a beneficial agronomic practice in order to increase the cation-exchange capacity of a highly weathered tropical soil. However, it may not have the expected effect of providing plant-available nutrient, since this anion is covalently and tenaciously bonded to the soil and is not readily extracted by plant roots.

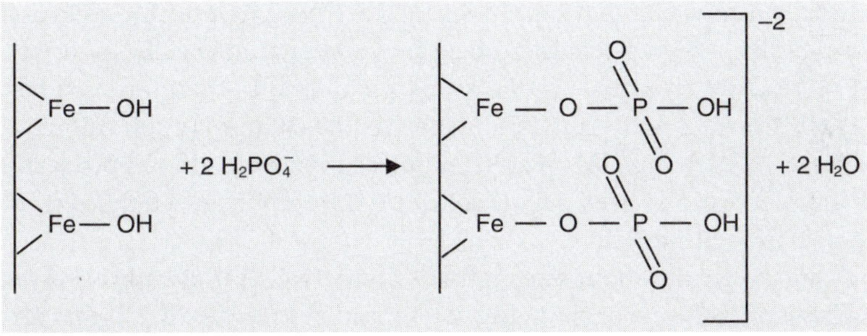

Fig. 18.3 Specific adsorption of phosphate on a hydrous iron oxide surface.

18.2.5 Soil pH

Soil pH depends on the nature and history of the soil. Soils rich in carbonate minerals tend to be somewhat alkaline. On the other hand, those containing large amounts of humic material are often, but not always, acidic. The acidity arises from microbial decomposition of the organic matter producing organic acids as a metabolic product, as well as from carbon dioxide released during respiration. Soils containing adsorbed iron and aluminium are also acidic as a consequence of the hydrolysis of the iron (III) and aluminium (III) cations. Prolonged leaching of the principal exchangeable metal cations, and their replacement by hydronium ion, also contributes to increasing acidity.

Soil pH is affected by changes in redox status. If a soil containing hydrous iron oxide is submerged, the iron (III) oxide is reduced as represented by the half reaction shown:

$$Fe(OH)_3 \text{ (s)} + 3H_3O^+ \text{ (aq)} + e^- \rightarrow Fe^{2+} \text{ (aq)} + 6H_2O \qquad (18.1)$$

Because hydronium ions are consumed in the reduction, there is an accompanying rise in pH. Flooded soils therefore tend to exhibit higher pH values than their upland counterparts.

Terminology commonly used to describe the acid–base status of soils is as follows:

pH	less than 4	strongly acid
	4 to 5	moderately acid
	5 to 6	slightly acid
	6 to 8	neutral
	8 to 9	slightly alkaline
	9 to 10	moderately alkaline
	greater than 10	strongly alkaline.

18.3 Soil profiles

The description of soil physical and chemical properties given in the previous section does not take into account the fact that soils in the field are not a monolithic mass of unchanging composition. Rather, they are characterized by large spatial variability in both the horizontal and vertical dimensions. As one scans a landscape, there are obvious variations across the Earth's surface—differing landforms, visible changes in soil colour, and changes in land use patterns. Although less visible, important variations in soil properties may exist on a much smaller scale of area. Likewise, there are important changes in physical and chemical properties with depth.

If one digs a pit, typically about one metre or more deep, it is usually observed that there is a series of horizontal layers of changing colour and/or texture. Such layers are referred to as *horizons*, and the assemblage of soil horizons is called a *soil profile*. There have been numerous classification systems which are all based on a description of the profiles but may also include assumptions about the soil-forming (pedogenetic) processes that the soil has undergone. One of the widely accepted systems, the Comprehensive Soil Survey System, was developed by the Soil Survey Staff of the US Department of Agriculture. Unlike most other taxonomic systems, this one is based exclusively on the observed properties of the soil in the field. It makes few or no assumptions about the processes that gave rise to those properties. The Comprehensive System has been widely adopted, but its adoption does not preclude using general and simpler terminology from other methods of classification. We cannot go into this important and extensive subject here.

It is helpful, however, to be aware of some of the widely (but not universally) accepted terminology used to describe certain features of soil profiles. To do this, we will examine three 'typical' examples from very different environmental situations. We will describe some of the relevant features, introduce basic terminology, and indicate aspects of the environmental features of these profiles.

18.3.1 A Canadian Shield Spodosol

The first example is of a Canadian Shield Spodosol also often called a Podzol. Spodosols are typical of humid, temperate regions usually under forest cover; they have formed on relatively acidic parent material, and are usually characterized by being coarse textured. A profile of such a soil from the boreal forest region of Montmorency, PQ, in central Canada, is shown in Fig. 18.4.[3]

Organic horizons[4] overlie the mineral soil. Two such horizons, O1 and O2, are distinguished in the present example; the former contains largely undecomposed litter from the forest canopy, while the latter consists of the same material in a form that has been partially or completely humified and therefore the original morphological features of the litter may not be recognizable.

The processes of decomposition of organic matter release organic acids and generates carbon dioxide—both processes contribute to acidification of the soil. In this case the pH is in the highly acidic category and this is due entirely to natural processes. Rainfall, even neutral rainfall, is thus acidified as it percolates through the coarse material. Moving downward into the upper horizon of mineral soil—the A horizon—it leaches out readily soluble components. Iron and aluminium solubility is greatly increased by the pore water acidity, while silicon solubility is unaffected by pH in the acid region. Therefore, the upper horizon of mineral soil becomes depleted in iron, aluminium, and other soluble elements

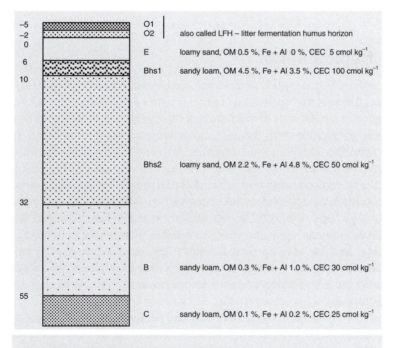

Fig. 18.4 The Montmorency soil profile. Horizon depth in cm given to the left of the profile. Forest floor at 0 cm.

while silica remains behind. The effect is clearly visible in the field in that the top mineral layer has a characteristic grey ashy appearance in contrast to the more highly coloured organic soil above and mineral soil below. The dark red-brown colour of the deeper mineral soil is to a large extent due to iron oxides, as will be explained below. Horizons where the main process is leaching are called eluvial horizons—symbolized as E horizons. Note that the pH of the E horizon in this Spodosol is still low, although somewhat higher than in the overlying organic layers.

Underlying the E horizon is a depositional layer which is called the B horizon. Here, a portion of the compounds transported from above are precipitated from solution due to a variety of causes including a rise in pH, a decline in microbial activity, and absorption. In a Spodosol, iron and aluminium are deposited as hydrous oxides in the B horizon along with organic matter. These compounds together give this part of the profile a characteristic dark brown or red-brown colour that contrasts with the overlying light coloured layer. In the particular soil being described here, the horizon is named Bhs where the 'h' refers to humic material and the 's' designates that iron and aluminium sesquioxides[5] have accumulated in the region. Below the Bhs horizon, the B horizon continues but is much less affected by depositional processes.

The C horizon is a relatively unaltered subsoil that overlies and grades into the bedrock.

In a general sense, then, a Spodosol is typically found in temperate forested regions and is characterized by an organic layer overlying a mineral soil, the surface of which is depleted in iron, aluminium, and organic matter. These constituents are deposited below. The pH of a Spodosol is in the acidic to very acidic range but rises somewhat with increasing depth.

18.3.2 A tropical Alfisol

The second soil profile example is taken from South India[6] and is in the general class of soils called Alfisols. The soil, named Tyamagondalu, is located near Bangalore in Karnataka State and is typical of many red soils in the southern Deccan Plateau. These soils have developed on weathered gneiss, are deep, clayey, and moderately acid to neutral. They support extensive cultivation of millet, some legumes, and rice. The profile is shown in Fig. 18.5.

The surface horizon is a loamy sand with pH about 6.8 and organic matter content of less than 1%. In the tropical environment of South India, the principal eluviated species are not iron and aluminium as in Spodosols but rather silicon, whose solubility is quite high in the neutral pH range. Over a lengthy period eluviation has generated a soil which, to considerable depth, is low in silica and contains relatively large proportions of iron and aluminium oxides and the clay mineral kaolinite. The small content of organic matter is characteristic of soils in the arid tropics. The surface horizon is designated as Ap where the suffix 'p' indicates soil—a plough layer—that has been altered by human activities, particularly cultivation.

Underlying the Ap horizon is a series of accumulation horizons designated as B21t, B22t, B23t, and B24t. B horizons are subdivided into a sequence B1, B2, and B3 where B1 indicates a transition between A and fully expressed B material, B2 is the true B horizon, and B3 is a transitional soil to that in the C horizon. In the present profile, all the soil in this horizon is in the B2 category. The second digits in this system are to denote morphological differences

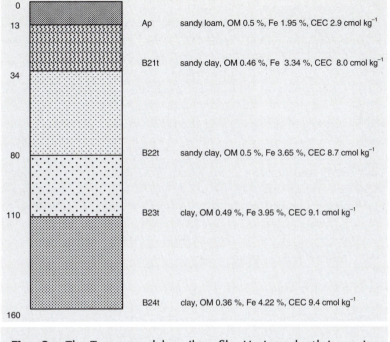

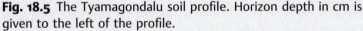

Fig. 18.5 The Tyamagondalu soil profile. Horizon depth in cm is given to the left of the profile.

within the B horizon observed in the field or by laboratory testing. The suffix 't' symbolizes that the transported material which has accumulated is silicate clay mineral, in this case kaolinite.

Therefore the Tyamagondalu soil contains a plough layer of relatively coarse material that is underlain by a series of layers where clay minerals accumulate. The entire profile is rich in iron and aluminium oxides, and kaolinite.

18.3.3 A subtropical Vertisol

The third soil profile example is a Vertisol from the Central Transvaal plateau of South Africa. This is a subtropical area characterized by hot summers and mild, dry winters. The total precipitation is about 600 to 800 mm y^{-1}. The topography is generally flat or a series of gently undulating ridges with dominant vegetation consisting of grasses and several varieties of trees (such as acacia) and bushes. Agriculture is centred on maize, sunflower, and fodder under dry-land conditions; wheat and tobacco when irrigated.

The soil has developed on basic igneous rocks such as norite. The Hartbeespoort soil[7] profile is shown in Fig. 18.6.

The A11, A12, and A13 horizons are heavy black clay soils distinguished from one another only by subtle structural differences and root content that declines from A11 to A13. Because the dominant clay mineral is montmorillonite, the soil has a large cation-exchange

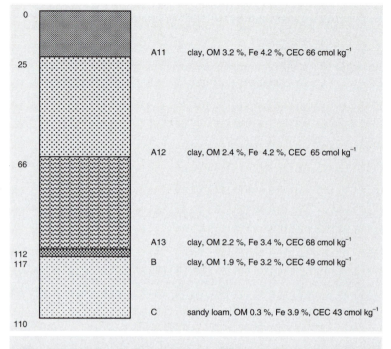

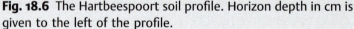

Fig. 18.6 The Hartbeespoort soil profile. Horizon depth in cm is given to the left of the profile.

capacity. Organic matter content is considerably lower than 1% and thus it makes little contribution to the CEC. (Note that the black colour of this and many similar soils in the tropics and subtropics is due to the nature of the parent material, not to the presence of organic matter.) The large content of montmorillonite also causes the soil to undergo a high degree of swelling and shrinking as the soil is wetted and dried. The cracks that form on drying provide a route by which surface material can slough off and fall down to a deeper part of the profile, resulting in a natural ploughing action. While Vertisols can be chemically rich, they are difficult to work because of the nature of the clay that makes them sticky when wet and very hard in the dry season.

In the Hartbeespoort soil, there is some free calcium carbonate throughout the profile and this is reflected in a soil pH that is in the slightly alkaline range.

The type of clay does not change with depth but the clay content declines, and this leads to smaller cation-exchange capacity values in the B and C horizons. They do, however, remain at a moderately (and surprisingly) high value.

The above three examples illustrate how physical and chemical properties of the horizons in a soil profile are used in order to classify individual soils. National soil surveys have been carried out in most countries, providing soil maps of varying detail. A general soil map of the world is provided in ref. 2 of the Additional reading list in Chapter 17. Table 18.6 lists the soil orders defined by the USDA Soil Taxonomy system and notes the principal features characteristic of each order.

Table 18.6 Soil orders according to the USDA system of soil taxonomy

Soil order	Principal characteristics
Alfisols	Soils with grey to brown surface horizons, medium to high exchangeable cation supply, and subsurface horizons of clay accumulation; usually moist but may be dry during warm seasons
Aridisols	Mineral soils in dry areas; may have a horizon where calcium carbonate, gypsum, or other salts accumulate
Entisols	Recent soils without pedogenic horizons
Histosols	Organic soils of various types; greater than 30% organic matter to a depth of greater than 40 cm
Inceptisols	Young soils formed by alteration of parent materials. Little evidence of accumulation; soils are usually moist
Mollisols	Soils with nearly black, humic-rich surface horizons and good supply of exchangeable, mostly divalent, cations
Oxisols	Highly weathered soils with a depositional horizon rich in hydrous oxides of aluminium and iron
Spodosols	Mineral soils with an eluvial layer underlain by a depositional horizon having accumulations of amorphous organic materials and aluminium and iron hydrous oxides
Ultisols	Soils that are usually moist with horizon of clay accumulation and low base saturation; formed in areas of warm, moist climate, usually under forest
Vertisols	Soils with high content of swelling clays and wide deep cracks at dry seasons

(From reference in Note 4 of this chapter).

18.4 Environmental properties of soils

Many of the environmental problems associated with soil are substantially physical ones. Soil erosion due to heavy cropping is a widespread phenomenon. In the corn belt across the mid-western United States, average losses of topsoil per ha have been estimated to be around 25 t each year. Serious erosion occurs to varying degrees in many other locations throughout the world.

Too little or too much water is also a problem. About one person in seven lives in arid or semi-arid regions and desertification is spreading in parts of North Africa, the Middle East, and the Indian subcontinent. On the other hand, waterlogging due to irrigation is a major issue near some of the world's great rivers and inland seas—the Nile, the Tigris–Euphrates, and the Indus rivers and the Caspian and Aral seas. While these problems have an apparently 'physical' basis, there are chemical aspects as well. We will look at some of the kinds of environmental chemistry issues associated with these and other problems in the terrestrial environment.

18.4.1 Leaching and erosion

Intensive land use has encouraged the application of chemical inputs to maximize the yield of crops and forests. Care is required to ensure that any chemical used serves its desired purpose and is not leached into groundwater or washed away as run off. In Chapter 20, we will consider in some detail the fate of organic biocides when applied to a crop or the soil. Here we will briefly examine the behaviour of inorganic nutrients.

The flux of water (rainfall or irrigation), soil texture, and nature of plant cover play major roles in determining extent of leaching. Because downward movement of chemical species is inhibited geochemically through nutrient binding to soil particle surfaces and biologically through uptake by the plant, leaching occurs to a smaller extent where the soils are fine textured and where plants are actively growing.

The physicochemical interactions between nutrient and soil are major determinants with respect to leaching. Cationic macronutrient species include ammonium, potassium, calcium, and magnesium ions added as ammonium salts, potassium chloride (potash), calcium carbonate (limestone), and calcium magnesium carbonate (dolomite). The positive ammonium and metal ions tend to be associated with the negatively charged exchange complex composed of clay minerals and organic matter. Soils with a large cation-exchange capacity therefore hold cations and prevent their leaching. At the same time, species held on a soil are in equilibrium with their dissolved counterparts, and so at least a small portion is present in the associated water. When there is competition with high concentrations of other cations (including hydronium ion) in the soil solution, additional nutrient ions are released from the solid phases and are then subject to leaching.

Ammonium is a special case because, under high pE conditions, it is readily oxidized to nitrate. Being an anion, the nitrate is not strongly bound to solid particles in most soils. It is then readily available to be taken up by plants. If, however, it is not utilized by plants or microorganisms, it then remains in solution.

To maximize production from high-yielding varieties of grains such as maize, heavy applications of nitrogen fertilizers are widely used. It has frequently been observed that a significant fraction of nitrate added or produced from inorganic fertilizers is not taken up by plants and is readily leached from the rooting zone, sometimes into groundwater. There is considerable variability in the amounts of nitrates leached from soils depending on the nature of the soil, amount of water, crop, tillage practices, and rates, types, and times of application of fertilizer. Nitrate reaching a surface water body can contribute to eutrophication although, as we have noted, more frequently phosphorus is the limiting nutrient. Nitrate also is toxic, particularly to young children, and drinking water standards have been set in various jurisdictions.

Box 18.1 Ammonium and nitrate

Nitrogen run-off and leaching from soils is a widespread environmental problem. As a major nutrient, and one obtained or produced at significant economic and environmental cost, losses into the hydrosphere mean reduced availability for its principal purpose—that of enhancing crop growth. Furthermore, in water, excessive nitrogen can be a cause of eutrophication, and its various species are toxic to some life forms.

As sources of nitrogen, both ammonium and nitrate, the common aquatic and terrestrial forms, can be utilized by plants. Ammonium is frequently the initial inorganic nitrogen form released into a soil–water system, because it is produced by mineralization of the organic nitrogen in plant and animal residues and because the element is added as an ammonium salt, liquid ammonia, or urea. In temperate regions, there are situations where animal manures are spread on fields during cold winter weather. Chemical and microbial reactions are slow, and the ammonium compounds are not used or transformed in any way. When the spring snow-melt occurs, large amounts of ammonium may then run off or percolate through the soil. In countries like the Netherlands and other European and North American countries where livestock are raised in large numbers within a limited area, disposal of manure on land is a major issue, especially in the winter.

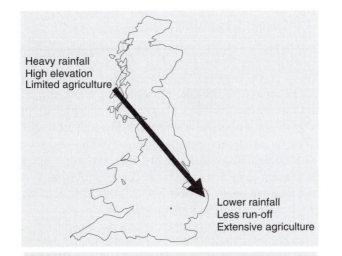

Heavy rainfall
High elevation
Limited agriculture

Lower rainfall
Less run-off
Extensive agriculture

Fig. 18b.1 Gradient in nitrate concentrations in rivers of England and Scotland. Concentrations increase in a south-easterly direction from levels that are typically $< 1\,mg\,L^{-1}$ in the north-eastern highlands to $> 5\,mg\,L^{-1}$ in the heavily agricultural south-eastern plains. (Goudie, A. S. and D. Brunsden, *The environment of the British Isles: an Atlas*, Clarendon Press, Oxford; 1994.)

However, during a normal warm growing season, in all but low pE environments, nitrification rapidly occurs, and at these times nitrate is the more common species. Where there is more nitrate than is required by the growing crop, the excess can run off or be leached through the soil. It is interesting to consider the features that lead to high nitrate loadings in streams and other surface waters. The situation in Britain is a good example. A map showing average levels in surface waters indicates a marked gradient of increasing nitrate concentration running from north-west to south-east. The highest values, typically from 5 to $9\,mg\,N\,L^{-1}$ or more, occur in the areas in and around East Anglia. Not surprisingly, low values are found in streams located in areas having high elevations where run-off is rapid, high rainfall which creates a dilution effect, and limited agricultural activity. In contrast, elevated levels of nitrate are found in the drier and intensively cultivated lowlands.

Ammonium increases the possibility of fish asphyxia and, in public water supplies, it reacts with chlorine to produce the chloramine compounds, reducing chlorine's effectiveness as a disinfectant. A maximum acceptable level of ammonium is $0.5 \, mg \, L^{-1}$. On the other hand nitrate, the principal nitrogen species in an oxygenated aqueous environment, can be toxic to humans. While it is rapidly excreted by adult kidneys, it accumulates in infants younger than 6 months and is reduced to nitrite where it combines with haemoglobin in the blood to form methaemoglobin, a form of the protein that is unable to carry oxygen. The result is the so-called *blue baby syndrome*, a sometimes fatal type of oxygen deprivation. Maximum acceptable nitrate levels in water are frequently set at $10 \, mg \, L^{-1}$.

The environmental behaviour of the second important plant nutrient, phosphorus, is quite different from that of nitrogen. We have seen that phosphorus, present in aqueous solution as the family of orthophosphate species, is strongly and specifically bound to particular soil minerals including aluminium and iron oxides and 1:1 clays. The fixation can be so strong that soluble phosphorus in run-off or leachate is almost negligible. It is when erosion occurs that losses of this nutrient become significant.

In many other instances, too, erosion can be a major contributor to loss of nutrients and other chemicals from the land. Species sorbed to soil particles are carried away in the run-off water and can later be released in soluble forms when they reach rivers or lakes. Table 18.7 gives concentrations of both soluble and particle-bound calcium and phosphorus measured in run-off from forested areas developed on Alfisols in western Nigeria. The amount of nutrient associated with suspended material in run-off can far outweigh the soluble nutrient and is especially strongly dependent on slope.

Table 18.7 Nutrient concentrations in water and soil particles in run-off from a forested Alfisol in western Nigeria

Slope	Soluble metal/$\mu g \, L^{-1}$		Particle-bound metal/$\mu g \, L^{-1}$	
	Calcium	Phosphorus	Calcium	Phosphorus
1	2.7	0.1	475	3.9
5	2.6	0.1	725	5.5
10	1.6	0.4	790	8.2
15	1.4	0.6	1135	14.7

From Lal, R., *Soil Erosion Problems on an Alfisol in Western Nigeria and their Control*, Monograph No. 1., IITA, Ibadan, Nigeria; 1976.

18.4.2 Reactions with acids and bases

As we noted earlier, mineral composition is the principal factor determining the intrinsic pH of a soil. Where carbonate minerals are present, soil pH tends to be in the range 7.5 to 8. On the other hand, soils containing significant quantities of exchangeable aluminium (III) and iron (III) or high in organic matter are generally somewhat acidic. Soil pH is also affected by acid inputs from a variety of sources—many of these associated with acid-generating reactions that occur naturally. One acid source involves microorganisms whose respiration produces carbon dioxide. It is for this reason that the soil atmosphere can be enriched in carbon dioxide resulting in reduced pH of the soil solution. Another contribution to reduced pH in soil solution is from the production of organic acids by microbes that degrade soil biomass. Each of these factors is influenced by climate and setting and the net result is a 'natural pH' for a particular soil.

There is also a variety of anthropogenic sources of acidity, but two stand out as being quantitatively especially important. Acid precipitation at various locations in the world supplies a large amount of hydronium ion to the soil. A mean annual precipitation pH of 4.2 is not unusual in some parts of northern Europe and eastern North America. Consider a situation where the yearly rainfall with average pH of 4.2 is 1000 mm (1 m) and the area is 1 ha. The total number of moles of acid being added to the 1 ha area of soil is then

$$M_{H_3O^+} = 1\,\text{m} \times 10000\,\text{m}^2 \times 1000\,\text{L}\,\text{m}^{-3} \times 10^{-4.2}\,\text{mol}\,\text{L}^{-1} = 630\,\text{mol}.$$

A second very common source of acid input into soil occurs due to the addition of nitrogen-containing fertilizers, where the added nitrogen is in the reduced form and is subject to nitrification. As an example, consider a situation in which 200 kg ha^{-1} nitrogen is added in the form of solid ammonium sulfate. This corresponds to 14 000 mol of ammonium. The initial nitrification reaction is

$$2NH_4^+ \,(\text{aq}) + 3O_2 + 2H_2O \overset{Nitrosomonas}{\rightarrow} 2NO_2^- \,(\text{aq}) + 4H_3O^+ \,(\text{aq}) \tag{18.2}$$

In our example, after the reaction is complete, 28 000 mol of hydronium ion has been generated *in situ* in the soil within the 1 ha area. The further oxidation of nitrite to nitrate does not lead to the formation of any additional acid or base.

There are many reactions potentially available to neutralize acidity that have been added to the soil or generated within it. Factors that determine which one is operative include the nature of the particular soil, the presence of vegetation, and the anions associated with the input acidity. We can divide the types of reactions that neutralize acidity into two categories—those that are geochemical and those that are biological.

18.4.3 Geochemical reactions that neutralize acidity

A soil that contains carbonate minerals neutralizes acid by dissolution of the solid carbonate species. For calcite (limestone) the reaction is

$$2CaCO_3 \,(\text{s}) + H_2SO_4 \,(\text{aq}) \rightarrow 2Ca^{2+} \,(\text{aq}) + 2HCO_3^- \,(\text{aq}) + SO_4^{2-} \,(\text{aq}) \tag{18.3}$$

As a consequence, the soil solution is enriched in calcium and hydrogen carbonate ions and depleted in hydronium ion. Soil pH, which is initially in the neutral or mildly alkaline ranges, is usually little affected by these reactions unless the supply of carbonate minerals is very small.

For soils containing little or no carbonate species, neutralization occurs due to cation-exchange reactions of the type shown:

$$\text{Soil}: K^+ + H_3O^+ \text{ (aq)} + NO_3^- \text{ (aq)} \rightarrow \text{Soil}: H_3O^+ + K^+ \text{ (aq)} + NO_3^- \text{ (aq)} \tag{18.4}$$

Such reactions serve to neutralize the solution acidity while the base saturation of the exchange complex is decreased. The extent to which the exchange reaction occurs depends on the original exchange capacity and base saturation of the soil. Because every soil has some exchange capacity, it has been suggested that the magnitude of the CEC along with the base saturation is a reasonable measure of acid-neutralizing capacity of soils—equivalent to alkalinity for a natural water body. Although CEC is important in many situations, other physical and chemical factors also come into play and must be considered in a detailed evaluation of soil resistance to acidification.

Soils containing iron and aluminium hydrous oxides are capable of removing sulfate (and with it hydronium ion) by specific complexation. One possible way to describe this process is

$$Al(OH)_3 \text{ (s)} + SO_4^{2-} \text{ (aq)} + 2H_3O^+ \text{ (aq)} \rightarrow AlOHSO_4 \text{ (s)} + 4H_2O \tag{18.5}$$

The reaction is a specific one between sulfate (but not nitrate) and solid aluminium or iron hydroxide, and while taking up sulfate it simultaneously eliminates hydronium species. The net result is diminished soil solution acidity and also diminished ionic concentration through removal of both positively and negatively charged species. The simple equation, however, describes only one possible means of interaction. An alternative removal process is through specific adsorption of the sulfate on to the amorphous hydrous oxide surface. Hydronium ion is then retained as a counter ion. Either process can be important ways of removing both sulfate and acidity simultaneously in tropical Oxisols and other red soils, as well as in the depositional horizons of Spodosols where aluminium and iron oxides accumulate. In many cases, these will be soils having an intrinsic low pH and these reactions provide a way in which acidic soils can neutralize some additional acidity.

The AlOHSO$_4$ is not very insoluble and there are situations where the soil solution is sufficiently dilute that sulfate and aluminium are released back into the aqueous phase. Depending on the solution pH and other factors, the aluminium ion goes into solution as Al^{3+}, $AlOH^{2+}$, $Al(OH)_2^+$, or is complexed with some other ligand such as F^-. When redissolution occurs, the elevated aluminium concentrations can have a serious detrimental effect on the growth of plants. If water carrying soluble aluminium species reaches lakes or rivers, the metal can also be toxic to a variety of aquatic species.

The three types of neutralization reactions described here occur sufficiently rapidly to cause the pH of an acidic solution to rise substantially during the time it takes to pass down the soil column. There are other reaction types, including some of the weathering reactions described in Chapter 17, that consume hydronium ions much more slowly. The reaction with orthoclase feldspar is a good example:

$$2KAlSi_3O_8 \text{ (s)} + 2H_3O^+ \text{ (aq)} + 7H_2O \rightarrow Al_2Si_2O_5(OH)_4 \text{ (s)} + 4H_4SiO_4 \text{ (aq)} + 2K^+ \text{ (aq)}$$
$$\tag{18.6}$$

While this reaction does occur, it takes place extremely slowly—too slow to be of any significance during an individual rainfall event.

18.4.4 Biological processes that neutralize acidity

Besides being under geochemical control due to processes that we have just looked at, acid inputs are simultaneously subject to biological control. Most importantly, acidity associated with nitrate (i.e. nitric acid) is influenced by the fact that nitrogen in the form of nitrate is a macronutrient for all plants. Therefore, where plants are growing, the nitrate is taken up into the roots by reactions that begin with an anion-exchange process at the root surface. (see p. 342) The exchanging anion is usually considered to be carbonate:

$$\text{Root}: CO_3 + 2NO_3^- \text{ (aq)} + H_3O^+ \text{ (aq)} \rightarrow \text{Root}: (NO_3)_2 + HCO_3^- \text{ (aq)} + H_2O \qquad (18.7)$$

The reaction shows that as nitrate is taken up, some of the hydronium ion goes toward neutralizing the released carbonate weak base, and thus the soil pH is buffered. Obviously, where acid is added to or generated in a fallow field or clear-cut forest, the biological control processes are limited.

Clearly, there are several possible abiotic and biotic reactions related to neutralization of acids that can occur within the soil. In all cases, both the soil and the soil solution are altered in the process. The effect may be to temporarily enhance the growth of plants by supplying nutrients such as nitrate or soil-derived ones such as calcium and magnesium. In other situations, potentially toxic elements like aluminium are dissolved and adversely affect plant growth.

Looking at the other side of the issue, we see that soils that are most likely to be unable to neutralize acid inputs are those having no free carbonate minerals, small CEC values, small content of iron and aluminium oxides, and ones where no vegetation is growing.

Box 18.2 Acid sulfate soils

Acid sulfate soils are characteristic of marine coastal plains in areas rich in organic matter. They are associated with brackish water, especially in areas covered by mangrove swamps. Over 2 million ha of soil in Vietnam are developed on pyrite-rich former marine deposits.

In submerged areas, organic debris from the mangroves accumulates, producing a dark-coloured organic muck and a highly reducing environment. The sea water provides a continuous source of sulfate which, under the anaerobic conditions, acts as an electron acceptor for the oxidation of the organic matter, as has been shown in Section 15.4. The reaction is mediated *via* the sulfur-reducing bacteria *Desulfovibrio* and, depending on pH, produces hydrogen sulfide whose odour is often detectable in such coastal areas.

$$SO_4^{2-} \text{ (aq)} + 2\{CH_2O\} + 2H_3O^+ \text{ (aq)} \rightarrow H_2S + 2CO_2 + 4H_2O \qquad (18b.1)$$

The low pE conditions also favour solubilization of iron (III) minerals by reducing the metal to iron (III), which then reacts with the reduced sulfur to form pyrite (FeS_2).

A possible overall representation of the complex set of reactions is

$$4Fe(OH)_3 + 8SO_4^{2-} \text{ (aq)} + 16H_3O^+ \text{ (aq)} \rightarrow 4FeS_2 + 15O_2 + 30H_2O \qquad (18b.2)$$

Over time this reaction proceeds and builds up a plentiful supply of pyrite that acts as a reservoir for the reduced sulfur.

The waterlogged soils have a pH that is usually in the slightly acidic region but if the soil is allowed to dry, the reverse of reaction 18b.2 occurs with the release of large quantities of sulfuric acid. As a result, the pH of the 'reclaimed' soil may fall as low as 1.5 or 2. The deposited iron (III) compounds give the fine-grained organic soil a yellow mottling and it is frequently called a 'cat clay'. The low pH results in high concentrations of soluble aluminium and these two factors alone make such soils completely unsuitable for agriculture. Manganese toxicity and phosphorus deficiency have also been observed.

For these reasons, coastal soils that have the potential to become acidic should not be drained. In the Mekong Delta, areas have been kept continually submerged and used for lowland rice culture. Previously drained soils can be treated with lime if the acidity is not too great. Where the pH is less than 3.5, reclamation requires uneconomically large quantities of lime and the soils are frequently abandoned.

18.4.5 Salt-affected soils

We saw in Chapter 5 that precipitation always contains small concentrations of many elements. When rain percolates freely through a well drained soil, some of these dissolved ionic species are retained at various depths by interaction with soil particles. At the same time, weathering and leaching can cause dissolution of elements from the soil. Soil pore water composition therefore is determined by a combination of removal and dissolution reactions. Overall, the drainage water usually contains only a very small concentration of ionic species, and there is no significant accumulation of salts in any part of the soil profile.

In contrast, where there is limited precipitation and high rates of evaporation, the downward movement of water may be insufficient to leach out all the salts that accumulate near the soil surface. Salt-affected soils are therefore common in arid and semi-arid regions of the world including parts of Tunisia, Iraq, Sudan, Pakistan, and Australia. Salinity and related problems arise when, over extended periods, evaporation and evapotranspiration from the soil exceed the downward percolation of rainfall or irrigation water. When the input water itself contains relatively high concentrations of salts, the possibility of the soils accumulating salts is enhanced. Typically, loss of water by upward movement away from the surface is greater than by downward movement when the water table is closer than 1 to 1.5 m below the surface.

A number of categories of salt-affected soils are classified with respect to their pH, electrical conductivity (EC) of the saturated extract, cation exchange capacity, exchangeable sodium percentage (ESP), and sodium absorption ratio (SAR). The latter two terms require some explanation.

Exchangeable sodium percentage is the fraction, expressed as a percentage of exchangeable sodium ions $[(Na^+)_E]$ compared to total exchangeable ions (CEC):

$$ESP = \frac{[(Na^+)_E]}{CEC} \times 100. \qquad (18.8)$$

A related parameter is the sodium adsorption ratio (SAR); it is used to characterize the sodium content of soil solutions and is defined as follows:

$$SAR = \frac{C_{Na^+}}{(C_{Ca^{2+}} + C_{Mg^{2+}})^{\frac{1}{2}}} \qquad (18.9)$$

The concentrations (mmol L^{-1}) C_{Na^+}, $C_{Ca^{2+}}$, and $C_{Mg^{2+}}$ are measured on aqueous extracts of the soil. Units of SAR are (mmol $L^{-1})^{\frac{1}{2}}$.

Electrical conductivity is a measure of total ionic concentration of the soil solution and is measured on a saturated extract.

- 'Normal' soils have EC values less than $4\,dS\,m^{-1}$ and ESP values below 15%.[8]
- Saline soils have EC values greater than $4\,dS\,m^{-1}$ with an ESP less than 15%. Because the soluble salts in saline soils are mostly neutral—made up of cations such as Ca^{2+} and Mg^{2+} and the anions Cl^- and SO_4^{2-}—the pH of saline soils is generally under 8.5.
- Sodic soils have EC values less than $4\,dS\,m^{-1}$ but ESP greater than 15%. The concentration of neutral salts is small and salts such as sodium carbonate are important. In water the carbonate ion hydrolyses, producing hydroxyl ion, and the pH of these soils is typically high, lying between 8.5 and 10.
- Saline–sodic soils are a fourth category of salt-affected soils. These soils have both high EC ($> 4\,dS\,m^{-1}$) and high ESP ($> 15\%$). The presence of a large concentration of neutral salts usually maintains the pH value at less than 8.5.

The high concentration of accumulated salts in soil leads to several environmentally significant consequences. Fine-grained soils have a large clay content and the clay minerals become dispersed or 'peptized', meaning that the individual particles remain as separate units. The problem is especially acute for sodic and sodic–saline soils, because it is the large (hydrated radius) monovalent sodium ion that contributes the most to dispersion of the clay minerals. The result is that the soil tends to lose any structure which it previously exhibited, and becomes highly impervious to the movement of water. With regard to plant growth, the elevated salt concentration requires an expenditure of energy as a physiological response by the plant in order to maintain a constant water potential gradient between the root and soil solution. As a result, plant growth is inhibited. Furthermore, high concentrations of particular ions like sodium in the soil solution can create a nutrient ion imbalance; for example, calcium deficiency is frequently observed in high-sodium soils. In some cases, toxic levels of certain elements—boron is a common example—may be reached in salt-affected soils. Where soils are alkaline, the hydroxyl ion contributes to toxicity along with the other factors.

Depending on the situation and the specific properties of the salt-affected soil in question, there are a number of technologies that can be applied for their reclamation.

Flushing the salts from a saline soil is perhaps the simplest remediation procedure. This requires using water that is itself low in salts and ensuring adequate drainage of the leached water. By flushing, neutral soluble salts are removed from the rooting zone and the conductivity moves toward a normal value. Because of low hydraulic conductivity, simple flushing cannot be used when the ESP of the soil is initially large.

For saline–sodic soils, a similar flushing procedure can be used, but the input water should contain a high concentration of calcium and/or magnesium ion in order to increase the soil permeability. As a result, exchangeable sodium ions are replaced by those of the divalent alkaline earth metals. Subsequent flushing using water with low ion concentrations can then be done without dispersion of the clays in order to bring the conductivity into the normal range.

Because sodic soils have a high concentration of sodium ion and are also alkaline as a result of the carbonate ion, it is necessary to remove both these species. One treatment is to add to the soil a large amount of calcium sulfate (gypsum) while maintaining continuously moist conditions. The calcium sulfate reacts with sodium carbonate to produce insoluble calcium carbonate and soluble (neutral) sodium sulfate. At the same time, a large fraction of the sodium on the exchange sites is replaced by calcium. Using low-conductivity water, subsequent flushing of the sodium sulfate from the soil is then possible.

An alternative treatment procedure makes use of elemental sulfur, which is worked into the surface soil where it is oxidized microbially to produce sulfuric acid. The acid again serves to convert sodium carbonate into sodium sulfate, which is then washed out of the soil.

It is important to note that soils, especially those in the sodic categories, can be irreversibly affected by excess salt concentrations. Such soils may be made so impervious to water movement that flushing is almost impossible. Attempts to enable some movement of water have been effected by reducing the bulk density through intensive tilling. This allows water-carrying calcium ions to penetrate into the macropores, resulting in a limited degree of leaching.

18.4.6 Trace metals in soils

In considering the issue of salt-affected soils, we were concerned with the accumulation of metal ions in the form of soluble salts near the soil surface. For the most part the metal ions were sodium, potassium, calcium, and magnesium all type A metals. In different environmental circumstances, other metal ions in the borderline and type B categories can also accumulate in the soil, although the amounts are usually much smaller than those of the alkali and alkaline earth families. Nevertheless, in some cases their availability for plant uptake or to be leached into groundwater is an important environmental issue.

Excessive concentrations of metals can be associated with naturally occurring ore deposits, but more usually are the result of human activities. In the next chapter, we will examine some specific cases where waste materials containing metals are disposed of on land. Here we will review the general principles that determine the fate of metal ions in the soil. With soil, we are always dealing with soil/water relationships, so all of the principles discussed in Chapter 13 are applicable in the present case as well.

With respect to water, we emphasized that all samples, even samples that are considered to be 'pure', have trace quantities of 'impurities'. Not surprisingly, this situation is even more dramatically expressed in soils, where it is observed that there is a background

Table 18.8 Concentrations ($\mu g\ g^{-1}$) of minor metals, including some semimetals in uncontaminated soils (values given as approximate means; bracketed values are ranges).

Element	World[a]	World[a]	World[b]		USA[a]	Canada[a]	
Arsenic			6	(0.1–40)			
Cadmium	0.5		0.06	(0.01–7)		<1	
Chromium	200	(100–300)	100	(5–3000)	53	43	(10–100)
Cobalt	8	(10–15)	8	(1–40)	10	21	(5–50)
Copper	20	(15–40)	20	(2–100)	25	22	(5–50)
Lead	10	(15–25)	10	(2–200)	20	20	(5–50)
Manganese	850	(500–1000)	850	(100–4000)	560	520	(100–1200)
Mercury	0.01				0.071	0.059	(0.005–0.1)
Nickel	40	(20–30)	40	(10–1000)	20	20	(5–50)
Selenium	0.01		0.5	(0.1–2)	0.45	0.26	(0.03–2)
Strontium	350		300	(50–1000)	240	210	(30–500)
Zinc	50	(50–100)	50	(10–300)	54	74	(10–200)

[a]Data reported in McKeague J. A. and M. S. Wolynetz, Background levels of minor elements in some Canadian Soils, *Geoderma* **24** (1980), 299–307.
[b]Data reported in Allaway, W. H., Agronomic controls over the environmental cycling of trace elements. *Advances in Agronomy*, **29** (1968), 235–74.

concentration of less common metals and other elements as well. The observed levels are variable all over the Earth's surface and depend on many factors. Table 18.8 gives some background concentration ranges and mean values that have been measured by various researchers on soils that are not considered to be 'contaminated' in any special way.

Where do the trace elements come from? To some extent, they are derived from the original minerals that were subject to weathering and produced the mineral soil. This does not necessarily mean that they reflect the composition of the associated bedrock, as many soils have been transported from other locations to their present site by wind, water, or ice. Likewise the organic components of the soil contain small amounts of many metals and they too may be derived from either local or distant sources. To these important 'original' sources of metals in the soil, other inputs are added. One of these additional sources is through deposition from the atmospheric aerosol. A detailed inventory[9] of the origin of metals in the atmosphere shows that important sources include wind-blown soil particles (dust), volcanoes, and volatile and particulate organic matter derived from forested areas and from the sea. The biogenic sources are especially important for many metals (accounting for more than 30% of the annual releases to the atmosphere) including most of those listed on Table 18.8. Industrial emissions are responsible for much of the flux of lead, cadmium, and zinc.

In soil, the metals are present in a variety of forms. In some cases they are structural components of soil minerals or minor constituents incorporated into soil organic matter. Metals can also be deposited by specific sorption processes on to surfaces of preexisting minerals of various types. An additional fraction of soil trace metals is accounted for by electrostatic ion retention on mineral or organic matter exchange sites. Finally, a (usually very small) concentration of metal species is found in the pore water of any soil.

Environmental issues regarding soil trace metals often centre around their mobility and this is related to the form in which it is present as well as the environmental situation. The most important environmental factors are the amount, chemical nature, and movement of water through the soil. Metal associated with the original inorganic material may not be readily available for uptake or leaching. Especially in the case of silicate species, such metals are resistant to weathering, tend to persist in the solid structure, and therefore do not achieve the mobility associated with being in solution. Aluminium present in alumino-silicate minerals such as the feldspars is essentially inert on a time-scale of years or decades. It is only over much longer time-scales that the weathering alterations we noted earlier become significant.

There are other associations involving the mineral phase, in particular metals chemically and specifically bonded to surfaces of various minerals. As has been frequently mentioned, the hydrous oxides of iron and aluminium are particularly active sorbents. Mobilization of metals in these associations can occur as a result of acid conditions or changes in the redox status of soils. Reducing conditions are especially significant in this context. When a soil that has previously existed in an oxidizing environment—under well aerated conditions—is flooded, oxygen movement is restricted and the pre-existing oxygen is used up in microbial aerobic reactions. Once the oxygen is depleted, other species can act as electron acceptors in conjunction with facultative anaerobes in the soil. We saw earlier that sulfate and nitrate were two such species. Solid mineral phases can serve the same purpose, manganese (IV) oxide and iron (III) oxide, both in hydrated forms are two important solid electron acceptors. Studies under controlled conditions have shown that a sequence of reduction events occurs when a soil is flooded.[10] The order and approximate pE range over which events occur are

$$O_2 \text{ (aq)} + 2H_3O^+ \text{ (aq)} + e^- \rightleftharpoons H_2O_2 \text{ (aq)} + 2H_2O \qquad pE = 6.3 \text{ to } 5.8 \qquad (18.10)$$

$$H_2O_2 \text{ (aq)} + 2H_3O^+ \text{ (aq)} + 2e^- \rightleftharpoons 4H_2O \qquad\qquad\qquad (18.11)$$

$$2NO_3^- \text{ (aq)} + 12H_3O^+ \text{ (aq)} + 10e^- \rightleftharpoons N_2 \text{ (g)} + 18H_2O \qquad pE = 4.2 \text{ to } 3.6 \qquad (18.12)$$

$$MnO_2 \text{ (s)} + 4H_3O^+ \text{ (aq)} + 2e^- \rightleftharpoons Mn^{2+} \text{ (aq)} + 6H_2O \qquad pE = 3.6 \text{ to } 3.1 \qquad (18.13)$$

$$Fe_2O_3 \text{ (s)} + 6H_3O^+ \text{ (aq)} + 2e^- \rightleftharpoons 2Fe^{2+} \text{ (aq)} + 9H_2O \qquad pE = 1.9 \text{ to } 1.4 \qquad (18.14)$$

$$SO_4^{2-} \text{ (aq)} + 8H_3O^+ \text{ (aq)} + 8e^- \rightleftharpoons S^{2-} \text{ (aq)} + 12H_2O \qquad pE = -0.7 \text{ to } -1.4 \quad (18.15)$$

The reactions are reversible, although oxidation occurs over a pE range that is approximately 0.8 units higher than for the corresponding reduction. The consequence of reactions 18.13 and 18.14 is that hydrated manganese and iron oxide minerals in soils are subject to reduction, a process that mobilizes the metals in the form of +2 species.

Manganese and especially iron are quantitatively important soil constituents, but trace elements are subject to similar mobilization/immobilization phenomena. Some of the phenomena are second order effects related to the behaviour of major elements. Arsenic is a good example.[11] Under high pE conditions, arsenic is present as $H_2AsO_4^-$ and $HAsO_4^{2-}$ species (see Chapter 10, problem 8) and these are strongly adsorbed by hydrous iron oxides in the soil. Under reducing conditions, solubilization of iron by reaction 18.14 releases, along with iron, arsenic that has been sorbed on the iron oxide surface. The reducing situation also converts the As(V) species into As(III) in the form of the weak acid H_3AsO_3, under most environmental conditions. Another factor controlling metal solubility is the precipitation of metal sulfides. When pE is low enough that sulfate is reduced to sulfide (reaction 18.15), the S^{2-} forms highly insoluble compounds with several metals including copper, nickel, and zinc, so that their already low solubility in soil pore water is suppressed even further. If the soil matrix reoxidizes, the trace metals are then released back into the associated solution.

Organic matter can act either to mobilize or to immobilize metals in soil. The solubility of metals that are structural components of organic matter, or that form strong complexes with it, is determined by the solubility of the associated organic matter. Often, decomposition to form products that are smaller and more soluble is an important factor in increasing the solubility of such metals. In temperate forest regions, the mineral soil is overlain by a regularly renewed layer of organic litter. The litter is subject to degradation and the products include small molecules like organic acids as well as a stable insoluble humic material. The organic acids act as ligands for many metals, enhancing movement in percolating water down through the soil profile. In part, this explains the formation of an eluvial layer in a Spodosol.

Some of the low molar mass fulvic acid material is itself soluble, giving the soil pore water a characteristic yellow or light brown colour. It is interesting that the solubility of soil humic material is larger when the pore water is neutral or mildly alkaline rather than acidic. Metals complexed with this soluble material are, of course, themselves made soluble in the high pH situation. In this respect, metal solubility may not follow the usual pattern of being greater in acid conditions.

There are also ways in which organic matter—the more insoluble, high molar mass humic fraction—acts as a reservoir to reduce the mobility of metals. Examples of this ability of soil humus to inhibit metal leaching in soils are well documented. Lead, released from combustion of leaded gasoline in vehicle engines, is deposited in soils on roadsides, and it has been observed that the metal accumulates in the surface of the soil in association with organic matter. Over many years, in most locations, only very limited movement downward has been observed.

The metals associated with the exchange complex, both mineral and organic, are a small fraction of the total metal content in most instances. Yet this fraction is readily available for plant uptake—a positive consequence when the metal is a required micronutrient, but negative if it exerts a toxic effect. The exchangeable metals are the most readily mobilized. They can be displaced by other ions that are present in large concentration in the soil solution, including the hydronium ion. It is this portion of soil metals that is made more soluble by acid inputs, as we noted earlier.

It is clear, then, that many factors control the ability of a metal to move down the soil profile in the soil water. Added to the chemical issues, the actual flux of water also plays a major role. Leaching of metals is therefore favoured where there is high rainfall and where soils are coarse textured, allowing water to move rapidly downwards. This was a factor explaining the development of highly leached tropical soils such as Oxisols, but it is also a factor that determines leaching of contaminant metals under various environmental conditions.

The main points

1 Besides being a medium for the production of food, fibre, and fuel, soils are also an important environmental agent.

2 The physical and chemical properties of soils both influence the ways in which they interact with chemicals that are added to or generated within the soil.

3 Soil texture and structure affect the flow of water through the soil, but also determine the extent of solid surface available to react with chemicals moving through the pores.

4 The chemical nature of the soil plays a major role in determining the types of interaction—retention or release of species—with the interstitial water. The ion-exchange properties associated with soil minerals and organic matter are especially important. Other geochemical properties and also biological reactions involving uptake by plants and decomposition are further factors that determine the fate of chemicals in the soil.

5 Saline conditions, acid additions, redox changes, metal contamination are some of the frequently encountered environmental issues related to soil behaviour.

Additional reading

1 Ellis, S. and A. Mellor, *Soils and Environment*, Routledge, London; 1995.

2 Greenland, D. J. and M. H. B. Hayes, eds, *The Chemistry of Soil Constituents*, John Wiley and Sons, Chichester; 1978.

3 McBride, M. B., *Environmental Chemistry of Soils*, Oxford University Press, New York; 1994.

4 Pitty A. F., *Geography and Soil Properties*, Methuen and Co, London; 1978.

Problems

1 The permeability of the Black Cotton soils (high in montmorillonite) of the Deccan Plateau in central India is very high (up to $20 \, cm \, h^{-1}$) at the beginning of the monsoon season, but soon becomes much less (below $1 \, cm \, h^{-1}$) as the rains continue. Suggest an explanation.

2 Discuss the extent and types of water contamination problems that are possible when septic tanks for sewage disposal are located in (a) sandy or (b) clayey soils.

3 Explain why clay-rich soils have desirable physical and chemical properties for use as liners for landfill sites.

4 Radiocaesium is one of the nuclides arising from the fallout from the Chernobyl reactor incident in 1986. Some of it rained out over the United Kingdom. In the soil, caesium is either immobilized or is taken up by plants that are then eaten by grazing animals. In the UK, controls come into force where levels in animal flesh are greater than $1000 \, Bq \, kg^{-1}$. In 1987, levels as high as four times this value were measured in some sheep. Highest

animal concentrations were measured where pasture was developed on upland soils, high in organic matter; lower levels were found where the sheep grazed on lowland soils, rich in clays. Suggest a reasonable explanation for these observations.

5 Consider the following data for a forest soil:

		Bulk density/g mL^{-1}	Particle density/g mL^{-1}
O	(−5 to 0 cm)	0.19	1.78
E	(0 to 8 cm)	1.08	2.61
B	(42 to 66 cm)	1.52	2.65

Comment on the significance of these values in terms of porosity and permeability in each horizon.

6 Salts are commonly spread on highways during winter to prevent build-up of ice. Sodium chloride is most commonly used, but calcium magnesium acetate $(Ca_{0.3}Mg_{0.7})(C_2H_3O_2)_2$ has been recommended as an alternative because it biodegrades, is less toxic to aquatic life and is less corrosive. There is concern, however, that it might increase the mobility of trace metals in roadside soils. What metals would be of concern and how could these two salts affect their mobility in soils?

7 With reference to an ion exchange medium, selectivity refers to the thermodynamic tendency to retain a particular species. The order of selectivity for alkali metal cations by most clays is

$$Cs^+ > K^+ > Na^+ > Li^+$$

Explain this in terms of the aqueous solution chemistry of these ions.

8 In soils of the eastern United States, the CEC (cmol (+) kg^{-1}) has been described as related to the OM (concentration as %) by the following equation:

$$CEC = 4.83 + 3.87\,OM$$

with $r = 0.73$ and $N = 57$ (r is the correlation coefficient and N is the number of data points). How does this relation correspond to the generalities presented in this chapter?

9 Logging of a forest by removing all the mature trees is a controversial forestry practice. Aside from issues such as the effect on species biodiversity and erosion, clear-cutting can alter chemical processes in the soil and even in the global environment. Explain how this practice could lead to increased nitrification and denitrification—and how this may affect soil acid–base properties and the stratospheric ozone concentration.

10 The following are chemical properties of two Venezuelan surface soils.
Predict their relative sensitivity to acidic inputs from rain or fertilizer and give reasons for your prediction.

	pH	OC	N	Clay	Ca	Mg	K	Na	Al	CEC/cmol kg^{-1}	BS/%
		%									
Machiques	6.0	0.75	0.08	7.2	0.3	1.4	0	0.01	0.1	3.7	44
Barinas	5.6	1.41	0.17	32.6	6.6	0.9	0.5	0.1	1	8.4	96

11 In a waterlogged soil, carbon dioxide from decomposition processes may increase, aqueous oxygen levels may decline to near zero due to impeded gas exchange, and the soil solution may accumulate significant concentrations of hydrogen sulfide and methane. Discuss possible effects of these properties on the behaviour of aluminium, iron, and manganese in the soil.

12 Traditional agricultural methods are used throughout the Zaire river basin and have been described.[12] Comment on the significance of the following practices in terms of soil chemistry and other related properties.

(a) Land with the thickest vegetation is cleared (by fire) before planting sorghum or manioc.

(b) Several crops are grown together or in sequence in the same area, so that the land is covered with vegetation over an extended period of time.

(c) Household waste, sod, and dry grass are incorporated into the soil before or after planting. Sometimes the compost is incinerated in mounds, later used for planting root crops such as yams.

(d) In some areas, nomadic communities are invited (sometimes even paid) to set up temporary animal corrals after which the site is used for cropping.

13 The planting of high-yielding hybrid varieties of grains is widespread and has contributed to increasing the global grain supply. A co-requirement is the application of large amounts of fertilizer to support the high levels of growth. For example, corn (maize) yields of greater than $20\,t\,ha^{-1}$ are obtainable under good agronomic conditions. This might require the addition of $200\,kg\,ha^{-1}$ of nitrogen in the form of urea. Calculate the amount $(kg\,ha^{-1})$ of limestone ($CaCO_3$) that should be added to the soil just to balance the acidity generated by this fertilizer, assuming complete nitrification of the urea.

Notes

1 Note that in soil terminology, clay may refer to either a size category or a class of minerals. To distinguish between these two usages, it is best to speak of clay-sized materials and clay minerals respectively.

2 Nommick, H., Ammonium chloride–imidazole extraction procedure for determining titratable acidity, exchangeable base cations and cation exchange capacity in soils. *Soil Science*, **118** (1974), 254.

3 Bentley, C. F., ed., *Photographs and Descriptions of Some Canadian Soils*, The University of Alberta, Edmonton; 1979.

4 Terminology used to define soil horizons is a complex and controversial subject that we cannot be concerned with here. In the occasional assignments of soil types employed in this book, we will use the US Soil Taxonomy system (Soil Survey Staff, *Keys to Soil Taxonomy*. AID, USDA, SCS, SMSS Techn. Monograph No. 19, 5th ed. Pocahontas Press, Inc., Blacksburg, VA; 1992).

5 The prefix 'sesqui' means one and a half, indicating that there are 1.5 oxygens for each metal in iron (III) and aluminium (III) oxides.

6 Murthy, R. S., L. R. Hirederur, S. B. Deshpande, and B. V. Venkata Rao, eds, *Benchmark Soils of India*, National Bureau of Soil Survey and Land Use Planning (ICAR), Nagpur; 1982.

7 Dudal R., *Dark Clay Soils of Tropical and Subtropical Regions.*, Food and Agriculture Organization of the United Nations, Rome; 1965.

8 The traditional units used to measure electrical conductivity are mmho cm^{-1} (a mho is a reciprocal ohm) which is equivalent to the SI unit decisiemens per metre (dS m^{-1}).

9 Nriagu, J. O., A global assessment of natural sources of atmospheric trace metals, *Nature*, **338** (1989), 47–9.

10 Patrick, W. H., Jr and A. Jugsujinda, Sequential reduction and oxidation of inorganic nitrogen, manganese, and iron in flooded soil. *Soil Sci. Soc. Am. J.*, **56** (1992), 1071–3.

11 Masscheleyn, P. H., R. D. Delaune, and W. H. Patrick, Jr, Arsenic and selenium chemistry as affected by sediment redox potential and pH. *J. Environ. Qual.*, **20** (1991), 522–7.

12 Miracle, M. P., *Agriculture in the Congo Basin*. The University of Wisconsin Press, Madison; 1967.

19

The chemistry of solid wastes

In any human society, bulk solid wastes are produced as a byproduct of the normal and fundamental activities of living. These wastes can be as rudimentary as food scraps, ash from fires, and excreta from humans and animals. In a modern, highly industrialized society, however, the wastes go much beyond the fundamental materials both in quantity and variety. The amounts of waste produced by intensive agriculture and by modern industry are staggering, to say nothing of waste generated by ordinary citizens in a wealthy consumer-oriented urban setting. Table 19.1 compares bulk solid waste products associated with various human activities.

The term *bulk* wastes distinguishes these large volume waste products from specific chemicals such as pharmaceuticals, hospital wastes, dyes, and chemical additives that are sometimes discarded, but over which detailed control protocols should be followed.

We use the word waste with some hesitation. Almost any substance that is discarded, and is therefore designated as waste, can also be thought of as a potential resource. Throughout human history, societies have found ways in which to use waste—organic residues as fertilizers, animal dung as fuel, inert materials as landfill. In the present era too, there is effort to discover new uses for materials that have served their primary purpose. This has given rise to the concepts of reuse and recycling. Alongside these efforts are ones to minimize the

Table 19.1 Bulk solid waste materials associated with various human activities

Human activity	Bulk solid waste materials
Food production	Plant residues, animal manures
Food consumption	Food wastes, packaging materials (paper, plastic, aluminum, steel, glass), sewage sludge
Shelter	Construction materials, soil and other landscaping wastes, ash from heat production
Transportation	Scrap metal, rubber, road and rail construction wastes
Consumable products	Metals, ceramics, polymers, paper, and other fibres

production of waste byproducts so that a reduced quantity of residual material remains to be discarded.

With or without the reduction/reuse/recycling efforts, some solid materials that must be discarded are inevitably an end product of many activities.

Disposal of bulk solid wastes

In the same way that liquid (aqueous) wastes from homes, factories, and agriculture are often discharged into water bodies, so also many solid waste materials are disposed of on land. The solid wastes cover a wide range of types—everything from animal manures, to municipal garbage, to waste steel from automobiles, to mine tailings. The waste materials inevitably interact chemically and physically with environmental components, but the time-scale over which such reactions occur may be short or long depending on the type of materials and their degree of exposure. Two important environmental components that play a central role in many reactions are oxygen and water and the solid wastes undergo various kinds of 'weathering' processes. The resulting products from weathering can include gases or liquids that become part of the atmosphere or hydrosphere. In most cases, residuals and reaction products remain with the solid phase of the waste itself or migrate into the soil on which the waste has been disposed.

Certain types of solid wastes are deliberately spread out over a large area of land in order to maximize the degree of mixing and interaction with the soil. Composted solids and manures are examples of this. Mixing is done so that the beneficial effects of the added solid are assimilated by the soil that receives it.

In contrast, for some wastes the degree of interaction is made as small as possible. Mixed urban waste, for example, can be compacted and then encapsulated by using synthetic membranes to isolate the garbage from the atmosphere, land, and groundwater. By minimizing free movement of air and water from outside the landfill, a different set of reactions takes place within the entombed material.

Because of the wide variety of solids that have been disposed of on land, it is best (once again) to consider the subject of solid wastes in terms of general principles. We will use some of the most common examples to illustrate these principles.

19.1 Solid wastes from mining and metal production

In total volume and mass, tailings are the major waste product from many mining operations. Throughout the world, these deposits cover hundreds of thousands of hectares of land. Tailings are the rejected rock produced during upgrading of a mined ore. In many mineral deposits, the original concentration of the ore mineral itself is very small—a few parts per million in the case of gold, for example—and much of the associated rock must be rejected. A common method of upgrading or *beneficiation* is by flotation. This process requires that the entire mined product be ground to small particle size and separated by gravity, aided by surfactants in an aqueous slurry. The *concentrate* is then sent for further processing and the *gangue* or *tailings* are piped out into a large receiving pond. This major portion of the originally mined rock comprises the tailings. Besides the flotation method, other physical and chemical processes are used depending on the nature of the mineral and the form in which it is found. The waste materials are variously referred to as tailings, spoils, or slimes.

Relatively coarse-grained tailings drain readily, leaving a dry sandy deposit that either fills a depression in the landscape or is built up into an artificial hill. Fine-grained deposits, and those that have an affinity for water, may not easily become dry and so stay partially wet or sometimes remain totally under water. Some types of tailings are deliberately kept submerged in a tailings pond.

The physical and chemical properties of tailings are highly variable depending on the nature of the ore mineral itself and of the host rock. Problems associated with disposal of tailings depend on the particular properties. Three commonly encountered situations will be discussed briefly here.

19.1.1 Benign tailings deposits

In some cases, the gangue minerals that form the bulk of the tailings are relatively inert. The diamond mines of South Africa represent such a case. There, the diamonds are found as clusters in funnel-shaped 'pipes', which are intrusions of kimberlite into the host rock. The kimberlite is variable in composition but usually consists of quite resistant ultrabasic igneous minerals including olivine, biotite, garnet, and ilmenite. The igneous intrusion contains sporadic inclusions of fossiliferous shales derived from earlier sedimentary deposits. There are many of these diamond-rich pipes around the country, with major sites being located near Kimberley and Pretoria.

Tailings from such mines are made up of finely ground rock fragments whose mineral composition reflects that of the kimberlite host rock. The minerals are ones that undergo only very limited and very slow chemical reactions after exposure to air and water. Because they are relatively inert, leaching of metals from them is minimal. As a result, chemical contamination of associated water bodies only rarely occurs.

However, if these tailings remain untouched as a massive deposit of inert material, they are unsightly and wind-blown dust is an aesthetic problem as well as a potential health hazard. For these reasons, major efforts have gone into stabilizing the deposits. The most appropriate method of stabilization is usually considered to be by vegetating them. To do this, the physical and chemical properties of the tailings as plant growth media must be considered.

Most tailings have relatively homogeneous particle size distributions, with particle size median values in the sand or large silt-size regions. Because intrinsic organic matter concentrations are close to zero, there are no agents for binding particles and so the deposit has little 'soil' structure. As a consequence, permeability is high, drainage is rapid, and there is very limited water-holding capacity. Where the tailings are dark coloured, their albedo is small and they are able to absorb solar radiation efficiently. Therefore, daytime temperatures in the surface material can be extremely high. Taken together these factors are responsible for a dry, hot medium that is not a hospitable site for growth of plants.

There is a second problem associated with trying to establish vegetation on benign tailings deposits. Tailings such as those from kimberlite diamond mines do not contain natural sources of the major plant nutrients such as nitrogen, phosphorus, potassium, and calcium. Furthermore, the fact that the mineral components are inert means that release of minor and trace nutrients into available forms is not sufficient to support the growth of most plants.

One solution to the combined physical and chemical problems is to incorporate organic matter (OM) into the surface of the tailings. Along with the organic amendment, chemical fertilizers are added at the time of planting and/or afterwards. The OM serves to improve the nutrient and water-holding capacity and can be a source of some nutrients, while fertilizers supplement the nutrient supply. Without the OM, cation-exchange capacity of the coarse-grained, unreactive material is so small that any soluble cations not immediately taken up by plants are leached from the deposit. Complexing components of the OM provide cation-exchange sites for retention of positively charged species. Furthermore, a portion of the complexing components go into solution and enhance the rate of chemical weathering. Once a growth cycle has been established, natural processes of litterfall and decay at least partially maintain the organic content of the 'soil'. The processes of soil formation then proceed at an accelerated rate.

19.1.2 Tailings from sulfide ore deposits

Tailings containing several per cent sulfur as sulfide minerals present additional problems for reclamation. The sulfide minerals, especially pyrite, FeS_2, are present in the host rock associated with several important ore minerals including copper, nickel, lead, zinc, and sometimes gold. Sulfides are also present in coal deposits and significant concentrations are found in the spoils associated with coal mining. Table 19.2 gives the composition of sulfide tailings associated with gold mining in South Africa and of spoils from a coal mine in Pennsylvania.

Table 19.2 Properties of waste material from two mines containing associated sulfide minerals

		Gold wastes, Witwatersrand, South Africa	Coal spoils, Pennsylvania, USA
Sand and gravel %		50	67
pH		2.5–3.1	3.3
Mg		110	127
Ca	available $\mu g\, g^{-1}$	1400	102
K		15	199
P		13	3
N total %		0.02	0.003
S total %		1.5–3.5	several %

From ref. 1, Additional reading list.

The gold and coal mine tailings suffer from similar problems to spoils defined above as benign; compared with good soils, they are too coarse grained and are low in nitrogen, phosphorus, potassium, and other nutrients. These properties alone make them a challenge with respect to establishing a vegetative cover. In addition, a major problem associated with the presence of sulfur in the tailings is the internal generation of sulfuric acid, giving them a very low pH.

There are several possible reactions that begin with pyrite, and most of them have a microbiological component. The reactions are related to those discussed in Chapter 15. Some that have been clearly identified are as follows.

The sulfur in pyrite is initially oxidized by oxygen in the air:

$$2FeS_2 + 7O_2 + 6H_2O \rightarrow 2Fe^{2+} (aq) + 4SO_4^{2-} (aq) + 4H_3O^+ (aq) \tag{19.1}$$

The acid-tolerant iron-oxidizing bacteria *Thiobacillus ferrooxidans* is then responsible for oxidation of the iron (II) released in the initial step:

$$4Fe^{2+} (aq) + O_2 + 4H_3O^+ (aq) \rightarrow 4Fe^{3+} (aq) + 6H_2O \tag{19.2}$$

The iron (III) that is formed acts as an oxidizing agent and further reacts with solid pyrite, without the need of molecular oxygen:

$$FeS_2 + 14Fe^{3+} (aq) + 24H_2O \rightarrow 15Fe^{2+} (aq) + 2SO_4^{2-} (aq) + 16H_3O^+ (aq) \tag{19.3}$$

Of reactions 19.2 and 19.3, the rate-determining step is the oxidation of iron (II) to iron (III). The sum of these two reactions gives reaction 19.4, which is the same as reaction 19.1. Therefore the three processes, taken together, describe a self-accelerating sequence of oxidation of pyrite:

$$2FeS_2 + 7O_2 + 6H_2O \rightarrow 2Fe^{2+} (aq) + 4SO_4^{2-} (aq) + 4H_3O^+ (aq) \tag{19.4}$$

Because aqueous iron (III) readily loses protons, unless the pH is extremely low (as it can be *within* the tailings deposit itself as described above), oxidation of iron (II) forms a solid product often given the approximate formula $Fe(OH)_3$. Deposits of this solid are frequently observed where the leachate has moved downstream from the highly acid tailings:

$$4Fe^{2+} (aq) + O_2 + 18H_2O \rightarrow 4Fe(OH)_3 + 8H_3O^+ (aq) \tag{19.5}$$

Other mechanisms for pyrite oxidation have also been suggested.[1] One of these begins with the dissolution of iron and the oxidation of sulfide to neutral sulfur:

$$2FeS_2 + O_2 + 4H_3O^+ (aq) \rightarrow 2Fe^{2+} (aq) + 2S_2^0 + 6H_2O \tag{19.6}$$

This is followed by oxidation of iron (II) (reactions 19.2 or 19.5) and of elemental sulfur, the latter reaction mediated by the bacteria *Thiobacillus thiooxidans*:

$$S_2^0 + 3O_2 + 6H_2O \rightarrow 2SO_4^{2-} (aq) + 4H_3O^+ (aq) \tag{19.7}$$

Each of these reactions (except reactions 19.2 and 19.6) is an acid-producing process. Similar reactions can be written for other sulfide minerals such as pyrrhotite (FeS) and chalcopyrite ($CuFeS_2$) that invariably are found in association with pyrite. Over a period of years, the sulfide minerals all oxidize, generating vast quantities of acidity and of sulfate ion. The amount of acid produced is such that leachate emanating from sulfide tailings deposits can have pH values of 1 or even lower. The sulfate-rich leachate is referred to as acid mine drainage. Being a highly acid solution, it solubilizes other metals so that concentrations of metals like copper, zinc, and lead can be very high. All these factors add to the hostile environment for plant, animal, and 'normal' microbial growth in the tailing deposits themselves and in nearby soil and water affected by the leachate.

Reclamation strategies for sulfide-bearing tailings must therefore take into account the essential requirement for reducing acidity of the environment. Liming is one possible amendment, but the limestone treatment must be repeated, as the production of acid is an ongoing process. Keeping the tailings submerged in a pond is also possible as this maintains the sulfide minerals in a stable reduced form. In spite of the difficulties, reclamation and revegetation of sulfide tailings have been carried out and self-supporting grasslands and tree plantations have been established at some sites.

It is interesting to note the similarities between sulfide tailings and acid sulfate soils (Box 18.2)—two environments that arise from quite different circumstances.

19.1.3 Red mud

Red mud is a particular type of waste associated with the production of alumina (Al_2O_3) from bauxite—an initial step in aluminium production. In the Bayer process, aluminium (as $Al(OH)_4^-$ or AlO_2^-) is solubilized by leaching finely milled crude bauxite with concentrated aqueous sodium hydroxide under controlled high temperature and pressure conditions. Alumina is later recovered after neutralizing the solution. The residual solids contain clay minerals, quartz, and insoluble iron and titanium oxides as well as unextracted aluminium oxide. This waste material, called red mud, has a chemical composition that is typically about 15% iron, 15% aluminium 8% titanium, 5% silicon, and contains variable amounts of sodium and calcium, mostly as hydrated oxides. The amount of red mud produced is between 0.5 and 2 t per t of alumina produced and throughout the world, the total annual production of red mud is about 4×10^7 t.

The red mud is pumped as a slurry from the bauxite plant. The viscous slurry contains 20 to 40% solids having a wide range of particle sizes, but a substantial fraction is in the silt to clay size range and complete settling occurs only very slowly. The high sodium content ensures that the clay is higly dispersed. Furthermore, the suspended solids are very hygroscopic and have considerable ability to retain the slurry water. These factors lead to one of the problems associated with red muds—that they do not dry out quickly when disposed of on land. One technique used to speed up the process is to spread a thin layer of the mud over a large area and allow it to dry before repeating the process. While drying, a crust of hardened material forms on the surface, inhibiting evaporation of the deposit underneath. For this reason, the crust is broken up periodically to expose the wet material below.

Another problem associated with the muds is that they are highly alkaline. Unlike acid mine drainage, leachate from alumina wastes does not solubilize most metals, but aluminium itself, because of its amphoteric nature, is an exception. As we have seen, aluminium is soluble in basic solution as the anionic hydroxy ($Al(OH)_4^-$) species (Fig. 16.2). The concentration of the element in leachate can therefore be very high and this, along with the alkaline nature of the solution, contributes to the toxic nature of the leachate.

The combined physical and chemical properties of red muds make disposal in tailing ponds a problem, and alternative disposal methods and uses for the material are being sought. Revegetation is possible only if the deposit is covered with a thick layer of soil or dredged sediment. Attempts to use sewage sludge as an amendment before growing plants have met with some success. Some red mud can be used for making bricks or ceramic products. It has also been shown to function somewhat effectively as a dephosphorizing agent in waste-water treatment.

19.2 Organic wastes

19.2.1 Direct disposal of animal wastes on land

A very common practice, observed worldwide amongst agriculturists, is the application of animal manure to soil as an amendment. Manure is a good source of organic matter, along with nutrients, especially nitrogen, phosphorus, and potassium. The composition of manure is variable, depending, in the first instance, on the species of animal, but also on diet and other conditions. Considering the common livestock species—cattle, goats, sheep, horses, pigs, and poultry—an average daily output of faeces and urine is approximately 60 kg per 1000 kg mass of animal. Average percentage of components in the excreta are approximately: total solids 10%, nitrogen 0.6%, phosphorus 0.1%, and potassium 0.3%. Concentrations are similar for other types of livestock, with the notable exception of poultry for which the nutrient values are somewhat higher than for the mammals. It is organic matter and nutrients that make the manure a particularly good soil amendment. Using it for this purpose also responds to the objective of returning to the soil some of the components that have been removed by grazing or harvesting of plants.

The spreading of animal manure brings up aesthetic issues and there are health concerns associated with the presence of pathological microorganisms. There are also environmental issues connected with the chemical behaviour of components of the manure. Before the nutrient elements can be utilized by growing plants, the organic matter must undergo decomposition, releasing the inorganic forms that are available for uptake. To encourage degradation, a well oxygenated environment is preferable. Even in optimum cases, however, it is likely that some of the nutrients remain in an unmineralized form beyond the growing season, and delayed release of the plant-available nutrient species may occur when there is no means by which it can be immobilized. Depending on the element, the hydraulic conductivity and chemical properties of the soil, and other environmental conditions, nutrients can move through the soil into surface or groundwater.

Potassium is retained as K^+ on the cation-exchange sites of soil materials that have adequate CEC values. Phosphorus is retained in various forms by its association with iron and aluminium hydrous oxides or as insoluble calcium species. Many problems occur, however, with nitrogen. In an oxygenated soil, ammonium that is released by mineralization undergoes nitrification, yielding nitrate as the thermodynamically stable species. As we have seen, most soils have only a small anion-exchange capacity and therefore very little ability to interact with the negatively charged nitrate species. Specific adsorption of nitrate is also not significant. Therefore nitrate tends to be mobile, and excess nitrogen released from manure may end up in water bodies. A particular situation that leads to high levels of nitrate leaching occurs where manure is kept outdoors in piles during winter prior to spreading at the beginning of the growing season. Nitrate leached from the pile is not retained by the soil, and there are no plants to act as a biological control. It can therefore move downwards below the water table; toxic levels in groundwater are sometimes observed in this situation. Similar nitrate problems can arise after the manure is spread on the field.

Leaching of phosphorus and potassium is less common, but these elements can be mobilized by erosion and transport of dried manure particles. We have already noted that non-point source contamination makes a major contribution to eutrophication of some surface water bodies.

Proper management can minimize chemical contamination of water from application of animal wastes. These include controlling the amounts of manure added—typically to about $30\,t\,ha^{-1}$ each year. This is equivalent to supplying one hectare of land with the waste from two cattle, or four or five hogs. Furthermore, to ensure maximum uptake, crops should be planted soon after application. Where heavy feeders like maize are used, there is reduced chance of leaching or run-off losses.

The total salt (chlorides of the alkali and alkaline earth elements) content of manure can be quite large, anywhere from 1 to 10% on a dry mass basis. Concentrations in leachate from the manure can therefore be substantial, and in some cases they are excessive, also having a deleterious effect on plant growth or on the groundwater quality. The relative concentrations of the four major cations are in the order calcium > potassium > magnesium > sodium. This order is similar to that frequently observed for cations on the exchange complex and therefore it does not upset the normal ionic balance. It is the total salt concentration that creates the salinity hazard. As might be expected, these problems are more intense in arid regions of the Earth.

19.2.2 Composting

For the disposal of bulk organic solid wastes, composting offers a number of attractive features. Composting is a process in which organic solids undergo decomposition to produce a relatively stable humus-like material. The process is microbiological, involving a suite of aerobic microorganisms including bacteria, actinomycetes, and fungi. It may be carried out on an individual or family-unit scale by placing the wastes in a pile or in a specially designed container. Alternatively, composting can be carried out on an industrial scale under carefully controlled conditions. With proper attention to quality control, the product is suitable for use as a soil amendment that acts as a source of organic matter as well as a source of small quantities of essential nutrients. A prime requirement is that there be an appropriate ratio of carbon:nitrogen in the compostable materials. It is also essential that the final product does not contain harmful components such as elevated concentrations of potentially toxic metals, organic chemicals, and pathogens.

While the production process is simple, careful control of conditions is necessary to ensure rapid and efficient synthesis of the stable product. The factors to be considered are:

● the nature of the organic material to be used in producing compost;
● the need to control environmental conditions for composting—temperature, aeration, and water supply;
● the time required to produce a mature and stable product.

The composting reactions are essentially aerobic, analogous to the activated sludge process in waste-water treatment. In these reactions, aerobic microorganisms grow and produce carbon dioxide and microbial biomass. In the process, other components of organic matter are converted into more stable species akin to humic materials. The microorganisms require water and oxygen and so compost must be kept moist and, at the same time, must be well aerated. This is done by supplying water when necessary, by preventing the mass of material from becoming too compacted, and/or by mechanically agitating it to allow the entry of air. As a consequence of the degradation reactions being exothermic, there is a temperature increase in the compost. The temperature should be allowed to rise to 50 or 60 °C so that decomposition occurs at a reasonably rapid rate. While aeration is essential, excessive aeration by agitation must be avoided so as not to release too much internally generated heat. Besides affecting the rate of decomposition, a final temperature of at least 60 °C should be reached and maintained for several hours or days to ensure that pathogenic microorganisms and enzymes are permanently inactivated.

The carbon:nitrogen ratio of the raw materials is critical for controlling the rate and extent of composting reactions. A large ratio means that nitrogen is limiting as a nutrient for microbial growth. When this occurs, decomposition is incomplete and the compost will be immature with partial degradation products—acetic, propionic, and n-butyric acid, all phytotoxic—remaining in the dormant material. Furthermore, when immature compost is added to soil as an amendment, in the initial stages heterotrophic microorganisms take up nitrogen from the soil, immobilizing it and therefore rendering it unavailable to plants. On the other hand, if the carbon:nitrogen ratio is too small, the excess nitrogen present is released by ammonification reactions as potentially toxic ammonia. Furthermore, the ammonia uses oxygen as it undergoes nitrification and this, along with the intrinsically large organic carbon content can result in a reducing (low pE) environment.

The carbon:nitrogen ratio is best controlled to a value around 30 by careful selection of proportions of high ratio (sawdust, straw, paper) and low ratio (food scraps, manures) solids. Soil or inorganic nitrogen fertilizers can also be added, if it is necessary to decrease the C:N ratio.

The range and median elemental composition of composted products is shown in Table 19.3.

Organic matter makes up most of the mass of the compost. Much of the OM is in the form of nitrogen-rich humic-like materials. Functional groups, especially carboxylic acids, give compost a high potential for ion-exchange sites. The sites are occupied by the usual cations but minor elements are also present. The trace metal content is often much larger on a mass fraction basis than in most soils (compare Table 19.3 with Table 18.8). Many of the trace metals are minor nutrients and their presence can add to the fertilizer value of compost.

Table 19.3 Compostion of selected composts on a dry mass basis

	C	N	P	K
Median/%	31	1.1	0.27	0.26
(Range/%)	(27–40)	(0.51–1.8)	(0.15–0.6)	(0.07–0.97)
Number of samples	4	6	6	6
	Ca	Mg	Na	S
Median/%	4.0	0.3	0.3	0.2
(Range/%)	(1.2–7.5)	(0.08–0.6)	(0.2–0.67)	(0.2–0.6)
Number of samples	5	5	4	5
	Cu	Ni	Mn	Zn
Median/$mg\,kg^{-1}$	230	110	500	930
(Range/$mg\,kg^{-1}$)	(100–630)	(0.76–190)	(400–600)	(500–1650)
Number of samples	6	3	3	6
	Hg	Pb	Cd	Cr
Median/$mg\,kg^{-1}$	4.5	600	7	220
(Range/$mg\,kg^{-1}$)	(4–5)	(9–900)	(0.04–100)	(2–270)
Number of samples	2	4	5	4

From He, X.-T., S. J. Traina, and T. J. Logan, Chemical properties of municipal solid waste composts, *Journal of Environmental Quality*, **21** (1992), 318–29.

In excessive amounts, however, these and other metals could be toxic to plants or to animals that consumed the plants.

As a soil amendment, compost has several virtues. Most important is that the large organic matter content can improve the physical properties of sand-rich or clay-rich soils. At the same time, the organic matter increases the exchange capacity of the soil. The possibility of toxicity due to trace metals or in particular cases due to organic contaminants must always be considered.

19.2.3 Sewage sludge

In our discussion of waste-water treatment, we saw that the process ends with two products—the treated water that is released, usually to a lake or river, and sewage sludge. Sludge has been disposed of in a variety of ways including disposal at sea, in landfills, and through incineration, but a particularly common and important means of disposal is to use it as a soil amendment on productive land.

Sewage sludge is a slurry made up of solid material, often about 1% of the total mass, with high residual water content. It is principally organic in composition, but also contains inorganic components derived from the original sewage as well as from any metal-based coagulants added during treatment. Depending on the sources, sewage sludge composition is highly variable. Typical values for major and minor elements in some sludge samples are given in Table 19.4. Several of the elements listed in the table are nutrients and these, along with the organic matter, suggest that sewage sludge can serve as an excellent soil amendment. Nitrogen is present in both free and combined forms with about one quarter to one half available as ammonia plus nitrate. The remainder is present as combined forms in organic molecules. The macronutrient content is somewhat greater than that of most composts. Sludge also builds up the soil organic matter content in the same manner as compost, animal, or green manures. It therefore improves soil structure and water retention and it can supply small but significant amounts of important nutrients to the soil.

Unfortunately, sludges sometimes contain relatively large concentrations of non-nutrient elements. These too are, in most cases, present in greater concentrations than they are in compost. Some of them are potentially toxic to plants growing on soil treated with sludge, and/or to animals (including humans) who might consume the plants.

Table 19.4 Median values for major and minor element content of dried sewage sludge from seven American States

Element	Content	Element	Content
Organic C	30.4%	Al	0.4%
Total N	2.5%	Cu	$850\,\mu g\,g^{-1}$
$NH_4^+ - N$	0.13%	Ni	$190\,\mu g\,g^{-1}$
$NO_3^- - N$	$194\,\mu g\,g^{-1}$	Mn	$200\,\mu g\,g^{-1}$
Total P	1.8%	Zn	$1800\,\mu g\,g^{-1}$
Total S	1.1%	Pb	$650\,\mu g\,g^{-1}$
K	0.24%	Cr	$910\,\mu g\,g^{-1}$
Na	0.12%	Cd	$20\,\mu g\,g^{-1}$
Ca	3.8%	Hg	$6\,\mu g\,g^{-1}$
Mg	0.46%		
Fe	0.8%		

Data from ref. 2, Additional reading list.

Industrial effluents can be a source of the elevated concentrations of toxic metals, although in many parts of the world, technological and legislative developments have diminished this problem. Households can be a major source as well; lead, copper, and zinc are all leached from domestic pipes, especially if the water is soft. Many consumer products such as phosphate-based detergents supply small amounts of metals to the waste-water stream.

Aerobic sludge is that which has been obtained directly from an aerobic system such as the aerator of an activated sludge plant. Prolonged aerobic digestion results in auto-oxidation of the synthesized biomass. Anaerobic sludge is obtained after the aerator sludge has undergone further reaction in the absence of air in order to produce methane. These processes have been described in Chapter 16.

Sludges are applied either as a slurry containing only about 1% solids, or in dewatered, partially dried form, where the water content is much smaller. There are obvious precautions, similar to those observed in the use of animal manures, that must be followed in applying sludge to agricultural land. Among these are

- consideration of proximity to residences and wells;
- maintaining a safe distance from surface water bodies—depending on slope and soil permeability;
- considering depth of the groundwater table—depending on soil permeability;
- application should be avoided on shallow soils, when bedrock is near the surface;
- a waiting period should be observed after application of sewage sludge before certain vegetable crops are planted, because some pathogenic organisms may have survived the treatment process.

The precautions related to potentially toxic metals require that sludge not be used on organic soils, on soils with low pH, and on land already containing high concentrations of the metals. Recommended applications take into account the metal content of the sludge. One set of recommendations specifies that there must be a minimum concentration ratio of

$$\frac{\text{nitrogen in the form of ammonia plus nitrate}}{\text{metal}}$$

before the sludge is considered acceptable. For example, the ratio is set at 500 for the semi-metal selenium. Suppose a sludge contains 0.28% ammonia plus nitrate nitrogen, and $3.3 \, \text{mg kg}^{-1}$ selenium. The calculated ratio for this sludge is $2800/3.3 = 850$. It is therefore acceptable for use according to the specified criterion for this one element. Another type of criterion assigns relative toxicity factors to the most abundant trace elements found in sludges. In the United Kingdom, the weighting is based on zinc equivalents, with copper and nickel factors of 2 and 8 respectively. In approximately 75% of cases, these three elements have been found to be the most likely to contribute to toxicity. The amount of the three metals added to a soil should not exceed $560 \, \text{kg ha}^{-1}$ (measured in zinc equivalents) over a 30 year period. For example, we can consider a sludge that contains $920 \, \text{mg kg}^{-1}$ zinc, $540 \, \text{mg kg}^{-1}$ copper, and $60 \, \text{mg kg}^{-1}$ nickel.

Zinc equivalent concentration $= (1 \times 920) + (2 \times 540) + (8 \times 60) = 2480 \, \text{mg kg}^{-1}$.

(Note that the unit $mg\,kg^{-1}$ is the same as $g\,t^{-1}$. The recommended application limit of zinc or zinc equivalents are

$$= 560\,kg\,(Zn\,eq)\,ha^{-1}$$

$$= 560\,000\,g\,(Zn\,eq)\,ha^{-1},$$

Since sludge contains $2480\,mg\,kg^{-1} = 2480\,g\,(Zn\,eq)\,t^{-1}(sludge)$,
Recommended application limit for sludge

$$= \frac{5.60 \times 10^5\,g\,(Zn)\,ha^{-1}}{2480g\,(Zn)\,t^{-1}(sludge)}$$

$$= 255\,t\,(sludge)\,ha^{-1}.$$

According to this criterion, a maximum of 225 t of sludge could be applied to 1 ha of land over a 30 year period.

As is expected, most of the metals added in sludge are originally present in organic forms, although some may be associated with mineral phases. In many soils, the trace metals are found in a wider variety of associations—with carbonate minerals, with iron and aluminium oxides, and incorporated into the silicate lattice. Metal associated with organic matter in sludge is released when the organic material decomposes in the field and most of it associates with some other phase in the soil. The metal is therefore quite immobile, and usually remains in the top layers of the soil and accumulates over time.

Salinity problems are not commonly associated with the application of sewage sludge. Much of the soluble salt content has been dispersed in the aqueous effluent from the treatment plant and the sludge itself does not have a large ionic concentration.

An attractive concept is to use two waste materials together, in a way that is an environmental benefit to both. An important example is the use of sewage sludge as an amendment for mine tailings. We emphasized that tailings are frequently deficient in water-holding capacity, nutrients, and exchange sites to hold the nutrients. Sludge, on the other hand, is rich in organic matter and contains at least small amounts of many nutrients. Therefore its application to tailings (usually some additional major nutrients like nitrogen and phosphorus are required as well) provides many of the needs of the tailings. This approach has been recommended for treatment of copper spoils.[2] Native shrubs and grasses were used in the revegetation protocol. Growth was greatly enhanced on spoils where sludge was added. Uptake of some of the important trace elements in the tailings and in the sludge gave high values in the shrub leaves. This indicates that in this and many other cases, the revegetated spoils could not be used to produce agricultural crops for human or animal consumption.

19.2.4 Small scale biogas synthesis

Similar to the anaerobic digestion of sewage sludge, a process that is carried out on an industrial scale, *biogas* can be synthesized at a domestic level in a relatively simple process. Biogas synthesis makes use of a substrate of biological waste material, particularly animal manures, to produce a mixture of methane and carbon dioxide. The methane is a convenient clean-burning fuel with many applications, principally for cooking and heating. The sludge material that is left behind retains many of the desirable properties of the original manure which make it suitable as an amendment for soils. A design of a simple biogas generator widely used in Chinese villages is shown in Fig. 19.1.

Biogas generation is a complex anaerobic process that can be divided conceptually into three steps, all of which involve bacterial mediation. The first step is solubilization and hydrolysis whereby solid organic matter made up largely of polysaccharides, proteins, and lipids is broken down into the component sugars, amino acids, glycerol, and fatty acids. Once in solution, the small molecules act as substrates for reactions producing acetic acid, carbonate species, and hydrogen—the latter two compounds being partially partitioned into the gas phase according to Henry's law. Smaller amounts of other substances like sulfide species and ammonia are also produced and relatively biorefractory compounds including aromatics and fatty acids remain unreacted at this stage. Finally methanogenesis occurs. Ammonia and hydrogen take part in reactions to further reduce the acetate and carbonate species *via* such reactions as

$$CH_3COO^- \, (aq) + H_2O \; \rightarrow \; CH_4 + HCO_3^- \, (aq) \qquad \Delta G = -31.0 \, kJ \qquad (19.8)$$

$$HCO_3^- \, (aq) + 4H_2 + H_3O^+ \, (aq) \; \rightarrow \; CH_4 + 4H_2O \qquad \Delta G = -135.6 \, kJ \qquad (19.9)$$

The end result is a mixture of gases containing, as major components, methane and carbon dioxide but with small amounts of hydrogen, ammonia, and hydrogen sulfide. The methane and carbon dioxide mixing ratios are typically in the ranges 60–70% and 30–40% respectively, with the heating value of the gas depending strongly on the relative amounts.

If the feedstock is predominantly cellulosic, relatively high ratios of carbon dioxide are produced while lipids and proteins favour methane. The reaction conditions must be carefully controlled in order to optimize production of high heat content biogas. The pH of the reaction mixture should be in a range 6.5 to 8.5. The alkalinity may be as high as 14 000 ppm calcium carbonate. Temperatures should be in the range between 20 and 60 °C—the optimum temperature for mesophilic bacteria is about 35 °C and for thermophiles about 55 °C. In winter, or to operate at higher temperatures, some of the produced biogas may be

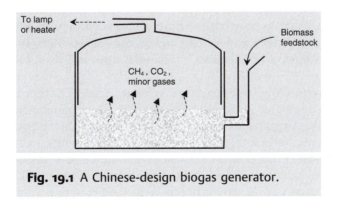

Fig. 19.1 A Chinese-design biogas generator.

required as supplemental heat for the digester, thus reducing the net production efficiency. Nitrogen is required to support the microbiological activity and so the carbon:nitrogen ratio of the feedstock is critical. As for composting, the optimum value of the ratio is about 30. Substances whose ratios are close to the optimum are animal manures (15 to 20), household food wastes (25), grasses and seaweed (15 to 20). Wood, even when finely chopped, is not suitable because of its ratio of 50. Urine has a value of 0.8. A near ideal mixture consists of animal wastes diluted by straw, as is found in cattle bedding mixtures.

The production of biogas is environmentally attractive because a clean-burning fuel is produced. Sulfur emissions remain low and the problem of particulate emissions is nearly eliminated. The residual sludge contains most of the nitrogen, phosphorus, potassium, and other nutrients of the original feedstock. The organic content is still substantial, being about 30% of the original value—most of it biorefractory lignin-related materials as well as bacterial cellular components. When added to soil, the clean-smelling sludge supplies nutrients as well as being a source of organic matter that enhances the physical and chemical exchange properties of the soil.

The process represents a compromise between using wastes exclusively for fuel or exclusively as a soil additive. Using typical figures, the extent of the compromise may be summarized as follows:

- One tonne of biomass (dry weight), added directly to the soil, provides it with 6 kg of nitrogen, 1.5 kg of phosphorus, and 3 kg of potassium.
- One tonne of biomass, when burned produces a theoretical 1.5×10^{10} J of energy.
- One tonne of biomass, when converted to biogas and sludge, depending on conditions generates about 200 to 1000 m^3 of biogas. For a 65 : 35 methane : carbon dioxide mixture with heating value of 2.4×10^7 J m^{-3}, the total energy content of the gas is between 4.8×10^9 and 2.4×10^{10} J. The lower number would be closer to the actual value in many cases. A slurry of sludge weighing 300 kg remains to be applied to the soil.

19.3 Mixed urban wastes

The material that is discarded by urban society is a complex mixture of many substances, and the nature of the mixture is very diverse depending on the situation. In high-income countries, the traditional dump (euphemistically called a 'sanitary landfill') has contained a wide range of organic wastes, paper, glass, metals, and plastics. Table 19.5 gives ranges of composition, determined in five studies, of material discarded from urban areas in the United States.

All of the materials in the table have some value in terms of nutrient content, energy content, and/or as a resource for recovery of metals or other commodities. Partly for this reason, in recent years wealthy societies have sought ways to use the discarded materials. Paper and aluminium, and in some cases a suite of other components, are separated prior to landfilling and recycled in other forms. Recycling is not a novel phenomenon in low-income countries. There has traditionally been a market for many of the materials, and systems have long been in place to enable collection for reuse and recycling.

Table 19.5 Composition of the solid waste stream as determined in several studies in the United States

	Median/%	Range/%
Food waste	8.8	8.0–22.8
Yard waste (grass, weeds, leaves etc.)	19.8	11.0–28.2
Other	8.9	7.8–16.2
Total organic	39.0	28.8–50.8
Newsprint	5.2	3.1–7.3
Cardboard	7.0	6.2–10.8
Mixed paper	21.8	5.6–32.7
Total paper	35.0	14.9–48.7
Ferrous metal	5.1	3.2–7.5
Aluminium	1.2	0.3–1.6
Other metal	0.2	0.2–0.7
Total metal	6.0	4.9–9.0
Plastics	7.3	7.0–9.4
Glass	5.0	2.4–8.3
Other inorganics	5.3	1.9–12.2
Total plastics + glass + other non-metal inorganics	18.0	16.1–27.3

Data reported in O'Leary, P. and P. Walsh, *Introduction to Solid Waste Landfills*. Course material for C240-A1 80 Solid Waste Landfils, a correspondence course from the Solid and Hazardous Waste Education Center, University of Wisconsin, Madison.

19.3.1 Landfilling

Disposing of mixed urban waste in a *sanitary landfill* (alternatively simply called a landfill or a dump) is one means by which urban garbage is handled. Landfilling may follow removal of components for composting or recycling, or it may be used for disposal of ash after an incineration process. In many cases, waste disposal is a one-step process in which all the collected material is placed directly in a dump. For public health and aesthetic reasons, a properly sited landfill should not be located in the immediate vicinity of population centres. Balancing this is the need to take into account economics related to transport of the garbage away from the population centre. To maximize the amount of garbage that can be disposed of in any site, the waste is usually compacted, often with heavy machinery, into a dense mass. What happens in this highly concentrated mixture of solid materials?

Most garbage sites contain substantial amounts of degradable organic substances—food wastes, wood, paper, and other fibres, in addition to more inert materials. Some of these materials will be present even when an efficient composting process precedes the landfilling operation. Degradation of the OM is again a microbial process, and as long as oxygen is present to act as an electron acceptor, the usual oxidation reactions proceed through various intermediates eventually producing carbon dioxide and water as the major products. Nitrogen is converted to nitrate. However, the compact nature of the deposit inhibits air transfer into the bulk of the mass, and the environment soon becomes anaerobic. The first stages of anaerobic decomposition lead to the production of small molar mass organic acids along with additional carbonate species and hydrogen gas. A possible reaction starting with the generic carbohydrate molecule is as follows:

$$4\{CH_2O\} \rightarrow CH_3CH_2COOH + CO_2 + H_2 \tag{19.10}$$

The pH of leachate in the landfill drops to around 4.5, sufficiently low that some metals are solubilized. Calcium and magnesium are the most common metallic ions found in leachate at this point. Under the low pH conditions, methane-producing bacteria do not function efficiently and little fermentation takes place. Eventually, however, methanogenic bacteria become dominant and the principal degradation process is fermentation leading to the formation of methane and carbon dioxide in a 1 : 1 molar (and volume) ratio:

$$2\{CH_2O\} \rightarrow CH_4 + CO_2 \tag{19.11}$$

The processes are similar to those involved in domestic biogas production and the proportion of methane is again higher for degradation of proteins and fats. The 'average' landfill generates these gases in an approximately equimolar ratio.

Assuming that the average degradation reaction is represented by reaction 19.11, $370\,m^3$ of methane would be produced from 1 t of waste. Because a typical mixed waste contains about 75% decomposable organic matter and because the decomposition is never complete, actual values of methane produced are often found to be approximately 25% (approximately $100\,m^3$) of this value. Nevertheless, this is a valuable energy resource that can be used to heat adjacent buildings, for a boiler in a factory, or for other purposes. Methane is of course a potent greenhouse gas, providing another reason why it should be collected and used as fuel. Engineered collection systems (Fig. 19.2) involve sealing the landfill to prevent escape of the gas into the surrounding soil, and installing pipes to transport the gas to the location where it is stored or used. The seal usually requires a synthetic polymer liner and capping system.

The degradation does not proceed cleanly to only gaseous products. A liquid leachate is also produced whose composition varies over the stages of decomposition described above. Figure 19.3 shows a sequence of this variation along with changes that take place in gas composition and production over a 2 year period. After the aerobic phase of decomposition is complete, the initial anaerobic processes produce a leachate with a large organic matter content as shown by the COD plot in the figure. This COD is due in large part to organic acids produced by the anaerobic decomposition. The low pH solution at this point can dissolve some metals, including those that are toxic and it can also contain a wide variety of refractory organic species. It is important, therefore, to contain the leachate. The synthetic liner serves this purpose, and along with it a thick clay underlayer is provided. A liquid collection system directs the leachate to the lowest point within the site where it can be pumped to the surface. There it is either recirculated through the landfill or removed for further treatment. In some cases, the treatment is done at a waste-water treatment plant where landfill leachate is added as a component of the influent stream.

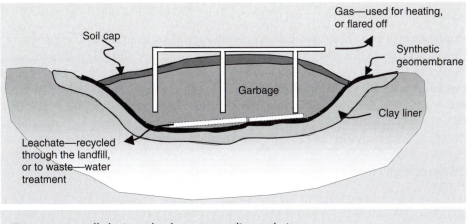

Fig. 19.2 A well designed urban waste disposal site.

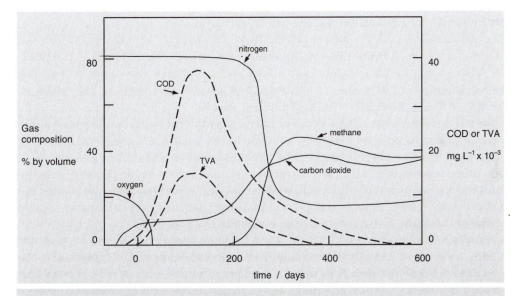

Fig. 19.3 Changes in gas composition(———)and leachate concentration(— —)over a 2 year period in a landfill. COD is chemical oxygen demand and TVA is the total volatile acids and is measured in terms of acetic acid. The total volume of gas evolved from the landfill reaches a maximum at about 380 days. The figure is based on one in Pohland, F. G., J. T. Derien, and S. B. Ghosh, Leachate and gas quality changes during landfill stabilization of municipal refuse, in *Proceedings of the 3rd International Symposium on Anaerobic Digestion*, R. L. Wentworth, ed., Boston, Dynatech, Cambridge, USA: 1983. pp. 185–201.

19.3.2 Incineration

Incineration of urban waste materials is an alternative to landfilling. There are several attractive features to this alternative. One is that it provides an efficient means for energy recovery from much of the waste, and it is a means of disposal that can be set up in an area immediately adjacent to a population centre. Second, it reduces the volume of waste considerably so that much less land is required for final disposal. Furthermore, it eliminates problems associated with methane generation and leachate that emanate from a landfill site.

We begin an examination of the environmental consequences by looking at the energy production issue. We noted that in a typical landfill, about $100 \, \text{m}^3$ (4500 mol) of methane could be recovered from 1 t of waste. When this is burned, the reaction is

$$CH_4 + 2O_2 \;\rightarrow\; CO_2 + 2H_2O \qquad \Delta H = -802.3 \, \text{kJ}. \qquad (19.12)$$

Therefore from 1 t of waste, it is possible to obtain $4500 \times 890\,300 = 4.0 \times 10^9$ J of energy, assuming an efficient collection and recovery system.

If, alternatively, the 1 t of waste is burned directly, about 85% is combustible (the combustible value is somewhat higher than that used for decomposition, because it includes plastics that break down in a landfill only very slowly). Again, using the generic carbohydrate formula, the combustion reaction can be written as

$$\{CH_2O\} + O_2 \;\rightarrow\; CO_2 + H_2O \qquad \Delta H = -440 \, \text{kJ}. \qquad (19.13)$$

In this situation, 1 t of waste $= 1 \times 10^6/30$ mol of generic carbohydrate and produces $(1 \times 10^6 \times 0.85 \times 440\,000)/30 = 1.2 \times 10^{10}$ J of heat energy.

There is therefore about a three-fold gain in maximum energy recovery when using direct combustion compared with recovering methane from a landfill and using it as fuel.

There are, however, other environmental consequences to consider. During incineration, most of the organic matter is converted into carbon dioxide and water. Depending on the nature of the waste and on the combustion conditions, other gases are also produced—sulfur dioxide, the NO_x compounds, polyaromatic hydrocarbons, and chlorinated organic substances are some that are of particular environmental concern.

While incineration reduces the mass of the mixed urban waste considerably, it does not eliminate it entirely. There are also residual solids called ash. Some of the ash (usually $< 1\%$) is emitted through the stack as fly ash. An incinerator which treats $20 \, \text{t} \, \text{h}^{-1}$ of waste can release 2 to $10 \, \text{kg} \, \text{h}^{-1}$ of fly ash. A much larger fraction of the solid component ($> 99\%$) is present as bottom ash that remains as residue in the incineration chamber after combustion is complete. Recovery of much of the fly ash is technically possible using technology described in Chapter 6. This residue and the bottom ash must be disposed of and both components present their own unique problems.

With regard to metals, at least a portion of the volatile metals such as cadmium, lead, and mercury are vaporized and emitted from the stack while more refractory elements tend to remain in the bottom ash. The fate of metals is not described in a simple way, however. Volatility depends on the form of the element, the nature of other materials in the waste stream, and the operating conditions of the incinerator.

Take the example of lead, an element that is present in waste materials in paints and discarded batteries, among other things. The metal (Pb^0) begins to volatilize significantly around 1000 °C; this temperature is usually achieved in the afterburner chamber of an incineration unit. Therefore, some of the lead is released as the element into the vapour phase. If the furnace is operating under pyrolysis (reducing) conditions, sulfur will be converted to sulfide (S^{2-}) that reacts with lead to generate lead (II) sulfide. This compound also becomes significantly volatile around 1000 °C. Under oxidizing conditions—an air-rich combustion mixture—some of the lead is converted to lead (II) oxide. The oxide is more volatile than either lead metal or lead (II) sulfide. A major effect on lead release occurs when there are chlorine-containing compounds in the waste stream; a common source is the widely used polymer, polyvinyl chloride (PVC). The reaction to produce lead (II) chloride is rapid and the compound is highly volatile so that losses of the metal are very rapid and complete. Other metals too have their own characteristic reaction possibilities, but in most cases the presence of chlorine enhances metal vaporization.

After leaving the combustion unit, the stack gases cool and further transformations of vaporized metals take place. One possibility is that metal species will homogeneously form nuclei which coagulate and grow to become a stable aerosol. A second possibility is that the metal, in free or combined form, will condense on the surface of other refractory fly-ash particles. Incorporation in the aerosol may also occur by a third mechanism wherein the metal *reacts at* rather than *condenses on* the surface of another particle.

It may be possible to take advantage of this last mechanism in order to design a means by which metals are efficiently removed from the stack gases. Uberoi and Shadman[3] tested various solid sorbents for their ability to capture lead emitted in a simulated flue gas in a laboratory reactor. The bed of sorbent was kept at an elevated temperature to ensure that the lead was in the vapour phase when it passed through this material. Alumina and kaolinite, among others, were found to be especially efficient in retaining lead, by reactions of the type

$$Al_2O_3 + PbCl_2 + H_2O \rightarrow PbO \cdot Al_2O_3 + 2HCl \tag{19.14}$$

Lead deposited by chemical reaction on such sorbents is not extractable by water, in contrast to metal that condenses rather than reacts to form an aerosol.

Retention of fly ash from municipal incinerators and other combustion sources is a very important issue. Much of the ash is well within the PM_{10} size range, and it can have a very high concentration of metals. The more volatile metals including cadmium, copper, lead, and zinc are especially concentrated. Lead and zinc have been found at levels between 1 and 10% in the particulates. Moreover, because these metals are condensed with the aerosol, they have been found to be readily available in physiological reactions and with respect to solubilization. For this reason, fly ash has been classified as a hazardous material in some jurisdictions and its disposal in landfills is proscribed.

We have emphasized the fact that metal concentrations are a major problem with fly ash. Organic contaminants are also very important. The compounds of greatest concern are chlorinated organic substances including chlorobenzenes (CB), chlorophenols (CP), polychlorinated biphenyls (PCB), furans (PCDF), and dioxins (PCDD). Some examples of the chlorinated compounds are shown in Fig. 19.4. Each of these examples represents a class of compounds with varying degrees and positions of chlorine substitution. This is indicated using the generic structure for PCBs. In total, there are 209 possible congeners and depending on the source of the PCBs, various congeners predominate. The PCB compounds are somewhat toxic and may be weakly carcinogenic; the more highly chlorinated PCBs have greater toxicity. Polychlorinated biphenyls have been an important industrial chemical with several uses, especially as liquid insulators for heavy electrical equipment. Because of their toxicity, production has been curtailed but much of the produced material is still in use. The PCBs can be destroyed by various means but when they find their way into landfills or conventional incinerators, uncontrolled release as leachate or in fly ash can occur.

Also shown in Fig. 19.4 are examples of the dibenzo-*p*-dioxins (PCDDs) and dibenzofurans (PCDFs). Again, there are many congeners of these compounds and as a group they are loosely referred to as dioxins. They are not an industrial product as such, but are produced and released in small amounts during combustion of chlorine-containing organic materials such as polyvinyl chloride (PVC) plastics. PCBs are themselves a precursor of dioxins and the product is much more toxic than the original PCB.

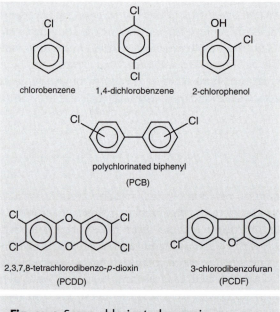

Fig. 19.4 Some chlorinated organic compounds that have been found in fly ash. A generic formula for PCBs is given. The 2,3,7,8-tetrachlorodibenzoparadioxin is considered to be a reference compound for dioxins in general.

As we have noted earlier, the polyaromatic hydrocarbons (PAHs) are another category of organic compounds emitted from combustion processes including incineration of solid waste. Unlike metals, which can only be released if they are already present in the waste material, many organic compounds can be formed in the incinerator from other organic materials. Chlorine is obviously a requirement for formation of chlorinated organics. It is available in various inorganic forms and also in some polymers such as the commodity product polyvinyl chloride.

On a mass basis, concentrations of the organic compounds in ash are much smaller than those of the metals. For the chlorinated substances maximum concentrations in fly ash may be as high as $1 \mu g\,g^{-1}$ for the individual species. In bottom ash the amounts are substantially less. The compounds are initially released in the vapour phase or are adsorbed on to smoke and dust particles. If they are not removed in the stack, their volatility ensures that they are carried into the atmosphere where they may condense or associate with other components of the aerosol. Eventually they settle as dry deposition or are washed out by rain into water or soil or onto the surface of vegetation.

Dioxins and the other chlorinated compounds have very limited solubility in water. The octanol/water partition coefficient (K_{ow}) for 2,3,7,8-TCDD is approximately 10^6–10^7 and the K_{oc} value is 2.3×10^4. The compounds therefore have a tendency to partition into soil, especially when the soil has a substantial organic content. Leaching tests indicate negligible solubility, although very small amounts of chlorophenols and chlorobenzenes are dissolved. If ash is deposited on the surface of the ground, the chlorinated compounds usually remain in the top 15 to 30 cm. In most cases, any vertical or lateral movement in pore water would occur *via* soluble organic matter or dispersed soil colloids rather than in true solution.

Polynuclear aromatic hydrocarbons (PAHs) are present in both fly ash and bottom ash in concentrations that are typically 100 to 1000 $ng\,g^{-1}$. Like the chlorinated compounds, they too have a strong affinity for organic matter as reflected in their K_{ow} values which fall in the range 10^4 to 10^7. This means that the PAH compounds also tend to associate with the soil phase, the tendency being greater when soil organic matter concentrations are large. Very little PAH dissolves directly in water, but where the water contains dissolved or particulate organic matter, there is a competition for association of the PAH between the aqueous phase organic matter and solid phase organic matter. The limited transport that does occur in soil is through association with organic material present in the soil pore water.

One set of specifications for municipal incinerator operations[4] recommends that maximum emissions from an urban waste incinerator be $11\,mg\,Rm^{-3}$ for total particulates, $7\,\mu g\,Rm^{-3}$ for cadmium, $76\,\mu g\,Rm^{-3}$ for lead, $56\,\mu g\,Rm^{-3}$ for mercury, and $0.14\,ng\,Rm^{-3}$ for TCDD. The unit Rm^{-3} is for a reference cubic metre, meaning that the conditions have been normalized to $25\,°C$, P^o, and 11% oxygen. In order to ensure compliance with emission standards, furnace design and operation must be evaluated carefully. In the particular study referred to, it was necessary to ensure that the combustion temperature consistently be at least $1100\,°C$ and must never fall below $1000\,°C$, that the residence time of gases in the furnace be at least 1 s, that there be good turbulence, and that oxidizing conditions are maintained by keeping at least 6% residual oxygen after burning.

The amount of solid residue at the bottom of the furnace is much greater than that of fly ash and depends on what, if any, prior removal of waste material has taken place.

Typically, 10 to 20% of the total mass remains after combustion of mixed urban wastes; it consists of 'siftings' that have fallen through the gratings in the furnace as well as the bottom ash. Assuming that efficient combustion has occurred, the slag-like material contains SiO_2, Al_2O_3, CaO, Fe_2O_3, Na_2O, K_2O, and MgO as its principal components. Small amounts of other elements are associated with the non-combustible solids and are often incorporated in a silicate matrix. The high temperature in the furnace volatilizes most of the low-boiling metals and organic materials, leaving much smaller concentrations than are present in fly ash. (The PAH compounds are an exception in that their concentration in bottom ash is comparable to or greater than in fly ash.) Even when present only in low concentration, leachability of undesirable chemicals from the residue is still a possibility. To test for this, standard leaching tests have been developed, such as the Sequential Batch Extraction Procedure (ASTM Standard No D4793–88). In this test, the ash is mixed with water in a mass ratio of 20 parts water to 1 part ash. The mixture is shaken for 18 h, separated, the aqueous phase saved, and the leaching procedure repeated for a total of five cycles. The combined leachate is then analysed for content of metal and organic contaminants. In most cases for mixed urban waste, chlorinated organic compounds of concern are extracted at levels near or below the detection limit. Most metals too are extracted in only very small concentrations. The residue is therefore a much more inert and benign product than is fly ash, and it can usually be used as construction material (for example for rail or road beds) or disposed in a landfill. Ferrous and non-ferrous metals are recovered from some incinerator facilities, especially in Europe.

The main points

1 Bulk solid wastes of a variety of types are often placed on or in the land for disposal. In some cases, effort is made to isolate them from the soil, in other cases they are deliberately incorporated into the soil.

2 Mine wastes give rise to a number of physical and chemical problems depending on their nature. The production of acid is a particular problem associated with tailings containing sulfide minerals—a problem that is usually treated by the application of lime. Many tailings deposits are stabilized by nutrients and other amendments that allow for revegetation of the deposit.

3 Organic residues of plant and animal origin have value as soil amendments and as energy sources. Three options are available for their disposal–application to land, combustion, or conversion to gaseous fuel. In the latter case, the residue can be applied to the land.

4 Urban wastes comprise a complex mixture of materials. Disposal must take into account their resource and energy values, and the environmental consequences of the various management strategies. Landfilling, composting, and incineration all have environmental consequences associated with them.

Additional reading

1 Bradshaw, A. D. and M. J. Chadwick, *The Restoration of Land*, Blackwell Scientific, Oxford; 1980.

2 Elliott, L. F. and F. J. Stevenson, eds, *Soils for Management of Organic Wastes and Waste Waters.*, Soils Science Society of America, Madison, Wisconsin USA; 1977.

3 Salomons, W. and U. Forstner, eds, *Chemistry and Biology of Solid Waste*, Springer, Berlin; 1988.

Problems

1 The 'Trelogan' tailings[5] occurs at a calcareous lead/zinc deposit in Britain and has properties as shown below.

Particle size	%
>2mm	6.7
2.0–0.2	14.6
0.2–0.02	29.7
0.02–0.002	24.8
<0.002	24.2

$pH = 7.0$
$CEC = 2.8\,cmol\,(+)\,k\,g^{-1}$
$Conductivity = 2.3\,dS\,m^{-1}$

Total analysis/$\mu g\,g^{-1}$:

N	126	Cu	205
P	160	Cd	267
K	1070	Ni	87
Ca	138 500	Pb	39 800
Mg	1500	Zn	95 000
F	185		

Describe the nature of any expected physical and chemical problems associated with revegetation of these tailings, and how the problems might be overcome.

2 Arsenic is often associated with sulfide minerals. Because of this, significant amounts of arsenic may be discarded along with the tailings from mining operations involving these minerals. Refer to the material provided in Chapter 10, problem 8 and indicate what species of arsenic would initially be present in the tailings, and what species might be generated over time.

3 The nickel pE/pH diagram is shown below. Included in setting up the diagram is a concentration of Ni^{2+} of 10^{-4} mol L^{-1} and a sulfur concentration of 10^{-5} mol L^{-1}. Predict the environmental fate of nickel discharged as the mineral nickel sulfide if it is disposed into a well aerated tailings pond or alternatively if it is placed at the bottom of a deep lake.

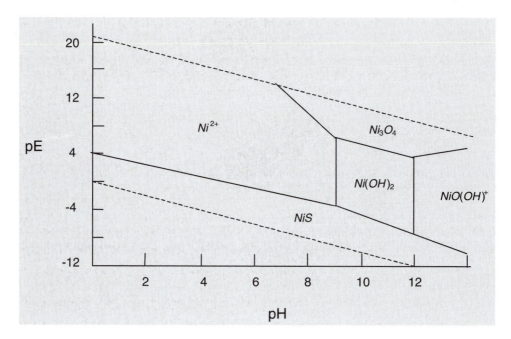

4 If the acidic leachate from sulfide tailings flows in a stream along the surface of the land, it is frequently observed that a red-brown deposit is found downstream on the rocks or sediment in the stream bed. Discuss the chemistry that might explain this observation.

5 Use values of the acid dissociation constants for aluminium aquo complexes to calculate the solubility of aluminium in red mud leachate that has a pH of 10.3.

6 Describe the significance of the following properties in assessing the quality of compost: concentration of carboxyl groups, carbon:nitrogen ratio, total phosphorus concentration, concentration of trace metals.

7 A possible classification of metals in incinerator ash is based on their leachability[6]:

Use your knowledge of metal association with solids to explain this classification system.

Ion exchangeable	Readily leached
Metals associated with oxides or carbonates	Can be leached under acidic conditions
Metals associated with iron and manganese	Can be leached under reducing conditions
Metals bound to sulfide or organic matter	Can be leached under strongly oxidizing conditions
Metals in the silicate lattice	Not leachable

Notes

1 Kittrick J. A., D. S. Fanning, and L. R. Hossner, eds, *Acid Sulfate Weathering*, SSA Special Publication Number 10, Soil Science Society of America, Madison Wisconsin; 1982.

2 Sabey, B. R. R. L. Pendelton, and B. L. Webb, Effect of municipal sewage sludge application on growth of two reclamation shrub species in copper mine spoils. *J. Environ. Qual.* **19** (1990), 580–6.

3 Uberoi, M. and F. Shadman, Sorbents for the removal of lead compounds from hot flue gases. *AIChE J.*, **36** (1990), 307–9.

4 Anonymous, *National Incinerator Testing and Evaluation Program: Environmental Characterization of Mass Burning Incineration Technology at Quebec City.*, Report EPS3/UP/5; 1988.

5 Data adapted from ref. 1, Additional reading list.

6 Fraser, J. L. and K. R. Lum, Availability of elements of environmental importance in incinerated sludge ash. *Environ. Sci. Technol.*, **17** (1983), 52–4.

20

Organic biocides

ONCE more, we are beginning an examination of a very large topic—a topic on which much research has been focused and which is the subject of thousands of books and papers. It is again appropriate to start with some definitions and a description of the scope of our discussion. We use the general term *biocide* according to its literal meaning *(bio* = life, *cide* = kill). An alternative name with a similar broad significance is *xenobiotic compound* *(xeno* = alien, *biotic* = to life). Both these terms are inclusive of substances inadvertently produced or released, yet toxic in varying degrees to life forms (category 1 in Table 20.1) The substances range from petroleum residues leaked from storage tanks to dioxins, small amounts of which form and are released during combustion and during manufacture of chlorophenol chemicals, as well as when chlorine-based bleaches are used in the pulp and paper industry. Also included in the biocide categories are substances synthesized deliberately to target and kill specific organisms (category 2 in Table 20.1). Pesticide is a general name for this latter class of compounds. Within the general class, specific types of pesticide may be identified—insecticides, bactericides, fungicides, herbicides, for example. When the pesticide is designed to eliminate all types of living organism, it is called a fumigant or sterilant.

The voluminous literature on all these substances includes material on their synthesis, mode of action, agricultural or industrial applications, and toxicological properties. However, in line with the goals set out at the beginning of the book, we limit ourselves to a discussion of what happens in the environment—air, water, and soil—after the compounds are synthesized and before they are taken up by living organisms. Given the very large number of compounds and the infinite range of soil, water, and atmospheric properties, there is still a great deal to say even in this defined area and we will restrict ourselves to general principles as illustrated by some specific examples. While the focus is on the soil in the terrestrial environment, many of the principles apply equally to sediment in the hydrosphere. Examples of xenobiotic compounds within some common classes are listed in Table 20.1.

The general question we wish to examine is 'What happens to biocides once they are released into the general environment?' A specific example of this question would be 'What is the fate of the herbicide glyphosate,[1] once is has been sprayed on a field as a broad-spectrum weed killer?' Aside from its toxicological effect on the target organism (assuming there are target organisms in the general case), there are two aspects involved in answering such questions. One is the *chemical stability* of the biocide—inversely related to how and at

Table 20.1 Examples of xenobiotic compounds (biocides)

Function	Chemical class	General formula of typical compound	Specific example
Category 1			
Petroleum residue	Hydrocarbon	C_nH_{2n+2}, CH_{2n}	Isooctane
Unwanted byproduct	Dioxin		2,3,7,8-TCDD
Category 2			
Insecticide	Organophosphorous		Parathion
Insecticide	Organochlorine	various	Methoxychlor
Herbicide	Triazine		Metribuzin
Herbicide	Carbamate		Aldicarb
Herbicide	Chlorophenoxy		2,4-D
Herbicide	Pyridylium		Paraquat

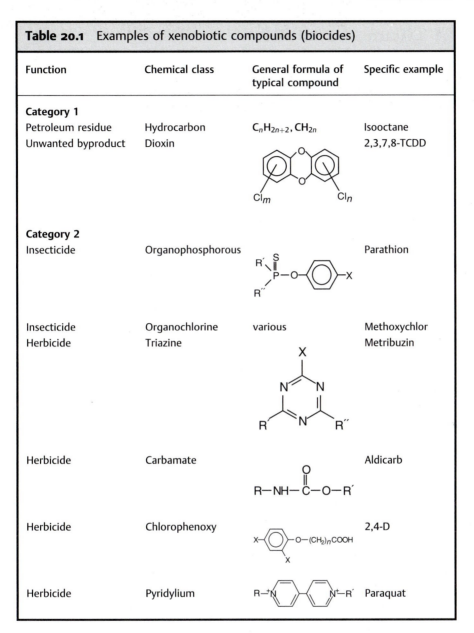

what rate it degrades. The other is *mobility* of the biocide—the mechanisms and rates at which it is transported through various compartments of the environment. The two aspects overlap. If degradation is rapid, then mobility becomes less of an issue. If transport is fast, then different degradation mechanisms may operate as the pesticide moves to a new environment. Degradation products can also have biocidal properties—even enhanced ones in some cases—and their transport properties are therefore also important. Keeping in mind the interconnections, we will look at the two issues separately and in sequence.

20.1 Chemical stability

Degradation of an organic molecule ultimately ends with the production of simple stable species like carbon dioxide, but there are intermediates of varying stability on the way to complete mineralization. Usually degradation reactions produce species with lower toxicity as in the case of dichlorodiphenyltrichloroethane (DDT). Dichlorodiphenyltrichloroethane degrades very slowly in water and soil by dechlorination reactions, and one of the products found under reducing conditions is dichlorodiphenyldichloroethene (DDE) (reaction 20.1):

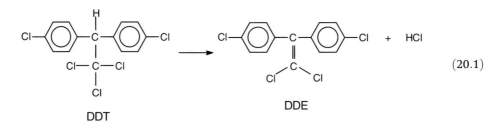

$$(20.1)$$

The product is less toxic than the parent compound. In fact, removal of chlorine from organochlorine molecules nearly always has a detoxifying effect.

Less frequently, such as with some organophosphorus compounds, degradation products exhibit enhanced toxicity. Organophosphorus compounds like parathion are properly called phosphorothioates because of the presence of the $P = S$ group. Inside cells, oxygenating enzymes convert this group to $P = O$ (reaction 20.2):

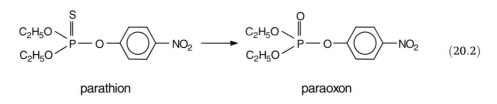

$$(20.2)$$

The phosphorus atom in the oxygenated product, paraoxon, then reacts with the acetylcholinesterase enzyme (AcE) forming a stable complex that inhibits the ability of AcE to catalyse acetylcholine hydrolysis during transmission of nerve impulses. This interference with the neurotransmission accounts for the activity of paraoxon as a pesticide and also as a serious potential health hazard to other organisms including humans.

Both paraoxon and parathion are subject to degradation *via* hydrolysis reactions, in which water acts as a nucleophile. These reactions are catalysed in organisms by various phosphatase enzymes (reaction 20.3). However, the presence of a sulfur attached to phosphorus in parathion instead of the oxygen in paraoxon reduces the tendency for reaction with the nucleophile (in addition to water, the nucleophile can also be a nucleophilic part of AcE) and therefore the sulfur analogue is less toxic. Consequently, it is of environmental concern when the more toxic oxygen-containing forms of these organophosphorus compounds are produced by oxidation outside the target organism. For this reason, it is important to consider the transport of the reaction products as well as of the biocide itself.

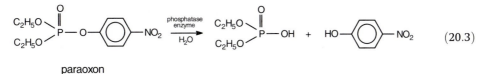

$$\text{paraoxon} \quad\quad\quad\quad\quad\quad\quad\quad\quad\quad\quad\quad\quad\quad\quad\quad\quad (20.3)$$

For the thousands of organic compounds to be considered, there are as many possible sets of reactions which could occur. However, certain types of functional groups are particularly susceptible to transformation and it is therefore convenient to document some of the common types of degradation processes.

20.1.1 Photolytic reactions

Photochemical degradation of biocides is a possibility only when the chemical is exposed to sunlight. In practice, this means that reactions occur during daytime and it is necessary that the chemical be in the gas phase, in or on atmospheric aerosols, in surface waters, or on the surface of plants or soil. Even when these conditions obtain, not every biocide undergoes photolytic decomposition. The further requirements are that the molecule absorb some portion of the radiation from the solar spectrum producing an excited state and that the quantum yield for subsequent decomposition be large compared to yields for other deactivation pathways.

Table 20.2 lists some approximate mean bond enthalpies in organic molecules. Also tabulated are the approximate wavelengths and associated energies for maximum absorption by common chromophoric groups. Recall that the solar spectrum at the Earth's surface has a low wavelength cut-off at about 285 nm (see pp. 43–4) because more energetic radiation has been almost completely absorbed by ozone in the stratosphere.

From this information, we could rule out the possibility of extensive photodegradation of an alkane, since very little radiation could be absorbed. We could also predict only limited degradation of naphthalene. Even though the molecule absorbs strongly at 286 and 312 nm, the energy required to break an aromatic C–C or C–H bond is greater than that taken up from the solar radiation. On the other hand, we might predict that the seed and soil fungicide fenaminosulf would be susceptible to photolysis because of the ability of the azo group to absorb and because of the relatively weak C–N bonds.

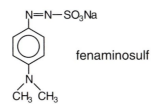

fenaminosulf

The description in the previous paragraph applies to *direct photochemistry*; this refers to the situation where a molecule absorbs radiation and this leads to dissociation or some other kind of chemical reaction. An example of direct photolysis is the photochemical

Table 20.2 *Upper part*—mean bond enthalpies for bonds in organic molecules.[a] *Lower part*—approximate wavelengths and associated energies of radiation absorbed by chromophoric groups (some of these groups also absorb at other wavelengths below 270 nm)

Bond in polyatomic molecule	Mean bond enthalpy/kJ mol^{-1}	
Carbon–carbon (single)	348	
Carbon–carbon (double)	613	
Carbon–carbon (aromatic)	518	
Carbon–hydrogen (alkane)	412	
Carbon–hydrogen (alkene)	440	
Carbon–hydrogen (aromatic)	431	
Carbon–fluorine	484	
Carbon–chlorine	338	
Carbon–bromine	276	
Carbon–iodine	240	
Oxygen–hydrogen	463	
Nitrogen–hydrogen	388	
Carbon–oxygen (single)	360	
Carbon–oxygen (double)	743	
Carbon–nitrogen (single)	305	
Nitrogen–nitrogen (double)	409	
Chromophore	**Wavelength/nm**	**Energy/kJ mol^{-1}**
Carbonyl	285	532
Nitro	280	427
Nitroso	300	399
	665	180
Nitrate	270	443
Azo	340	351
Phenol	270	443
Naphthalene	286	419
	312	384

The following chromophores absorb radiation only at wavelengths less than 270 nm (equivalent to energy of 443 kJ mol^{-1}): alkane, alkene, alkyne, benzene, carboxyl, alcohol, ether, ester, amino, amido, nitrile. Values reported in Atkins, P., *Physical Chemistry*, 6th edn, W. H. Freeman and Company, New York; 1997.

degradation of trifluralin in an alkaline aqueous solution leading to the formation of N-dealkylated products (reaction 20.4).

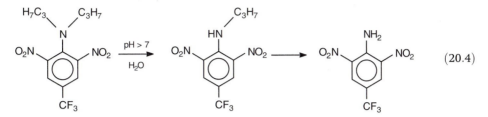

$$\text{rate} = f\,[\text{trifluralin}] \tag{20.5}$$

For this and other direct photolytic reactions, the reaction rate obeys first order or pseudo first order kinetics:

where f, the photochemical rate constant, is a property of the solar flux as well as of the molecule itself as shown in eqn. 2.17.

Another type of photochemical process is *indirect photolysis*. In this situation, a sensitizer molecule is radiatively excited. If the excited species is sufficiently long lived, it is able to transfer energy or an electron, hydrogen atom, or proton to another molecule in the solution. In this way, without absorbing radiation directly, the receptor molecule can be activated so as to take part in a subsequent chemical reaction. There are specific examples of sensitized photodegradation such as the case of rotenone, a natural product extracted from the derris root, which is susceptible to excitation by sunlight. The excited state rotenone is able to transfer its surplus energy to aldrin and other organochlorine compounds leading to their degradation. In a more general sense, a number of important environmental species, including some mineral surfaces and humic material, are capable of acting as sensitizers. The rate of an indirect photolytic reaction is given by

$$\text{rate} = f\,[\text{sensitizer}]\,[\text{receptor}] \tag{20.6}$$

but when the sensitizer is present at much higher concentration than the receptor, the reaction becomes pseudo first order:

$$\text{rate} = f_1\,[\text{receptor}] \tag{20.7}$$

where $f_1 = f\,[\text{sensitizer}]$.

The ability to take part in indirect photochemical reactions extends the possibility of photolysis to biocides which would, by themselves, be stable to solar radiation.

20.1.2 Non-photolytic reactions

Degradation in the soil/water matrix by a variety of thermal reaction processes also occurs. When these processes are purely chemical they are referred to as being *abiotic*; as long as the required reactants are present, such reactions can take place in a sterile environment. Others are chemical reactions which are mediated by the autochthonous or zymogenous microorganisms present in the particular environmental situation. The microorganisms may degrade the biocide as a primary substrate from which they derive energy and/or nutrient requirements. Alternatively, the biocide may be co-metabolized along with the principal substrates required by the autochthonous population. Both such processes are called *biotic* processes. A number of reactions take place both biotically and abiotically and, depending on circumstances, one or other mechanism predominates. It is sometimes difficult to distinguish between mechanisms and to determine which one is operative in a given situation.

Some of the most important types of chemical reactions involving biocides are indicated below and in each case one or more examples is given.

Hydrolysis

Hydrolysis is a general term to describe a nucleophilic reaction where water interacts with a substrate molecule replacing a portion (the leaving group) of the molecule with OH. A general equation for hydrolysis is

$$RX + H_2O \rightarrow ROH + HX \tag{20.8}$$

This type of reaction proceeds by either purely chemical or microbiological mechanisms.

1. Ethers (reaction 20.9), esters (reaction 20.10), and thioesters (reaction 20.10, where C=S replaces C=O) undergo hydrolysis as follows:

$$R - O - R' + H_2O \longrightarrow R\,O\,H + R'OH \tag{20.9}$$

$$R - \overset{\overset{\displaystyle O}{\|}}{C} - O - R' + H_2O \longrightarrow R - \overset{\overset{\displaystyle O}{\|}}{C} - OH + R'OH \tag{20.10}$$

2,4-dichlorophenoxyacetic acid (2,4-D) is one of the most widely used herbicides. It contains both carboxylic acid and ether functional groups. Hydrolysis takes place as shown, involving ether C–O bond scission:

$$\text{Cl}-\underset{\text{Cl}}{\bigcirc}-\text{O}-CH_2-COOH + H_2O \longrightarrow \text{Cl}-\underset{\text{Cl}}{\bigcirc}-OH + HOCH_2COOH \tag{20.11}$$

Parathion and malathion (eqn 20.12) are potent organophosphorus insecticides which also undergo hydrolysis reactions of this type.

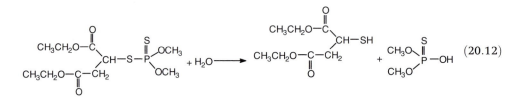

$$(20.12)$$

Note that further hydrolysis of the diester could occur.

2. An amide is hydrolysed to an acid and an amine. Depending on pH of the solution, the products may exist as a free acid and amine (ammonium) salt or a free amine and carboxylate salt or an equilibrium mixture of all these species.

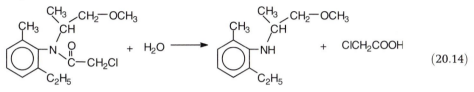

$$\text{(20.13)}$$

Metolachlor, a selective pre-emergence herbicide, undergoes this type or hydrolysis in an abiotic process:

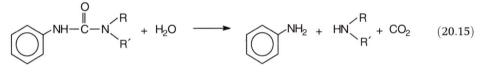

$$\text{(20.14)}$$

metolachlor

3. Phenylurea compounds are hydrolysed to give two amines: an aniline and an aliphatic amine:

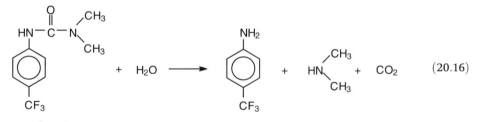

$$\text{(20.15)}$$

An example of this reaction is the case of the pesticide fenuron, a herbicide which is used to control a broad spectrum of weeds including deep-rooted grasses and woody plants:

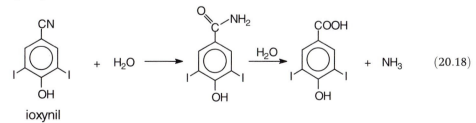

$$\text{(20.16)}$$

fenuron

4. A nitrile is hydrolysed to produce, in succession, an amide and then a carboxylic acid:

$$R-CN + H_2O \longrightarrow R-\overset{\displaystyle O}{\overset{\displaystyle \|}{C}}-NH_2 \xrightarrow{\ H_2O\ } RCOOH + NH_3 \qquad \text{(20.17)}$$

The selective herbicide ioxynil (and its bromine analog bromoxynil) undergoes such hydrolytic processes

$$\text{(20.18)}$$

ioxynil

5. A carbamate hydrolyses to generate an amine, an alcohol, and carbon dioxide:

$$R—NH—\overset{\displaystyle O}{\overset{\|}{C}}—O—R' \ + \ H_2O \ \longrightarrow \ RNH_2 + R'OH + CO_2 \tag{20.19}$$

The hydrolysis reaction for carbaryl, a very widely used horticultural insecticide, occurs as follows. In this case, the aromatic alcohol 1-naphthol is formed:

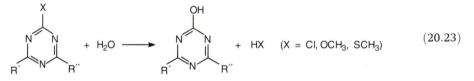

$$(20.20)$$

6. The thiocarbamate family likewise reacts to form amines, thiols, and carbon dioxide:

$$R—NH—\overset{\displaystyle O}{\overset{\|}{C}}—S—R' \ + \ H_2O \ \longrightarrow \ RNH_2 + R'SH + CO_2 \tag{20.21}$$

This is illustrated by the hydrolysis of benthiocarb, a selective pre-emergence herbicide used especially against annual grasses and broad leaf weeds in rice fields:

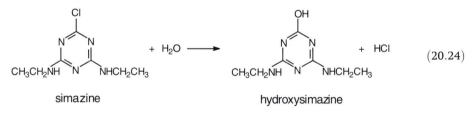

$$(20.22)$$

7. A well known example of hydrolysis occurs with the triazine herbicides:

![triazine hydrolysis reaction]

$$(20.23)$$

An example is the hydrolysis of simazine:

![simazine hydrolysis reaction, simazine + H₂O → hydroxysimazine + HCl]

$$(20.24)$$

Oxidation

Oxidation reactions are extremely important in terms of the total degradation of organic species in the environment. For oxidation to occur, an appropriate oxidant must be available; the nature of the oxidant depends on the environmental circumstance. In surface soils and water, where the pE is high, there is a plentiful supply of oxygen and in addition, more powerful oxidants produced by photochemical processes may be present. These include the hydroxyl radical, hydrogen peroxide, ozone, and singlet oxygen ($O(^1D)$). While the concentration of each of these species is normally small compared with ground-state dioxygen ($O_2(^3\sum_g^-)$), they can play a dominant role in bringing about oxidation of otherwise refractory molecules. Under the variety of anoxic conditions, other less powerful oxidants like nitrate and sulfate are potential electron acceptors for the oxidation of biocides. When all oxidants are depleted, reductive degradation becomes operative.

Some of the types of oxidation reactions are as follows.

1. Alkanes or aliphatic hydrocarbon substituents are oxidized, often at the terminal group, to produce alcohols, aldehydes, and then carboxylic acids:

$$RCH_3 \longrightarrow RCH_2OH \longrightarrow RCHO \longrightarrow RCOOH \qquad (20.25)$$

Various microorganisms using unique mechanisms are capable of bringing about parts or all of this sequence. Subterminal oxidation is also possible.

2. Alkenes are oxidized through a number of processes resulting in the production of ketones, alcohols, and carboxylic acids. When attack occurs at the double bond, the following reaction sequence occurs from alkene to epoxide to 1,2-diol to α-hydroxy acid and finally to a carboxylic acid that has one less carbon than the original alkene.

$$(20.26)$$

3. Branched chain hydrocarbons, while somewhat more stable than straight chain molecules, undergo similar types of biological oxidation processes. Alicyclic hydrocarbons are highly resistant to oxidative degradation in most situations.

4. Aromatic hydrocarbons are also highly resistant to oxidation reactions as was observed in our earlier discussion of polyaromatic hydrocarbons in the atmosphere. Nevertheless, oxidation does occur with the rate and extent strongly influenced by the nature of substituents on the molecule. Halogens, nitro-, and sulfonate groups are stabilizing compared with hydroxy, methoxy, and carboxylate groups. The number and position of substituents is also important.

One mechanism for the oxidation of benzene itself involves the initial formation of an epoxide or oxirane which is subsequently converted to the diol with concomitant rearomatization of the benzene ring:

$$(20.27)$$

Further oxidation can lead to ring fission with the production of a dicarboxylic acid:

The rate of oxidative degradation of petroleum residues is of considerable current interest with respect to leakage from gasoline tanks and oil spills associated with extraction or transport of crude petroleum. Many of the generalizations regarding hydrocarbon decomposition have been found to apply in a study[2] of the microbial degradation of petroleum drilling cuttings. The cuttings, containing fuel oil, calcium salts and a small amount of emulsifier had a pH of 9.1, an organic matter content of 12.4% and a carbon nitrogen ratio of 103. They were applied to a silty clay agricultural soil, of pH 5.1, organic matter 2.62%, and carbon:nitrogen ratio of 8.4. If the soils had been sterilized, there was little or no degradation over a 9 month period, but in non-sterilized soils, about 75% of the original mass of hydrocarbon was degraded during this time. Amongst the classes of hydrocarbon, there were very large differences in decomposition rate, with essentially all the low molar mass alkanes (those with less than 27 carbons) being degraded within 16 days. There was slower decomposition of branched chain and aromatic fractions. The residual unreacted material was an unresolved complex mixture of mostly high molar mass hydrocarbons.

5. Particular functional groups of biocide molecules are also subject to oxidative processes, although biotic hydrolysis reactions more often initiate a degradative sequence. A thioether sulfur atom can be oxidized in two steps, first to a sulfoxide and then to a sulfone (reaction 20.28):

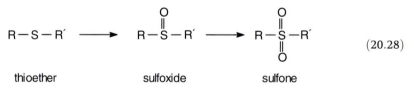

$$\tag{20.28}$$

Aldicarb is a potent and broad spectrum insecticide, acaricide (active against spiders) and nematicide (active against nematodes), which is susceptible to such oxidation (reaction 20.29). As in the case of parathion, the oxidized analogue is more toxic than the original compound:

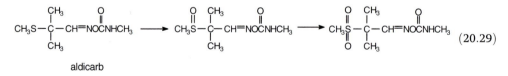

$$\tag{20.29}$$

6. Another very important oxidative reaction is β oxidation of fatty acid side chains, through a ketone intermediate to an acid with two fewer carbons in the chain:

$$R-CH_2-CH_2-COOH \longrightarrow R-\overset{\overset{\displaystyle O}{\|}}{C}-CH_2-COOH \longrightarrow R-COOH \tag{20.30}$$

An example of this reaction is conversion of MCPB (2-(4-chloro-2-methylphenoxy) butryic acid), a herbicide used for post-emergence weed control in cereals and pasture, to MCPA (4-chloro-2-methyl-phenoxyacetic acid), as shown:

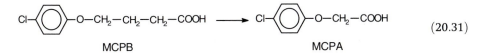

$$(20.31)$$

Reduction

Reduction reactions occur in low pE environmental situations such as anoxic groundwater and flooded soils. Some important and well defined biotic degradation processes take place in these situations.

1. Dehalogenation is a principal process of degradation of refractory halogenated organic compounds. While oxidative decomposition of such compounds is slow, the reduction reactions are frequently faster in some waste water treatment processes, advantage is taken of the increased rate in order to enhance the degradation of halogenated compounds present in a waste stream. It is also a possible means by which halogenated compounds are detoxified in a sanitary landfill. Zero-valent iron may be used as a reductant to increase the rate of reaction. Two general mechanisms of dehalogenation have been identified. Hydrogenolysis takes the following form:

$$R-X + H^+ + 2e^- \longrightarrow R-H + X^- \qquad (20.32)$$

Dichlorodiphenyltrichloroethane (DDT), the well known insecticide, is reduced to DDD (dichlorodiphenyldichloroethane) in this way.

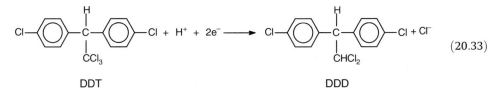

$$(20.33)$$

Pentachlorophenol, a common wood preservative, is similarly reduced to trichlorophenol.

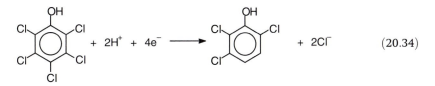

$$(20.34)$$

2. A second type of reductive dehalogenation reaction is called vicinal dehalogenation because the halogen leaving groups are arranged vicinally in the compound.

$$
\begin{array}{cc}
X & X \\
| & | \\
R-CH-CH-R' + 2e^- \longrightarrow R-CH{=}CH-R' + 2X^-
\end{array}
\qquad (20.35)
$$

This mechanism has been observed for lindane in flooded soils and in anaerobic sludge:

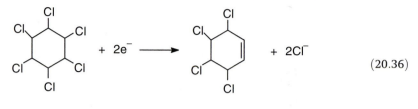

$$+ 2e^- \longrightarrow \qquad + 2Cl^- \tag{20.36}$$

lindane

3. Reduction of nitro groups is a multistep process which, if carried to completion, has an overall reaction represented by:

$$R-NO_2 + 6H^+ + 6e^- \longrightarrow R-NH_2 + 2H_2O \tag{20.37}$$

The insecticide fenitrothion and the herbicide trifluralin undergo such reactions under reducing conditions:

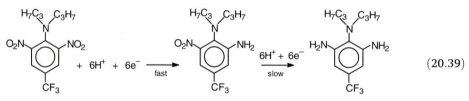

fenitrothion

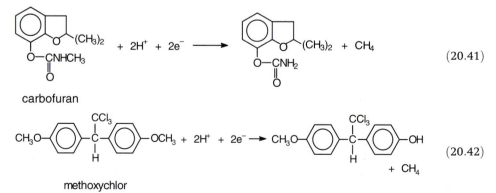

trifluralin

4. A third type of reduction involves dealkylation or dealkoxylation:

$$\begin{array}{cc} R\text{--}X\text{--}R' & +2H^+ + 2e^- \longrightarrow \\ (X = O, S, NH) \end{array} \quad \begin{array}{c} R\text{--}XH + R'H \\ (R' = \text{-}(CH_2)_n CH_3 \text{ or } O(CH_2)_n CH_3) \end{array} \tag{20.40}$$

This reaction is operative in the case of carbofuran, a carbamate insecticide, and methoxychlor, a close analogue of DDT.

carbofuran

methoxychlor

20.1.3 Rates of degradative reactions

It is important to have information concerning the rate of degradation (inversely related to molecular stability) of a particular xenobiotic. The chemical nature of the substance is obviously of key importance, but rate also depends on the availability of other reactants as well as additional environmental factors. Photolytic, biotic, and abiotic thermal processes all can contribute to breakdown of the molecule; the extent to which each of these occurs varies greatly in each environment, even between microenvironments, and so specific studies are required for individual situations. In this context, great care is needed in transferring information from laboratory experiments to that which would apply in the field.

The nature of the biocide

Very broad conclusions regarding the persistence of xenobiotic classes may be drawn from present data. Thus, Table 20.3 summarizes some of this information for several important classes of pesticide compounds. In the table, persistence ($t_{3/4}$) is defined loosely as the approximate time taken for 75% of the pesticide to degrade in the field.

Temperature

For any abiotic thermal reaction, temperature affects the rate according to the usual Arrhenius relation. A similar situation obtains in the case of biotic reactions, but each microorganism has its own optimum temperature. Below $0\,°C$, most microbiological processes essentially cease. High-temperature situations pertain to extreme environments such as dark soil under intense solar radiation where surface soil temperatures can be high enough ($> 50\,°C$) to kill off some of the mesophilic population.

Moisture

Moisture is essential for many abiotic and most biotic reactions. Almost all microorganisms require an abundance of moisture to flourish, and degradation rates increase with increasing moisture in soil up to field capacity. As soil pores are filled with water, oxygen

Table 20.3 Persistence of pesticides in the environment

Pesticide class	Persistence ($t_{3/4}$)/months
Chlorinated hydrocarbons	$16->24$
Urea, trazine	$3->18$
Benzoic acid, amide	3–12
Phenoxy, toluidine, nitrile	1–6
Organophosphorus	0.2–3
Carbamate, aliphatic acid	0.5–3

From Edwards C. A., *Persistent pesticides in the environment*, 2nd edn, CRC press, Cleveland; 1973.
The $t_{3/4}$ value is the time taken for 75% of the pesticide to degrade in the field.

transfer becomes limited and anoxic, reducing conditions set in with accompanying microbial switch-over to facultative and anaerobic species. The actual degradation mechanism may then change substantially.

Soil/water properties—pH

Hydrolysis by nucleophilic substitution can be catalysed by either acids or bases and therefore there is a strong pH dependence (Fig. 20.1). In the example shown, the degradation of atrazine is very slow at intermediate pH values such as are found in most environments. At pH 6, the pseudo first order rate constant of approximately 10^{-10} s^{-1} corresponds to a half-life of over 200 y. However, hydrolysis of atrazine does occur under apparently neutral conditions. In part this is because the bulk pH of the water body or soil solution does not indicate the true availability of protons or hydroxyl ions to the pesticide. The microenvironment provided by soluble species or solid phases must also be taken into account.

Soil/water properties—organic matter

Natural organic matter (NOM) is a source of hydrogen ions, through dissociation of acidic groups, and is therefore one environmental component which determines the bulk pH as well as the local availability of protons. In solution or as a solid, NOM can interact strongly with many other organic compounds as shown in Chapter 12. The complexed species then show altered—either enhanced or diminished—reactivity. Probably the most significant effect of NOM on degradation is with respect to biotic processes. A soil rich in organic matter usually has a large microbial population. Heterotrophic organisms, by definition, require presynthesized organic compounds, and as they make use of the NOM they can also co-metabolize the foreign compound. Zymogenous species which make use of a specific chemical also contribute to degradation.

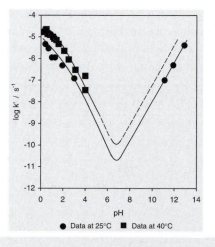

Fig. 20.1 The pH/rate constant profile for the triazine pesticide, atrazine. The figure is redrawn from some of the experimental and interpolated data provided in Plust, S.J., Loehe, J.R., Feher, F.J., Benedict, J.H., and Herbrandson, H.F., *J. Org. Chem.*, **46**, 3661–3665 (1981).

Soil/water properties—inorganic species

Dissolved metal ions are known to catalyse some nucleophilic substitution reactions. By forming a complex through an electron-donating atom like oxygen, electrons are withdrawn from the molecule making it more susceptible to attack by a nucleophile such as water. Reaction 20.43, the hydrolysis of an amide, is an example.

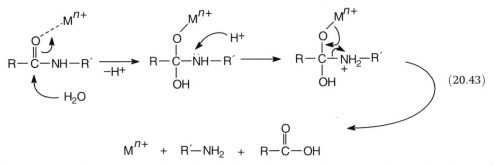

(20.43)

Note again, in the sequence shown the final products may be present in varying degrees of protonation depending on bulk or local pH.

Surface-bound metals can effect the same enhancement. For example, it has been shown that organophosphorus compounds hydrolyse more rapidly in the presence of hydrous iron and aluminium oxides such as are found in abundance in some tropical soils as well as in the depositional horizon of Spodosols and Brunisols.

Besides being available to complex xenobiotics, certain soil minerals, including hydrous iron (III) oxide (as limonite or goethite), are oxidizing agents, enhancing the oxidation of organic material in the soil. On the other hand, minerals like iron sulfide (pyrite) provide sites for reduction at the interface between the solid and the associated water. Natural organic matter, especially relatively undecomposed material with a small O : C ratio, is also essentially a reducing agent which can contribute to anoxic conditions and a reducing environment.

20.1.4 Kinetic calculations

In calculating rate of degradation of biocides and the kinetic parameters associated with a particular reaction, it is common to assume that the process is first order. In many cases this is a reasonable assumption because the xenobiotic compound is usually present in very small concentration compared with other reactants (water and organic matter, for example).

An example of a quantitative assessment is taken from a study of a number of pesticides, including one case where the degradation kinetics for the organophosphorus insecticide fenthion[3] were examined.

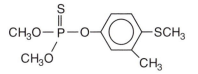

fenthion

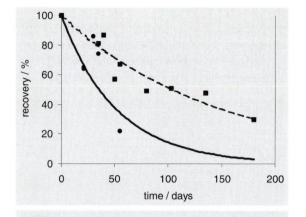

Fig. 20.2 Degradation kinetics for fenthion in a laboratory experiment where the pesticide had been dissolved in filtered river water. Filled squares, 6 °C Filled circles, 22 °C. Data redrawn from the reference in Note 3.

Figure 20.2 shows the degradation behaviour measured as the percentage of fenthion remaining (recovery) against time at two temperatures (6 °C and 22 °C) in the laboratory. The data were obtained using river water with pH 7.3 that had undergone filtration through a 0.7 μm filter. The experiment was carried out in darkness so that photolytic decomposition did not occur.

Some qualitative observations regarding the plotted data are as follows.

1. The scatter of data is characteristic of results obtained on water and soil after extraction and gas chromatographic analysis. Exponential best-fit lines are drawn through the points.

2. As would be expected, the rate of degradation increased with temperature. As the samples were not sterilized, it is not possible to determine whether the principal degradation mechanism was abiotic or biotic.

3. Additional data were obtained on samples that had not been filtered, in order to determine whether the solid materials stabilized the chemical against degradation or enhanced its decomposition. The results were inconclusive.

Using the original data points and assuming that the degradation follows first order kinetic behaviour ($C_t = C_0 e^{-kt}$) the value of the rate constant k can be estimated from a plot of $\ln(C_t/C_0)$ vs. time t (Fig. 20.3). The plot shows a linear relation (with considerable scatter) and the slope of the best-fit straight line is measured. Its value is equal to $-k$, with units of reciprocal time. For a first order reaction, the half-life is independent of the original concentration of the reactant and is given by $t_{1/2} = (\ln 2)/k$.

In the case of fenthion in filtered river water, the values of the rate constants are therefore 0.0066 days^{-1} and 0.0194 days^{-1} with $t_{1/2}$ then 105 days and 36 days at 6 °C and

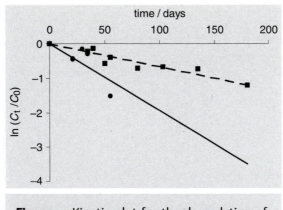

Fig. 20.3 Kinetic plot for the degradation of fenthion at (filled squares) 6 °C and 22 °C (filled circles). The slopes of the lines are − 0.0066 days^{-1} (R^2=0.882) and − 0.00194 days^{-1} (R^2=0.632) respectively.

22 °C respectively (our calculations are based on the reported data). The qualitative conclusions regarding increasing decomposition rates with increasing temperature are confirmed by these calculations.

Through the Arrhenius relation ($k = Ae^{-Ea/RT}$) which relates the rate constant to the frequency factor A and the activation energy for the process E_a, data obtained at the two temperatures can be used to calculate the value of E_a. In the present instance, E_a in filtered water was 20 kJ mol^{-1}. This value is lower than the range of usual activation energies for ester hydrolysis—typically between 40 and 30 kJ mol^{-1}However, scatter in the data used for the calculations is such that it would be unwise to suggest that the two results are significantly different.

20.2 Mobility

The second factor related to environmental behaviour of biocides is their mobility. A number of mechanisms are responsible for their migration, or attenuation, at the surface and in the subsurface of the soil (Fig. 20.4).

20.2.1 Aqueous transport

Xenobiotic compounds move through the soil within both the gas and liquid phases. In most cases aqueous phase transport is the most important process. In water, chemicals are carried downward by percolating rain and irrigation waters, upward by capillary action in poorly drained arid situations, and laterally on slopes and in groundwater aquifers.

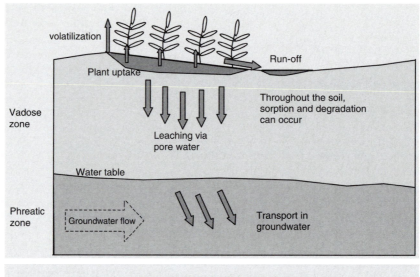

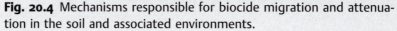

Fig. 20.4 Mechanisms responsible for biocide migration and attenuation in the soil and associated environments.

Whether the biocide moves readily or tends to remain fixed in association with the soil depends on properties of both the particular chemical and the soil. In this regard, a frequently calculated property is the octanol: water partition coefficient (K_{OW}) which was described in Chapter 14. As was shown there, this parameter can be related empirically to two other closely connected coefficients, the K_{OM} and K_{OC}, describing the distribution between organic matter and water. These latter parameters are themselves related to the fundamental distribution coefficient K_d. which is the ratio of the concentration of the compound in soil to its concentration in the associated water, by eqns 20.44 and 20.45:

$$K_{OM} = K_d f_{OM} \qquad (20.44)$$

and

$$K_{OC} = K_d f_{OC} \qquad (20.45)$$

where f_{OM} and f_{OC} are the fractions of organic matter and organic carbon respectively in the soil. The important assumption here is that the organic matter fraction of the soil is solely reponsible for sorption and retention of the xenobiotic compound. This is not true in cases where specific interactions between soil minerals and certain species play an important role in retention. An example of this is the strong bonds formed between the carboxylate, amine, and phosphonate groups of the herbicide glyphosate (Fig. 20.5) with iron in hydrous oxide minerals in the soil. In spite of exceptions like glyphosate, the relations described above are widely used and often approximately valid.

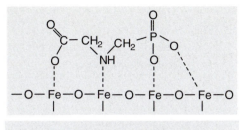

Fig. 20.5 Bonding between glyphosate and an iron oxide mineral surface.

Consider the three distribution constants which we have described. K_d is used to determine the distribution of the xenobiotic compound between soil and water *in a particular situation*. On the other hand, the K_{OW} is *a fundamental property* of the xenobiotic itself.

The intermediate parameter, the K_{OC} (or K_{OM}), is particularly useful; while it is a property of an individual chemical, it can be readily converted to a K_d value if the organic matter content of the soil is known.

Table 20.4 lists K_{OC} ranges, corresponding K_d (assuming 1% organic carbon) and R_f values. The K_{OC} has been calculated using concentrations in $\mu g\ g^{-1}$ (in soil) and $\mu g\ mL^{-1}$ (in water) and therefore has units of $mL\ g^{-1}$. The R_f function[3] is borrowed from chromatography and is a measure of the fractional transport of the compound compared with the water solvent. When $K_{OC} = 0$, $R_f = 1$ and there is no interaction with soil. Consequently, the compound moves freely with water. When K_{OC} is very large R_f approaches 0, signifying that the compound is completely immobilized. The following derivation and example illustrate how predictions can be made regarding the rate of movement of a biocide in the aqueous phase in soil.

$$R_f = \frac{\text{rate of movement of solute}}{\text{rate of movement of aqueous phase}} \tag{20.46}$$

Table 20.4 Distribution coefficients and mobility properties of various classes of organic compounds in soil

$K_{OC}/mL\ g^{-1}$	$K_d/mL\ g^{-1}$	R_f	Mobility	Class (typical)
0–50	0–0.5	1–0.7	Very high	Aliphatic acids
50–150	0.5–1.5	0.7–0.4	High	Carbamates
150–500	1.5–5	0.4–0.2	Medium	Benzoic acids
500–2000	5–20	0.2–0.05	Low	Triazines
2000–5000	20–50	0.05–0.02	Slight	Organophosphates
> 5000	> 50	< 0.02	Immobile	Organochlorines

Assuming equilibrium, and considering a particular volume of the soil/water system, the definition of R_f is equivalent to

$$R_f = \frac{\text{amount of solute in the aqueous phase}}{\text{total amount of solute}}$$

$$= \frac{x_m}{x_m + x_s} \qquad (20.47)$$

where $x_m =$ the amount of solute in the aqueous phase and $x_s =$ the amount of solute sorbed to the soil.

$$\frac{1}{R_f} = \frac{x_m + x_s}{x_m} = 1 + \frac{x_s}{x_m} \qquad (20.48)$$

The ratio x_s / x_m is the ratio of the amount of solute sorbed on soil to that in the aqueous phase *in a given volume of the column at any particular time*. This fraction is related to the K_d value by taking into account the density of soil particles (ρ) and the porosity of the soil (f_p):

$$\frac{x_s}{x_m} = K_d \times \rho \times \left(\frac{1 - f_p}{f_p}\right) \qquad (20.49)$$

$$R_f = 1 \Big/ \left(1 + K_d \times \rho \times \left(\frac{1 - f_p}{f_p}\right)\right) \qquad (20.50)$$

using the definition of R_f in eqn 20.46, rate of movement of the solute $= R_f \times$ rate of movement of the aqueous phase. Consider a situation where groundwater is moving at a rate of 2.3 cm h^{-1} in a soil with fractional porosity 0.27, consisting of particles whose density is 2.6 g m^{-1}. The water contains a solute and, in this situation, the K_d value is 10 mL g^{-1}.

$$R_f = 1/(1 + 10 \times 2.6 \times 0.73/0.27) = 0.014$$

Therefore, the rate of movement of the solute is $0.014 \times 2.3 = 0.032$ cm h^{-1}.

What are the factors contributing to transport properties?

1. The xenobiotic compound. Structural features of the molecule define its relative tendency to remain associated with soil or sediment particles or to move with the aqueous phase. Molecular factors contributing to retention (large K_{OW}, K_{OC}, and K_d) include hydrocarbon components, halogenation, being non-ionizable or cationic, and having the ability to form covalent bonds with soil minerals. Hydrophilicity (small K_{OW}, K_{OC}, and K_d) is favoured in the case of ionizable anionic species and where polar oxygen and nitrogen groups are present.

2. The soil. We have emphasized that, to a large extent, it is organic matter that is responsible for retention and this is the reason for the use of the coefficients K_{OM}, or K_{OC}. Retention is in part due to the hydrophobic nature of much of the organic matter complex, and its interaction with hydrophobic structures in the xenobiotic organic compound. Usually to a lesser extent, retention is also due to interaction between the compound and mineral surfaces. In the case of cationic species, however, the latter interaction may be very significant.

20.2.2 Vaporization

Mobility of pesticides has another aspect besides transport through porous media in the aqueous phase. Transport in the gas phase is a second means by which biotic compounds move vertically or horizontally. Gas-phase transport in soil does occur, but is usually not quantitatively as significant as transport *via* water. What can be very significant, however, is transport across the landscape, in the atmosphere above the soil surface.

The principal molecular property defining the ability of a pure compound to vaporize is its vapour pressure. For neutral, covalent compounds, this parameter depends in large part on the molar mass and polarity of the material; small molar mass and a non-polar structure favour volatility. For many xenobiotic compounds, the vapour pressure at 25 °C ranges from 10^{-6} to 1 Pa. (For comparison, recall that the vapour pressure of water is about 3000 Pa at the same temperature.) Exceptions with higher vapour pressures include low molar mass hydrocarbons such as some petroleum residues. The fact that the vapour pressure of many compounds of interest is very small does not mean that evaporative losses are necessarily negligible. In an open environment with even minimal air movement, the atmospheric partial pressure of the compound is almost zero except in the microlayer just above the pesticide surface. Therefore, there can be a strong thermodynamic tendency for evaporation or sublimation.

$$\text{Biocide (l or s)} \rightarrow \text{Biocide (g)} \tag{20.51}$$

However, the extent of evaporation is under kinetic not thermodynamic control. Environmental factors affecting the rate of vaporization include placement of the biocide (i.e whether on the surface or incorporated into the soil), temperature, air currents (wind), and the nature of the surface on which the biocide is deposited. Chemical interactions between a biocide and soil, the same ones which inhibit transport in water, can greatly reduce the equilibrium vapour pressure and also retard the rate of solid to vapour transition.

If a plot of vapour pressure against soil moisture content is made, assuming a constant concentration of biocide in the soil, there is an increase in vapour pressure until a constant value is reached that is characteristic of the pure compound (Fig. 20.6). Where water content is small, the forces retaining the condensed phase biocide on the soil particles act to reduce the extent of evaporation, but as moisture content increases, water molecules occupy adsorptive sites, releasing the biocide so that it may more readily vaporize. The plateau region occurs when monolayer coverage of all active sites has been achieved, and the compound is freed from the influence of the soil matrix.

When considering vaporization from a solid surface, vapour pressure of the pure chemical is the most important molecular property. However, it is frequently the case that evaporation takes place from an aqueous solution, as when herbicides are used to control water weeds or applied in wetland paddy culture. For soluble species, the equilibrium vapour pressure is controlled by relations defined by Henry's law.

With plants, sorption is due to retentive chemical forces on the plant surface and also to absorption into the epidural layer of the plant tissue. Recently[4] it has been postulated that evaporation from plant surfaces and also from the organic fraction of soil is best described by a dimensionless octanol/air coefficient K_{OA}, exactly analogous to the octanol/water coefficient, K_{OW}. The reasoning behind this is, again, that plant material has both

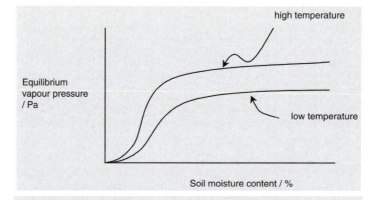

Fig. 20.6 Equilibrium partial pressure of an organic compound in soil with varying water content. At low values of soil moisture, retention of the compound is due to interaction with the soil. The plateau values are characteristic of the vapour pressure of the solute from a saturated aqueous solution.

hydrophilic and hydrophobic character, well represented by the amphiphilic solvent octanol. Values of K_{OA} for chlorobenzenes, PCBs, and DDT range from 10^4 to 10^{12}. Such large values indicate a strong tendency for these compounds to remain associated with growing crops in the terrestrial environment. This is consistent with the observed accumulation of organochlorines in plants and soil. Between $-10\,°C$ and $+20\,°C$, the value of log K_{OA} increases linearly with $1/T$. The major increase in volatility with temperature is critical in determining that these classes of compounds, when released in tropical or temperate (during the warm growing season) regions, vaporize to some extent, are carried elsewhere by global air currents and then condense in cold areas. Even in the absence of local sources, significant concentrations of organochlorines have been measured in the polar regions.

20.3 Leachability

The leachability of a biocide is its tendency to remain chemically stable while moving in the aqueous phase to a new location. This is an important parameter because leachable compounds can potentially move to other, perhaps more sensitive, parts of the environment. Taking chemical stability and mobility together, attempts have been made to characterize the leachability of organic chemicals in a simple manner. An example of this approach is the 'groundwater ubiquity score',[5] *GUS index*, defined in the following way:

$$\text{GUS} = \log_{10}(t_{1/2}^{\text{soil}}) \times (4 - \log_{10}(K_{OC})) \tag{20.52}$$

where $t_{1/2}^{\text{soil}}$, is the half-life in days for degradation in the soil and K_{OC} is the sorption coefficient in mL g^{-1} as defined above. It is important that the $t_{1/2}^{\text{soil}}$ be based on field experiments, as any measure of degradation rates is highly influenced by experimental conditions. As such, the value of $t_{1/2}{}^{soil}$ includes loss by biotic and abiotic degradation, as well as by vaporization. The $t_{1/2}{}^{soil}$ is the stability term (a large value indicates potential leachability) and the K_{OC} is the mobility term (a small value favours leachability). The values used in Table 20.5 are averages from a number of situations. It has been suggested that a GUS score of less than 1.8 indicates a species which is not prone to leaching (either because it degrades rapidly or is strongly retained by the solid matrix), while a score of greater than 2.8 is characteristic of highly leachable compounds. Table 20.5 lists selected pesticides along with their K_{OC} and $t_{1/2}{}^{soil}$ and GUS values.

As an example, the herbicide picloram (a derivative of picolinic acid) has a GUS index as follows:

$$\text{GUS} = \log(206) \times (4 - \log 26)$$
$$= 2.31 \times (4 - 1.42)$$
$$= 6.0.$$

picloram

Table 20.5 Properties related to leachability of selected pesticides

Pesticide	Class	$t_{1/2}{}^{soil}$/d	K_{OC}/mL g^{-1}	GUS
Picloram	Picolinic acid	206	26	5.98
Atrazine	Triazine	74	107	3.68
Carbofuran	Carbamate	37	55	3.54
Metolachlor	Amide	44	99	3.29
Simazine	Triazine	56	138	3.25
Aldicarb	Carbamate	7	17	2.34
Oxamyl	Dithiocarbamate	8	26	2.33
Carbaryl	Carbamate	19	423	1.76
Toxaphene	Organochlorine	9	96 000	1.23
Trifluralin	Dinitroaniline	83	8000	0.66
Chlorpyrifos	Organophosphate	54	6100	0.37
Heptachlor	Organochlorine	109	13 000	− 0.25
Dieldrin	Organochlorine	934	12 000	− 0.25
Chlordane	Organochlorine	37	19 000	− 0.45
DDT	Organochlorine	38 000	210 000	− 6.09

Data summarized from Gustafson, R. L.; see Note 5.

The large GUS value indicates that picloram is prone to leaching. This is because it has little affinity for soil organic matter (small K_{OC}) and is also chemically stable in the soil (large $t_{1/2}{}^{soil}$).

In contrast, the organochlorine heptachlor has a very large K_{OC} value and in spite of its chemical stability, it has a GUS index indicating only limited tendency to leach into groundwater:

$$\text{GUS} = \log(109) \times (4 - \log 13000)$$
$$= -0.23.$$

Another compound which is relatively resistant to leaching is the carbamate carbaryl:

$$\text{GUS} = \log(19) \times (4 - \log 423)$$
$$= 1.8.$$

In this case its resistance to leaching has more to do with the fact that it readily degrades. The undecomposed material is itself moderately mobile.

Many other attempts have been made to model the physical, chemical and biological behaviour of biocides in the field. The task is challenging, given the broad range of compounds and especially because each environmental situation is complex and unique. Of all the variables, perhaps the most unpredictable one is the microbiological activity which, in turn, depends on the specific ecological situation. When depth is included as a variable in developing models for persistence and mobility, the changes in soil physical, chemical, and microbiological properties add a new dimension to the complexity. A further element required in refined models takes account of the fact that retention is not necessarily defined by a single relation. In some situations, rapid and nearly reversible retention is observed, presumably due to surface adsorption. In the same system, slower, irreversible processes can also take place and this is attributed to *intraparticle diffusion*. Intraparticle diffusion implies that, over time, the biocide migrates into the interior of the solid phase through micropores or within the solid matrix itself. Material that has moved away from the surface is unable to exchange readily with the surrounding solution. This inhibits the mobility of the solute and has implications for both chemical clean-up and bioremediation of contaminated soils.

In spite of the limitations, indices and models are important in making qualitative and semi-quantitative predictions regarding the behaviour of particular biocides in specific soil environments.

The main points

1 Organic biocides include many types of synthetic compounds foreign to the soil/water environment. Some are materials inadvertently discharged into the environment, while others, of which pesticides are the most common example, have been manufactured and applied because of their particular toxicological properties. In both cases, we are interested in their persistence and mobility in the environment.

2 Biocides degrade by either abiotic (including photolytic) or biotic processes. Their stability depends on the chemical structure, the availability of reactants and microorganisms required for decomposition, and a variety of environmental factors. The degradation products have their own unique toxicological properties.

3 Mobility in air and water is related to the distribution of the chemical between the different phases. Like chemical stability, distribution also depends on the structure of the compound as well as physical and chemical properties of the soil and air.

4 Leachability is a function of both stability and mobility. Stable, mobile compounds are potentially leachable, while readily degradable biocides and/or those which have a large affinity for the solid phase are less likely to move far from their point of application.

Additional reading

1 Matolcy, G., M. Nadasy. and V. Andriska, *Pesticide Chemistry. Studies in Environmental Science 32*, Elsevier, Amsterdam; 1988.

2 Grover, R., ed., *Environmental Chemistry of Herbicides, Volume 1*, CRC Press, Inc., Boca Raton; 1988.

3 Morrill, L. G., B. C. Mahilum, and S. H. Mohiuddin, *Organic Compounds in Soils: Sorption, Degradation and Persistence*, Ann Arbor Science Publishers Inc./The Butterworth Group, Ann Arbor; 1982.

4 Borner, H., ed., *Pesticides in Ground and Surface Water. Chemistry of Plant Protection 9*, Springer, Berlin; 1994.

5 Larson, R. A. and E. J. Weber, *Reaction Mechanisms in Environmental Organic Chemistry*, Lewis Publishers, Boca Raton; 1994.

Problems

1 Consider the structure of the pesticide metolachlor and suggest how it might be bound to organic matter in the soil.

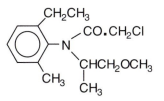

metolachlor

2 Compare the chemical forces with respect to the retention of the pesticides dieldrin and malathion by soil organic and mineral phases.

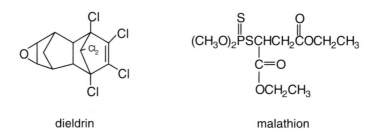

dieldrin   malathion

3 Qualitatively predict the relative extent of downward movement of the herbicides aldicarb and trifluralin if they are applied to soil containing 3.6% OM just before a rain storm which causes the water to penetrate 5 cm into the soil.

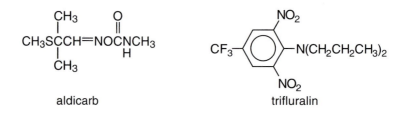

aldicarb   trifluralin

4 The presence of a halogen atom at an odd-numbered carbon, with respect to a primary substituent site on an a benzene ring, enhances the stability of the compound in comparison to a situation where the halogen is on an even-numbered carbon. Explain.

5 The $t_{1/2}$ for the hydrolytic degradation of the carbamate insecticide carbaryl is reported to be 31 days at 6 °C and 11 days at 22 °C. Calculate the activation energy for the reaction. Write an equation for the hydrolysis of this pesticide.

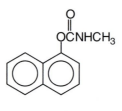

carbaryl

6 At 22 °C, the organophosphorus compound diazinon has a degradation $t_{1/2}$ of 80 days in river water, and 52 days in the same water after filtration. In contrast, the triazine cyprazine has corresponding $t_{1/2}$ values of 190 days and 254 days. Suggest reasons why filtration increases the $t_{1/2}$ in one case and decreases it in the other.

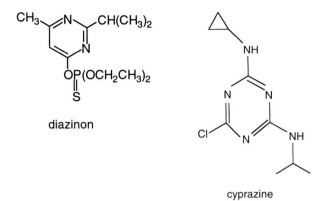

diazinon

cyprazine

7 Predict degradation processes which might occur for the herbicide metribuzin.

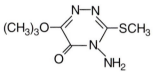

metribuzin

Notes

1 Glyphosate HOOC-CH_2-CH_2-NH-CH_2-PO(OH)$_2$ is a recently developed non-selective herbicide that is absorbed by plant foliage, effectively destroying many deep rooted annuals, biennials, and perennials.

2 Chaineau, C-H., J-L. Morel and J. Oudot, Microbial degradation of fuel oil hydrocarbons from drilling cuttings, *Environ. Sci. Technol.*, **29** (1995), 1615–21.

3 Lartiges, S. B. and P. P. Garrigues, Degradation kinetics of organophosphorus and organonitrogen pesticides in different waters under various environmental conditions, *Environ. Sci. Technol.*, **29** (1995), 1246–54.

4 Harner, T. and D. Mackay, Measurement of octanol–air partition coefficients for chlorobenzenes, PCBs and DDT, *Environ. Sci. Technol.*, **29** (1995), 1599–1606.

5 Gustafson R. L., Groundwater ubiquity score: a simple method for assessing pesticide leachability. *Environ. Toxicol. Chem.*, **8** (1989), 339–57.

21

The future Earth

We began our survey of environmental chemistry by summarizing the early stages of the Earth's natural history. The Earth was formed when gases on the periphery of the solar nebula cooled and contracted to form planets. The newly-formed planet was a solid sphere with a core made up of iron alloys and a crust of metal oxides and silicates. During the Earth's five billion year history, major changes occurred within and above the crust. Early on, water was formed and eventually came to cover two thirds of the Earth's surface at an average depth of over 3 km in the great oceans. Changes in the atmosphere also took place. Most significantly, free oxygen gas was released from its combination with other elements in the crust. The presence of atmospheric oxygen was essential as a support for some living organisms, while at the same time, other organisms maintained and regulated its supply. And so, the world as we know it took shape and became a place where microorganisms, plants, and animals of all kinds could thrive and support one another. In fact, all the inorganic and living components of the Earth, through interactions and cycles, have been acting together to make our planet what it is. We have studied some of the major interactions and cycles.

Human life is short in the context of this more than five billion years of geological time, and we see the Earth's composition and processes in the light of what we observe at this point in its long history. It is easy to think that the natural world has always been the same. But obviously this has not been true in the past, and changes in our surroundings will continue to occur. Volcanoes erupt, spewing gases into the atmosphere and altering the global climate. Fresh mineral material is spread over the continents and under the oceans. The continents continue to shift, causing cracking and upwelling of the crust. The highlands are eroded by rainfall; physical and chemical weathering of the eroded material alters the shape and composition of the individual particles. Sediment fills the lakes, bringing in nutrients so that they eutrophy and become wetlands.

All these events are occurring today and will continue to occur. The Earth adjusts to these changes. Species evolve or become extinct in response to their changing surroundings. The ability of living and inorganic components to modify the environment and to adapt to evolving conditions leads some people to consider the Earth itself as an individual organism.[1]

In this book we have tried to emphasize that environmental chemistry begins with a knowledge of the natural processes that have taken place and are ongoing now: the processes within the atmosphere, the hydrosphere, the geosphere and the biosphere, and the processes that move across boundaries, linking all parts of the environment together.

Humans (*Homo sapiens sapiens*) have been inhabitants of the Earth for the past 200 000 years. We live here, and by being alive we are participants in many of the natural processes and cycles. Our lungs take in oxygen and expel carbon dioxide. Our bodies require biomass and nutrients to survive and they release wastes that are then recycled through water, soil, plants and other animals. But we do more than just breathe and eat. Each of the civilizations that humans have created over millennia have had a broader impact on their surroundings. At this stage in human history, that impact is very great indeed.

There are two reasons for this—the number of human beings and the *environmental footprint* that each human creates and leaves behind as a legacy for the future world. Both factors are important. In some parts of the world—in particular much of southern Asia, Africa, South and Central America, it is the numbers of people that are most significant. In other regions—North America, and Australia, for example—it is the environmental footprint of each person that is more important. We saw (Table 8.6) that, in 1996, annual per capita energy use in North America was 325 GJ, while in Africa corresponding usage was 13 GJ. If we think of energy as the common currency of physical development, these figures imply that the average North American has a far greater impact on his/her surroundings than the average African. This statement is broadly true and it is certainly the case that many developments in industry and agriculture leave a large imprint on the global environment. Gases are released from factories and vehicles; liquid and solid wastes are discharged into water and on land—sometimes partially treated, sometimes untreated. Every year, new synthetic compounds are developed and, inadvertently or deliberately, some are released into the air or water or onto the soil where they interact with ongoing natural processes. All these impacts put together are the source of the growing environmental footprint associated with modern humans at the end of the twentieth century. Multiply this by the growing human population—five times greater than it was even one hundred years ago, and it becomes clear that the human influence on the environment is massive.

The chemistry of the atmosphere has been changing constantly since the beginning of Earth history. Just prior to the last glacial period, which began about 150 000 years ago, the atmospheric carbon dioxide mixing ratio as measured in Antarctic ice cores rose from about 200 to 290 ppmv, over a 20 000 year stretch of time. Recently there has been an increase of similar magnitude—from about 275 to 365 ppmv—but the present jump has taken place over a mere 150 years. Simultaneously, the past century has been a period during which there were substantial increases in atmospheric mixing ratios of other trace gases like methane and nitrous oxide. To add to this, new human-produced gases—the chlorofluorocarbons, halons, sulfur hexafluoride and many others—have been developed and released. While present in only trace quantities, the new compounds are able to interact with the 'natural' components and with solar radiation in the troposphere and stratosphere. The effects of trace gases on the ozone cycle in the stratosphere are well documented and large increases in the flux of solar UV-B radiation reaching parts of the Earth have been clearly measured. Details about how growth in greenhouse gas mixing ratios will affect the climate are debatable, but the fact that there has been unprecedented rapid growth cannot be questioned. This is evidence of the significant footprint which humans are contributing to the global environment. Whatever the impact may be, it is one that will affect the entire world, not just the areas where the chemicals are released. That is the nature of atmospheric environmental problems. They know no boundaries as we have seen in additional examples from literature on acid precipitation and on Arctic pollution.

Contaminated water too does not keep within well-defined boundaries, although its mobility may be somewhat less than that of air. The global water resource experiences the footprint of human activities and the impact is also energy-related in many cases. We have become familiar with the periodic occurrence of oceanic oil spills from super tankers in various places around the world. Perhaps it is surprising that the well-publicized major accidents account for only about 10 to 15% of petroleum that is discharged into the oceans. A larger proportion comes from bilge pumping (an illegal practice), land-based industrial sources and runoff from roads and parking lots in urban areas. There are also inputs into the oceans from the atmosphere.

Oil slicks are the most obvious evidence of petroleum contamination. Natural processes of volatilization, dissolution, abiotic and biotic degradation operate to make the slick disappear, especially its low molar mass components. The more resistant compounds are emulsified *via* oceanic turbulence and some of this emulsion eventually forms tar balls. The tar balls have small surface areas and are relatively inert so they degrade only slowly and can be carried great distances before sinking into the sediment or washing up on shore. Over long periods of time they break down into smaller particles that can be resuspended, degraded or be taken up by organisms in the water or sediment column. Some of the compounds, in particular aromatic and polyaromatic hydrocarbons, have toxic and carcinogenic properties and interfere with various metabolic reactions in many marine species. All these processes take place throughout the oceans, often far from the site of the original spill.

A recent study[2] in the Black Sea describes the types and amounts of petroleum residues found in water and sediment and creates a picture that is typical of situations around the world. The western part of the Black Sea, shared by Turkey, Bulgaria, Rumania, and Ukraine, receives petroleum-related contaminants from discharges released by industries along the Danube, Dnieper, and Dniester Rivers. Added to these are direct inputs from marine transport, crude oil spills, and the atmosphere. High concentrations of aliphatic and aromatic (including polyaromatic) hydrocarbons have been found in the river estuaries, but substantial amounts are also found offshore in both the sediment and dissolved phases, usually associated with combustion-derived particulate material.

The solid Earth has undergone major changes over the planet's five billion year history. We discussed the processes by which soils, the basis of agriculture and forestry, have formed (and are still forming) from massive rock and other materials. It is usually only over long periods of time that productive soils form from geologically-derived materials. And of course, there has always been a balance between soil formation and loss due to wind and water. In some cases, the transported materials are deposited elsewhere on land, replenishing the agricultural resource in productive areas like the lower reaches of the Yellow River in Northern China, or the Delta of the Nile in Egypt. Eventually, some soil is delivered to the oceans where it is permanently lost to agriculture. Sometimes the human footprint intersects these natural processes in a dramatic way, as when there is deforestation of headlands leading to enhanced erosion. In other cases, great dams are built, changing not only the water flow and topography but also the chemical processes related to eutrophication and rate of flushing of solid and dissolved species in the river.

Disposal sites for solid waste material now occupy thousands of hectares of land. In some cases, these areas are carefully monitored but from time to time accidents occur, bringing into focus the mark that human activities leave on the solid Earth.

In southern Spain, 45 km northwest of Seville, a deposit of copper-zinc-lead ore abundant in pyrite has been mined off and on since the Roman period.[3] In recent years, a profitable mining operation (the Los Frailes mine) for the three base metals has been established and a tailings dam was built with a capacity of about 70 million to deal with the waste rock. The tailings slurry has a pH of 2 to 4 and after settling, the liquid is pumped into a treatment facility to recover the metals before the wastewater is discharged into the Rio Agrio. This river empties into the Rio Guadiamar further downstream.

Below the tailings dam, the river flows through a rich farming area that produces citrus, peaches, sunflower, wheat, maize, cotton and olives, before reaching marsh lands that surround a National Park. The marshes are breeding grounds for more than 250 species of migratory birds and the area is protected to preserve the unique fauna and flora of the area.

On April 26, 1998, the tailings dam burst sending a 2 m wall of slurry coursing down the rivers. The amount of slurry released is uncertain but may have been as much as 5 million m^3, containing up to 2 million t of solids. Approximately 2000 ha of agricultural land were completely inundated with slurry and a further 2000 ha were affected less severely. Eighty percent of the tailings settled out along the first 13 km of the river bed and its surroundings.

Initially, farmers sustained great losses due to the burial and destruction of standing crops. What will be the long-term effects? At least in part, natural processes—the same ones that are responsible for formation of soils—have already been able to 'neutralize' some of the impact from the sudden release of the heavy slurry. The pH of the river returned to near normal within 10 days. However, zinc and lead levels remained high for some time and it is not certain whether production of food crops can resume in the near future. Humans were called upon to assist and 3 million m^3 of tailings have now been removed from the river banks.

The catalogue of environmental issues is a long one and, once again, out of this list we have chosen only a few examples showing how physical and chemical changes have been imposed on our surroundings by human activities. Some issues are acute and obviously require immediate attention, others are more subtle and may begin to become visibly serious only after a very long time. The question then becomes—how should we respond to these perceived and actual human impacts? Discussions centre around two different approaches. One is to look for the scientific/technical solution for each issue either in response to damage already observed or in anticipation of problems that might occur. Better control devices to minimize or eliminate emissions, materials that degrade photo-lytically or biologically, integrated pest management that includes crop rotation, biolog-ical control agents and only selective use of pesticides—all are examples of using improved science to minimize impacts. There is also considerable current interest in *green chemistry* for industrial manufacturing processes.[4] This term refers to the redesigning, where pos-sible, of manufacturing technologies in a way that they employ more benign chemicals—water rather than organic compounds as solvent, for example—so that the byproducts of manufacturing are themselves relatively harmless materials. Each of these approaches requires creative thinking and can markedly improve the environment by minimizing the direct impact we make.

But modification of technology cannot be the sole approach. For one thing, almost all technology involves some input of energy. For example, one experimental approach for reducing harmful smog-producing emissions in urban areas is to design vehicles that run on hydrogen–oxygen fuel cells. The energy-generating reaction that powers the vehicle is

$$2H_2 \text{ (g)} + O_2 \text{ (g)} \rightarrow 2H_2O \text{ (g)} \qquad \Delta G = -457 \text{ kJ}$$

Indeed, as combustion products, such fuel-cell engines emit mostly water vapour (but also small amounts of hydrogen peroxide and nitric oxide, if air is used as the oxygen source) and would go a long way toward reducing the load of volatile hydrocarbons in the atmosphere of a large city. However, it is necessary to produce hydrogen for the fuel cell, and this is often accomplished by reversing the energy-producing reaction. One method involves electrolysis of water and the required energy is frequently provided by a coal-fired power station operating at around 30% efficiency. An environmental advantage associated with centralizing the production of fuel at the power station is that excellent emission control practices may be followed, and whatever emissions are released can be dispersed at a high altitude remote from heavily populated areas. But, inevitably, control devices at power stations require material inputs and will produce their own waste products requiring disposal. As always, carbon dioxide is almost unavoidably, a released product.

Aside from scientific/technical solutions, the other mechanism that humans must consider for minimizing our environmental footprint is to reduce substantially the activities that impact the environment. This 'solution' must be especially directed toward the 25% of people, mostly in wealthy countries, who consume 75% of the world's resources. It involves consumption of less materials, reduced use of transportation (particularly the personal motorized vehicle), lessened dependence on non-renewable energy for providing a comfortable living environment and many other ways of curtailing resource depletion. It is clear that the huge imbalance in energy use between, for example, North America and Africa cannot be explained away and justified on the basis of climate and size of country.

Clearly, in these closing words, we have been moving beyond the field of environmental chemistry and beyond the purpose of this book. And so, we return to our stated goal which was to provide a description of the chemical basis for understanding our surroundings, the global environment. Whether one is dealing with ideas of human health, with design of devices and processes for controlling emissions, or setting policy, the importance of understanding the underlying nature of an issue cannot be overemphasized.

Notes

1 Lovelock, J. E. *Gaia: a New Look at Life on Earth*. Oxford University Press, Oxford; 1979, 1987, 157 pp.

2 Maldonado, C., Bayona, J. M., and Bodineau, L. Sources, distribution and water column processes of aliphatic and polycyclic aromatic hydrocarbons in the northwestern Black Sea water. *Environ. Sci. Technol.*, **33** (1999), 2693–702.

3 Sassoon, M. Los Friles aftermath. *Mining Environ. Management*, July 1998, pp. 8–12.

4 Anastas, P. and Williamson, T. C. *Green Chemistry*. Oxford University Press, Oxford; 1998.

Appendices

Appendix 1 Properties of the Earth

Mass of the Earth	$5.98 \times 10^{24}\,kg$
Mass of the atmosphere	$5.27 \times 10^{18}\,kg$
Mass of the oceans	$1.37 \times 10^{21}\,kg$
Mass of fresh water (surface)	$1.27 \times 10^{17}\,kg$
Mass of pore and groundwater	$9.5 \times 10^{18}\,kg$
Mass of ice	$2.9 \times 10^{19}\,kg$
Mass of atmospheric water	$1.3 \times 10^{16}\,kg$
Mass of living OM (dry wt, carbon)	$8 \times 10^{14}\,kg$
dead OM (dry wt, carbon)	$3.5 \times 10^{15}\,kg$
Average radius of the Earth	6378.2 km
Total area of Earth's surface	$5.10 \times 10^{14}\,m^2$
Area of continents	$1.48 \times 10^{14}\,m^2$
Continental area covered by ice	$1.72 \times 10^{13}\,m^2$
Volume of oceans	$1.35 \times 10^{18}\,m^3$

Appendix 2

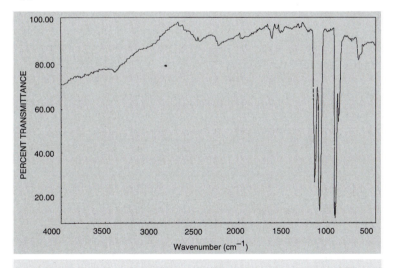

Fig. A.1 Infrared absorption spectrum of CFC-12. The relation between wave number and wavelength is: wave number $(cm^{-1}) = 10\,000/$wavelength (μm) (© BIO-RAD Laboratories, Sadtler Division, 2000).

Appendix 3 The elements

Atomic number	Element	Symbol	Atomic mass Dalton	Crustal mg kg^{-1}	Abundance Oceanic† mol L^{-1}	Freshwater mol L^{-1}	Atmospheric ppmv
1	hydrogen	H	1.00794	1520			0.53
2	helium	He	4.002602	0.008	1×10^{-9}		5.2
3	lithium	Li	6.941	20	2.5×10^{-5}	1.7×10^{-6}	
4	beryllium	Be	9.01218	2.6	7×10^{-12}		
5	boron	B	10.81	10	4.2×10^{-4}	1.7×10^{-6}	
6	carbon	C	12.011	480	2.0×10^{-3}		365
7	nitrogen	N	14.00674	25			209 500
8	oxygen	O	15.9994	466 000			
9	fluorine	F	18.99840	950	7.0×10^{-5}	5.3×10^{-6}	
10	neon	Ne	20.1797	7×10^{-5}	1×10^{-7}		18
11	sodium	Na	22.98977	28 300	0.481	2.2×10^{-4}	
12	magnesium	Mg	24.3050	20 900	5.5×10^{-2}	1.6×10^{-4}	
13	aluminium	Al	26.98154	81 300	3.5×10^{-8}	1.9×10^{-6}	
14	silicon	Si	28.0855	272 000	1×10^{-6}	1.9×10^{-4}	
15	phosphorus	P	30.97376	1000	5×10^{-8}	1.3×10^{-6}	
16	sulfur	S	32.066	260	2.9×10^{-2}		
17	chlorine	Cl	35.4527	130	0.561		
18	argon	Ar	39.948	1.2	1.2×10^{-5}		9300
19	potassium	K	39.0983	25 900	1.01×10^{-2}	3.4×10^{-5}	
20	calcium	Ca	40.078	36 300	1.06×10^{-2}	3.6×10^{-4}	
21	scandium	Sc	44.95591	16	1×10^{-11}	8.9×10^{-11}	
22	titanium	Ti	47.88	5600	1×10^{-8}	2.1×10^{-7}	
23	vanadium	V	50.9415	160	3×10^{-8}	2.0×10^{-8}	

24	chromium	Cr	51.9961	~100	3×10^{-9}	1.9×10^{-8}	
25	manganese	Mn	54.93805	950	2×10^{-9}	1.5×10^{-7}	
26	iron	Fe	55.847	50 000	1.7×10^{-9}	7.2×10^{-7}	
27	cobalt	Co	58.93320	20	1×10^{-10}	3.4×10^{-9}	
28	nickel	Ni	58.70	80	2×10^{-9}	3.8×10^{-8}	
29	copper	Cu	63.546	50	1.3×10^{-9}	1.6×10^{-7}	
30	zinc	Zn	65.39	75	8×10^{-10}	4.6×10^{-7}	
31	gallium	Ga	69.723	18	4×10^{-10}	1.3×10^{-9}	
32	germanium	Ge	72.61	1.8	3×10^{-12}		
33	arsenic	As	74.9216	1.5	2.0×10^{-8}	2.3×10^{-8}	
34	selenium	Se	78.96	0.05	4×10^{-13}	2.5×10^{-9}	
35	bromine	Br	79.904	0.37	8.6×10^{-4}	2.5×10^{-7}	
36	krypton	Kr	83.80	1×10^{-5}	1×10^{-10}		1.14
37	rubidium	Rb	85.4678	90	1.5×10^{-6}	1.8×10^{-8}	
38	strontium	Sr	87.62	370	9.4×10^{-5}	6.9×10^{-7}	
39	yttrium	Y	88.90585	30	1×10^{-10}	7.9×10^{-9}	
40	zirconium	Zr	91.224	190	1×10^{-10}		
41	niobium	Nb	92.90638	20	1×10^{-12}		
42	molybdenum	Mo	95.94	1.5	1.1×10^{-7}	5.2×10^{-9}	
43	technetium*	Tc	98.9062				
44	ruthenium	Ru	101.07	$\sim 1 \times 10^{-3}$			
45	rhodium	Rh	102.90550	$\sim 2 \times 10^{-4}$			
46	palladium	Pd	106.42	$\sim 6 \times 10^{-4}$	2×10^{-13}		
47	silver	Ag	107.8682	0.07	1×10^{-12}	2.8×10^{-9}	
48	cadmium	Cd	112.411	0.11	1×10^{-11}		
49	indium	In	114.82	0.049	9×10^{-13}		
50	tin	Sn	118.710	2.2	2×10^{-11}		
51	antimony	Sb	121.75	0.2	$\sim 3 \times 10^{-9}$	8.2×10^{-9}	
52	tellurium	Te	127.60	$\sim 5 \times 10^{-3}$	1.5×10^{-12}		
53	iodine	I	126.90447	0.14	3.7×10^{-7}	5×10^{-8}	
54	xenon	Xe	131.29	2×10^{-6}	8×10^{-10}		0.086
55	caesium	Cs	132.9054	3	2.3×10^{-9}	2.6×10^{-10}	

Appendix 3 (Continued)

Atomic number	Element	Symbol	Atomic mass Dalton	Crustal mg kg-1	Abundance Oceanic mol L⁻¹	Freshwater mol L⁻¹	Atmospheric ppmv
56	barium	Ba	137.327	500	3.5×10^{-8}	4.4×10^{-7}	
57	lanthanum	La	138.9055	32	1.6×10^{-11}	3.6×10^{-10}	
58	cerium	Ce	140.115	68	4×10^{-11}	5.7×10^{-10}	
59	praseodymium	Pr	140.90765	9.5	3×10^{-12}	5.0×10^{-11}	
60	neodymium	Nd	144.24	38	1.3×10^{-11}	2.8×10^{-10}	
61	promethium*	Pm	(145)				
62	samarium	Sm	150.36	7.9	3×10^{-12}	5.3×10^{-11}	
63	europium	Eu	151.965	2.1	7×10^{-13}	6.6×10^{-12}	
64	gadolinium	Gd	157.25			5.1×10^{-11}	
65	terbium	Tb	158.925			6.3×10^{-12}	
66	dysprosium	Dy	162.50			3.0×10^{-10}	
67	holmium	Ho	164.930			6.1×10^{-12}	
68	erbium	Er	167.26			2.4×10^{-11}	
69	thulium	Tm	168.934			5.9×10^{-12}	
70	ytterbium	Yb	173.04			2.3×10^{-11}	
71	lutetium	Lu	174.97			5.7×10^{-12}	
72	hafnium	Hf	178.49				
73	tantalum	Ta	180.948				
74	tungsten	W	183.85			1.6×10^{-10}	
75	rhenium	Re	186.21				
76	osmium	Os	190.2				
77	iridium	Ir	192.2				
78	platinum	Pt	195.08				
79	gold	Au	196.9665			2.0×10^{-11}	

80	mercury	Hg	200.59	3.5×10^{-10}
81	thallium	Tl	204.38	
82	lead	Pb	207.2	4.8×10^{-9}
83	bismuth	Bi	208.980	
84	polonium	Po	(209)	
85	astatine	At	(210)	
86	radon	Rn	(222)	
87	francium	Fr	(223)	
88	radium	Ra	226.025	
89	actinium	Ac	227.028	
90	thorium	Th	232.038	
91	protactinium	Pa	231.036	
92	uranium	U	238.03	1×10^{-9}
93	neptunium	Np	237.048	
94	plutonium	Pu	(244)	
95	americium	Am	(243)	

*These are radioactive elements and are not found naturally except for trace quantities in uranium deposits

†Ocean concentrations are averages; in some cases values refer to the surface water

Source: Most of the values for terrestrial and oceanic concentrations are from J. Emsley, *The Elements*, 2nd edn, Clarendon Press, Oxford; 1991. Freshwater values are reported as mean concentrations in river water in S.M. Libes, *An Introduction to Marine Biogeochemistry*, John Wiley and Sons Inc., New York; 1992.

Appendix 4   Properties of air and water

Air (dry, P°)

Average molar mass (troposphere)	28.96	dalton
Density (0 °C)	1.293	$kg\,m^{-3}$
(20 °C)	1.205	$kg\,m^{-3}$
Viscosity (0 °C)	1.7×10^{-2}	$g\,m^{-1}s^{-1}$
(20 °C)	1.9×10^{-2}	$g\,m^{-1}s^{-1}$

Water

Molar mass	18.015	dalton
Density (0 °C)	999.87	$kg\,m^{-3}$
(20 °C)	998.23	$kg\,m^{-3}$
Viscosity (0 °C)	1.79	$g\,m^{-1}s^{-1}$
(20 °C)	1.00	$g\,m^{-1}s^{-1}$

Appendix 5   Area, biomass, and productivity of ecosystem types

Type	area/$10^{12}\,m^2$	mean plant biomass/ $kg\,C\,m^{-2}$	productivity/ $kg\,m^{-2}y^{-1}$
tropical forest	24.5	18.8	0.83
temperate forest	12.0	14.6	0.56
boreal forest	12.0	9.0	0.36
woodland and shrubland	8.0	2.7	0.27
savanna	15.0	1.8	0.32
grassland	9.0	0.7	0.23
tundra and alpine meadow	8.0	0.3	0.065
desert scrub	18.0	0.3	0.032
rock, ice, sand	24.0	0.01	0.015
cultivated land	14.0	0.5	0.29
swamp and marsh	2.0	6.8	1.13
lake and stream	2.5	0.01	0.23
open ocean	332.0	0.0014	0.057
upwelling zones	0.4	0.01	0.23
continental shelf	26.6	0.005	0.1
algal bed and reef	0.6	0.9	0.90
estuaries	1.4	0.45	0.81

Source: J. Harte, *Consider a Spherical Cow*, University Science Books, Mill Valley, Ca; 1988, 283pp.

Appendix 6   SI prefixes and fundamental geometric relations

atto	a	10^{-18}
femto	f	10^{-15}
pico	p	10^{-12}
nano	n	10^{-9}
micro	μ	10^{-6}
milli	m	10^{-3}
centi	c	10^{-2}
deci	d	10^{-1}
hecto	h	10^{2}
kilo	k	10^{3}
mega	M	10^{6}
giga	G	10^{9}
tera	T	10^{12}
peta	P	10^{15}
exa	E	10^{18}

Circumference of a circle $= 2\pi r = \pi d$
Surface area of a circle $= \pi r^2$
Surface area of a sphere $= 4\pi r^2$
Volume of a sphere $= \frac{4}{3}\pi r^3$

Appendix 7   Fundamental constants

Avogadro's number	N_A	6.0221367×10^{23}	mol^{-1}
Boltzmann's constant	k	1.38066×10^{-23}	$J\,K^{-1}$
Faraday's constant	F	9.6485309×10^{4}	$C\,mol^{-1}$
Gas constant	R	8.314510	$J\,K^{-1}\,mol^{-1}$
		0.082057	$L\,atm\,K^{-1}\,mol^{-1}$
Planck's constant	h	$6.6260755 \times 10^{-34}$	$J\,s$
Speed of light (vacuum)	c	2.99792458×10^{8}	$m\,s^{-1}$
Standard Pressure	P^{o}	1.01325×10^{5}	Pa

Appendix 8   Thermochemical properties of selected elements and compounds

	ΔH°_f/kJ mol^{-1}	ΔG°_f/kJ mol^{-1}	S°/J mol^{-1} K^{-1}
Aluminium			
Al (s)	0	0	+28.33
Al^{3+} (aq)	−531	−485	−321.7*
Al$_2$O$_3$ (s, corundum)	−1675.7	−1582.3	+50.92
Al$_2$O$_3$·3H$_2$O (s, gibbsite)	−2586.67	−2310.41	+136.90
Calcium			
Ca (s)	0	0	+41.42
Ca^{2+} (aq)	−542.83	−553.58	−53.1*
CaO (s)	−635.09	−604.05	+39.75
CaCO$_3$ (s, calcite)	−1206.92	−1128.84	+92.6
Carbon (including some common organic compounds)			
C (s, graphite)	0	0	+5.740
C (g)	+716.68	+671.29	+158.99
CO (g)	−110.53	−137.15	+197.57
CO$_2$ (g)	−393.51	−394.36	+213.63
CO$_2$ (aq)	−413.80	−385.98	+117.6*
H$_2$CO$_3$ (aq)	−699.65	−623.08	+187.4*
HCO$_3^-$ (aq)	−691.99	−586.84	+91.2*
CO$_3^{2-}$ (aq)	−677.14	−527.86	−56.9*
CCl$_4$ (l)	−135.44	−65.28	+216.40
CS$_2$ (l)	+89.70	+65.27	+151.34
HCN (aq)	+107.1	+119.7	
CN$^-$ (aq)	+150.6	+172.4	+94.1*
CH$_4$ (g, methane)	−74.81	−50.75	+186.16
CH$_3$ (g, methyl radical)	+145.69	+147.92	+194.2*
C$_2$H$_6$ (g, ethane)	−84.68	−32.92	+229.49
C$_2$H$_4$ (g, ethene)	+52.26	+68.08	+219.45
C$_2$H$_2$ (g, ethyne)	+226.73	+209.17	+200.83
C$_3$H$_8$ (g, propane)	−104.5	−23.4	+269.9
C$_4$H$_{10}$ (g, n-butane)	−126.5	−17.15	+310.1
C$_5$H$_{12}$ (g, n-pentane)	−146.5	−8.37	+348.9
C$_6$H$_6$ (l, benzene)	+49.0	+124.7	+172
C$_6$H$_6$ (g, benzene)	+82.9	+129.7	+269.2
C$_{10}$H$_8$ (s, naphthalene)	+78.53		
CH$_3$OH (l, methanol)	−238.66	−166.35	+126.8
C$_2$H$_5$OH (l, ethanol)	−277.69	−174.89	+160.7
C$_6$H$_5$OH (l, phenol)	−165.0	−50.9	+146.0*
HCOOH (l, formic acid)	−424.72	−361.42	+128.95
CH$_3$COOH (aq, acetic acid)	−485.76	−396.56	+159.8
CH$_3$COO$^-$ (aq, acetate)	−486.01	−369.39	+86.6*
HCHO (g, formaldehyde)	−108.57	−102.55	+218.66

Appendix 8   (Continued)

	$\Delta H^\circ_f/kJ\,mol^{-1}$	$\Delta G^\circ_f/kJ\,mol^{-1}$	$S^\circ/J\,mol^{-1}\,K^{-1}$
CH_3CHO (l, acetaldehyde)	−192.30	−128.20	+160.2
CH_3CHO (g, acetaldehyde)	−166.19	−128.91	+250.2
CH_3COCH_3 (l, acetone)	−248.1	−155.4	+200.4*
$C_6H_{12}O_6$ (s, β-D-glucose)	−1268	−910	+212*
Chlorine			
Cl_2 (g)	0	0	+222.96
Cl (g)	+121.68	+105.70	+165.09
Cl^- (aq)	−167.16	−131.24	+56.5*
HCl (g)	−92.31	−95.30	+186.80
HCl (aq)	−167.16	−131.23	+56.5*
Hydrogen			
H_2 (g)	0	0	+130.58
H (g)	+217.97	+203.26	+114.60
H_2O (l)	−285.83	−237.18	+69.91
H_2O (g)	−241.82	−228.59	+188.72
H_2O_2 (l)	−187.78	−120.42	+109.6
H_3O^+ (aq)	−285.83	−237.13	+69.91
Iron			
Fe (s)	0	0	+27.28
Fe^{2+} (aq)	−89.1	−78.9	−137.7*
Fe^{3+} (aq)	−48.5	−4.7	−315.9*
Fe_2O_3 (s, hematite)	−824.2	−742.2	+87.40
Fe_3O_4 (s, magnetite)	−1118.4	−1015.4	+146.4
Nitrogen			
N_2 (g)	0	0	+191.50
N (g)	+472.70	+455.58	+153.19
NO (g)	+90.25	+86.55	+210.65
N_2O (g)	+82.05	+104.20	+219.74
NO_2 (g)	+33.18	+51.29	+239.95
N_2O_5 (g)	+11.3	+115.0	+355.7
HNO_3 (aq)	−207.36	−111.25	+146.4*
NH_3 (g)	−46.11	−16.42	+192.34
NH_3 (aq)	−80.29	−26.57	+111.3*
NH_4^+ (aq)	−132.51	−79.31	+113.4*
NH_2CONH_2 (s, urea)	−333.51	−197.44	+104.60
Oxygen			
O_2 (g)	0	0	+205.03
O (g)	+249.17	+231.75	+160.95
O_3 (g)	+142.7	+163.2	+238.82
OH^- (aq)	−229.99	−157.28	−10.75*

Appendix 8   (Continued)

	$\Delta H°_f/\text{kJ mol}^{-1}$	$\Delta G°_f/\text{kJ mol}^{-1}$	$S°/\text{J mol}^{-1}\text{K}^{-1}$
Sulfur			
S (s, rhombic)	0	0	+31.80
SO_2 (g)	−296.83	−300.19	+248.11
SO_3 (g)	−395.72	−371.08	+256.65
HSO_4^- (aq)	−887.34	−755.99	+131.8*
SO_4^{2-} (aq)	−909.27	−744.60	+20.1*
H_2S (g)	−20.63	−33.59	+205.68
H_2S (aq)	−39.7	−27.86	+121*
HS^- (aq)	−17.6	+12.08	+62.08*
SF_6 (g)	−1209	−1105.4	+291.71

Source: The National Bureau Standards Tables of Chemical Thermodynamic Properties. J. Phys. and Chem. Reference Data, 11, Supplement 2 (1982) as reported in W.G. Breck, R.J.C. Brown, and J.D. McCowan, *Thermochemical Tables to Accompany Chemistry for Science and Engineering*, McGraw-Hill Ryerson Ltd., Toronto, 1989. These values have been calculated using $P = P° = 101\,325$ Pa. Some additional values* were obtained from P. Atkins, Physical Chemistry, 6th edn., W.H. Freeman and Co., New York; 1998 and are calculated with $P = 1$ bar $= 100\,000$ Pa.

Appendix 9   Mean bond enthalpies $\Delta H/\text{kJ mol}^{-1}$ at 298 K

	H	C	N	O	F	Cl	Br	I	S	P	Si
H	436										
C	412	348 s									
		612 d									
		838 t									
		518 a									
N	388	305 s	163 s								
		613 d	409 d								
		890 t	946 t								
O	463	360 s	157	146 s							
		743 d		497 d							
F	565	484	270	185	155						
Cl	431	338	200	203	254	242					
Br	366	276			219	193					
I	299	238			210	178	151				
S	338	259			496	250	212		264		
P	322									201	
Si	318		374	466							226

Source: P. Atkins, *Physical Chemistry*, 6th edn., W.H. Freeman and Co., New York; 1998.
Note: s = single, d = double, t = triple, a = aromatic bond.

Appendix 10 Dissociation constants for acids and bases in aqueous solution at 25 °C

Acid	Protonated Species	K_a	pK_a	Base	Deprotonated species	K_b	pK_b
Acetic acid	CH_3COOH	1.8×10^{-5}	4.75	Acetate	CH_3COO^-	5.6×10^{-10}	9.25
Aluminium (III)	$Al(H_2O)_6^{3+}$	7.2×10^{-6}	5.14	Hydroxyaluminum (III)	$Al(OH)^{2+}$	1.4×10^{-9}	8.86
Ammonium	NH_4^+	5.6×10^{-10}	9.25	Ammonia	NH_3	1.8×10^{-5}	4.75
Arsenic acid	H_3AsO_4	5.8×10^{-3}	2.24	Dihydrogen arsenate	$H_2AsO_4^-$	1.7×10^{-12}	11.76
Dihydrogen arsenate	$H_2AsO_4^-$	1.10×10^{-7}	6.96	Hydrogen arsenate	$HAsO_4^{2-}$	9.1×10^{-8}	7.04
Hydrogen arsenate	$HAsO_4^{2-}$	3.2×10^{-12}	11.50	Arsenate	AsO_4^{3-}	3.1×10^{-3}	2.50
Arsenious acid	$As(OH)_3$	5.1×10^{-10}	9.29	Dihydrogen arsenite	$H_2AsO_3^-$	2.0×10^{-5}	4.71
Boric acid	$B(OH)_3$	7.2×10^{-10}	9.14	Borate	$B(OH)_4^-$	1.4×10^{-5}	5.86
Carbon dioxide*	CO_2	4.5×10^{-7}	6.35	Hydrogen carbonate	HCO_3^-	2.2×10^{-8}	7.65
Hydrogen carbonate	HCO_3^-	4.7×10^{-11}	10.33	Carbonate	CO_3^{2-}	2.1×10^{-4}	3.67
Formic acid	$HCOOH$	1.8×10^{-4}	3.75	Formate	$HCOO^-$	5.6×10^{-11}	10.25
Hydrofluoric acid	HF	3.5×10^{-4}	3.46	Fluoride	F^-	2.9×10^{-11}	10.54
Hydrogen cyanide	HCN	4.9×10^{-10}	9.31	Cyanide	CN^-	2.0×10^{-5}	4.69
Hydrogen sulfate	HSO_4^-	1.0×10^{-2}	2.00	Sulfate	SO_4^{2-}	1.0×10^{-12}	12.00
Hydrogen sulfide	H_2S	1.0×10^{-7}	7.00	Hydrogen sulfide ion	HS^-	1.0×10^{-7}	7.00
Hydrogen sulfide ion	HS^-	1.1×10^{-12}	11.96	Sulfide	S^{2-}	9.1×10^{-3}	2.04
Hypochlorous acid	$HClO$	3.0×10^{-8}	7.52	Hypochlorite	ClO^-	3.3×10^{-7}	6.48
Iron (III)	$Fe(H_2O)_6^{3+}$	6.3×10^{-3}	2.19	Hydroxyiron (III)	$FeOH^{2+}$	1.6×10^{-12}	11.80
Methylammonium	$CH_3NH_3^+$	2.2×10^{-11}	10.66	Methylamine	CH_3NH_2	4.5×10^{-4}	3.34
Nitrous acid	HNO_2			Nitrite	NO_2^-		

Acid		K_a	pK_a	Conjugate base		K_b	pK_b
Phenol	C_6H_5OH	1.3×10^{-10}	9.89	phenate	$C_6H_5O^-$	7.7×10^{-5}	4.11
Phosphoric acid	H_3PO_4	7.1×10^{-3}	2.15	Dihydrogen phosphate	$H_2PO_4^-$	1.4×10^{-12}	11.85
Dihydrogen phosphate	$H_2PO_4^-$	6.3×10^{-8}	7.20	Hydrogen phosphate	HPO_4^{2-}	1.6×10^{-7}	6.80
Hydrogen phosphate	HPO_4^{2-}	4.2×10^{-13}	12.38	Phosphate	PO_4^{3-}	2.4×10^{-2}	1.62
Silicic acid	$Si(OH)_4$	2.2×10^{-10}	9.66	Trihydrogen silicate	$H_3SiO_4^-$	4.6×10^{-5}	4.34
Sulfurous acid	H_2SO_3	1.72×10^{-2}	1.76	Hydrogen sulfite	HSO_3^-	5.81×10^{-13}	12.24
Hydrogen sulfite	HSO_3^-	6.43×10^{-8}	7.19	Sulfite	SO_3^{2-}	1.56×10^{-7}	6.81

*Note the relation between aqueous carbon dioxide and carbonic acid (pp. 226–7)

Source: Most values are taken from P. Atkins, *Physical Chemistry*, 6th edn., W.H. Freeman and Co., New York; 1998.

Appendix 11   Standard redox potentials in aqueous solutions (For simplicity all equations use H^+ as a substitute for the hydronium ion H_3O^+)

Reduction half reaction	E^o/v	pE^o	$pE^o(w)$
$O_3 + 2H^+ + 2e^- \rightarrow O_2 + H_2O$	+2.075	+35.1	+28.1
$H_2O_2 + 2H^+ + 2e^- \rightarrow 2H_2O$	+1.763	+29.8	+22.8
$MnO_4^- + 4H_3O^+ + 3e^- \rightarrow MnO_2 + 6H_2O$	+1.692	+28.6	+19.3
$2HClO + 2H^+ + 2e^- \rightarrow Cl_2 + 2H_2O$	+1.630	+27.6	+20.6
$Cl_2 + 2e^- \rightarrow 2Cl^-$	+1.396	+23.6	+23.0
$Cr_2O_7^{2-} + 14H^+ + 6e^- \rightarrow 2Cr^{3+} + 7H_2O$	+1.36	+23.0	+6.67
$2NO_3^- + 12H^+ + 10e^- \rightarrow N_2 + 6H_2O$	+1.25	+21.1	+12.7
$O_3 + H_2O + 2e^- \rightarrow O_2 + 2OH^-$	+1.24	+21.0	+28.0
$MnO_2 + 4H^+ + 2e^- \rightarrow Mn^{2+} + 2H_2O$	+1.230	+20.8	+6.80
$O_2 + 4H^+ + 4e^- \rightarrow 2H_2O$	+1.229	+20.8	+13.8
$NO_3^- + 4H^+ + 3e^- \rightarrow NO + 2H_2O$	+0.955	+16.1	+6.77
$NO_3^- + 3H^+ + e^- \rightarrow HNO_2 + H_2O$	+0.940	+15.9	−5.11
$ClO^- + H_2O + 2e^- \rightarrow Cl^- + 2OH^-$	+0.89	+15.0	+22.0
$NO_3^- + 10H^+ + 8e^- \rightarrow NH_4^+ + 3H_2O$	+0.882	+14.9	+6.15
$NO_3^- + 2H^+ + 2e^- \rightarrow NO_2 + H_2O$	+0.837	+14.2	+7.15
$Fe^{3+} + e^- \rightarrow Fe^{2+}$	+0.771	+13.0	+13.0
$CH_2O + 4H^+ + 4e^- \rightarrow CH_4 + H_2O$	+0.411	+6.94	−0.06
$O_2 + 2H_2O + 4e^- \rightarrow 4OH^-$	+0.40	+6.76	+13.8
$SO_4^{2-} + 8H^+ + 6e^- \rightarrow S + 4H_2O$	+0.353	+5.96	−3.37
$SO_4^{2-} + 9H^+ + 8e^- \rightarrow HS^- + 4H_2O$	+0.248	+4.20	−3.68
$CO_2 + 8H^+ + 8e^- \rightarrow CH_4 + 2H_2O$	+0.170	+2.87	−4.13
$2H^+ + 2e^- \rightarrow H_2$	0.00	0.00	−7.00
$CO_2 + 4H^+ + 4e^- \rightarrow [CH_2O] + H_2O$	−0.071	−1.20	−8.20
$S + 2e^- \rightarrow S^{2-}$	−0.50	−8.45	−8.45
$2H_2O + 2e^- \rightarrow H_2 + 2OH^-$	−0.828	−14.0	−7.00

Sources: D.C. Harris, *Quantitative Chemical Analysis*, 4th edn, W.H. Freeman and Co., New York (1995) and W. Stumm and J.J. Morgan, *Aquatic Chemistry*, John Wiley and Sons, New York; 1981.

Index